Adhesion and Adhesives

Adhesion and Adhesives

SCIENCE AND TECHNOLOGY

A. J. Kinloch

Department of Mechanical Engineering
Imperial College of Science and Technology
University of London

CHAPMAN & HALL
London · Glasgow · New York · Tokyo · Melbourne · Madras

Published by Chapman & Hall, 2-6 Boundary Row, London SE1 8HN, UK

Chapman & Hall, 2-6 Boundary Row, London SE1 8HN, UK

Blackie Academic & Professional, Wester Cleddens Road, Bishopbriggs, Glasgow G64 2NZ, UK

Chapman & Hall Inc., One Penn Plaza, 41st Floor, NY 10119, USA

Chapman & Hall Japan, Thomson Publishing Japan, Hirakawacho Nemoto Building, 6F, 1-7-11 Hirakawa-cho, Chiyoda-ku, Tokyo 102, Japan

Chapman & Hall Australia, Thomas Nelson Australia, 102 Dodds Street, South Melbourne, Victoria 3205, Australia

Chapman & Hall India, R. Seshadri, 32 Second Main Road, CIT East, Madras 600 035, India

First edition 1987
Reprinted 1990, 1994

Printed in Great Britain by the University Press, Cambridge

ISBN 0 412 27440 X

British Library Cataloguing in Publication Data

Kinloch, A.J.
Adhesion and adhesives: science and technology

1. Adhesives.
I. Title
620.1'99 TA455.A34
ISBN 0-412-27440-X

Library of Congress Cataloging-in-Publication Data

Kinloch, A.J.
Adhesion and adhesives.
Includes bibliographical references and indexes.
1. Adhesives. 2. Adhesion. I. Title
TP968.K46 1987 668'.3 86.34333
ISBN 0-412-27440-X

∞ Printed on acid-free text paper, manufactured in accordance with ANSI/NISO Z39.48-1992 and ANSI/NISO Z39.48-1984 (Permanence of Paper).

To my parents, my wife Gillian
and my children Ian, Elizabeth and David

Contents

Preface

Over the last decade, or so, the growth in the use of adhesives, especially in ever more technically demanding applications, has been rapid and many major developments in the technology of adhesives have been reported. This growth has also led to attention being focused on somewhat more basic studies of the science of adhesion and adhesives, and in recent years our level of fundamental knowledge concerning the formation and mechanical performance of adhesive joints has increased dramatically. Such studies have, of course, been aided greatly by the development of the tools at the disposal of the investigators. For example, specific surface analytical techniques, such as X-ray photoelectron and secondary-ion mass spectroscopy, and the increasingly sophisticated methods of stress analysis and fracture mechanics have been put to good use in furthering our understanding of the science of adhesion and adhesives. The present book attempts to review the multidisciplined subject of adhesion and adhesives, considering both the science and technology involved in the formation and mechanical performance of adhesive joints.

The author would like to thank his friends and colleagues for useful discussions and help in the preparation of this book. I am particularly grateful to P. Cawley, J. Comyn, W. A. Lees, A. C. Roulin-Moloney, W. C. Wake, J. G. Williams and R. J. Young who have read and commented on various chapters and P. Farr for preparing the diagrams. I would also like to thank Professor H. H. Kausch and Dr A. C. Roulin-Moloney for inviting me to spend a summer sabbatical at the École Polytechnique Fédérale de Lausanne where this book was completed. Finally, I would like to thank my family for their patience, understanding and assistance in so many ways.

A. J. Kinloch
Imperial College of Science and Technology
September 1986

Acknowledgements

The author wishes to thank the following publishers and companies for permission to reproduce illustrations from previous publications:

American Cynamid – Fig. 6.4. (© American Cynamid)
Applied Science Publishers – Fig. 3.4.
British Leyland – Figs 1.13 and 1.14
British Aerospace – Figs 1.15, 1.16 and 1.17.
Butterworths – Fig. 6.34.
Chapman and Hall Ltd – Figs 7.21, 7.22b, 8.17 and 8.18.
Ciba-Geigy Plastics – Figs 1.10, 1.12 and 1.20.
Gordon and Breach – Fig. 3.3. (© Gordon and Breach)
Imperial Metal Industries Ltd (R. O. Summerfield) – Fig. 1.18.
Institute of Metal Finishing – Fig. 3.2.
Loctite – Figs 1.2, 1.3, 1.4 and 1.5.
Permabond Adhesives Ltd/JCB Ltd – Fig. 1.6.
Permabond Adhesives Ltd/Delo GmBH – Fig. 1.7.
Permabond Adhesives Ltd/Lotus Cars Ltd – Fig. 1.8.
Permabond Adhesives Ltd/Foden Ltd – Fig. 1.9.
Permabond Adhesives Ltd/David Brown Tractors Ltd – Fig. 1.11.
Permabond Adhesives Ltd – Fig. 1.19.
Plenum Press – Fig. 2.5.
National Adhesives and Resins Ltd – Fig. 1.1.
The Royal Society – Fig. 7.20.

A number of other figures in the book are based upon already published material and, in these cases, an appropriate acknowledgement appears in the caption.

CHAPTER ONE

Introduction

An *adhesive* may be defined as a material which when applied to surfaces of materials can join them together and resist separation. Adhesive is the general term and includes cement, glue, paste, etc. and these terms are all used essentially interchangeably. Various descriptive adjectives are often applied to indicate certain characteristics. For example, to indicate the physical form of the adhesive, e.g. *liquid adhesive*, *liquid two-part adhesive*, *film adhesive*; its chemical form, e.g. *epoxy adhesive*, *cyanoacrylate adhesive*, *polychloroprene adhesive*; to indicate the type of materials bonded, e.g. *metal-to-metal adhesive*, *paper adhesive*, *wood adhesive*; or to show the conditions of use, e.g. *solvent-based adhesive*, *cold-hardening*, or *-curing*, *adhesive*, *hot-melt adhesive*.

The term *adhesion* is used when referring to the attraction between the substances. As will become evident later, the level of adhesion forces which are operating across an interface cannot usually be measured by mechanical tests. For example, the measured energy for interfacial fracture is generally orders of magnitude higher than that arising solely from the *intrinsic adhesion* forces, such as molecular van der Waals' forces or covalent bonds, which may be operating across the interface.

The materials being joined are commonly referred to as the *substrates* or *adherends*. The latter term is sometimes employed when the materials are part of a joint, but in the present text only the term substrate will be used. In the following chapters the importance of considering the adhesive joint from the viewpoint of being an '*adhesive system*' will be emphasized. The term adhesive system is employed to indicate that, for many aspects of the performance of adhesive joints, other factors besides the physical and chemical properties of the selected adhesive may be of importance. For example, parameters such as the type of alloy selected for the substrate and the form of any surface pretreatment or primer chosen for the substrates may be extremely important. Indeed, in many instances, such parameters may be the main factors in determining whether the adhesive joint will perform adequately, especially with respect to attaining the desired service-life from the bonded structure.

Also, of course, the exact details of the *joint design*, e.g. the geometric features of the joint and the way in which the applied loads are transmitted from one substrate to the other, will greatly influence the observed mechanical behaviour of the bonded joint.

As a means of joining materials adhesives have been used by mankind for many centuries. However, it is only in the last 50 years, or so, that the science and technology of adhesion and adhesives has really progressed significantly and the major advances that have been made may be traced from the middle of the 1940s. The main reason for this is that the adhesives employed in nearly all the technically demanding applications are based upon synthetic polymers. As discussed later, such materials possess the balance of properties that enables them to adhere readily to other materials and to have an adequate strength so that they are capable of transmitting the applied loads or forces from one substrate to the other. The use of adhesives in technically demanding applications has provided the spur for the research and development of new, improved materials and identified the need for, and supported, studies on the more fundamental aspects of the underlying science. The wide range of synthetic polymers, and ancillary products such as hardeners, stabilizers, toughening additives, etc., which have become available over the last few decades, has enabled the adhesives technologist to develop specific adhesive formulations to meet the manufacturing and performance requirements of very diverse applications for both industrial and domestic applications.

Notwithstanding the above comments, solely meeting the manufacturing and performance requirements of a given industry is generally an insufficient reason for any industry to adopt a new technology, with all the teething problems that such an adoption usually implies. However, a driving force for the development and continual growth of the adhesives market is the many advantages that they offer compared with other, more traditional, techniques for fastening materials, such as brazing, welding, riveting, bolting, etc.

Advantages that adhesives can offer include:

(a) The ability to join dissimilar materials; e.g. the joining of metals, plastics, rubbers, fibre-composites, wood, paper products, etc.
(b) The ability to join thin sheet-material efficiently. This is a major use of adhesives for joining both metallic and non-metallic materials. Compared to metallic substrates, adhesives, being based largely upon organic polymers, do not possess anywhere near the level of tensile fracture strengths exhibited by most metals but when used to join relatively thin sheets of metal their strengths are usually more than adequate.
(c) An improved stress distribution in the joint which imparts, for example, a very good dynamic-fatigue resistance to the bonded component.
(d) The fact that they frequently represent the most convenient and cost-effective technique. Indeed, the bonding operation can often be auto-

mated. This obviously removes the necessity for any human operator to mix together the various components of the adhesive, if required, apply the adhesive to the correct location and repeat the procedure correctly many hundreds of times a day. Also, as demonstrated by the vehicle manufacturing industries, adhesives may be readily adapted to robotic assembly techniques.

(e) An increase in design flexibility which enables novel design concepts to be implemented and allows a wider choice of materials to be available to the designer. A good example of these advantages is honeycomb structures where a honeycomb core is faced with two adhesively bonded sheets of material to produce a structure with a very good stiffness-to-weight ratio. The materials may be metallic, non-metallic or combinations of both.

(f) An improvement in the appearance of the fastened structure; for example, if adhesives are used instead of spot-welding then the smooth, blemish-free, appearance of the bonded structure is more appealing to the consumer.

(g) An improvement in corrosion resistance. The above comparison to the spot-welded component serves as a good example where the use of a well-selected adhesive system will inherently result in far less corrosion.

However, as with any technology, there are some disadvantages:

(a) To attain long service-life from adhesive joints in very severe, hostile environments may often require the use of a surface pretreatment process for the substrates being joined.

(b) Compared to other fastening techniques such as welding, riveting, etc. the upper service-temperatures that adhesives can withstand are limited.

(c) As remarked above, the strength and toughness of adhesives in tension or shear is relatively low compared to many metals. Hence, whilst adhesives are very effective at joining thin sheets of metal, they are not typically used for joining thick metallic components, unless the bonded area is large or the adhesive is kept in compression.

(d) Non-destructive test methods for adhesive joints are relatively limited compared to those used with other fastening methods.

To illustrate these various points and show some of the very wide uses of adhesives the reader is referred to Figs 1.1 to 1.20. From these applications it is obvious that adhesives now play a most important role in the fastening of materials. They clearly demonstrate that, whilst uses of adhesives in the home, particularly by the do-it-yourself enthusiast, represent a valuable service to mankind and a vital market for the adhesives supplier, modern adhesives are not confined to such domestic applications. Indeed, adhesives are currently employed in industrial applications of all kinds and the applications illustrated reveal that adhesives are used in some of the most critical areas in engineering structures and in some of the most demanding environments, and that they have a good history of meeting the performance and service-life requirements.

If the various stages in the formation of adhesive joints are considered then several important requirements have to be met during the bonding operation. The first stage in the formation of an adhesive joint is the attainment of interfacial contact between the adhesive and substrates and this aspect is considered in Chapter 2. Having obtained intimate interfacial contact then some form of intrinsic adhesion forces needs to be established between the adhesive and substrates, and such intrinsic adhesion forces will hold the materials together throughout the service-life of the joint. Strictly, the exception to this statement is when solely mechanical interlocking occurs between the adhesive and substrates, since then neither the attainment of intimate interfacial contact nor the establishment of intrinsic adhesion forces need to be proposed. The types of intrinsic adhesion forces that may operate across an interface, which in the present text is taken to include mechanical interlocking at the interface for the sake of conciseness and clarity, are reviewed in Chapter 3.

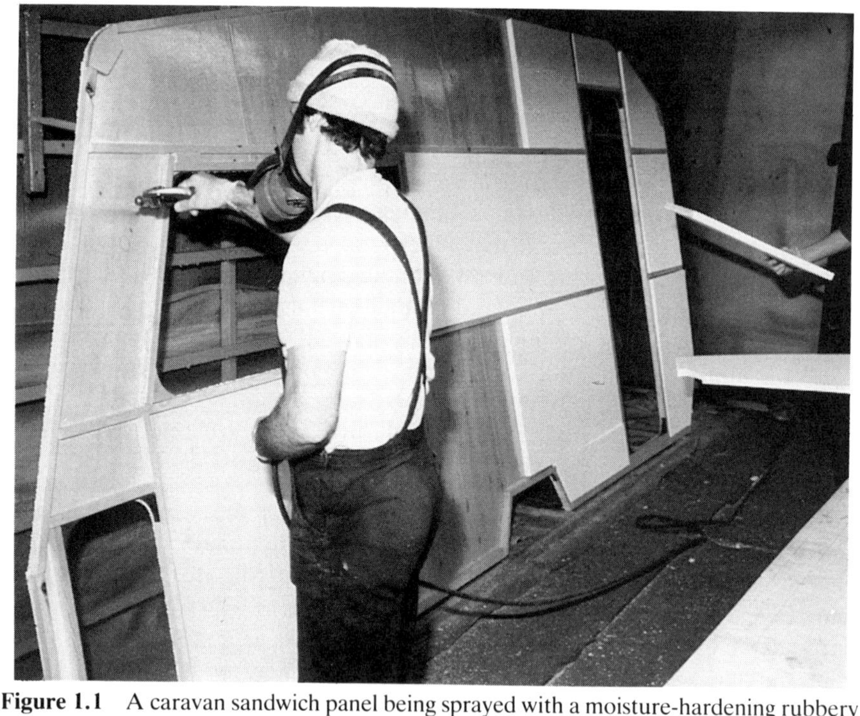

Figure 1.1 A caravan sandwich panel being sprayed with a moisture-hardening rubbery polyurethane adhesive, in a solvent, prior to bonding on the inner skin. The absence of mechanical fasteners greatly increases the production rate and results in a more pleasing appearance.

Figure 1.2 Ice cream and fruit dishes are assembled using an acrylic adhesive which is hardened by the application of an ultraviolet light source.

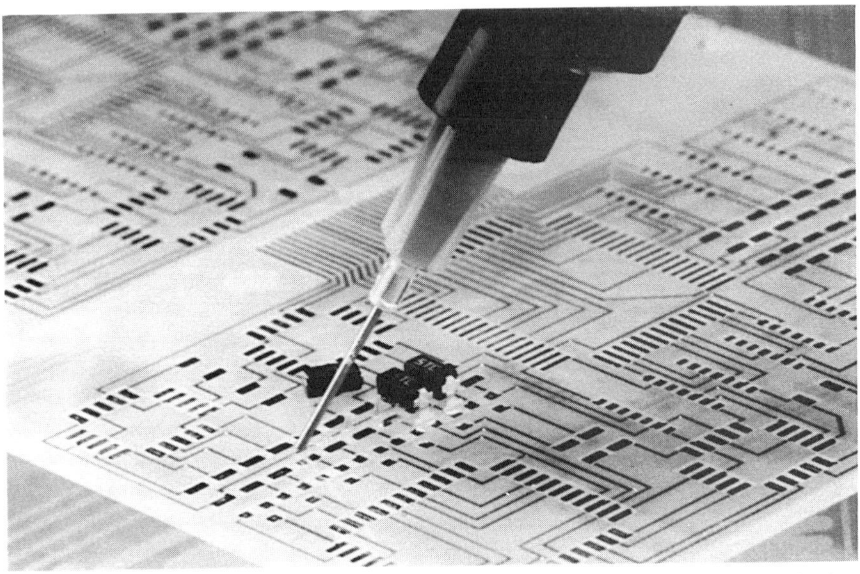

Figure 1.3 Electronic microchips may be mounted onto the printed circuit board using a very fast, room-temperature (via ultraviolet light) or hot-temperature hardening acrylic adhesive. Electronic applications invariably involve a large number of interfaces between different materials and the use of adhesives in the electronics industry is one of the most exciting growth areas for adhesive bonding.

Figure 1.4 Fast, room-temperature hardening, single-part cyanoacrylate adhesives are widely used for bonding metallic, plastic, rubber and wooden substrates, such as in attaching the rubber recoil pad to this hardwood gun stock.

Figure 1.5 Cyanoacrylate adhesives are also used to assemble loudspeakers.

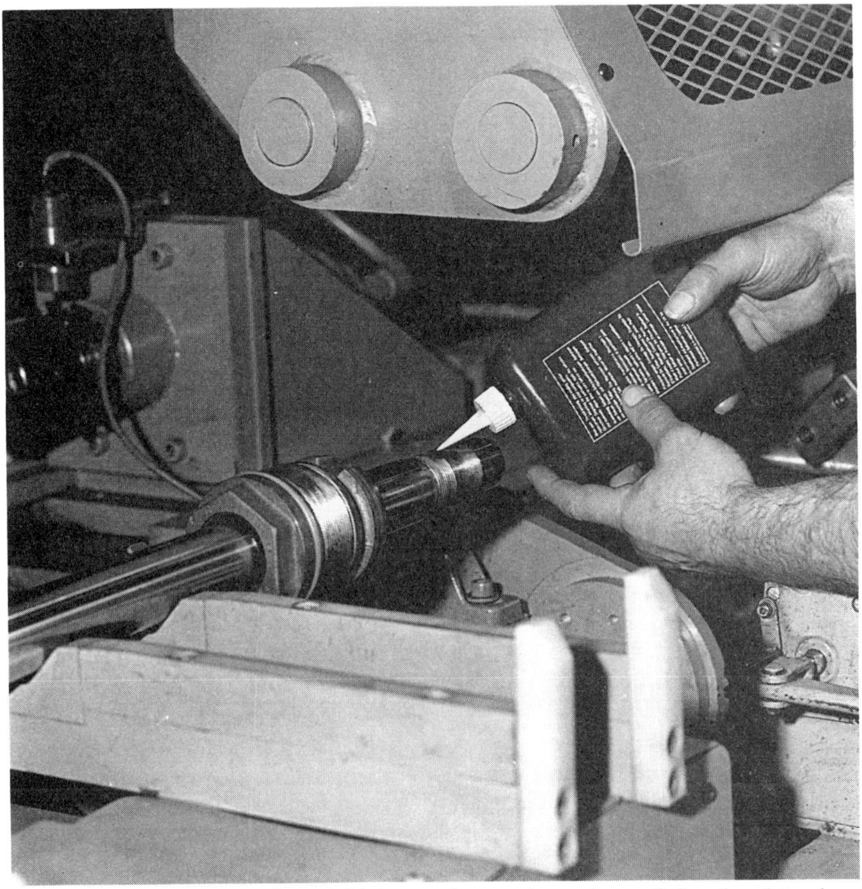

Figure 1.6 In the production of this heavy-duty engine, prior to fitting the piston head to the threaded end of the piston rod, the thread is coated with an anaerobic acrylic adhesive to ensure that the piston is held firmly in position. Such adhesives harden in the absence of oxygen.

The discussions in Chapters 2 and 3 emphasize that some form of surface pretreatment is often required in order for the bonded component to meet the necessary levels of service performance and life. Therefore, Chapter 4 reviews this very important aspect of adhesives technology in some detail. To achieve the earlier requirement of attaining intimate interfacial contact the adhesive had to be in a 'liquid' form at some stage during the bonding operation. However, the adhesive almost invariably now has to harden in order to be able to withstand the stresses and strains that may be applied to the joint. This aspect is considered in Chapter 5. However, unlike the other chapters, this chapter is not intended as an in-depth review, principally because such a review would

Figure 1.7 This gear is part of the drive mechanism of a steel rolling mill. After many years of use, the shaft cracked and could not be repaired by welding. Instead the fractured end was cut away and a new driving collar fitted. The shaft and collar were bonded together with an anaerobic acrylic adhesive.

take the present book to at least a two-volume set and, in any event, the chemistry and formulation of adhesives are well-covered in several recent books. References to such literature are given in Chapter 5 and this chapter is therefore intended to give a broad introduction to the chemical types of adhesives to enable the reader, unfamiliar with adhesives technology, to follow more readily points made in the other chapters of the present book. Thus, Chapters 2 to 5 inclusive deal with aspects of the formation of adhesive joints.

However, it must be appreciated that the details of the design of an adhesive joint, the way in which loads are applied to it and the service environment that it must withstand, will all affect its mechanical performance and life expectancy. These aspects are considered in Chapters 6, 7 and 8. In Chapters 6 and 7 the design and the mechanical performance of adhesive joints are discussed. In order to present a clearer picture of these aspects, the different, but complementary, approaches that have been adopted to examining the problem of joint design and mechanical performance are reviewed in two chapters: Chapter 6

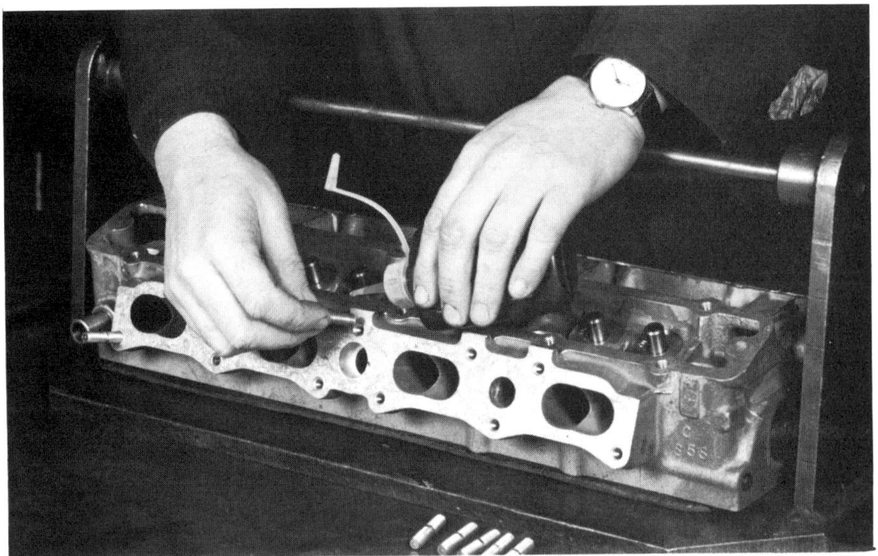

Figure 1.8 In this high-performance car engine the studs used to secure the inlet manifold to the cylinder head are retained in position with an anaerobic acrylic adhesive. Although of high strength, the grade chosen will allow maintenance and ready removal of the studs should this be necessary. Because the studs are a loose fit, and hand fitted, and retention is ensured by the adhesive, no stresses are generated into the aluminium cylinder head. This means that the sections employed may be of minimal thickness as there is no danger of cracks being initiated by overtightened studs. The studs of the exhaust manifold (rear face of head) are treated in a similar manner. Note that, when the turbo modified engine is being built these studs are fabricated from stainless steel.

Figure 1.9 The doors of this 'Foden S10 Fleetmaster' are constructed from bonded aluminium alloy panels using a two-part, room-temperature hardening acrylic adhesive.

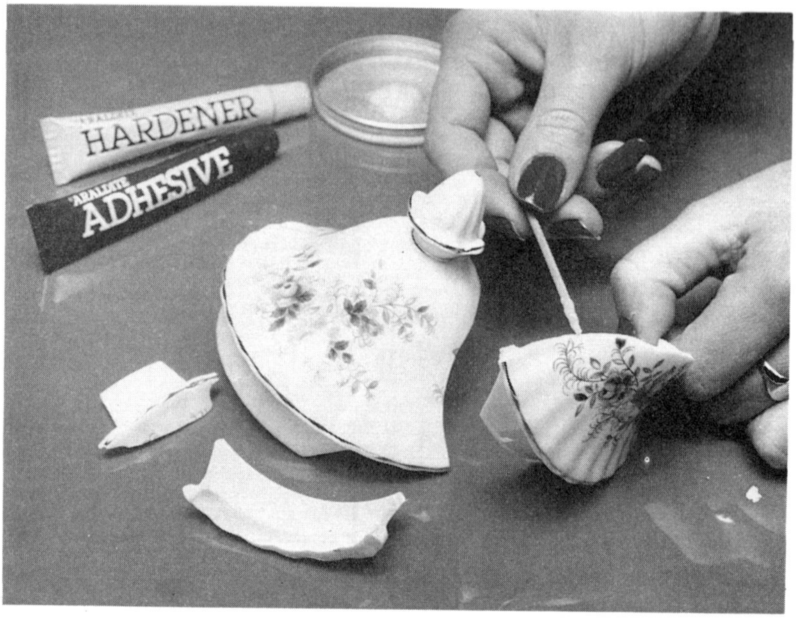

Figure 1.10 A two-part, room-temperature hardening epoxy adhesive is a favourite retail pack for the 'do-it-yourself' domestic consumer.

Figure 1.11 The steel bonnet of a tractor and its various stiffeners are bonded with a rubber-toughened, single-part, heat-cured, epoxy edhesive.

Figure 1.12 A modified single-part, heat-cured, epoxy adhesive is used in the production-line assembly of steel car bodies, and here the roof section is about to be bonded into place.

Figure 1.13 Automatic and robotic application of adhesives is a fast expanding field and offers many advantages to general engineering industry. The photograph shows the robotic application of an adhesive in the assembly of components in a car production line.

Figure 1.14 The British Leyland 'Energy Conservation Vehicle Mark 3 (ECV3)' is probably one of the most innovative cars ever built. A significant and unique feature is its aluminium alloy monocoque body which relies entirely on a toughened epoxy adhesive for its structural integrity.

Figure 1.15 The 'British Aerospace 146'. Like all modern aeroplanes adhesives are used extensively in its construction, for bonding both metallic and non-metallic materials together.

Figure 1.16 In the 'British Aerospace 146' one of the largest components is the wing-skin assembly which is manufactured using aluminium alloy. The stiffening stringers are bonded onto the skin using a modified phenolic adhesive. The two gentlemen are conducting non-destructive tests on the bonded joints.

essentially considers the nature and magnitude of stresses in adhesive joints, as ascertained using stress-analysis techniques, and the effects of such stresses on the behaviour of joints, whilst Chapter 7 is concerned with the use of continuum fracture mechanics methods. Chapter 6 also contains a brief review of the non-destructive testing of adhesive joints. In Chapter 8 the factors which influence the service-life of an adhesive joint are discussed; in particular consideration is given to the effects of stressing the joint under dynamic and static fatigue loads and of exposing it to an hostile environment.

As may be appreciated from the above comments, the science and technology of adhesion and adhesives is a truly multidisciplined subject. A detailed consideration of the many theoretical and experimental techniques which have been used, and are therefore mentioned, is well beyond the scope of the present book. However, a bibliography of books for the various main disciplines which are relevant to the present discussions is given at the end of the present chapter and further references are given in the respective chapters which follow.

Finally, it is hoped that the present book will do justice to the exciting developments that have occurred, and are still occurring, in the field of the science and technology of adhesion and adhesives.

Figure 1.17 Another major use of structural adhesives in the aerospace industry is in making honeycomb sandwich panels where the face skins are bonded to the honeycomb core. This example is from a section of the rudder of Concorde and consists of aluminium alloy core and skins. More recently one has seen the increased use of non-metallic core and skin materials, i.e. carbon-fibre reinforced plastic (CFRP) skins.

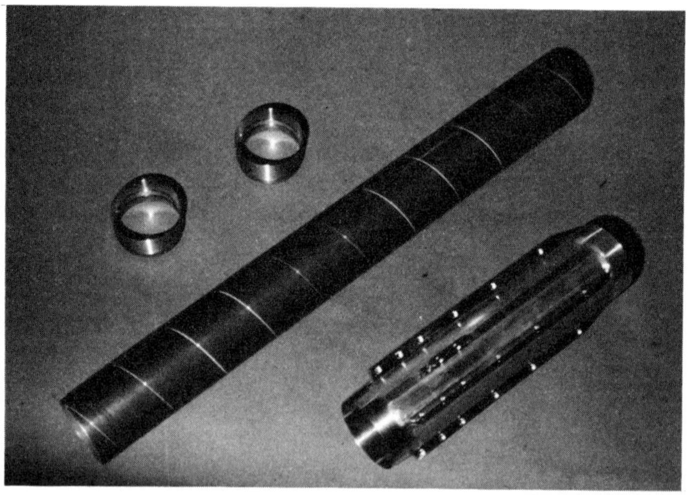

Figure 1.18 These 'pressure vessels' are rocket motors, and are manufactured by helically winding a number of layers of metal strip coated with an adhesive onto a mandrel and then heating the tube to harden the adhesive. End fittings and wing rings are then bonded into and over the tube respectively to produce a finished rocket-motor case. Two major advantages of this method of manufacture, compared to welded tubes, are high specific strength and low cost of the finished product.

Figure 1.19 Adhesives are widely used in the construction of many civil engineering projects, such as the bonding of aluminium panels.

Figure 1.20 Segmental bridge construction takes advantage of room (cold)-temperature hardening, epoxy-based adhesives for bonding and sealing the concrete units.

BIBLIOGRAPHY OF GENERAL BACKGROUND BOOKS

Surfaces and surface analysis

1. Adamson, A. W. (1967) *Physical Chemistry of Surfaces*, Wiley-Interscience, New York.
2. Brewis, D. M. (ed.) (1982) *Surface Analysis and Pretreatment of Plastics and Metals*, Applied Science Publishers, London.
3. Briggs, D. and Seah, M. P. (eds) (1983) *Practical Surface Analysis by Auger and Photoelectron Spectroscopy*, Wiley-Heyden, Chichester.
4. Clark, D. T. and Feast, W. J. (eds) (1977) *Polymer Surfaces*, John Wiley, New York.
5. Czanderna, A. W. (ed.) (1975) *Methods of Surface Analysis*, Elsevier, New York.
6. Hansma, P. K. (ed.) (1981) *Tunnelling Spectroscopy: Capabilities, Applications, and New Techniques*, Plenum Press, New York.
7. Jaycock, M. J. and Parfitt, G. D. (1981) *Chemistry of Interfaces*, Ellis Horwood, Chichester.
8. Lee, L. H. (ed.) (1977) *Characterisation of Metal and Polymer Surfaces*, (2 vols), Academic Press, New York.
9. Margenau, H. and Kestner, N. (1971) *Theory of Intermolecular Forces*, 3rd edn., Pergamon Press, Oxford.
10. Pauling, L. (1960) *The Nature of the Chemical Bond*, 3rd edn., Cornell University Press, Ithaca, New York.

Polymer science

11. Aklonis, J. J., Macknight, W. J. and Shen, M. (1972) *Introduction to Polymer Viscoelasticity*, Wiley-Interscience, New York.
12. Andrews, E. H. (ed.) (1979) *Developments in Polymer Fracture*, Applied Science Publishers, London.
13. Billmeyer, F. W. (1971) *Textbook of Polymer Science*, Wiley-Interscience, New York.
14. Ferry, J. D. (1970) *Viscoelastic Properties of Polymers*, John Wiley, New York.
15. Flory, P. J. (1953) *Principles of Polymer Chemistry*, Cornell University Press, Ithaca, New York.
16. Hertzberg, R. W. and Manson, J. A. (1980) *Fatigue of Engineering Plastics*, Academic Press, New York.
17. Kausch, H. H. (1978) *Polymer Fracture*, Springer-Verlag, Berlin.
18. Kinloch, A. J. and Young, R. J. (1983) *Fracture Behaviour of Polymers*, Applied Science Publishers, London.
19. Lenz, R. W. (1967) *Organic Chemistry of High Polymers*, Wiley-Interscience, New York.
20. Margerison, D. and East, G. C. (1967) *An Introduction to Polymer Chemistry*, Pergamon Press, Oxford.
21. Odian, G. (1970) *Principles of Polymerization*, McGraw-Hill Inc., New York.
22. Treloar, L. R. G. (1958) *The Physics of Rubber Elasticity*, Oxford University Press, Oxford.
23. Ward, I. M. (1971) *Mechanical Properties of Solid Polymers*, Wiley-Interscience, New York.
24. Young, R. J. (1981) *Introduction to Polymers*, Chapman and Hall, London.

Stress and fracture analysis

25. Broek, D. (1974) *Elementary Engineering Fracture Mechanics*, Noordhoff, Leyden.
26. Knott, J. F. (1973) *Fundamentals of Fracture Mechanics*, Butterworths, Guildford.
27. Williams, J. G. (1973) *Stress Analysis of Polymers*, 2nd edn., Longman, Harlow.
28. Williams, J. G. (1984) *Fracture Mechanics of Polymers*, Ellis Horwood, Chichester.
29. Zienkiewicz, O. C. (1971) *The Finite Element Method in Engineering Science*, McGraw-Hill, London.

Miscellaneous

30. Comyn, J. (ed.) (1985) *Polymer Permeability*, Applied Science Publishers, London.
31. Crank, J. (1973) *The Mathematics of Diffusion*, Clarendon Press, Oxford.
32. Crank, J. and Park, G. S. (eds) (1968) *Diffusion in Polymers*, Academic Press, New York.
33. Hollahan, J. R. and Bell, A. T. (eds) (1974) *Techniques and Applications of Plasma Chemistry*, Wiley-Interscience, New York.

CHAPTER TWO

Interfacial contact

2.1 INTRODUCTION

As discussed in Chapter 1, it has been recognized for many years that the establishment of intimate molecular contact is a necessary, though sometimes insufficient, requirement for developing strong adhesive joints. This means that the adhesive, and primer if one is employed, needs to be able to spread over the solid surface, and needs to displace air and other contaminants that may be present on the surface.

An adhesive which conforms ideally to these conditions must:

(a) when liquid exhibit a zero or near zero contact angle;
(b) at some time during the bonding operation have a viscosity that should be relatively low, e.g. no more than a few centipoises;
(c) be brought together with the substrate in a rate and manner that should assist in the displacement of any trapped air.

In order to assess the ability of a given adhesive/substrate combination to meet these criteria it is necessary (i) to consider wetting equilibria, (ii) to ascertain values of the surface free energies of the adhesive and substrate and the free energy of the adhesive/substrate interface, (iii) to examine the kinetics of the wetting process and (iv) to consider details of the bonding operation.

2.2 SURFACE TENSION

Surface tension is a direct measurement of intermolecular forces. The tension in surface layers is the result of the attraction of the bulk material for the surface layer and this attraction tends to reduce the number of molecules in the surface region resulting in an increase in intermolecular distance. This increase requires work to be done, and returns work to the system upon a return to a normal configuration. This explains why tension exists and why there is a surface free energy.

The most common type of physical attractive forces are the van der Waals forces, and they can be attributed to different effects:

(a) dispersion (or London) forces arising from internal electron motions which are independent of dipole moments and
(b) polar (or Keesom) forces arising from the orientation of permanent electric dipoles and the induction effect of permanent dipoles on polarizable molecules.

The dispersion forces are usually weaker than the polar forces but they are universal and all materials exhibit them. Another type of force that may operate is the hydrogen bond, formed as a result of the attraction between a hydrogen atom and a second, small and strongly electronegative atom such as a fluorine, oxygen or nitrogen atom.

The various types of physical attractive forces are listed in Table 2.1 together with estimates of the range of magnitude of their respective bond energies.

Table 2.1 Types of physical attractive forces and typical bond energies [1,2]

Type	Bond energy (kJ/mol)
Permanent dipole–dipole interactions:	
Hydrogen bonds involving fluorine	Up to 40
Hydrogen bonds excluding fluorine	10–25
Other dipole–dipole (excluding hydrogen) bonds	4–20
Dipole-induced dipole	Less than 2
Dispersion (London) forces	0.08–40

2.3 WETTING EQUILIBRIA

2.3.1 Theoretical considerations

Wetting may be quantitatively defined by reference to a liquid drop resting on a solid surface as shown in Fig. 2.1. The tensions at the three-phase contact point are indicated such that lv is the liquid/vapour point, sl is the solid/liquid point and sv is the solid/vapour point. The Young equation [3,4], relating these tensions to the equilibrium contact angle, θ, may be written as:

$$\gamma_{sv} = \gamma_{sl} + \gamma_{lv} \cos\theta \quad (2.1)$$

The term γ_{sv} represents the surface free energy of the solid substrate resulting from adsorption of vapour from the liquid and may be considerably lower in

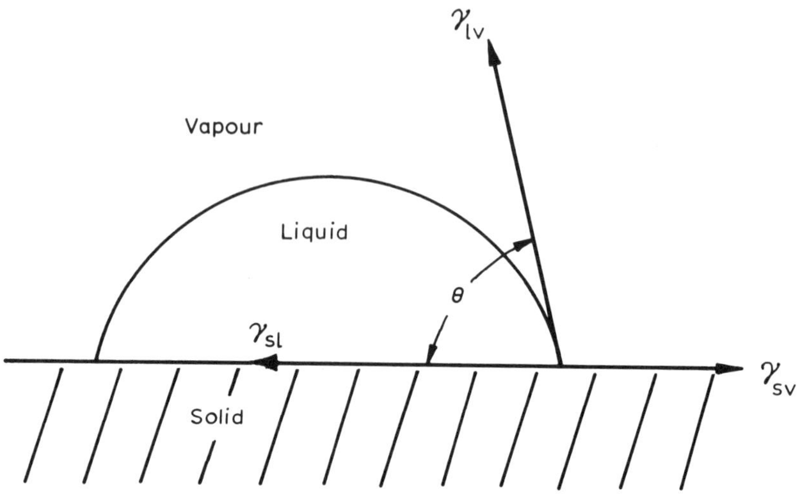

Figure 2.1 A liquid drop resting at equilibrium on a solid surface.

value than the surface free energy of the solid *in vacuo*, γ_s. This reduction in the surface free energy of the solid when covered by a layer of vapour has been defined by the concept of an equilibrium spreading pressure, Π_s, such that when the vapour obeys the ideal gas law [5]:

$$\Pi_s = \gamma_s - \gamma_{sv} = RT \int_0^{p_0} \Gamma \, d(\ln p) \tag{2.2}$$

where p is the vapour pressure, p_0 is the equilibrium vapour pressure, R is the gas constant, T is the absolute temperature and Γ is the surface concentration of adsorbed vapour. Thus Equation 2.1 may be rewritten as:

$$\gamma_s = \gamma_{sl} + \gamma_{lv} \cos\theta + \Pi_s \tag{2.3}$$

When $\theta > 0°$ the liquid is nonspreading, but when $\theta = 0°$ the liquid wets the solid completely and spontaneously spreads freely over the surface at a rate that depends upon such factors as the liquid viscosity and roughness of the solid surface, as discussed later. Thus for spontaneous wetting to occur:

$$\gamma_{sv} \geq \gamma_{sl} + \gamma_{lv} \tag{2.4}$$

$$\gamma_s \geq \gamma_{sl} + \gamma_{lv} + \Pi_s \tag{2.5}$$

These criteria may also be expressed by defining a parameter termed the equilibrium spreading coefficient, S, where:

$$S = \gamma_{sv} - \gamma_{sl} - \gamma_{lv} \tag{2.6}$$

or alternatively

$$S = \gamma_s - \gamma_{sl} - \gamma_{lv} - \Pi_s \qquad (2.7)$$

Hence, a liquid will spread spontaneously and completely wet a solid surface when $S \geq 0$. It is also possible, of course, to make a liquid spread across a solid surface even when $\theta > 0°$, but this requires the application of a pressure or a force to the liquid forcibly to spread it over the solid surface.

However, before these concepts may be of use in adhesives technology the surface and the interfacial free energies need to be evaluated and the kinetics considered. Since this frequently involves measuring contact angles, experimental aspects will be considered next.

2.3.2 Experimental considerations

For a planar surface, the equilibrium contact angle can be measured very simply from the profile of either a sessile liquid drop (Fig. 2.2a) or a bubble (Fig. 2.2b) resting on the surface, and these methods are known as the sessile-drop and captive-bubble techniques, respectively [6,7]. The contact angle may be determined by drawing a tangent to the profile at the point of three-phase contact after the drop profile has been enlarged either by image projection or photography. It may also be measured directly using a telescope fitted with a goniometer eyepiece, or indirectly by measuring the angle at which light from a point source is reflected from the surface of a liquid drop at its point of contact with the solid [8]. The author has employed this latter technique frequently and found it to be a rapid and reproducible method, but it is limited to values of $\theta < 90°$. Other indirect techniques include the measurement of drop dimension [9,10] and the use of interference microscopy [11,12]. In the former technique the contact angle may be readily calculated from the height and base diameter of the drop, provided the drops are sufficiently small that the gravitational distortion of the drop can be considered negligible. For relatively large drops, i.e. drops with volumes larger than about 10^{-10} m^3, the relationships between drop size, profile shape and contact angle have been considered by Padday [9]. The interference microscopy method is particularly useful for determining contact angles less than about 10°, which are difficult to measure accurately by the tangent techniques.

Considering curved surfaces, then the contact angles of various liquids on the cylindrical surfaces of fibres have been determined using the tensiometric (or Wilhelmy) plate method [13,14] or by calculating the value of θ from measurements of drop size and fibre diameter [15,16].

Another aspect to be considered when measuring contact angles is the phenomenon of contact angle hysteresis. This term describes the observation that different values of θ may be obtained depending upon whether the liquid drop is advanced or withdrawn across the solid surface. Contact angle

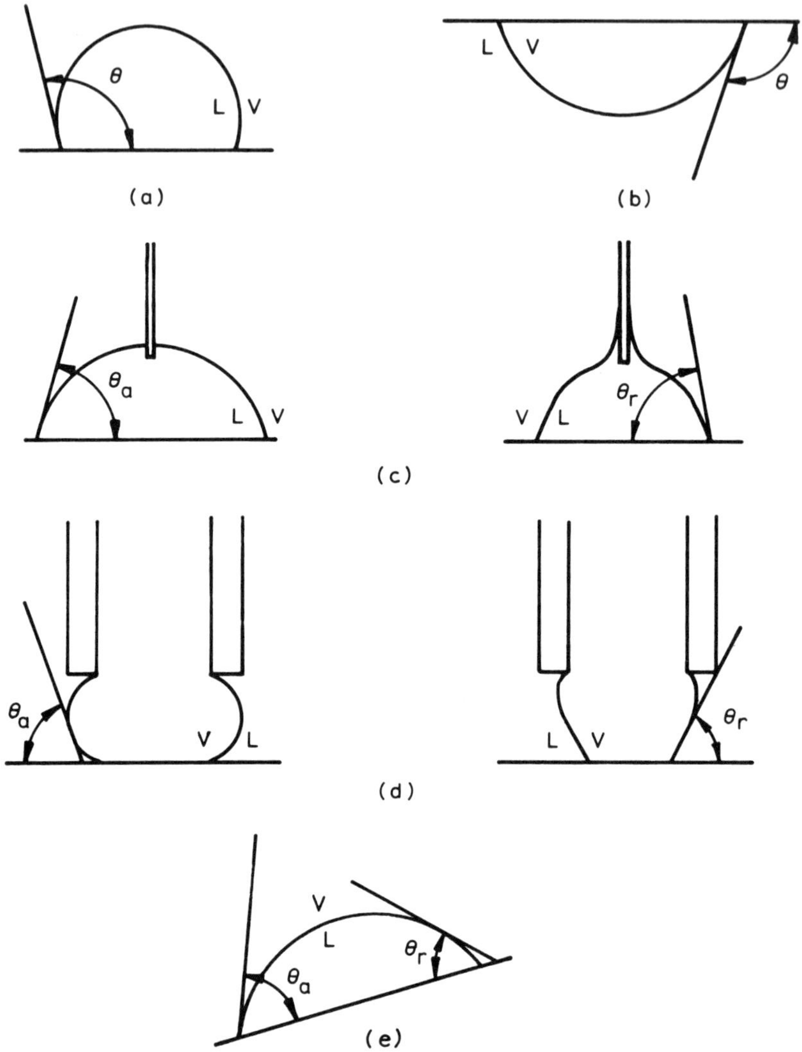

Figure 2.2 Drop and bubble configurations for measurement of equilibrium, advancing and receding contact angles [6]. (a) Equilibrium sessile drop, (b) equilibrium pendant bubble, (c) advancing and receding drop, (d) advancing and receding bubble, (e) drop on tilted plate. L = liquid, V = vapour.

hysteresis arises from solid surfaces seldom being either ideally smooth or chemically homogeneous [17–19]. The common methods of determining the advancing and receding angles are shown in Fig. 2.2c–e. In Fig. 2.2c the liquid is advanced or withdrawn across the surface by increasing or decreasing the size of the liquid drop, which is typically dispensed by a syringe. As discussed by Rance [6], it is best to let the needle remain in the drop during measurements to avoid unnecessary vibration or distortion of the drop; this does not affect the contact angle. The equivalent captive-bubble technique is shown in Fig. 2.2d, and in Fig. 2.2e the tilted drop technique is illustrated. In the latter method the advancing and receding angles can both be measured from a single drop in the limit when the drop begins to move as the tilt angle of the plate is increased. By far the most commonly determined values are the equilibrium advancing values of the contact angle.

Finally, it should be noted that instead of employing a contact angle method Papirer [153] and Schultz *et al.* [154] have recently used the technique of inverse gas phase chromatography to characterize the surfaces of low and high-energy fibres and fillers. In this technique the surface free energy (and acid–base properties) are determined by the injection of specific 'probes' (e.g. ether, chloroform, benzene, acetone, etc.) into the column containing the solid. The retention times of these probes measured at 'infinite dilution', or near zero surface coverage, enable the interactions between the probes and the solid to be ascertained in the absence of interactions between the probe molecules themselves.

2.3.3 Effect of surface roughness

Wenzel [20] has shown that surface roughness may change the apparent advancing contact angle, θ_f, observed for a given liquid on a rough solid, compared to the angle, θ_s, observed on a smooth surface. This change in the contact angle may be expressed by:

$$\cos\theta_f = r_f \cos\theta_s \tag{2.8}$$

where r_f is the roughness factor or the ratio of the actual area to the projection area of the solid. If on a smooth surface θ_s is less than 90°, then roughening the surface will result in θ_f being even smaller. This will obviously increase the apparent surface free energy of the solid surface and consequently also increase the extent of wetting. However, if for a smooth surface θ_s is greater than 90°, roughening the surface will increase the contact angle θ_f still further and therefore decrease the degree of wetting. Other, more sophisticated models of substrate surface topography have since been developed [17–19,21–26] to explain features such as contact angle hysteresis. These models have led to the Wenzel relationship being re-examined and Hitchcock *et al.* [27] have reported that increasing the roughness of the substrate may only increase the degree of wetting if the liquid exhibits a comparatively low contact angle or the surface texture is very rough, e.g. an abraded surface. They found that an empirical relationship existed between the enthalpy of a liquid and the extent to which its wetting behaviour was altered by substrate roughening. This enabled theoretical predictions to be made about the behaviour of liquids.

More recently Carre and Schultz [28] have used the Wenzel equation to determine the value of the roughness factor, r_f, for various types of solid surfaces. They measured the contact angle of a liquid on the rough surface coated with a thin gold film and the contact angle of the same liquid on a perfectly flat gold surface prepared by metallizing a clean and smooth glass plate. The value of r_f is then calculated from Equation 2.8. Table 2.2 shows the coefficients of roughness obtained for various aluminium substrates which have been subjected to different pretreatments. Thus, this novel method enables the surface roughness to be accounted for in subsequent wettability studies.

Table 2.2 Roughness factors, r_f, for various aluminium surfaces

Solid surface	r_f
Smooth glass plate (reference)	1.0
Anodized aluminium	1.47
Sealed anodized aluminium	1.08
Phosphated aluminium	1.01

2.4 SURFACE AND INTERFACIAL FREE ENERGIES

A distinction may be made between low-energy and high-energy solid surfaces. Organic compounds, such as polymers, belong to the first group and their surface free energies are usually less than 100 mJ/m^2. Metals, metal oxides and ceramics belong to the second group and their surface free energies are typically greater than 500 mJ/m^2.

2.4.1. Low-energy surfaces

2.4.1.1 The critical surface tension

The first approach to the characterization of low-energy, polymeric solid surfaces was an empirical one developed by Zisman and co-workers. He and his colleagues [29–33] established that, for low-energy solids and a series of liquids, a rectilinear relationship frequently existed between the cosine of the contact angle, $\cos\theta$, and the surface tension of the wetting liquid, γ_{lv}. This is illustrated for various polymeric materials in Fig. 2.3. From such observations Zisman defined a critical surface tension of wetting, γ_c, by the value to which γ_{lv} extrapolated as $\cos\theta$ tends to unity, i.e. as θ tends to 0°. Thus γ_c is the surface tension of a liquid which will just spread on the surface giving a zero contact angle. Zisman and subsequent workers [33] have found a similar behaviour for a wide range of homologous series of liquids and solid surfaces. Indeed, Zisman found that even when $\cos\theta$ was plotted against γ_{lv} for a series of non-homologous liquids the points fell close to a straight line. However,

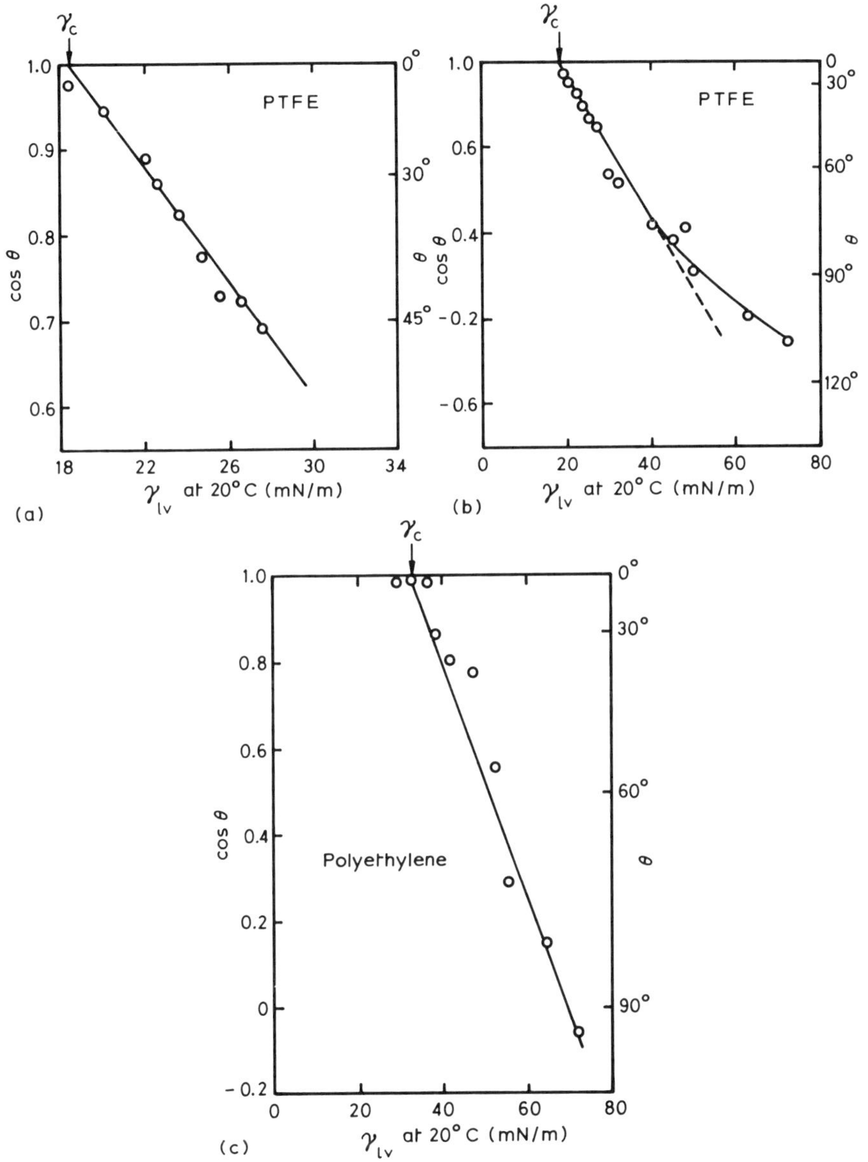

Figure 2.3 Zisman plots for various low-energy polymeric surfaces. (a) Polytetrafluoroethylene (PTFE) with *n*-alkanes as the liquid series, (b) PTFE with a wide range of liquids, (c) polyethylene with the liquid series commonly used by Zisman.

Table 2.3 Values of surface free energies of plastics, rubber and composites obtained at room temperature by various methods [6,28,35–42,51]

Solid surface	Critical surface tension, γ_c (mN/m)	γ_s^D from Equation 2.16 (mJ/m^2)	Values from Equation 2.18 (mJ/m^2)			Surface tension from molten polymers (mN/m)
			γ_s^D	γ_s^P	γ_s	
Polyhexafluoropropylene	16.2–17.1	18.0	11.7	0.7	12.4	—
Polytetrafluoroethylene	18.5	19.5	18.6	0.5	19.1	26.5
Poly(vinylidene fluoride)	25	—	23.2	7.1	30.3	—
Poly(vinyl fluoride)	28	—	31.3	5.4	36.7	—
Poly(chlorotrifluoroethylene)	31	30.8	31.4	2.1	33.5	31.1
Polyethylene	31	35.0	31.3	1.1	32.4	35.7
Polypropylene	31	30.2	—	—	—	—
Polystyrene	32.8	44.0	38.4	2.2	40.6	40.7
Poly(vinyl chloride)	39	—	40.0	1.5	41.5	—
Poly(methyl methacrylate)	39	—	35.9	4.3	40.2	—
Poly(vinylidene chloride)	40	—	42.0	3.0	45.0	—
Nylon-6,6	42.5	—	33.6	7.8	41.4	—
Poly(ethylene terephthalate)	43	—	41.8	3.3	45.1	—
Typical amine-cured epoxy	—	—	41.2	5.0	46.2	—
Rubber-toughened epoxy	—	—	37.2	8.3	45.5	—
Phenol–resorcinol resin	52	—	—	—	—	—
Urea–formaldehyde resin	61	—	—	—	—	—
Styrene–butadiene rubber	—	—	27.8	1.3	29.1	—
Acrylonitrile–butadiene rubber	—	—	26.5	9.5	36.0	—
Carbon fibre reinforced plastic (abraded)	—	—	27.4	30.6	58.0	—

some deviations were observed for solid surfaces having very low values of γ_c (where $\gamma_c < 20$ mN/m), and this feature may be seen in some of the data shown in Fig. 2.3b.

The widespread occurrence of the rectilinear relationship between $\cos\theta$ and γ_{lv} led Zisman to use the critical surface tension to characterize and compare the wettabilities of a variety of low-energy surfaces. Some typical values of γ_c for polymeric solids are shown in Table 2.3. The critical surface tension is an extremely useful parameter for the characterization of low-energy solids. Lee [43] has even derived an empirical relationship between γ_c and the glass transition temperature. However, Zisman has always emphasized that γ_c is not the surface free energy of the material, but only an empirical parameter the relative values of which act as one would expect the surface free energy to behave. Many subsequent authors have assumed that these two parameters are equal without comment. The relationships between γ_c, γ_{sv} and γ_s for a low-energy solid are discussed below.

In more recent work Dann [44] has re-examined the empirical observations of Zisman. He measured the contact angles of homologous and non-homologous series of liquids on a number of polymeric surfaces and found that different liquid series gave different values of critical surface tension, γ_c, for a particular polymeric surface. In fact Fox and Zisman [30] had also previously observed this result for paraffin and hexatriacontane surfaces when they employed different liquid series. They correctly attributed these observations to the different values of solid/liquid interfacial tensions and stated that there could be different values of γ_c depending upon the solid/liquid interfacial tension. Dann also noted that the relationship between the cosine of the contact angle and the surface tension of the liquid was normally a curve of a power form rather than a linear one. He employed the concepts of Good, Girifalco and Fowkes [45–47], which are discussed in detail below, to explain this feature. Basically, the reasons why Zisman frequently observed linear relationships were, firstly, that the hydrocarbon liquids he employed possessed only a narrow range of surface tensions. Secondly, only dispersion (or London) force components contributed to their total surface tensions, i.e. no polar contribution was involved. Thirdly, the non-homologous liquids commonly used by Zisman had the unusual feature that the dispersion force component decreased as the total surface free energy increased, and it was this feature that caused the plot of $\cos\theta$ against γ_{lv} to approximate to a linear relationship. Dann recommended that, when determining values of γ_c, long extrapolations should be avoided and that upon reporting values of γ_c the type of liquid series employed should be quoted. Finally, Dann demonstrated that when a liquid series possessing only dispersion forces is used, the critical surface tension equals the dispersion force component of the surface free energy of the solid and consequently provides a method for determining the true surface free energy of the solid.

2.4.1.2 The Good–Girifalco approach

Good and Girifalco [45,46] have expressed the interfacial free energy in terms of the surface free energies of the pure phases by analogy to the Berthelot relation [48]. This relation is for the attractive constants between like molecules A_{aa} and A_{bb} and that between unlike molecules A_{ab} and may be expressed by:

$$\frac{A_{ab}}{(A_{aa}A_{bb})^{1/2}} = 1 \tag{2.9}$$

For generality Good and Girifalco set their relationship to Φ so that Φ possessed a maximum value of unity and was defined in terms of molecular geometrics and forces of interaction across adjacent phases [49]. Thus when expressed in terms of surface free energies for two phases a and b, the interfacial free energy between the phases, γ_{ab}, is given by:

$$\gamma_{ab} = \gamma_a + \gamma_b - 2\Phi(\gamma_a\gamma_b)^{1/2} \tag{2.10}$$

where γ_a and γ_b are the surface free energies of the phases a and b. This geometric mean average has previously been used successfully by Hildebrand and Scott [50] for predicting the intermolecular energy of solutions. Equations 2.3 and 2.10 may be combined to yield:

$$\gamma_s = \frac{[\gamma_{lv}(1+\cos\theta)+\Pi_s]^2}{4\Phi^2\gamma_{lv}} \tag{2.11}$$

Various workers [51–55] have suggested that the spreading pressure, Π_s, for liquid on a surface with which it makes finite contact, is negligible, which means that a relatively high-energy liquid is not adsorbed and hence cannot reduce γ_s of a low-energy solid.

Thus Equation 2.11 becomes:

$$\gamma_s = \frac{\gamma_{lv}(1+\cos\theta)^2}{4\Phi^2} \tag{2.12}$$

Setting $\theta = 0°$, it is evident that the relation of γ_s to γ_c is:

$$\gamma_c = \Phi^2\gamma_s \tag{2.13}$$

Hence, from contact angle measurements, liquid surface tensions and a knowledge of Φ [54,56], the value of γ_s may be ascertained, and from Equation 2.10 the value of γ_{sl} may also be found.

2.4.1.3 *Additivity of components*

Fowkes has developed a similar argument for the interfacial free energy but adopted a slightly different approach. Fowkes [47] first proposed that the surface free energy of a pure phase, γ_a, could be represented by the sum of the contributions arising from different types of force components. He identified [57] at least seven components but Schultz *et al.* [58] have suggested that the surface free energy may be generally expressed by two terms, namely, a dispersion and a polar component, such that:

$$\gamma_a = \gamma_a^D + \gamma_a^P \tag{2.14}$$

where γ_a^D is the dispersion force component and γ_a^P is the polar force component. This latter term incorporates all the non-dispersion force components. Fowkes then proposed that the geometric mean of the dispersion force components is a reliable prediction of the interaction energies at the interface caused by dispersion forces. Thus, for interactions involving only dispersion forces:

$$\gamma_{ab} = \gamma_a + \gamma_b - 2(\gamma_a^D \gamma_b^D)^{1/2} \tag{2.15}$$

Fowkes [59,60] confirmed the validity of these equations by experimental and theoretical approaches, and has extended these concepts to solid/liquid systems by combining Equation 2.3 with Equation 2.15 to give:

$$\cos\theta = \frac{-\gamma_{lv} + 2(\gamma_s^D \gamma_{lv}^D)^{1/2}}{\gamma_{lv}} - \frac{\Pi_s}{\gamma_{lv}} \tag{2.16}$$

where γ_s^D and γ_{lv}^D are the dispersion force components to the solid and liquid surface free energies, respectively.

Tamai *et al.* [61] and Dann [62] introduced an energy term, $-I_{ab}$, into Equation 2.14 to account for the stabilization by non-dispersion forces. However, γ_s^P cannot be evaluated merely by the introduction of this term. Consequently, Owens and Wendt [63] and Kaelble and Uy [64], by analogy with the work of Fowkes, have proposed that I_{ab} may be expressed by the relation $2(\gamma_a^P \gamma_b^P)^{1/2}$. This expression may be introduced into Equation 2.15 to give:

$$\gamma_{ab} = \gamma_a + \gamma_b - 2(\gamma_a^D \gamma_b^D)^{1/2} - 2(\gamma_a^P \gamma_b^P)^{1/2} \tag{2.17}$$

Considering a solid/liquid system, this relationship may be combined with

Equation 2.3 to eliminate the interfacial free energy. Assuming the spreading pressure to be negligible, then:

$$1+\cos\theta = \frac{2(\gamma_s^D)^{1/2}(\gamma_{lv}^D)^{1/2}}{\gamma_{lv}} + \frac{2(\gamma_s^P)^{1/2}(\gamma_{lv}^P)^{1/2}}{\gamma_{lv}} \tag{2.18}$$

Fowkes [60] and Dann [44] have evaluated γ_{lv}, γ_{lv}^D and γ_{lv}^P for many liquids using the concepts previously discussed, and some typical liquids used in contact angle studies together with their values of γ_{lv}, γ_{lv}^D and γ_{lv}^P are given in Table 2.4. Consequently, if the contact angles of two liquids on a solid surface are measured, simultaneous equations may be formed from Equation 2.18 and solved to obtain values of γ_s^D and γ_s^P. The total surface free energy of the solid surface, γ_s, is then simply the sum of these components. This approach has been discussed in detail by Kaelble [65] and Sherriff [66]. Values of γ_s, γ_s^D and γ_s^P for various low energy surfaces are given in Table 2.3.

The results in Table 2.3 indicate that the values of surface free energies for the different polymers evaluated from the approaches outlined above are in good agreement with values obtained from other methods. However, the concepts outlined above have their critics. Fowkes [67], Wake [68] and Panzer [69] have

Table 2.4 Surface free energies of liquids used for contact angle measurements

	Surface free energy (mJ/m^2)		
Liquid	γ_{lv}^D	γ_{lv}^P	γ_{lv}
Water	22.0	50.2	72.2
Glycerol	34.0	30.0	64.0
Formamide	32.3	26.0	58.3
Di-iodomethane	48.5	2.3	50.8
Ethan-1,2-diol	29.3	19.0	48.3
1-Bromonaphthalene	44.6	0.0	44.6
Dimethylsulphoxide	34.9	8.7	43.6
Tricresylphosphate	36.2	4.5	40.7
Pyridine	37.2	0.8	38.0
Dimethylformamide	32.4	4.9	37.3
Polyglycol E-200	28.2	15.3	43.5
Polyglycol 15-200	26.0	10.6	36.6
2-Ethoxyethanol	23.6	5.0	28.6
Hexadecane	27.6	0.0	27.6
Tetradecane	26.7	0.0	26.7
Dedecane	25.4	0.0	25.4
Decane	23.9	0.0	23.9
Octane	21.8	0.0	21.8
Hexane	18.4	0.0	18.4

objected to the use of a relation, such as the geometric mean, to predict the polar force interactions. They consider that polar force interactions within liquids or solids cannot be predicted because of interaction energies observed between them. Their criticisms are based upon experimental evidence and theoretical considerations. For example, water may form hydrogen bonds with the π orbitals of polystyrene but there are no hydrogen bonds within polystyrene. This aspect is discussed in more detail below when acid–base interactions are reviewed (Section 2.4.1.4). They also commented that the flexibility of the macromolecular chains means that the polar groups will have specific orientations according to the nature of the liquid present on the surface of the polymer. Indeed, the work of Baszkin *et al.* [70] does provide evidence for this effect. Again this aspect will be discussed in detail later (Section 2.4.3). Lipatov and Feinerman [71] have agreed with these objections but consider that they are equally true for the dispersion force interactions and thus invalidate the whole approach, as expressed in Equations 2.14 and 2.15. They also concurred with Bikerman [72] who has commented that the additivity hypothesis contradicts both electrodynamics and interaction theory for liquids. Yet others [73,74] have accepted the basic additivity approach for deducing both the dispersion and polar force interfacial interactions but examined the possibility of using other relationships, such as the harmonic mean, rather than the geometric mean. However, the concept of employing the geometric mean for dispersion and non-dispersion forces has received experimental [58,75] and theoretical [76,77] justification. To summarize, whilst the concept of the additivity of surface force components has obviously yet to be unequivocally established, it has proved to be extremely useful in understanding and predicting adhesion phenomena. It appears that, providing the liquid series chosen can interact with the solid surface in a similar way to the adhesive of interest, then the values of γ_s^D and γ_s^P are of considerable help in adhesives technology, as will be shown later.

2.4.1.4 Donor–acceptor interactions

It is some 60 years ago that Lewis [78] proposed that bases are electron donors and acids are electron acceptors. More recently, Fowkes and co-workers [79–83] have treated hydrogen bonds as a subset of Lewis acid–base interactions and suggested that interfacial acid–base bonds between the adhesive and the substrate may often play a key role in adhesives technology.

Essentially, they have accepted that interfacial interactions between polar (Keesom) forces may possibly be predicted from a geometric mean relation, but consider that such interfacial interactions are usually negligible. In their view, more important are the acid–base interactions (which include hydrogen bonds) and they considered that these could not be predicted from a geometric mean relation. To deduce the contributions acting across an interface from such forces, Fowkes and colleagues have employed the expression developed by Drago *et al.* [84,85], who proposed that the enthalpy, ΔH^{AB}, of acid–base interactions could be described by an equation of the form:

$$-\Delta H^{AB} = C^{A}C^{B} + E^{A}E^{B} \tag{2.19}$$

where C^{A} and E^{A} are the two constants which are needed to characterize the acid and, similarly, C^{B} and E^{B} characterize the base. It was shown that from Equation 2.19 the extent of acid–base interactions could be determined to within ± 5%, or less, of experimental observations. As will be discussed in Chapter 3, the predictive capability of the approach embodied in Equation 2.19 is, as yet, somewhat limited in adhesives technology due to the lack of values of C and E for polymers and potential substrates.

However, Huntsberger [86] has questioned whether the interfacial interactions arising from polar forces, as embodied in Equations 2.14 and 2.17, can really be considered to be negligible. He has compared the predictions of the interfacial forces between a basic polyimide and various liquids from assuming either the geometric mean relation for the polar forces or the acid–base approach. He concluded that the contribution from polar forces was not always negligible and was reasonably well predicted by the geometric mean relation. Further, the contribution calculated from Equation 2.19 for the acid–base interactions was not always consistent with the known acidity/basicity of the liquids employed. Thus, choosing between the polar force approach and the acid–base approach is currently difficult. Polar contributions are probably included in the values of the acid–base constants in Equation 2.19 and, on the other hand, acid–base interactions are undoubtedly included in the polar force interactions deduced from Equations 2.14 and 2.17.

2.4.2 High-energy surfaces

As stated previously, metals, metal oxides and ceramics may be classified as high-energy solid surfaces and their surface free energies are typically greater than about 500 mJ/m^2, although it should be noted that there is still some debate [51,87–91] as to whether clean oxide-free metals, as opposed to metal oxides, are truly high-energy surfaces. The work of Schrader [90,91] on the contact angles of water on copper, silver and gold in ultrahigh vacuum suggests that clean metal surfaces are hydrophilic, i.e. water will spontaneously spread over their surfaces with a zero contact angle. However, from using Equations 2.7 and 2.15, this requires values for γ_s^D of 243 mJ/m^2 or greater which is at variance with some of the reported values (see Table 2.5). This apparent discrepancy may arise from the estimated values of γ_s^D given in Table 2.5 being in error, Equation 2.15 being invalid or, in the case of copper and silver, the water drop reacting with the clean metal surface to form an oxide layer, which

Table 2.5 Values of surface free energies for some high-energy surfaces [28,51,92,97–100,103,112–115]

Material	γ_s (pendant bubble) (mJ/m^2)	γ_s (sessile drop) (mJ/m^2)	γ_c (mN/m)	γ_s (theoretical) (mJ/m^2)	Surface free energy (mJ/m^2)		
					γ_s^D	γ_s^P	γ_s
Lead				442	99		
Copper				1360	60		
Nickel				1770			
Platinum		2672	3120	1915			
Mercury				319	200		
Graphite			924	1250	128		
Silver				890	74		
Gold					120 to 464		
Fe_2O_3				1357	107		
TiO_2 (anatase)					100		
Silica					78	209	287
Al_2O_3 (sapphire)	574	638	905	577–690	100		
Al_2O_3 (anodized)					125	44	169
BeO			640	1107			
Mica					30	90	120

Note: The values quoted are at the test temperature given in the original reference, thus some of the apparent discrepancies arise from different test temperatures, with the exception of the γ_s^D and γ_s^P which are at 20 °C.

has a high γ_s^P value, and so leads to the surface being hydrophilic. Schrader found that exposure of clean copper or silver to even small amounts of moisture resulted in the evolution of hydrogen, thus lending support to the last explanation. However, this matter has still to be completely resolved.

Many techniques have been used to determine the values of surface free energies of high-energy surfaces. Several workers [92–97] have calculated values of γ_s theoretically from, for example, the energy of vaporization. Some typical values are shown in Table 2.5 and, as may be seen, values of γ_s of metals and metal oxides at their melting points range from about 440 mJ/m^2 for lead to 1900 mJ/m^2 for platinum.

Considering experimental studies, then several techniques have been employed. Rasmussen and Nelson [98] have developed an X-ray radiographic method to obtain profiles of molten ceramic pendant bubbles in tungsten. The value of γ_s of the ceramic was then calculated from the shape of the pendant bubble. For example, a value of γ_s for Al_2O_3 of 574 ± 68 mJ/m^2 was determined, in good agreement with other values [94,95] using a similar method and with theoretical estimates [101]. As might be expected, many workers [102–108] in this area have used the sessile-drop method (Section 2.3.2). For high-energy surfaces, which are prone to surface contamination, contact angles are more reliably measured in a clean environment under vacuum. Studies of this type have been conducted by Murr [107,108] who has measured the contact angle of sub-micron liquid metal particles on ceramic substrates at temperatures greater than 1200 °C from scanning electron micrographs. He has also employed mercury as the test liquid and measured the contact angle of mercury on various metallic substrates using the scanning electron microscope, and typical micrographs are shown in Fig. 2.4. Such studies not only enable the contact angle to be determined accurately but also permit an energy-dispersive characteristic mercury X-ray map to be obtained. Thus, interfacial changes due to alloying or diffusion of atoms across the interface may be readily observed. This aspect is most important since it has been suggested [109] that, in the wetting of ceramic materials by molten metals, the contact angle is invariably small when the metal reacts with the oxide on the ceramic, but when there is no reaction the contact angle is large. Indeed, the strange shape of the drop profile for mercury on the zinc substrate shown in Fig. 2.4a is due to zinc diffusing into the mercury drop and vice versa, as revealed by the X-ray analysis for mercury (Fig. 2.4b).

However, since the wetting of high-energy materials is such an important area, many wetting studies have been reported in the literature, but the possibility of interfacial reactions affecting the measured contact angle must be raised as a cautionary note. In some studies [99] the contact angle of a sessile drop of a liquid salt or metal on the solid surface has been measured and, from a separate determination of the interfacial free energy, the value of γ_s for the solid deduced from Equation 2.1. Alternatively, Rhee [104] has adopted the

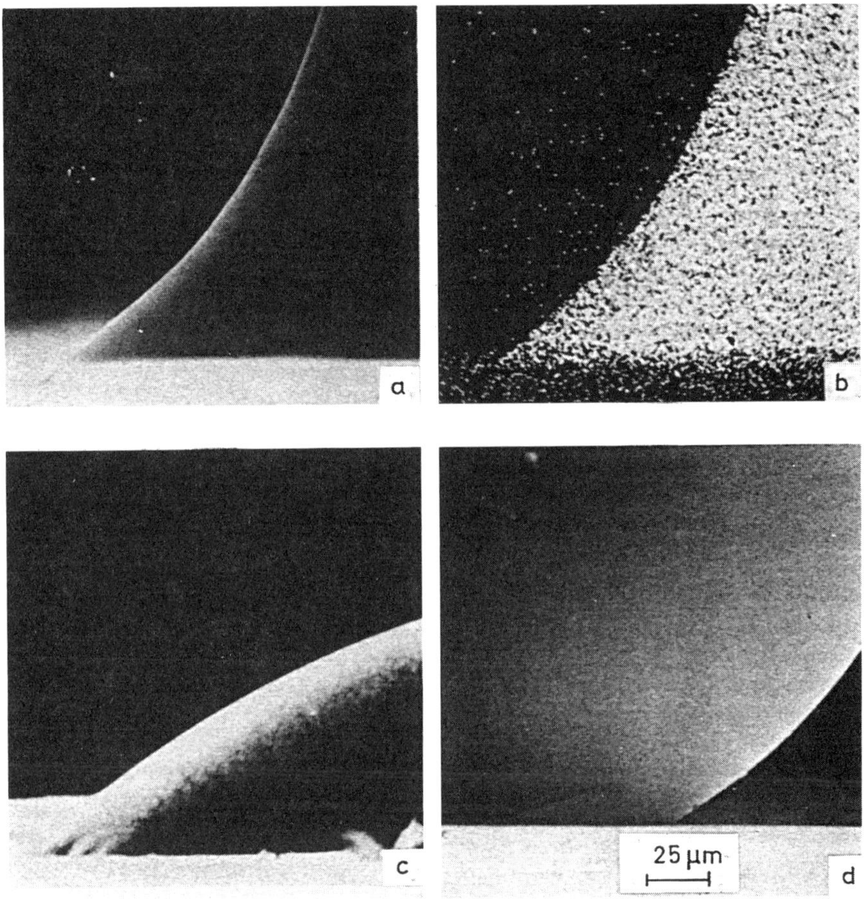

Figure 2.4 Contact angles of mercury/solid substrate systems viewed in the scanning electron microscope [108]. (a) Hg/Zn, (b) energy-dispersive, characteristic mercury X-ray map of (a), (c) Hg/Pd, (d) Hg/Zr.

Zisman critical surface tension approach, as discussed above for low-energy solids, and for example obtained a linear relationship between $\cos\theta$ and γ_{lv} for a BeO 'solid surface' and a series of Al–Mg alloy 'liquids' at 650 °C. A value for γ_c of 640 mN/m was reported, but this involved a very long extrapolation. Valentine [105] has not only criticized this work on these grounds but has also shown that the behaviour of binary molten metal alloys on ceramic substrates is often unpredictable (as, by analogy, is the behaviour of mixed liquids on low-energy surfaces) and therefore considers that this technique is usually inapplicable. Rhee [103] subsequently suggested a somewhat different experimental method. He reported that the value of γ_c may be found by using a single liquid metal and varying the temperature. For several ceramic surfaces he found a Zisman-type plot where the cosine of the contact angle increased linearly with increasing temperature, i.e. with decreasing liquid surface energy, γ_{lv}, for a given ceramic surface. However, further work [106] has clearly demonstrated that this temperature variant, contact angle method cannot be generally used to determine surface free energies of high-energy materials. Values of γ_c for cases where the above methods appear to be reasonably reliable are listed in Table 2.5. The values quoted are, of course, for a given (often very high) test temperature, but temperature coefficients of the surface free energy have been reported [92,99,103] and so room-temperature values may be ascertained; typical coefficients are in the range -0.1 to -0.3 mJ/m^2 per K increase in temperature.

Several methods have been proposed to enable the value of the dispersion force component, γ_s^D, to the solid surface free energy to be determined. Schonhorn [92] has commented that from theoretical considerations for the noble metals (Cu,Fe,Co,Ni,Pb,W,Tl,Al,Sn) and alkali metals, which possess cubic or tetragonal crystal structures, then:

$$\gamma_s^D/\gamma_s \cong 0.07 \tag{2.20}$$

Whilst for metals such as Hg,Sb,Cd,Zn,Mg and Bi, i.e. those with hexagonal close-packed or rhombohedral crystal structures, the relation is:

$$\gamma_s^D/\gamma_s \cong 0.4 \tag{2.21}$$

Experimental values have been reported for γ_s^D by Fowkes [51] and Schultz *et al.* [28,58,110]. Fowkes has combined Equations 2.2 with 2.15 to give:

$$\gamma_s^D = (\Pi_s + 2\gamma_{lv}^2)/4\gamma_{lv}^D \tag{2.22}$$

Values of the equilibrium spreading pressure, Π_s, for various adsorbates were obtained from adsorption studies and, from a knowledge of γ_{lv} and γ_{lv}^D for the adsorbate, the value of γ_s^D for the high-energy solid surface was calculated. The method of Schultz and co-workers is more direct and is based upon the measurement of the contact angle of water on the solid which is immersed in an *n*-alkane. They have studied mica and reported values for γ_s^D and γ_s^P of 30 and 90 mJ/m^2, respectively. These values give a total surface free energy of 120 mJ/m^2 which is in excellent agreement with the value reported by Bailey and Price [111] from cleavage experiments. Values of γ_s^D and γ_s^P for some high-energy surfaces are given in Table 2.5.

Finally, it must be noted that the values of critical surface tensions and surface free energies listed in Table 2.5 are for *clean* surfaces. In the typical bonding environment this is, of course, not the usual situation. The surfaces may be clean from the technologist's viewpoint but they are certainly not truly clean as understood by the surface scientist. This aspect, and its implications, are considered later in Section 2.6.2.

2.4.3 Orientation at interfaces

In attaining intimate molecular contact at the interface another aspect which should be considered [116–118] is the conformation of the adhesive or primer molecule when it is adsorbed onto the substrate surface. The conformation of the adsorbed molecule will be a function of many parameters, including the chemical and physical nature of the adsorbing species and the substrate surface, the concentration and molecular weight of the adsorbing molecules, the temperature, etc.

Considering those aspects most relevant to the present discussions, then small low molecular weight molecules, such as adhesives (or primer) monomers or oligomers, will typically adsorb in the most favourable conformation from the energetics standpoint. This may lead to preferential adsorption of one particular component from a given formulation, which typically contains a mixture of resin, hardener, catalysts, etc. This aspect, and its implications, are discussed further in Section 8.3.3.5.

If the adhesive possesses a relatively high molecular weight, then the molecules may adopt a variety of conformations giving, for example, multisite adsorption for one molecule. In between the adsorbed sites the polymer chain will typically form 'loops', with the polymer chains at the ends of the macromolecules acting as 'tails'. The length of the loops and tails, and their orientation to the surface, will be a function of the adsorption energy. At low energies long loops or tails are favoured and give rise to a conformation extended in the direction normal to the substrate surface, whereas, at high adsorption energies, attachment at a large number of sites per molecule results and small loops and tails predominate, which leads to a flattened conformation. Silberberg [117] has commented that if adsorption involves such conformation changes then these will in general not lower the tendency to establish intimate molecular contact, but will slow down the process, i.e. the kinetics will be changed. Implications of such multisite adsorption are discussed further in Sections 3.5.4 and 8.3.3.2.

Early work by Schonhorn [119] showed that orientated *monolayers* of amphipathic molecules, such as stearic acid, could be employed as extremely effective adhesives in the bonding of polyethylene to aluminium; indeed, so effective were these adhesives that joint strengths often exceeded the cohesive strength of the polyethylene! Multilayer adsorption lowered the joint strength not because of less intrinsic adhesion of the amphipathic molecule to the substrate, but because the relatively thick layer possessed low cohesive strength and thus behaved as a *weak boundary layer*. Chemisorption was thought to occur for the stearic acid on the aluminium oxide [120,121], and this has indeed been shown [122] to be the case by more recent work which employed Auger spectroscopy. This technique clearly revealed the formation of aluminium stearate. At the other interface, Wake [123] has speculated that some diffusion may have occurred across the interface between the alkane chains of the stearate and the polyethylene.

More recent elegant work by Baszkin *et al.* [70,75,124] has demonstrated the effect that orientation at the interface may have upon wettability and measured joint strengths. They quantitatively determined the surface densities of the polar sites on a polyethylene substrate as the surface was progressively oxidized, employing a technique based upon the adsorption of radioactive calcium ions. They showed that as the temperature was raised to between 80 and 85 °C there was a decrease in wettability. This arose from an increase in chain mobility which led to the redistribution of polar groups into the bulk, and that the remaining surface density of polar groups depended upon the nature of the liquid in contact with the polyethylene. This loss of polar groups at elevated temperatures was mirrored by a loss in joint strength.

2.4.4 Applicability to adhesives technology

Sharpe and Schonhorn [125] have emphasized the importance of wetting and proposed that one of the most important factors influencing adhesive joint strength is the ability of the adhesive to spread spontaneously on the substrate when the joint is initially formed. They developed a criterion from Equation 2.4 for the case when the adhesive will spontaneously spread on the substrate. By ignoring the interfacial free energy and using Zisman's critical surface tension concepts, they proposed that the γ_c of the adhesive must be less than that of the substrate. From Table 2.3 it may be seen that this means, for example, that high joint strengths should not result when polyethylene substrates are bonded using an epoxy–amine adhesive, and this is in agreement with experimental observations. However, this criterion predicts that polyethylene should adhere strongly when it is used as a hot-melt adhesive to bond cured epoxy–amine solid substrates, again this is in accord with experimental observations. This arises because, in the former case, γ_c (polyethylene substrate) $< \gamma_{lv}$ (epoxy–amine adhesive), and hence wetting is predicted to be limited. In the latter case γ_c (cured epoxy–amine substrate) $> \gamma_{lv}$ (polyethylene adhesive) and so spontaneous and complete wetting occurs. Thus, the Sharpe and Schonhorn criterion essentially proposes that a mobile, liquid adhesive with small or zero contact angle is of prime importance. Such an adhesive will spread readily, flow into crevices and achieve true interfacial contact with little opportunity for the formation of voids which may act as a source of inherent weakness. Further work [126–129] has since suggested that an additional requirement is that the interfacial free energy, γ_{sl}, should be kept as low as possible.

However, Huntsberger [86,130,131] has argued that the above criterion is incorrect. Considering the free energy, W_F, change upon formation of unit area of interface, then W_F is given by:

$$W_F = \gamma_{sl} - \gamma_{sv} - \gamma_{lv} \tag{2.23}$$

Substituting from Equation 2.1, and considering a rough solid surface and so introducing the roughness factor, r_f, (Section 2.3.3, Equation 2.8) gives:

$$W_F = -\gamma_{lv}(1 + r_f \cos\theta_s) \tag{2.24}$$

Thus Equation 2.24 suggests that a decrease in free energy accompanies wetting for all cases except when $r_f \cos\theta_s < -1$. Hence, Huntsberger has concluded that for most practical adhesive systems where the adhesive will exhibit a contact angle with the substrate of less than 90°, and therefore a positive $\cos\theta_s$, that thermodynamic equilibrium will correspond to a completely wetted state. However, whilst this is obviously true for the local area of interface formed, the adhesive will not spontaneously spread and completely wet the substrate if the value of the contact angle is more than 0°. Indeed, as might be expected, a low value of contact angle is often associated with the maximum rate of spreading.

These discussions emphasize the importance of wetting as a kinetic process. For, although the thermodynamics may indicate the establishment of intimate molecular contact, the kinetics of wetting may be the determining factor. Such aspects are reviewed next.

2.5 KINETICS OF WETTING

So far wetting has been considered from the thermodynamic viewpoint but the kinetics of wetting are also important. Indeed, although the thermodynamics may indicate the establishment of intimate molecular contact, the kinetics of wetting may be the determining factor.

2.5.1 Surface tension gradients

Bascom *et al* [132,133] have employed interference microscopy and ellipsometry and showed that if the contact angle is zero, surface tension gradients may exist at the spreading front which may assist or hinder spreading depending upon their direction. These surface tension gradients arise from thermal gradients or, in the case of liquids containing a more volatile component of different surface free energy, such as a trace impurity, they may arise from a concentration gradient. This effect is illustrated in Fig. 2.5. At the leading edge there is a thin (either one of two monomolecular layers) primary

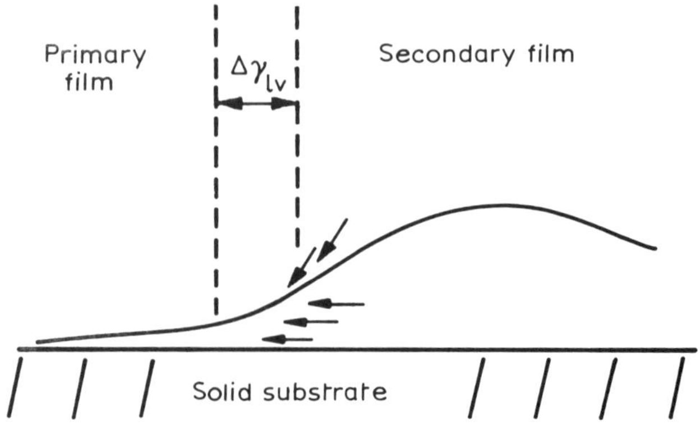

Figure 2.5 Schematic diagram of the surface and bulk flow patterns at the leading edge of a spontaneously spreading film [132].

film which is followed by a transition region in which thermal or concentration gradients create a surface flow that drags the underlying liquid forward. The rate of flow in the transition region may exceed the rate of gravity flow in the much thicker secondary film so that a ridge of liquid develops. Spreading rates resulting from this gradient effect are relatively low: typically about 0.1 μm/s for a polyisobutylene fluid on polished steel.

2.5.2 Dynamic contact angles

When an adhesive is applied to the substrate surface it is usually forcibly spread across the surface. Now if a liquid, having a viscosity greater than a few tens of centipoise, is forced rapidly across a solid surface then the liquid near the surface cannot keep up with the advancing front [132,134,135]. This leads to the dynamic contact angle, θ_d, being greater than the equilibrium value, and for liquids with viscosities in the range 10 to 1000 cp the dynamic angle is best described by the Fritz equation:

$$\tan\theta_d = m\left(\frac{\eta v_s}{\gamma_{lv}}\right)^n \tag{2.25}$$

where η is the viscosity, v_s is the velocity, and m and n are constants. In Fig. 2.6 calculated [132] values of θ_d are plotted against viscosity, η, assuming a spreading velocity of 100 mm/s and a liquid surface energy, γ_{lv}, of 30 mJ/m^2. Hence, under such circumstances the extent of wetting is obviously more limited than would be predicted from simple contact angle measurements like those described in Section 2.3.2.

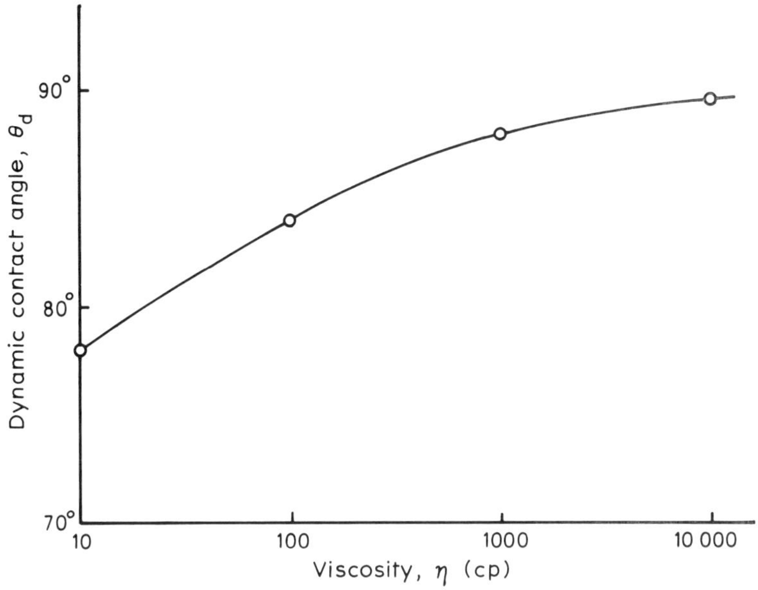

Figure 2.6 Dynamic contact angle, θ_d, versus the liquid viscosity, η [132].

2.5.3 The influence of surface roughness

The roughness of the substrate surface may also influence the kinetics of wetting.

Firstly, a liquid forming a contact angle of less than 90° with a solid surface may spread along fine pores, scratches and other inhomogeneities by capillary action, even though it will not spontaneously wet a planar surface. Bascom *et al.* [133] have reported that random surface scratches may increase the spreading rate of some liquids by as much as 50% and that open capillaries filled well ahead of the diffusional advance of the primary film. Similar observations have been recorded by Cottington *et al.* [136]. Cheever [137] has studied zinc phosphate conversion coatings and treated the situation theoretically as a capillary in which the capillaries assume the particular shape of slits. A mathematical and physical model was derived from Poiseuille's equation (see below) which adequately described the flow of liquid on zinc phosphated steel substrates, and gave the capillary pressure which was generated in the zinc phosphate coating to be about 0.7 MPa. However, spreading rates from this effect and the surface tension gradients are not high and may be overshadowed when a liquid adhesive is forcibly spread over such a surface. Nevertheless, Bascom and Patrick [132] have suggested that such effects may play a role in the redistribution of the adhesive after its initial application.

Secondly, de Bruyne [138] has modelled various types of substrate surface topography and obtained quantitative expressions for the relationship between the extent of wetting and the driving pressure, taking any back pressure in the case of closed pores into account. For a closed cylindrical pore the depth of penetration, l_p, is given by:

$$l_p = l\left(1 - \frac{p_a r_o}{2\gamma_{lv}\cos\theta + p_a r_o}\right) \tag{2.26}$$

where p_a is atmospheric pressure, l is the length and r_o is the radius of the cylindrical pore. Obviously, the smaller the value of r_o the greater the proportion of the pore's length which is filled at equilibrium. Also, considering the effect of contact angle and shape of cavity, then Fig. 2.7 shows the capillary penetration to be expected, under atmospheric pressure and against the back pressure of trapped air, as a function of the contact angle made by the penetrating liquid on the solid surface. In the case of the cylindrical pore, at low values of θ the penetration is good, but falls off rapidly as the value of θ increases. However, the limited penetration into the 'ink-bottle' pits is especially marked. This effect will be heightened if the liquid is forced rapidly across the surface, since, as mentioned above, a higher, dynamic, contact angle will be observed. Further, the above discussions assume equilibrium conditions are appropriate but, since adhesives usually harden in a relatively short time, the kinetics of penetration also need to be considered. As reviewed by Packham [139], the kinetics may be expressed by reference to Poiseuille's law which, for a Newtonian liquid flowing into a tube of radius r_o under the influence of a driving pressure p_r, gives l_p as:

$$l_p \frac{dl_p}{dt} = \frac{r_o^2 p_r}{8\eta} \tag{2.27}$$

where t is time. If the driving pressure is taken to be the capillary pressure, i.e. $2\gamma_{lv}\cos\theta/r_o$, then Equation 2.27 may be integrated, assuming θ and η are constant to give the Washburn equation which shows the penetration, l_p, is given by:

$$l_p^2 = \frac{r_o \gamma_{lv} \cos\theta}{2\eta} \cdot t \tag{2.28}$$

Thus, the extent of penetration, l_p, is proportional to the square root of time. Also, this equation reveals that the penetration will increase as the viscosity of the adhesive is decreased, as the adhesive surface free energy is increased and as its contact angle approaches zero. Obviously, these last two requirements

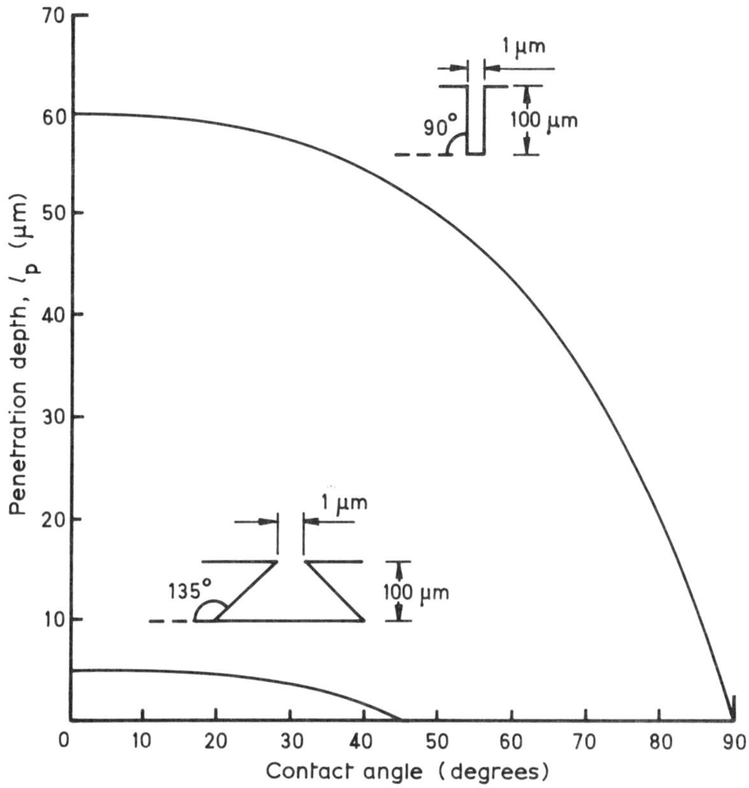

Figure 2.7 Comparison of the penetration of a liquid into cylindrical and 'ink-bottle' pits [138].

represent conflicting properties. Good [4] has shown that if the walls of the pore immediately ahead of the penetrating liquid are free from adsorbed vapour of the liquid, an additional force associated with the spreading pressure, Π_s, will draw the liquid into the tube and Equation 2.28 should be modified to:

$$l_p^2 = \frac{r_o(\gamma_{lv}\cos\theta + \Pi_s)}{2\eta} \cdot t \tag{2.29}$$

As commented by Packham [139], with a polymeric adhesive, evaporation from the liquid might well be slow, thus preventing the formation of a fully saturated band of adsorbed vapour ahead of the penetrating liquid. Hence, the value of l_p will be intermediate between that given by Equations 2.28 and 2.29. Considering technological applications, then the above discussions indicate that external pressure to assist the capillary driving pressure and heat to lower the viscosity and the contact angle (see below) will all aid wetting and penetration. Indeed, the adhesive, or primer if employed, is sometimes applied from a solvent which will ensure a low viscosity. Although, particularly in the case of structural adhesives, the problem of adequate solvent removal frequently inhibits the use of solvents.

2.5.4 Time–temperature considerations

Various workers [140–144] have examined the kinetics of wetting of molten polymers on substrates by following the change in the contact angle of molten polymers as a function of time at different temperatures. The value of θ obviously decreases with increasing time, until equilibrium is reached; and equilibrium is reached in a shorter time at higher temperatures. Further, the rate of wetting is independent of the mass of the drop. Schonhorn *et al.* [140] have demonstrated that the ratios of the cosine of the contact angle, θ_t. at a time, t, to the cosine of the contact angle, θ_∞, at infinite time, could be superimposed using a time–temperature equivalence to give a single master curve. Newman [143] and Cherry and Holmes [142] have reported that this time–temperature dependence may be expressed by:

$$\cos\theta_t = \cos\theta_\infty[1 - k_1\exp(-k_2 t)] \tag{2.30}$$

where k_1 and k_2 are constants and where [141]:

$$k_2 = \frac{\gamma_{\mathrm{lv}}}{\eta L} \tag{2.31}$$

where η is the viscosity of the molten polymer and L is a parameter with dimensions of length [142,144], independent of temperature and related to the adhesive/substrate interfacial interactions. Thus, this approach confirms that rapid wetting is favoured by the spreading adhesive exhibiting a low contact angle and possessing a high surface tension and a low viscosity; obviously for a low-energy substrate these first two requirements are antithetical, as pointed out earlier. On a practical level, Equations 2.30 and 2.31 may be used at least to give a guide to the time necessary to attain equilibrium wetting and the effect of raising the temperature on the kinetics of wetting.

Cherry and Holmes also concluded that the spreading of a molten polymer across a metal surface could be described in terms of an activated rate process in which the activation energy barriers were the same as those which oppose

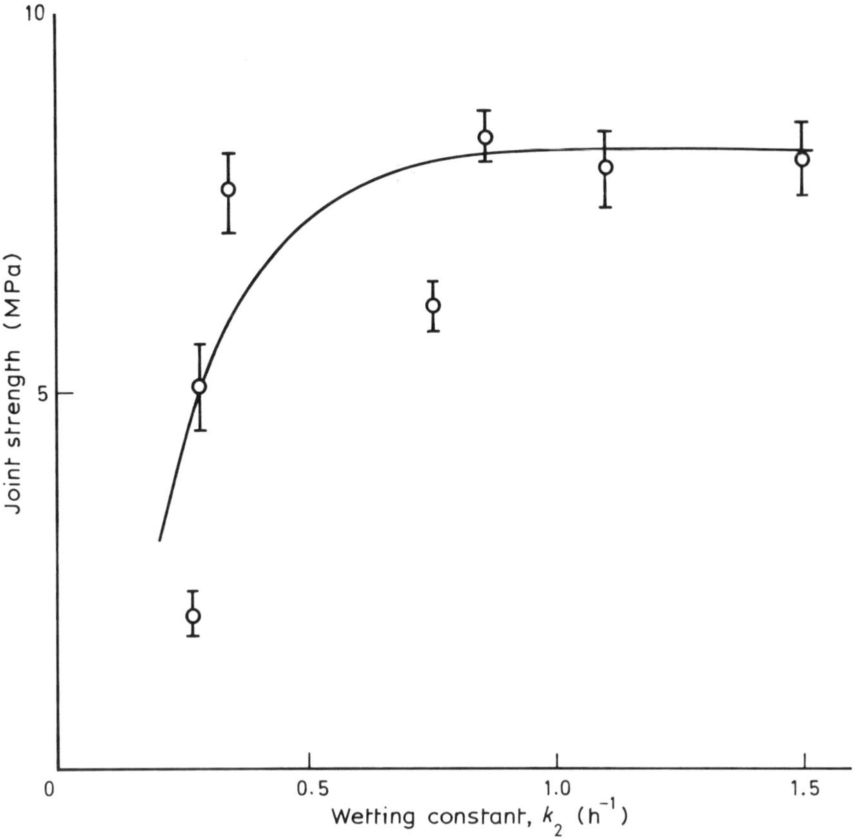

Figure 2.8 Joint strength versus wetting constant, k_2, for a series of thermosetting resin/polyethylene joints [144].

viscous flow. The driving force for this derives from the action of surface forces during the replacement of a solid/vapour interface by the solid/liquid interface. In a subsequent work [144] the parameter k_2 was termed the 'wetting constant' and values of k_2 were determined for a series of epoxy–urea–melamine adhesives bonding polyethylene substrates. The resulting joint strengths were also measured and the correlation between the wetting constant, k_2, and the joint strengths are shown in Fig. 2.8. It was generally thought that high joint strengths were associated with high values of the wetting constant, rather than being associated with a low contact angle, emphasizing the importance of wetting as a kinetic process. It is noteworthy that, like the Washburn equation (Equation 2.28), this approach emphasizes the importance of the ratio γ_{lv}/η in determining the kinetics of wetting, rather than simply just the viscosity, η, of the adhesive and, these parameters will, of course possess different temperature dependencies. However, in practical adhesives technology the formulator will have far greater control over the adhesive's viscosity than its surface free energy.

The time–temperature dependence of the contact angle, as expressed by Equation 2.30, may be combined with Poiseuille's law, Equation 2.27, and integrated to yield [139]:

$$l_{\mathrm{p}}^2 = \frac{r_{\mathrm{o}}\gamma_{\mathrm{lv}}\cos\theta_\infty}{2\eta} \cdot \left[t - \frac{k_1}{k_2} + \frac{k_1 \exp(-k_2 t)}{k_2}\right] \tag{2.32}$$

Thus, the value of the contact angle at time t, being greater than the equilibrium value, θ_∞, leads to a lower extent of penetration than if the equilibrium value had been relevant from the onset of the wetting process. This equation also reveals that a higher temperature, which will lead to the equilibrium value being attained in a shorter time, will therefore aid penetration, as indeed intuitively expected.

2.6 THE BONDING OPERATION

2.6.1 Air entrapment

Bascom and Cottington [145] have studied the mechanism and effects of air entrapment during the preparation of joints with structural film adhesives. Optical microscopy studies on glass/adhesive joints revealed that initially a thin film of air was trapped between the adhesive and glass substrate which usually extended over about half of the interfacial area. As the temperature was raised, to effect cure of the adhesive, this air drew up into bubbles that were eventually displaced into the adhesive layer. However, incomplete displacement occurred if the adhesive had a contact angle greater than zero on the substrate surface or the adhesive did not become sufficiently fluid during the heat cure. Such air-filled voids may be eliminated if the joint is bonded in an autoclave, where there is hydrostatic pressure high enough to compress entrapped air to a negligible volume and force it to dissolve into the adhesive or by employing a 'vacuum release' technique [145]. This latter method simply involves starting the cure in vacuum (about 650 Pa) and subsequently releasing the vacuum at the temperature at which the adhesive's viscosity is at a minimum. Air voids may obviously act as stress raisers and, indeed, increases in the peel strength of aluminium/epoxy adhesive joints of up to 30% were reported upon complete void removal using the vacuum release technique.

2.6.2 The bonding environment

Other aspects of the bonding operation are the environmental conditions employed and these can be particularly important when bonding metallic substrates which are, of course, usually covered by an oxide layer. As discussed in Section 2.4.2, metals and metal oxides are classified as high-energy solids since in the ultra-clean state they have surface free energies, γ_s, typically greater than 500 mJ/m^2. Thus, it would seem at first sight that such a surface would be readily wetted by organic adhesives having low surface free energies (typically less than about 70 mJ/m^2, see Table 2.3) once any machine oil, protective greases or other contaminants remaining from the material's production had been removed. However, even in the absence of these obvious impurities a high-energy surface adsorbs water vapour from the atmosphere [146,147], and other contaminants such as nitrogen and organic adsorbates (e.g. hydrocarbons [148]), which significantly lower the surface free energy of the substrate, i.e. to a value denoted by γ_{sv}. This effect may hinder spreading of the adhesive and limit the degree of interfacial contact.

This aspect is clearly illustrated by the work of Bernett and Zisman [149] who measured the critical surface tension, γ_c, of various 'clean' hydrophilic high-energy surfaces using a series of organic liquids at two extreme values of relative humidity (r.h.), 0.6 and 95% r.h. Their studies included oxides of nickel, iron, copper, aluminium, etc. and they found that all the surfaces were converted to ones of low γ_c (about 46 and 38 mN/m, respectively) and, further, that the values of γ_c were virtually independent of the actual substrate. The high-energy substrates had therefore a surface covering of adsorbed water, and possibly organic test liquid, the concentration of which was only a function of the relative humidity. The measured value of γ_c reflected the nature of this surface covering, rather than the nature of the underlying substrate. This may also be stated by taking $\gamma_c \cong \gamma_{sv}$ so that from Equation 2.2 the reduction of the surface free energy is given by the spreading pressure, Π_s:

$$\Pi_s = \gamma_s - \gamma_c \tag{2.33}$$

Gledhill *et al.* [150] have recently extended this earlier work by examining the effect of a range of relative humidities on the wettability of mild steel surfaces of different rugosities and have also considered the implications of such data on the strength of joints prepared under different humidities. From the contact angle measurements and, using Equations 2.18 and 2.14, the value of γ_{sv} for a polished and grit-blasted steel surface was deduced as a function of relative humidity. The results are shown in Fig. 2.9. For the polished surface there is a linear correlation and the results are in good agreement with the values previously reported by Bernett and Zisman [149]. For the rougher, abraded surfaces there is a considerably greater variation in the value of γ_{sv} with

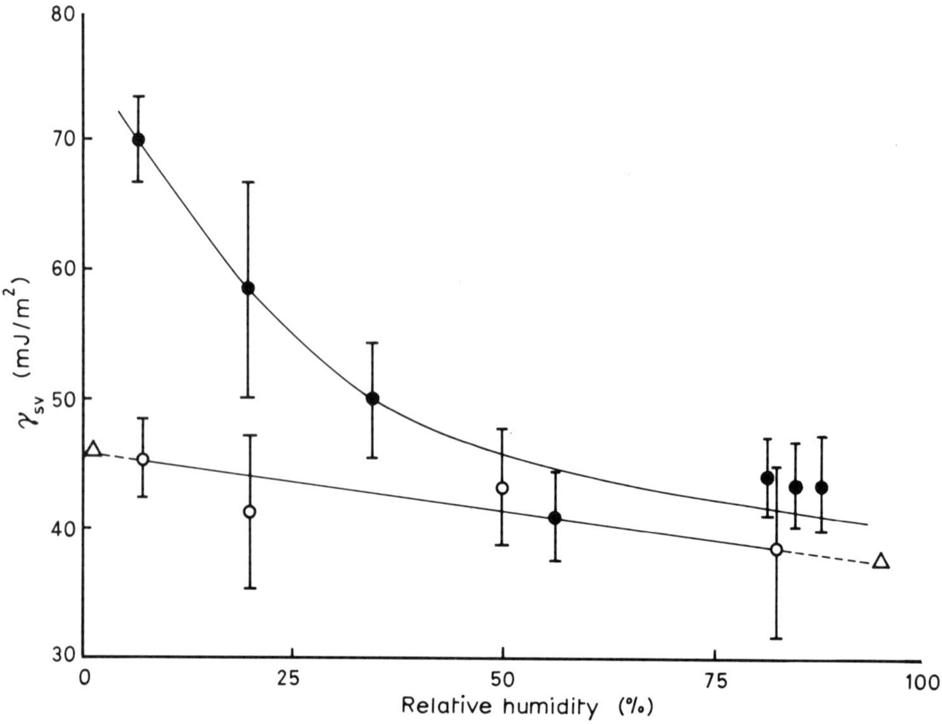

Figure 2.9 Surface free energy, γ_{sv}, of abraded and polished mild steel surfaces as a function of relative humidity [150]. ● Abraded steel surface; ○ polished steel surface; △ γ_c values for a polished steel surface, after Bernett and Zisman.

humidity and, as might be expected from Section 2.3.3, the value of γ_{sv} (rough) is greater than γ_{sv} (polished). Now from the values of γ_{sv}^D and γ_{sv}^P a wettability envelope may be constructed. For wetting to occur, Equation 2.4 stated that:

$$\gamma_{sv} \geq \gamma_{sl} + \gamma_{lv} \tag{2.4}$$

Substituting for γ_{sl} from Equation 2.17 into Equation 2.4 and re-arranging, the condition for wetting just to occur becomes:

$$\gamma_{lv} - (\gamma_{sv}^D \gamma_{lv}^D)^{1/2} - (\gamma_{sv}^P \gamma_{lv}^P)^{1/2} = 0 \tag{2.34}$$

Thus, knowing values of γ_{sv}^D and γ_{sv}^P for a solid surface enables Equation 2.34 to be solved to yield values of γ_{lv}^D and γ_{lv}^P for liquids which would just completely wet and spread spontaneously on the solid. The diagram which may be constructed from such data is termed a wettability envelope. Wettability envelopes for the abraded surfaces in 7 and 56% r.h. environments are shown in Fig. 2.10, together with typical γ_a^D and γ_a^P values for liquid amine-cured

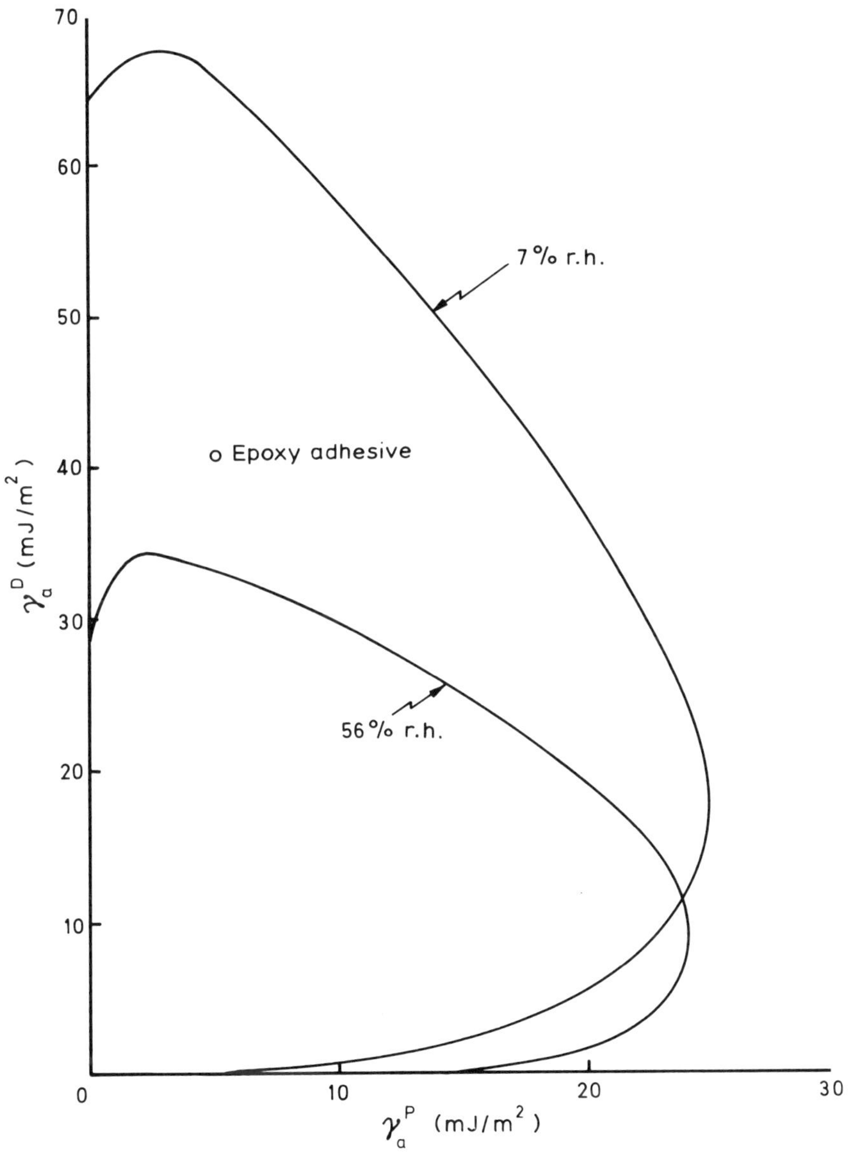

Figure 2.10 Wettability envelopes for abraded mild steel surfaces at various humidities [150].

epoxy adhesive (Table 2.3). As may be seen, the γ_a^D and γ_a^P values for this adhesive lie inside the 7% r.h. envelope but outside the 56% r.h. envelope. This indicates complete wetting of the steel surface by the epoxy adhesive at the lower humidity but non-spontaneous spreading at the higher humidity. When adhesive joints were prepared in these respective relative humidity environments the lower humidity resulted in significantly stronger joints. Thus the enhanced spreading and wetting of the steel surface by the adhesive at the lower humidity, as indicated by the wettability envelopes, was directly reflected in higher joint strengths. However, the particular epoxy adhesive used possessed a high viscosity and poor wetting characteristics and it is unlikely that similar improvements would be universally expected.

It should be noted that, in a few instances, the presence of moisture in the bonding, and storage, environment may also result in the uncured adhesive absorbing water which subsequently affects the cohesive strength of the cured adhesive [151,152]. This effect is particularly pronounced in the case of structural adhesive formulations based upon epoxy resins cured with dicyandiamide, and which contain additives such as inorganic thixotropic fillers, toughening rubbers and cure catalysts. It has been reported that the moisture content of the uncured adhesive should be below about 0.3 to 0.8% in order to achieve the optimum joint strength. Drying the adhesive to attain this value was recommended for those adhesives where the cure reaction was especially sensitive to the presence of absorbed moisture. This effect has been suggested to arise from the dicyandiamide which is dissolved in the resin or absorbed water, rather than being present as a separate solid phase, undergoing a hydrolysis reaction to produce guanylurea. The guanylurea will act as a catalyst for the cure reaction but possibly lead to a lower glass transition temperature for the cured adhesive.

Absorbed moisture in certain substrates, e.g. fibre composites, may also adversely affect the performance of the adhesive joint, but these aspects are best considered in Section 4.2.5 when surface pretreatments for such materials are reviewed.

From the above discussions it is obviously important to ensure that the bonding environment is such that adsorbed contamination is kept to a minimum; the lower the extent of such contamination, the more readily will it be displaced by the adhesive which may then achieve intimate molecular contact with the solid substrate. If multilayer contamination is not displaced by the adhesive it will act as a weak boundary layer of material at the interface and low joint strengths will result. It is of interest to note that the presence of adsorbed hydrocarbons probably assists this displacement process. This arises because the common adhesives are usually polar in character and thus will have a thermodynamic tendency to displace such non-polar contamination, especially if it is only physisorbed. The adhesives, such as certain epoxies and acrylics (Chapter 5), which are formulated to displace specific types of oils, protective waxes or pressing lubricants which may be covering the metal substrates rely upon this thermodynamic driving force, as well as their solvating power. It is also important in the case of some adhesive types and certain substrate materials to ensure that they do not absorb too high a moisture content either during storage or bonding.

Thus, on a practical level, attention should be given to establishing clean, controlled areas for adhesive bonding. Considering adhesive storage, in most cases a controlled humidity level is not necessary but a storage area where the storage conditions recommended for the various adhesive types can be met is a worthwhile investment. In the bonding shop a high level of cleanliness should always be maintained and care should be taken to exclude obvious contaminants such as silicones and aerosol sprays.

2.7 CONCLUDING REMARKS

This chapter has considered the various aspects which influence the attainment of intimate molecular contact between the liquid adhesive and the substrate materials which are to be bonded. All the reported work in this area is in agreement that this is a necessary, though by itself an often insufficient, requirement for obtaining strong and stable adhesive joints. However, the exact conditions and criteria needed to obtain such interfacial contact have yet to be firmly established. For example, it is not well established whether the thermodynamics of wetting really do play an important role in practical adhesives technology, where mechanical energy is often supplied to assist wetting. Alternatively, it may be that the kinetics of spreading the adhesive over the substrate surfaces are the major factor in attaining good intermolecular contact at the interface. Nevertheless, whatever the exact details, the available evidence suggests that both these aspects are important and that the relatively simple concepts and theories discussed in the present chapter do give the adhesives technologist a good guide as to both the important parameters involved in attaining adequate interfacial contact and the way in which changes in these parameters affect the degree and kinetics of wetting.

Assuming that intimate intermolecular contact across the interface has been attained, we next need to consider the nature of the intrinsic adhesion forces which act across the interface and will hold the joint together throughout its service life. Thus, Chapter 3 will discuss the mechanisms of adhesion.

REFERENCES

1. Pauling, L. (1960) *The Nature of the Chemical Bond*, Cornell University Press, New York.
2. Good, R. J. (1967) in *Treatise on Adhesion and Adhesives*, Vol. 1 (ed. R. L. Patrick), Marcel Dekker, New York, p. 15.
3. Young, T. (1805) *Trans. Roy. Soc.* **95**, 65.

4. Good, R. J. (1973) in *Aspects of Adhesion – 7* (eds D. J. Alner and K. W. Allen), Transcripta Books, London, p. 182.
5. Bangham, D. H. and Razouk, R. I. (1937) *Trans. Faraday Soc.* **33**, 1459.
6. Rance, D. G. (1985) in *Industrial Adhesion Problems* (eds D. M. Brewis and D. Briggs), Orbital Press, Oxford, p. 48.
7. Neumann, A. W. and Good, R. J. (1979) *Surf. Colloid Sci.* **11**, 31.
8. Fort, T. and Patterson, H. T. (1963) *J. Colloid Sci.* **18**, 217.
9. Padday, J. F. (1972) *Proc. Roy. Soc. Lond.* **A330**, 561.
10. Johnson, R. E. and Dettre, R. H. (1969) *Surf. Colloid Sci.* **2**, 85.
11. Sawicki, G. C. (1978) in *Wetting, Spreading and Adhesion* (ed. J. F. Padday), Academic Press, London, p. 361.
12. Callaghan, I. C., Everett, D. H. and Fletcher, A. J. P. (1983) *J. Chem. Soc. Faraday Trans.* **79**, 2723.
13. Penn, L. S., Bystry, F. A. and Marchionni, H. J. (1983) *Polymer Composites* **4**, 26.
14. Drzal, L. T., Mescher, J. A. and Hall, D. L. (1979) *Carbon* **17**, 375.
15. Carroll, B. J. (1976) *J. Colloid Interf. Sci.* **57**, 488.
16. Akutin, M. S., Kerber, M. L. and Stal'nova, I. O. (1973) *Zavod Lab.* **39**, 716.
17. Johnson, R. E. and Dettre, R. H. (1964) in *Advances in Chemistry Series, 43* (ed. R. F. Gould), American Chemical Society, Washington, p. 112.
18. Johnson, R. E. and Dettre, R. H. (1964) in *Advances in Chemistry Series, 43* (ed. R. F. Gould), American Chemical Society, Washington, p. 136.
19. Neumann, A. W. and Good, R. J. (1972) *J. Colloid Interf. Sci.* **38**, 341.
20. Wenzel, R. N. (1936). *Ind. Eng. Chem.* **28**, 988.
21. Cassie, A. B. D. and Baxter, S. (1944) *Trans. Farad. Soc.* **40**, 546.
22. Bartell, F. E. and Shephard, J. W. (1953) *J. Phys. Chem.* **57**, 211.
23. Bartell, F. E. and Shephard, J. W. (1953) *J. Phys. Chem.* **57**, 455.
24. Oliver, J. C., Huh, C. and Mason, S. G. (1977). *J. Adhesion* **8**, 233.
25. Oliver, J. C., Huh, C. and Mason, S. G. (1980) *Colloids and Surfaces* **1**, 79.
26. Eick, J. D. and Good, R. J. (1975) *J. Colloid Interf. Sci.* **53**, 235.
27. Hitchcock, S. J., Carroll, N. T. and Nicholas, M. G. (1981) *J. Mater. Sci.* **16**, 714.
28. Carre, A. and Schultz, J. (1983) *J. Adhesion* **15**, 151.
29. Fox, H. W. and Zisman, W. A. (1950) *J. Colloid Sci.* **5**, 514.
30. Fox, H. W. and Zisman, W. A. (1952) *J. Colloid Sci.* **7**, 109.
31. Fox, H. W. and Zisman, W. A. (1952) *J. Colloid Sci.* **7**, 428.
32. Shafrin, E. G. and Zisman, W. A. (1960) *J. Phys. Chem.* **64**, 519.
33. Zisman, W. A. (1964) in *Advances in Chemistry Series, 43* (ed. R. F. Gould), American Chemical Society, Washington, p. 1.
34. Shafrin, E. G. (1977) in *Handbook of Adhesives* (ed. I. Skeist), Van Nostrand, New York, p. 67.
35. Wu, S. (1970) *Polymer Reprints* **11**, 1921.
36. Wu, S. (1974) *J. Macromol. Sci. Rev. Macromol. Chem.* **C10**, 1.
37. Kaelble, D. H. (1971) *Physical Chemistry of Adhesion*, John Wiley and Sons, New York, p. 164.
38. Andrews, E. H. and Kinloch, A. J. (1973) *Proc. Roy. Soc.* **A332**, 385.
39. Good, R. J. (1979) *J. Colloid Interf. Sci.* **59**, 398.
40. Crane, L. W., Hamermesh, C. L. and Maus, L. (1976) *SAMPE J.* **12**, 6.
41. Lewis, A. F. and Gounder, R. T. N. (1981) in *Treatise on Adhesion and Adhesives*, Vol. 5 (ed. R. L. Patrick), Marcel Dekker, New York, p. 325.
42. Gledhill, R. A. and Kinloch, A. J. (1974) *J. Adhesion* **6**, 315.
43. Lee, L. H. (1968) *J. Appl. Polymer Sci.* **12**, 719.
44. Dann, J. R. (1970) *J. Colloid Interf. Sci.* **32**, 302.

45. Good, R. J. and Girifalco, L. A. (1957) *J. Phys. Chem.* **61**, 904.
46. Good, R. J. and Girifalco, L. A. (1960) *J. Phys. Chem.* **64**, 561.
47. Fowkes, F. M. (1963) *J. Phys. Chem.* **67**, 2538.
48. Berthelot, D. (1898) *Compt. Rend.* **126**, 1703.
49. Good, R. J. and Elbing, E. (1970). *Ind. Eng. Chem.* **62**, 54.
50. Hildebrand, J. H. and Scott, R. L. (1950) *Solubility of Non-electrolytes*, Reinhold, New York.
51. Fowkes, F. M. (1964) *Ind. Eng. Chem.* **56**, 40.
52. Graham, D. P. (1965) *Ind. Eng. Chem.* **57**, 4387.
53. Whalen, J. W. (1968) *J. Colloid Interf. Sci.* **28**, 443.
54. Good, R. J. (1975) *J. Colloid Interf. Sci.* **52**, 308.
55. Good, R. J. (1975) in *Adsorption at Interfaces* (ed. K. L. Mittal), American Chemical Society Symposium Series 8, American Chemical Society, Washington, p. 28.
56. Wu, S. (1970) *J. Phys. Chem.* **74**, 632.
57. Fowkes, F. M. (1968) *J. Colloid Interf. Sci.* **28**, 493.
58. Schultz, J., Tsutsumi, K. and Donnet, J. B. (1977) *J. Colloid Interf. Sci.* **59**, 277.
59. Fowkes, F. M. (1964) *Ind. Eng. Chem.* **56**, 40.
60. Fowkes, F. M. (1967) in *Treatise on Adhesion and Adhesives*, Vol. 1 (ed. R. L. Patrick), Marcel Dekker, New York, p. 352.
61. Tamai, Y., Makuuchi, K. and Suzuki, M. (1967) *J. Phys. Chem.* **71**, 4167.
62. Dann, J. R. (1970) *J. Colloid Interf. Sci.* **32**, 321.
63. Owens, D. K. and Wendt, R. C. (1969) *J. Appl. Polymer Sci.* **13**, 1740.
64. Kaelble, D. H. and Uy, K. C. (1970) *J. Adhesion* **2**, 50.
65. Kaelble, D. H. (1971) *Physical Chemistry of Adhesion*, Wiley-Interscience, New York, p. 153.
66. Sherriff, M. (1976) *J. Adhesion* **7**, 257.
67. Fowkes, F. M. (1972) *J. Adhesion* **4**, 155.
68. Wake, W. C. (1978) *Polymer* **19**, 291.
69. Panzer, J. (1973) *J. Colloid Interf. Sci.* **44**, 142.
70. Baszkin, A., Nisnino, M. and Ter-Minassian-Saraga, L. (1976) *J. Colloid Interf. Sci.* **54**, 317.
71. Lipatov, Y. and Feinerman, A. (1979) *Adv. Colloid Interf. Sci.* **11**, 195.
72. Bikerman, J. J. (1971) *J. Adhesion* **3**, 19.
73. Wu, S. (1973) in *Recent Advances in Adhesion* (ed. L. H. Lee), Gordon and Breach, London, p. 45.
74. Cortes, J. (1973) *J. Colloid Interf. Sci.* **45**, 209.
75. Baszkin, A. and Ter-Minassian-Saraga, L. (1978) *Polymer* **19**, 1083.
76. Good, R. J. (1979) *J. Colloid Interf. Sci.* **59**, 398.
77. Barton, A. F. M. (1982) *J. Adhesion* **14**, 33.
78. Lewis, G. N. (1923) *Valence and the Structure of Atoms and Molecules*, Chemical Cataloguing Co., New York, p. 142.
79. Fowkes, F. M. and Maruchi, S. (1977) *Org. Coatings Plastics Chem.* **37**, 605.
80. Fowkes, F. M. and Mostafa, M. A. (1978) *Ind. Eng. Chem. Prod. Res. Dev*, **17**, 3.
81. Fowkes, F. M. (1983) in *Physicochemical Aspects of Polymer Surfaces*, Vol. 2 (ed. K. L. Mittal), Plenum, New York, p. 583.
82. Fowkes, F. M. (1984) *Rub. Chem. Technol.* **57**, 328.
83. Fowkes, F. M., Tischler, D. O., Wolfe, J. A., Lannigan, L. A., Ademu-John, C. M. and Halliwell, M. J. (1984) *J. Polymer Sci., Polymer Chem. Ed.* **22**, 547.
84. Drago, R. S., Vogel, G. C. and Needham, T. E. (1971) *J. Am. Chem. Soc.* **93**, 6014.

85. Drago, R. S., Parr, L. B. and Chamberlain, C. S. (1977) *J. Am. Chem. Soc.* **99**, 3203.
86. Huntsberger, J. R. (1981) *J. Adhesion* **12**, 3.
87. White, M. L. (1964) *J. Phys. Chem.* **68**, 3083.
88. Erb, R. A. (1965) *J. Phys. Chem.* **69**, 1306.
89. Bernett, M. K. and Zisman, W. A. (1970) *J. Phys. Chem.* **74**, 2309.
90. Schrader, M. E. (1970) *J. Phys. Chem.* **74**, 2313.
91. Schrader, M. E. (1974) *J. Phys. Chem.* **78**, 87.
92. Schonhorn, H. (1967) *J. Phys. Chem.* **71**, 4578.
93. Livey, D. T. and Murray, P. (1956) *J. Am. Ceram. Soc.* **39**, 363.
94. Benson, C. C., Freeman, P. I. and Dempsey, E. (1963) *J. Am. Ceram. Soc.* **46**, 43.
95. Bryant, P. J., Gutshall, P. L. and Taylor, L. H. (1964) *Wear* **7**, 118.
96. Reiss, H., Frisch, H. L. and Lebowitz, J. L. (1959) *J. Chem. Phys.* **31**, 369.
97. Bondi, A. (1953) *Chem. Rev. Am. Chem. Soc.* **52**, 417.
98. Rasmussen, J. J. and Nelson, R. P. (1971) *J. Am. Ceram. Soc.* **54**, 398.
99. Kingery, W. D. (1959) *J. Am. Ceram. Soc.* **42**, 6.
100. Von Wartenberg, H., Wehner, G. and Saran, E. (1936) *Nachr. Ges. Wiss Goettingen, Math. Phys. Kl., Fachgruppe 2* **2,** 65.
101. Sokolov, O. K. (1963) *Izv. Akad. Nauk. SSSR, Met. Gorn. Delo*, **4**, 59.
102. Bartlett, R. W. and Hall, J. K. (1965) *Am. Chem. Soc. Bull.* **44**, 444.
103. Rhee, S. K. (1973) *Mater. Sci. Eng.* **11**, 311.
104. Rhee, S. K. (1974) *Mater. Sci. Eng.* **16**, 45.
105. Valentine, T. M. (1977) *Mater. Sci. Eng.* **30**, 205.
106. Valentine, T. M. (1977) *Mater. Sci. Eng.* **30**, 211.
107. Murr, L. E. (1978) *ASTM, Special Technical Pub.* **640**, 82.
108. Murr, L. E. (1981) *Mater. Res. Sci.* **14**, 107.
109. Bikerman, J. J. (1978) *ASTM, Special Technical Pub.* **640**, 97.
110. Schultz, J., Tstutsumi, K. and Donnet, J. B. (1977) *J. Colloid Interf. Sci.* **59**, 272.
111. Bailey, A. I. and Price, A. G. (1970) *Disc. Faraday Soc.* 118.
112. Harkins, W. D. and Loeser, E. H. (1950) *J. Chem. Phys.* **18**, 556.
113. Eberhart, J. G. (1966) *Trans. Met. Soc. AIME* **236**, 1362.
114. Boyd, G. E. and Livingstone, H. K. (1942) *J. Am. Chem. Soc.* **64**, 2373.
115. Shartsis, L. and Spinner, S. (1951) *J. Res. Nat. Bur. Std* **46**, 385.
116. Stromberg, R. R. (1967) in *Treatise on Adhesion and Adhesives*, Vol. 1 (ed. R. L. Patrick), Marcel Dekker, New York, p. 69.
117. Silberberg, A. (1971) *Pure Appl. Chem.* **26**, 583.
118. Takahashi, A. and Kawaguchi, M. (1982) *Adv. Polymer Sci.* **46**, 1.
119. Schonhorn, H. (1963) *J. Polymer Sci.* **A1**, 2343.
120. Schonhorn, H. (1963) *J. Colloid Sci.* **18**, 445.
121. Schonhorn, H. and Frisch, H. L. (1973) *J. Polymer Sci. Polymer Phys. Ed.* **11**, 1005.
122. Barraud, A., Rosilio, C. and Ruaudel-Teixier, A. (1980) *Thin Solid Films* **68**, 7.
123. Wake, W. C. (1982) *Adhesion and the Formulation of Adhesives*, Applied Science Pub., London, p. 96.
124. Baszkin, A., Nisnino, M. and Ter-Minassian-Saraga, L. (1977) *J. Colloid Interf. Sci.* **59**, 516.
125. Sharpe, L. H. and Schonhorn, H. (1964) in *Advances in Chemistry Series, 43* (ed. R. F. Gould), American Chemical Society, Washington, p. 189.
126. Mittal, K. L. (1975) in *Adhesion Science and Technology* (ed. L. H. Lee), Plenum Press, New York, p. 129.
127. Mittal, K. L. (1977) *Polymer Eng. Sci.* **17**, 467.

128. Iyengar, Y. and Erickson, D. E. (1967) *J. Appl. Polymer Sci.* **11**, 2311.
129. Dyckerhoff, Von G. A. and Sell, P. J. (1972) *Die Angew. Mackrolmol. Chemie* **21**, 169.
130. Huntsberger, J. R. (1967) in *Treatise on Adhesion and Adhesives*, Vol. 1 (ed. R. L. Patrick), Marcel Dekker, New York, p. 119.
131. Huntsberger, J. R. (1978) *Adhesives Age* **21 (12)**, 23.
132. Bascom, W. D. and Patrick, R. L. (1974) *Adhesives Age*, **17 (10)**, 25.
133. Bascom, W. D., Cottington, R. L. and Singleterry, C. R. (1964) in *Advances in Chemistry Series, 43* (ed. R. F. Gould), American Chemical Society, Washington, p. 355.
134. Fritz, G. Z. (1965) *Z. für Angew. Physik* **19**, 374.
135. Schwartz, A. M. and Tegada, S. B. (1972) *J. Colloid Interf. Sci.* **38**, 359.
136. Cottington, R. L., Murphy, C. M. and Singleterry, C. R. (1964) in *Advances in Chemistry Series, 43* (ed. R. F. Gould), American Chemical Society, Washington, p. 341.
137. Cheever, G. D. (1968) in *Interface Conversions for Polymer Coatings* (eds P. Weiss and G. D. Cheever), Elsevier, New York, p. 150.
138. De Bruyne, N. A. (1956) *Aero Research Technical Notes* No. 168, p. 1.
139. Packham, D. E. (1983) in *Adhesion Aspects of Polymeric Coatings* (ed. K. Mittal), Plenum, New York, p. 19.
140. Schonhorn, H., Frisch, C. and Kwei, T. K. (1966) *J. Appl. Phys.* **37**, 4967.
141. Cherry, B. W., Muddaris, S. El., and Holmes, C. M. (1969) *J. Australasian Inst. Met.* **14**, 174.
142. Cherry, B. W. and Holmes, C. M. (1969) *J. Colloid Interf. Sci.* **38**, 174.
143. Newman, S. (1968) *J. Colloid Interf. Sci.* **26**, 209.
144. Cherry, B. W. and Muddaris, S. El. (1970) *J. Adhesion* **2**, 42.
145. Bascom, W. D. and R. L. Cottington (1972) *J. Adhesion* **4**, 193.
146. Bowden, F. P. and Throssell, W. R. (1951) *Proc. Roy. Soc.* **A209**, 297.
147. Bailey, A. I. and Kay, S. M. (1967) *Proc. Roy. Soc.* **A301**, 47.
148. Gettings, M., Baker, F. S. and Kinloch, A. J. (1977) *J. Appl. Polymer Sci.* **21**, 2375.
149. Bernett, M. K. and Zisman, W. A. (1968) *J. Colloid Sci.* **28**, 243.
150. Gledhill, R. A., Kinloch, A. J. and Shaw, S. J. (1977) *J. Adhesion* **9**, 81.
151. Pike, R. A., Lamm, F. P. and Pinto, J. P. (1982) *J. Adhesion* **13**, 229.
152. Doduik, H., Drori, L. and Miller, J. (1984) *J. Adhesion* **17**, 33.
153. Papirer, E. (1986) in *Composite Interfaces* (eds. H. Ishida and J. Koenig), North-Holland, London, p. 203.
154. Schultz, J., Lavelle, L. and Martin, C. (1987) *J. Adhesion*, to be published.

CHAPTER THREE

Mechanisms of adhesion

3.1 INTRODUCTION

The previous chapter considered the various aspects involved in the attainment of intimate molecular contact at the adhesive/substrate interface. As discussed, the attainment of such interfacial contact is invariably a necessary first stage in the formation of strong and stable adhesive joints. The next stage is the generation of intrinsic adhesion forces across the interface, and the nature and magnitude of such forces are extremely important. They must be sufficiently strong and stable to ensure that the interface does not act as the 'weak link' in the joint, either when the joint is initially made or throughout its subsequent service life. The various types of intrinsic forces which may operate across the adhesive (or primer)/substrate interface are commonly referred to as the *mechanisms of adhesion*, and they are discussed in this chapter.

As shown in Chapter 2, the molecular forces in the surface layers of the adhesive and substrate greatly influence the attainment of intimate molecular contact across the interface and such molecular forces are now frequently the main mechanism of adhesion, and this is called the adsorption theory of adhesion. However, this is only one of the four main mechanisms of adhesion which have been proposed, namely:

(a) mechanical interlocking,
(b) diffusion theory,
(c) electronic theory,
(d) adsorption theory.

Some years ago many workers searched for 'the mechanism of adhesion', but more recently it has become generally accepted that, whilst the adsorption theory has the widest applicability, each of the others may be appropriate in certain circumstances and often make a contribution to the intrinsic adhesion forces which are acting across the interface. As discussed below, much of the confusion that has arisen in the literature concerning the mechanisms of adhesion has undoubtedly been caused by the test methods commonly employed to measure the strengths of adhesive joints not being well suited to ascertaining the magnitude of the intrinsic adhesion forces which are acting across the adhesive/substrate interface. They introduce geometrical and loading factors which are difficult to analyse, and the measured joint strength typically includes contributions from rheological energy losses in the adhesive and substrates. Thus, although the intrinsic adhesion forces influence the joint strength they are usually completely obscured by other contributions. Information concerning the magnitude of such forces may therefore usually only be obtained indirectly, or by the use of especially developed continuum fracture mechanics approaches.

The four main mechanisms of adhesion which have been proposed will now be considered in turn. Examples will be given in this chapter to illustrate the points which are discussed, and more examples are to be found throughout the text since subsequent chapters will draw heavily upon the ideas reviewed below.

3.2 MECHANICAL INTERLOCKING

3.2.1 Introduction

This theory essentially proposes that mechanical keying, or interlocking, of the adhesive into the irregularities of the substrate surface is the major source of intrinsic adhesion. One example where mechanical interlocking is of prime importance is in the use of mercury amalgam for filling tooth cavities. The dentist drills out the tooth material to give a relatively large 'ink-bottle' pit, ideally with an undercut angle of about 5°. Unfortunately, this frequently requires removing sound, as well as decayed, tooth material. The mercury amalgam paste is then forced into this large cavity. Mechanical interlocking is the main mechanism of adhesion which operates and ensures that the required service life of approximately seven to ten years is often achieved. More recent developments in restorative materials has led to ones which do rely upon secondary and primary interfacial forces (Section 3.5) and such truly adhesive materials offer many benefits: the elimination of the need to remove sound tooth material being especially noteworthy!

However, the attainment of good adhesion between smooth surfaces exposes the mechanical interlocking theory as not being of wide applicability. For example, the elegant work of Tabor *et al.* [1,2] who studied the adhesion between two perfectly smooth mica surfaces and Johnson *et al.* [3] who examined the adhesion to optically smooth rubber surfaces, clearly demonstrates that adhesion may be attained with smooth surfaces. Also, detailed examination of surfaces roughened by typical industrial pretreatment methods, for example, grit-blasted metallic substrates (Fig. 3.1), usually reveals little

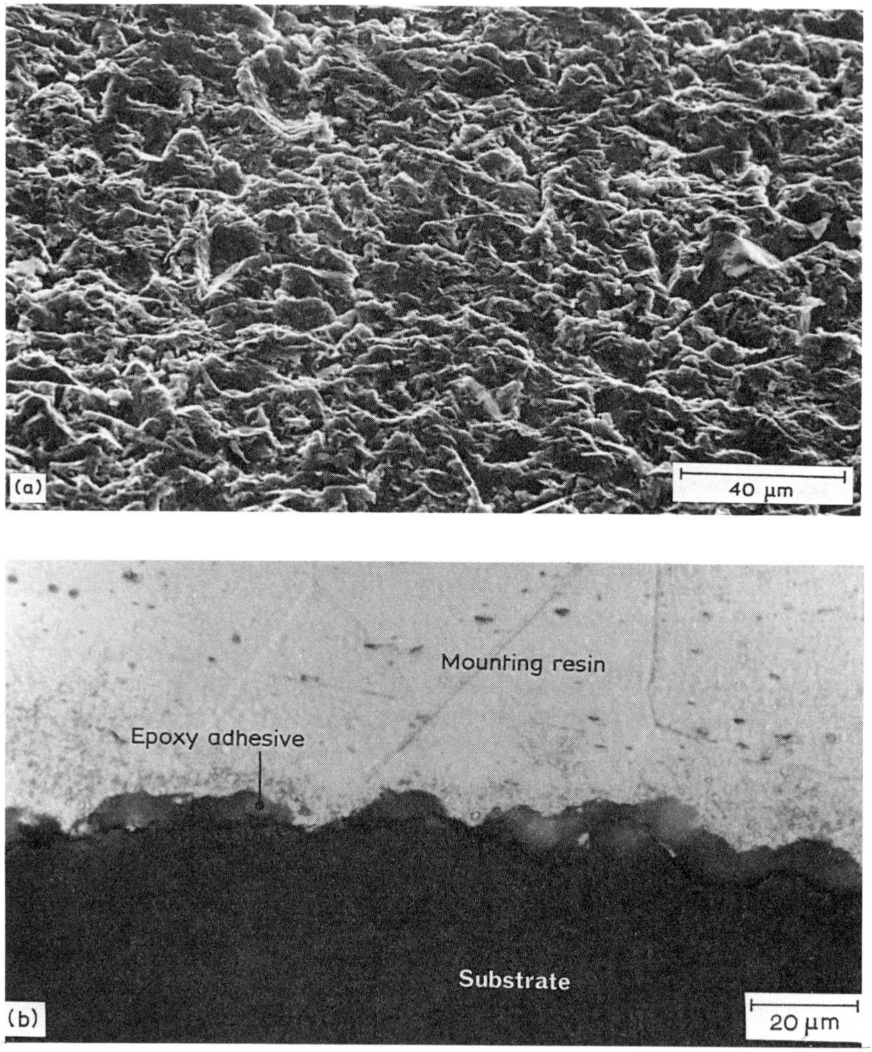

Figure 3.1 (a) Scanning electron micrograph of an abraded mild steel surface. (b) Optical micrograph of a section cut normal to an epoxy/abraded steel interface. (c) Talysurf profilometer trace for an abraded steel surface.

indication of cavities such as 'ink-bottle' pits which would enable mechanical interlocking to be the major mechanism of adhesion. In the case of the anodization of aluminium alloys a deep porous topography is produced (see Section 8.3.3.3) and, with many of the more open porous structures, the adhesive (or primer) typically penetrates to virtually the bottom of the pores,

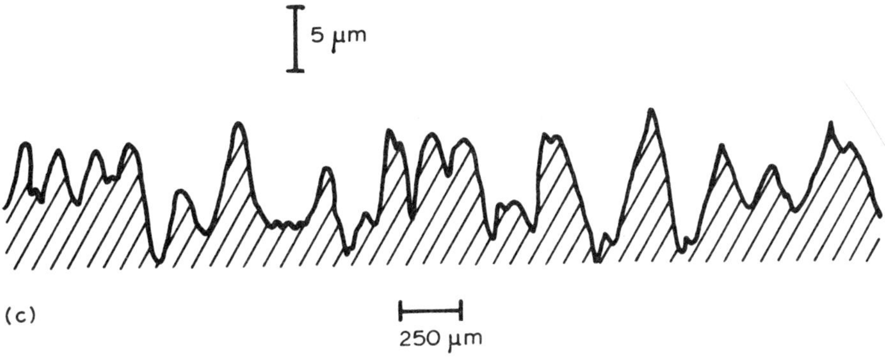

Figure 3.1—continued

and so a 'composite' interfacial region is created. This composite region will have a modulus and strength intermediate between that of the polymeric adhesive and the oxide and this would be expected to be beneficial from the viewpoint of joint strength and toughness, as discussed later (unless the oxide itself is too brittle as typically for oxides produced by a sulphuric acid anodizing technique.) However, considering the role of mechanical interlocking, it is firstly debatable whether mechanical interlocking really does occur [4] and secondly, even if it does, to assess its contribution to the strength and stability of the interface. These aspects are also considered further in Section 8.3.3.2, since it is of relevance to the stability of the interface in the presence of hostile environments.

Notwithstanding the above comments, there is much work to be found in the literature which convincingly demonstrates that increasing the surface roughness of the substrate may increase the measured strength of the adhesive joint. The important question that must be asked is: does such an observation arise from a contribution from mechanical interlocking to the intrinsic strength of the interface, or from other causes?

3.2.2 Plating of plastics

An area where mechanical interlocking may contribute significantly to the intrinsic adhesion is the metal plating of plastics where a chemical pretreatment of the plastic substrate is employed prior to plating. Some workers [5–9] have argued that the adhesion of metal plating to polymeric substrates is a function of the surface topography. However, others [9–12] have emphasized the role that increased oxidation of the plastic surface, commonly induced by the pretreatments employed prior to plating, plays, thus emphasizing the importance of surface force interactions (Section 3.5). A balanced view emerges from the quantitative experiments of Perrins and Pettett [13] who examined the adhesion of electroplated copper to polypropylene. They concluded that the adhesion was controlled by mechanical interlocking. Excavation and undercutting of the cavities in the polymer surface was dependent upon the presence of amorphous material in the polymer which was more readily attacked by the etching acid than the crystalline polymer spherulites. However, for optimum results then interfacial chemical (copper oxide/hydroperoxide) bonds had also to be established. Further, their work enabled the contributions to the intrinsic adhesion arising from mechanical interlocking and chemical interfacial effects to be distinguished. Following the theories of Andrews and Kinloch [14,15] and Gent and Schultz [16], Wake [17] has suggested that the effects of mechanical interlocking and interfacial chemical effects could be multiplied to give the following expression for the optimum measured joint strength:

$$\begin{aligned}\text{Optimum joint strength} = {} & (\text{constant}) \\ & \times (\text{mechanical interlocking component}) \\ & \times (\text{interfacial chemical component})\end{aligned} \tag{3.1}$$

Wake deduced that the values of the various terms for the present experiments were:

Constant	0.23
Mechanical interlocking component	2.9
Interfacial chemical component	5.9

Thus, the interfacial chemical component appears twice as important as the mechanical effect. However, as commented by Wake [17], in the experiments of Perrin and Pettett they could remove the interfacial component completely by selecting the appropriate surface treatment, but the mechanical component could only be minimized. Hence, the earlier remark that the adhesion was *controlled* by mechanical interlocking.

3.2.3 Mechanically roughened substrates

It is often observed that some form of mechanical abrasion will increase the measured joint strength [18]. However, as noted above, such treatments do not generally result in a surface topography with cavities suitable for establishing mechanical interlocking with an adhesive. This can be readily seen on the 'micro' level from electron micrographs (Fig. 3.1a) and on the 'macro' level from optical micrographs of sections taken normal to the surface (Fig. 3.1b) and from 'Talysurf' profilometry measurements (Fig. 3.1c).

One effect of mechanical abrasion in an industrial environment is usually to ensure that the substrates are free from release agents, machine oil, grease, scale, etc., which obviously aids subsequent wetting and removes any weak boundary layer which would fail prematurely when the joint was loaded. Also, as discussed in Chapter 2, surface roughness may aid both the thermodynamics and kinetics of wetting by the adhesive. It has also been suggested that increases in joint strengths might be attributed to the extra interfacial bonding area available, typically about 5 to 30%, or to changes in the stress distribution in the interfacial regions of the joint. This last feature may essentially promote plastic deformation in the adhesive in the interfacial regions, as discussed in more detail below. Alternatively, although somewhat more unlikely, it may prevent small cracks, flaws, etc., which are located at, or close to, the interface from aligning and rapidly propagating along any line of interfacial weakness, as might be envisaged in the case of a smooth planar interface. Thus, it appears unlikely that mechanically abraded surfaces offer much scope for mechanical interlocking, but such a pretreatment may well be beneficial for other reasons.

3.2.4 Chemically roughened substrates

Whilst mechanical abrasion does not appear to produce a surface topography suitable for mechanical interlocking to occur to any significant extent, chemical pretreatments can result in such surfaces. However, care is needed to ensure that the resulting surface layer is not too weak or friable and, even with the advent of modern surface-specific analytical equipment, the question of whether the treatment has induced chemical, as well as topographical, changes remains a difficult point to resolve.

Arrowsmith [19] has studied the adhesion of electroformed copper and nickel foils (35 μm thick) to glass cloth reinforced epoxy laminates, as used in printed circuit boards. He most elegantly altered the electroforming conditions for the metallic substrates so as to give various surface topographies, and he also measured the peel strength of the different joints. The results are summarized in Table 3.1 and, as may be seen, the 3 μm high-angle pyramid plus dendritic structure on the copper foil and club-headed nodular structure on the nickel foil give the highest peel strengths. These surface topographies are shown in Fig. 3.2 and can obviously give rise to true mechanical interlocking. Indeed, Arrowsmith commented that the slope of the pyramids is important for the copper foils since the dendrites grown on their slopes must stick out sideways in order to give the most effective mechanical interlocking with the epoxy resin (see Table 3.1).

A series of papers by Packham and co-workers [20–26] also emphasizes the improvements in joint strength that may be observed when a 'microfibrous' surface topography is created. In their studies on the adhesion of polyethylene (used as a hot-melt adhesive) to metallic substrates they found that high peel strengths were obtained when a very rough, microfibrous, oxide surface was produced on the substrate. Fig. 3.3a is a scanning electronic micrograph of the

Table 3.1 Peel adhesion of electroformed copper foil to epoxy laminates [19]

Surface topography of copper foil		Peel energy	
Description	Diagrammatic representation	kJ/m^2	lbf/in
Flat		0.66	3.8
Flat + 0.3 μm dendrites		0.67	3.8
Flat + 0.3 μm dendrites + oxidation		0.77	4.4
3 μm high-angle pyramids		1.0	5.9
2 μm low-angle pyramids + 0.3 μm dendrites		1.3	7.4
2 μm low-angle pyramids + 0.3 μm dendrites + oxidation		1.5	8.8
3 μm high-angle pyramids + 0.3 μm dendrites + oxidation		2.4	13.5
Nickel foil with club-headed nodular structures		2.3	13.1

surface of copper treated in a proprietary oxidizing solution which resulted in high joint strength. When the fibres were damaged to reduce the capability for mechanical interlocking, without detectably having changed the chemical nature of the substrate, the joint strength dropped markedly. However, if the chemical nature of the oxide was changed by electrolytic reduction, with a minimum change in topography from that shown in Fig. 3.3a, the strength was still substantial. In a more recent study [26] similar observations have been recorded upon bonding zinc substrates using epoxy adhesives: the fully dendritic zinc oxide (Fig. 3.3b), which was by far the most microfibrous, gave the best results.

3.2.5 The role of localized energy dissipation

The enhancement of joint strength that may result from increasing the rugosity of the substrate surface, as detailed above, for example, by Packham and co-workers, need not necessarily arise either from a mechanical interlocking mechanism, or indeed from an increase in surface area for bonding or from

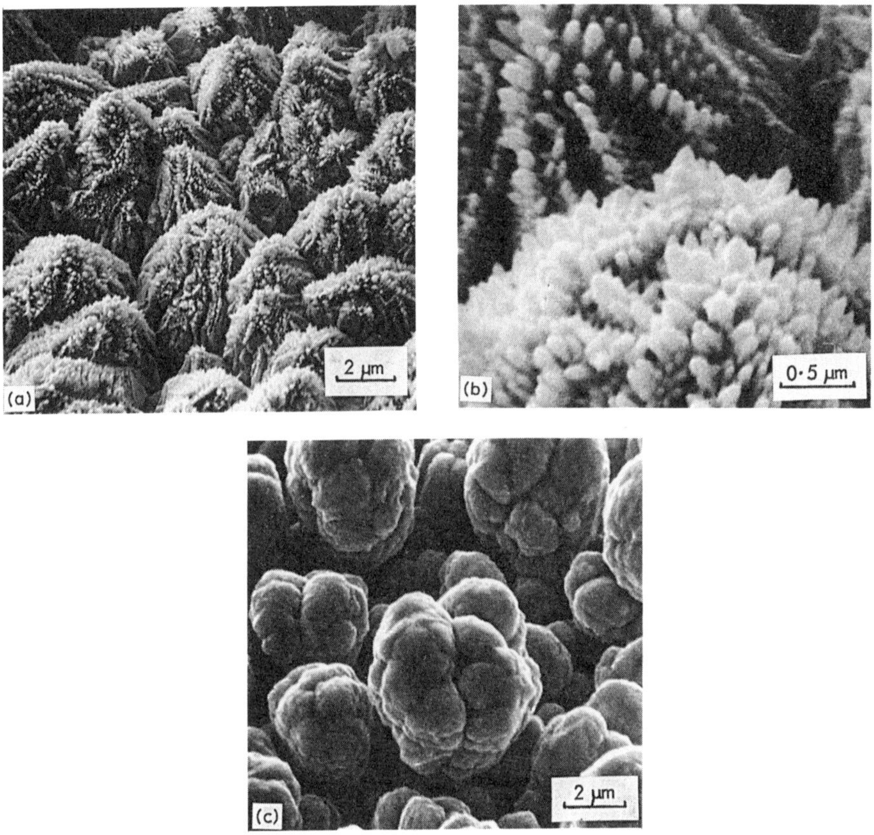

Figure 3.2 (a) Dendrites of second-stage electroformed copper after plating on high-angle first-stage pyramids. (b) Central field of (a). (c) Club-headed nodules on electroformed nickel foil [19].

improved wetting. The measured adhesive joint strength basically reflects the value of two parameters [14–16,27]: (a) the intrinsic adhesion and (b) the energy dissipated viscoelastically and plastically in the highly strained volume around the tip of the propagating crack and in the body of the joint, e.g. in bending a flexible substrate. The latter term usually dominates the measured joint strength, and also gives rise to the rate and temperature dependence of joint strengths. Several workers [23–26,28,29] have suggested that the importance of high surface rugosity is to increase the energy dissipated viscoelastically and plastically during joint failure, and Evans and Packham [23] and Wang and Vazirani [29] have applied the theory of the strength of

Figure 3.3 (a) Scanning electron micrograph of the surface of copper oxidized at 90 °C in Ebonol C solution for 10 min [22]. (b) Scanning electron micrograph of zinc electrode deposited from a potassium hydroxide solution giving a fully dendritic topography [26].

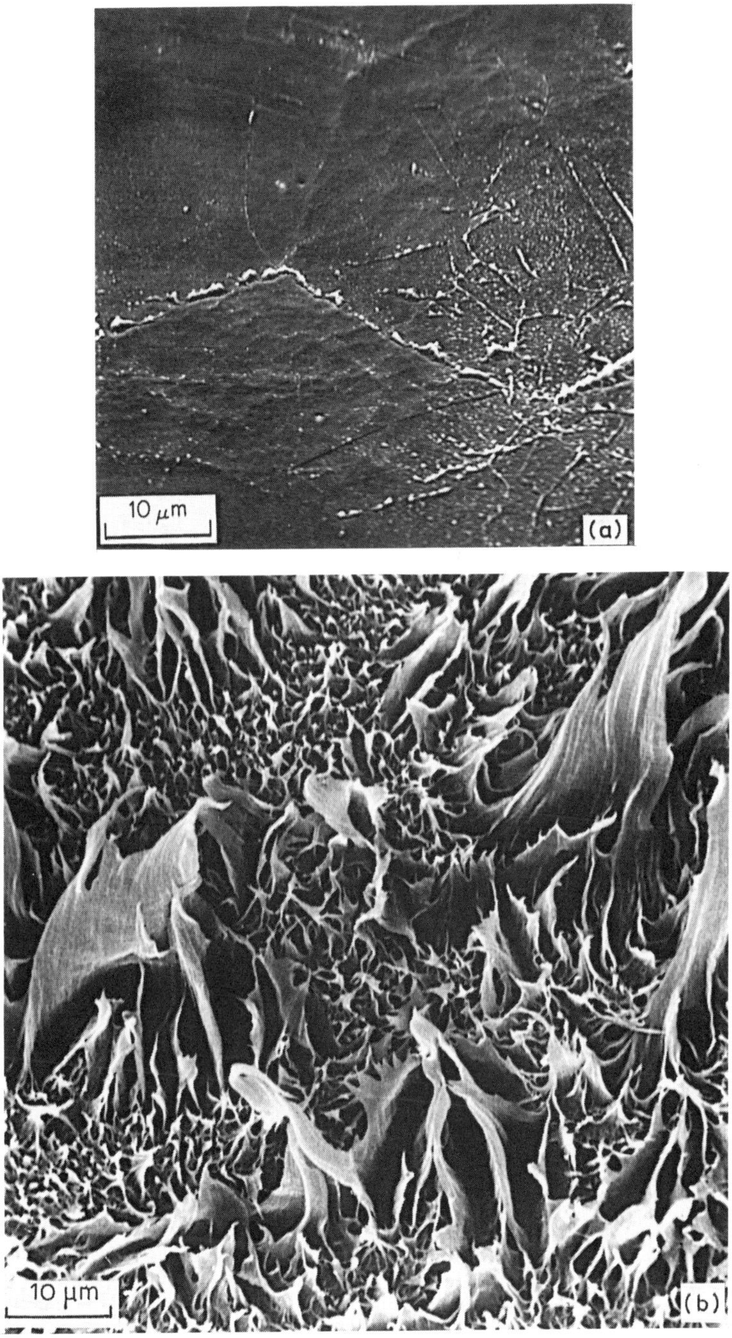

Figure 3.4 Scanning electron micrographs of the fracture surface of polyethylene/copper joints showing the polyethylene remaining on: (a) a smooth, chemically polished copper substrate, (b) a micro-fibrous copper substrate [24].

fibre-reinforced composite materials to the problem. Evans and Packham treated the polyethylene/fibrous oxide interfaces as a composite with discontinuous fibres in a resin matrix. When such a composite is stressed, stress has to be transferred from matrix to fibre and back again. This leads to tensile stress in the fibres which is a maximum in the centre, and to a shear stress between fibre and matrix which is largest at the ends. Thus, by analogy and providing the microfibres are sufficiently strong, it is to be expected that the presence of microfibres on the substrate would lead to high shear stresses around the ends of the fibres. Hence, joint failure would occur by plastic deformation of the polymer, initially around the fibre tips, and then, as the stress concentration is relieved, further into the bulk of the polymer. Thus, a much larger volume of polymer will be plastically deformed during fracture, compared to that volume deformed when bonding to a chemically similar but smooth surface, and this large amount of plastic deformation accounts for the high joint strength. Direct evidence for this proposal may be seen in Fig. 3.4 which shows scanning electron micrographs of the fracture surfaces of polyethylene/copper joints. Only in the case of the copper substrate where there was a microfibrous topography, generated by an alkaline chlorite pretreatment, is there evidence of extensive plastic deformation in the polyethylene adhesive.

3.2.6 Summary

Therefore, to summarize, in certain instances mechanical interlocking may contribute to the intrinsic adhesion mechanisms. However, the substrate has to be specifically pretreated to obtain the appropriate topography for this to occur and this appears to happen in only a few instances. The frequently observed increases in measured joint strengths with increasing surface rugosity are most usually attributable to other factors such as efficient removal of weak surface layers, improved interfacial contact (e.g. better wetting and a greater area for interfacial contact) and the enhancement of energy dissipative mechanisms in the adhesive.

3.3 DIFFUSION THEORY

3.3.1 Introduction

Voyutskii [30–32] is the chief advocate of the diffusion theory of adhesion which states that the intrinsic adhesion of polymers to themselves (*autohesion*), and to each other, is due to mutual diffusion of polymer molecules across the interface. This requires that the macromolecules, or chain segments of the polymers (adhesive and substrate) possess sufficient mobility and are mutually soluble. This latter requirement may be restated by the condition that they possess similar values of solubility parameter. The solubility parameter, δ_s, may be defined by:

$$\delta_s = \left(\frac{\Delta H_v - \mathbf{R}T}{V}\right)^{1/2} \tag{3.2}$$

where ΔH_v is the molar heat of vaporization, $\mathbf{R}$ is the gas constant, T is the temperature (K) and V is the molar volume. Hence, the solubility parameter is an index of the compatibility of two components, e.g. if an amorphous polymer and a solvent have similar values then they should form a solution. (The added need for the polymer to be amorphous is included since if the polymer possesses a significant degree of crystallinity then the free energy of crystallization makes it more resistant to dissolving in the solvent. The concept of a solubility parameter does not take this aspect into account.)

3.3.2 Modelling interfacial diffusion

Voyutskii's experimental evidence for his proposals are the effects of contact time, temperature, polymer type, molecular weight and viscosity on the measured joint strength. He argues that the functional dependence of joint strength on some of these parameters is similar to that expected for a diffusion process and therefore that the adhesion is a result of diffusion. These experimental observations have been reinforced by the quantitative models for the diffusion theory developed by Vasenin [33–35]. He began from Fick's first law which relates the amount of material, w, diffusing in the x-direction across a plane of unit area normal to the concentration gradient, $\mathrm{d}c/\mathrm{d}x$, and the time, t, by:

$$\mathrm{d}w = -D_f \mathrm{d}t \cdot \mathrm{d}c/\mathrm{d}x \tag{3.3}$$

where D_f is the diffusion coefficient. However, this can only be directly applied to steady-state diffusion where the concentrations at points within the system do not vary with time. This is clearly not the case for the penetration of a segment of polymer chain into the surface regions of a polymeric substrate. The build-up or decay of a diffusing species is given by Fick's second law which can be derived from the first. Vasenin used Fick's laws therefore to model the depth of penetration of the diffusing molecule. He first assumed that as the diffusing species interdiffused into the surface regions during the time of contact, t_c, the reduction in the value of the diffusion constant with time would be of the form:

$$D_f = D_d t_c^{-\beta} \tag{3.4}$$

where D_d is a constant which characterizes the mobility of the macromolecules and β is a constant which determines the rapidity of change of the coefficient of diffusion, D_f, with time and is of the order of 0.5 in value. Next he deduced that the depth of penetration, l_p, could be expressed by:

$$l_p = \frac{(\pi D_d t_c^{1/2})^{1/2}}{k_3} \tag{3.5}$$

where k_3 is a constant which characterizes the stiffness, bond length and valency angles along the polymeric molecules and Vasenin did indeed express the value of k_3 in terms of such molecular constants. Next he deduced that the number, N_c, of molecular chains crossing the phase boundary between the adhesive and substrate was given by:

$$N_c = \left(\frac{2\mathbf{N}\rho}{M}\right)^{2/3} \tag{3.6}$$

where **N** is Avogadro's number, ρ is the density and M is the molecular weight of the polymer. Finally, he assumed that the measured peeling energy, P, to rupture the interphase region was proportional to both the depth of penetration, i_p, and the number of chains crossing the phase boundary, N_c, so that the value of P is given by:

$$P = k_4\left(\frac{2\mathbf{N}\rho}{M}\right)^{2/3} D_d^{1/2} t_c^{1/4} \tag{3.7}$$

where k_4 is a constant which may be expressed in terms of molecular characteristics of the polymers involved. Equation 3.7 therefore relates the measured peel energy, P, to the time of contact, t_c, allowed between the materials and to the molecular weight, M, of the polymers and has been used successfully to predict the autohesion of polyisobutylene. In Fig. 3.5 the measured peeling energy, P, is shown as a function of the time of contact, t_c, for polyisobutylenes of different molecular weights. The agreement between the experimental results and theoretical predictions is very good with the value of P obeying the $t_c^{1/4}$ dependency. However, it should be noted that the value of D_d could not be absolutely ascertained and was obtained by fitting one set of experimental data. Also, in Fig. 3.6 the dependence of P upon the molecular weight of the polyisobutylene is illustrated at three different contact times; again good agreement between experiment and theory is demonstrated.

It is interesting to consider that, if the diffusion coefficient is taken to be 10^{-14} cm^2/s, Vasenin theory predicts, for instance, that it takes about 100 h (3.6×10^5 s) for the molecular segments of one polyisobutylene sheet to penetrate a depth of 10 μm into the other. However, Voyutskii and Vasenin point out that a much smaller penetration can still theoretically result in high joint strengths and, for example, that an interpenetration of macromolecules of between 1 and 2 nm may result in a five- to nine-fold increase in joint strength.

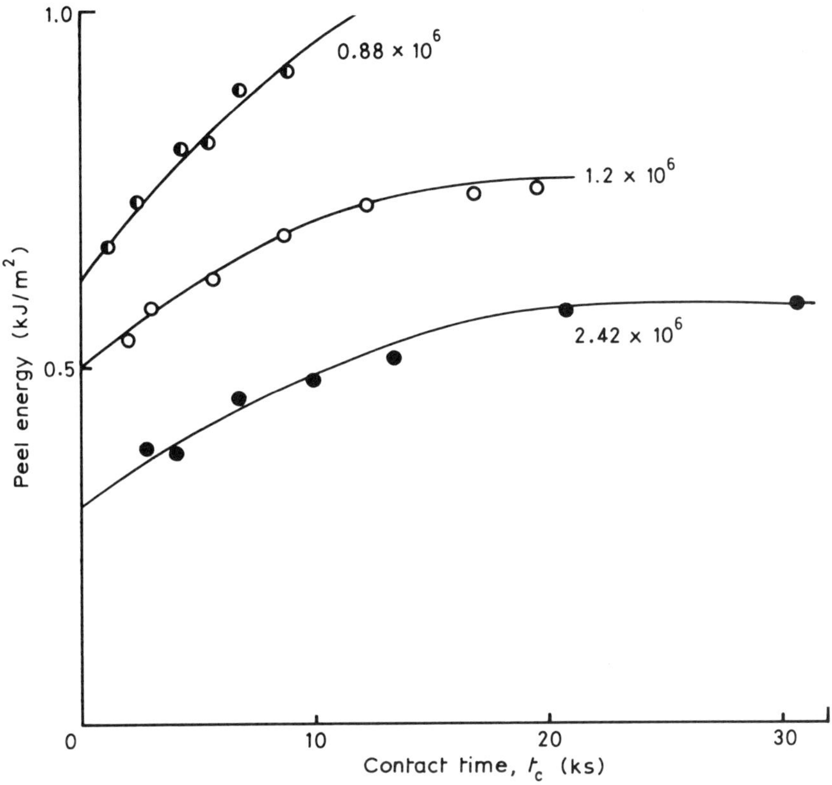

Figure 3.5 Relation between the measured peel energy and contact time for the autohesion of polyisobutylenes of various molecular weights. The curves are calculated from Equation 3.7 and the points are experimentally determined [33].

Jud *et al.* [148] have reached a similar conclusion and an important factor may be to ensure that the degree of interpenetration is sufficient for the diffusing polymer segments to establish an entanglement network structure.

Campion [36,37] has criticized the Vasenin model on the grounds that it is not very sensitive to changes in molecular shapes and structures. For example, he comments that Vasenin's model cannot readily account for the poor levels of autohesion exhibited by rubbers based upon ethylene–propylene copolymer. To explain such differences Campion proposed that, for some elastomers, free volume or 'hole' formation is facilitated by certain polymer chain structural features to provide a means for this apparently more rapid diffusion. He developed a simple ranking model which correctly ranked the degree of autohesion of various elastomers from a consideration of their microstructure. This is illustrated in Fig. 3.7 where a packing arrangement of four participating chains is shown for natural rubber (Fig. 3.7a) and ethylene–propylene–diene rubber (Fig. 3.7b). In both cases the chains have moved co-operatively to form

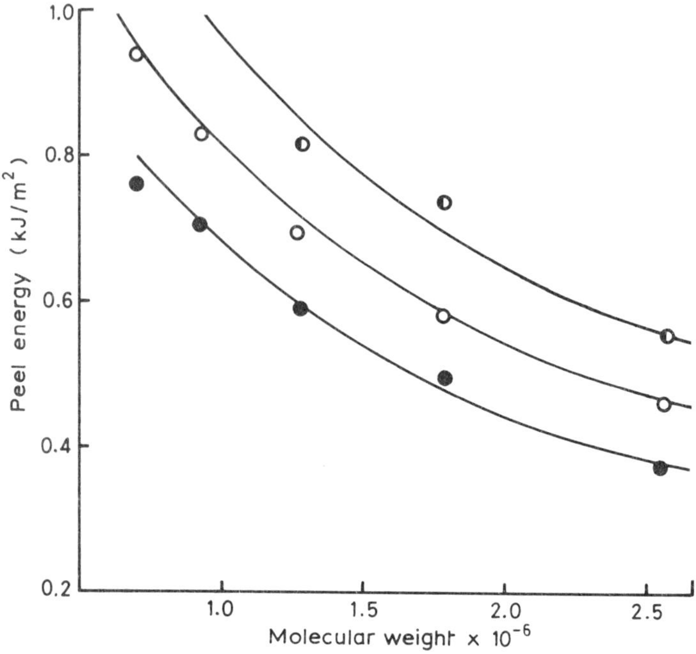

Figure 3.6 Relation between the measured peel energy and molecular weight for autohesion of polyisobutylenes as a function of contact time. The curves are calculated from Equation 3.7 and the points are experimentally determined [33].
◐ t_c = 15 mins; ○ t_c = 1 h; ● t_c = 5 h

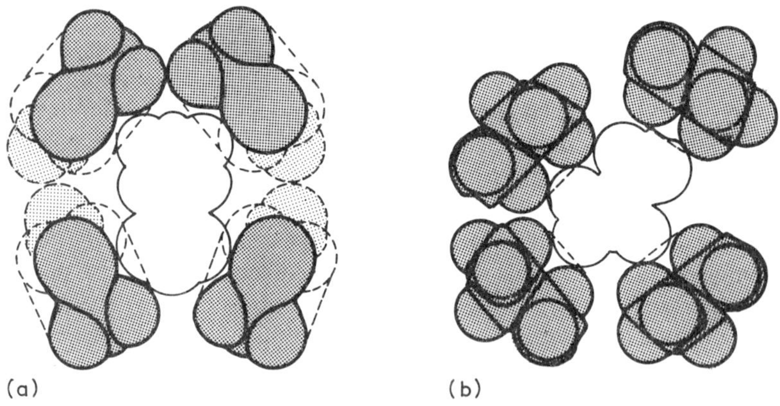

Figure 3.7 Hole formation via a packing arrangement of four participating polymer chains [37]. (a) Natural rubber, (b) ethylene–propylene–diene rubber. The holes are just large enough to allow diffusion of another polymer molecule of the same chemical species. However, the distance between the participating chains forming the hole needs to increase by 15% for natural rubber but by 27% for ethylene–propylene–diene rubber, compared to their respective close-packed chain conformations.

a hole large enough to accept an incoming molecular segment of the same respective chemical type, which is also depicted in outline in the figures. Obviously, for the given size of penetrating molecule, if the four chains were any closer together then neither hole would be sufficiently large to allow autohesion via a diffusion mechanism to occur. However, at these critical hole sizes the four natural rubber chains are closer together. Consequently, if the same extent of co-operative movements was considered for both natural and ethylene–propylene–diene rubbers, then a greater number of holes large enough to accept diffusing chain-ends would exist for natural rubber, irrespective of packing arrangement chosen. This is because of the more favourable molecular structure of the natural rubber for forming such holes. Indeed, ethylene–propylene–diene rubber was found to possess the least favourable molecular structure for hole formation; hence suggesting an explanation for its poor level of autohesion. However, the Campion model is not quantitative and has not been developed further.

3.3.3 Direct experimental evidence

Direct evidence for interdiffusion in compatible polymers does exist. Radiometric studies [38,39] have demonstrated the presence of macromolecular diffusion. The diffusion coefficients were found to be of the order of 10^{-11} to 10^{-14} cm^2/s, which Voyutskii [40] argues is completely adequate for the formation of an intrinsically strong interface between the polymers after a contact time of only a few seconds, as discussed above. Further work [41–43], using techniques of optical microscopy, including ultraviolet light employing luminescence analysis [43], has indicated that in compatible, non-polar polymers the interphase region where interdiffusion has occurred may be about 10 μm deep but in cases where the solubility parameters, δ_s, of the two polymers were appreciably different then no interdiffusion zone could be detected.

3.3.4 Criticisms of the diffusion theory

One criticism of the Vasenin model discussed in Section 3.3.2 is that it takes no account of the energy dissipation that occurs viscoelastically and plastically during a fracture test. Vasenin partially recognized this problem and commented that Equation 3.7 should only be considered approximate. However, Equation 3.7 does provide a very good expression for the experimental data as may be seen from Figs 3.5 and 3.6. Recalling that the constants were evaluated by fitting the equation to the experimental data, it is probable that the viscoelastic and plastic energy contributions to the observed peeling energy are 'contained' in these constants.

Another, and far more fundamental criticism, has been raised by Anand and his colleagues. They consider [44–49] that the dependence of the measured joint strength on parameters such as time of contact and polymer molecular weight, may readily be explained by their effect on the kinetics of wetting. Thus, essentially, they believe that the increases in joint strength that are recorded as a function of t_c and M arise from an increasing degree of interfacial contact occurring and that the mechanism of adhesion is the formation of secondary (van der Waals) forces across the interface (Section 3.5). In response to the direct experimental evidence for interdiffusion, especially in the case of autohesion of elastomers, Anand [49] believes that, whilst it does take place, its contribution to the intrinsic adhesion is minimal compared to that from the formation of interfacial secondary bonds. Quantitatively determining the separate contributions to the intrinsic adhesion from interdiffusion and physisorption is obviously extremely difficult and the above disagreement on the level of the respective contributions has yet to be resolved.

3.3.5 Welding of plastics

An area where interdiffusion appears to play an important role in plastics technology is in the welding of plastics [e.g. 50–52] when the plastics materials possess similar solubility parameters. The polymer chains in the surface layers are given sufficient mobility to interdiffuse either by the application of heat to the regions to be bonded or by applying a suitable solvent. Solvent welding is a technique in which the adhesion of plastic components is promoted by the temporary presence of a solvent, in the absence of an extraneous adhesive. The solvent is usually applied to one of the surfaces which is applied after a short interval to the other surface where it is held under pressure, often heat also being applied to the interfacial zone. A requirement of the solvent is that it strongly plasticizes the surface of the polymers, which results in a large increase in free volume and hence in the chain mobility of the polymer in the interfacial region, increasing the rate and extent of interdiffusion of the polymer chains. There should also be no major structural restriction on polymer chain movement such as crosslinks and a significant degree of crystallinity, which confines this joining technique typically to glassy thermoplastics such as polycarbonate, acrylic polymers, polysulphate, etc.

3.3.6 Polymer/metal interfaces

Although it might be argued that the penetration of polymer molecules into the pores of an anodized metal surface is a form of interfacial diffusion, the present text will not consider such a process to represent an example of the diffusion theory. However, it has been reported [53] that interdiffusion does occur across polymer/metal interfaces when certain metals are evaporated or sputtered onto polymeric substrates. From using the surface analytical techniques of elastic recoil detection combined with Rutherford backscattering, this was suggested to occur in the case of copper/polyimide interfaces, since both techniques indicated an interphase region of about 0.1 μm. Nevertheless, X-ray photoelectron spectroscopy revealed that chemisorption had also occurred and a polymer/metal interfacial complex had been formed (see Section 3.5.5). The authors concluded that the chemisorption provided the intrinsic adhesion forces but that this mechanism was enhanced by the interdiffusion of copper atoms into the polyimide surface regions.

Another example comes from the work of Yasuda and co-workers [54]. They have used the term 'atomic interfacial mixing' to describe their work on creating an interdiffused, interphase region between polymers and metallic substrates by ion implantation and plasma polymerization techniques. The purpose of this study was to improve the intrinsic adhesion and they considered that this might be achieved in two ways. Firstly, an injected atom such as carbon in the surface regions of the metal could act as a seed site for attachment and growth of a plasma-deposited polymer. Secondly, interfacial bonding may result from effects which indirectly involve the injected atoms. For example, surface structural damage might provide additional types of adsorption sites, such as surface vacancies at sputtered atom sites. Their limited work provided some evidence that improved intrinsic adhesion could be attained via these concepts. However, it appears that, as above, any increases in the intrinsic adhesion result largely from improved adsorption on the metallic substrate by the polymer, e.g. a greater number of adsorption sites and/or stronger adsorption forces, rather than directly by a diffusion mechanism.

3.3.7 Summary

In summary, the interdiffusion of polymer chains across a polymer/polymer interface requires the polymers (adhesive and substrate) to be mutually soluble and the macromolecules or chain segments to have sufficient mobility. These conditions are usually met in the autohesion of elastomers and in the solvent welding of compatible, amorphous plastics. In both these examples interdiffusion does appear to contribute significantly to the intrinsic adhesion. However, where the solubility parameters of the materials are not similar, or one polymer is highly crosslinked, crystalline or below its glass transition temperature, then interdiffusion is an unlikely mechanism of adhesion. In the case of polymer/metal interfaces it appears that interdiffusion can be induced and an interphase region created. But this effect enhances the interfacial adhesion by improving the adsorption of the polymeric material rather than by a classic diffusion mechanism.

3.4 ELECTRONIC THEORY

3.4.1 Introduction

If the adhesive and substrate have different electronic band structures there is likely to be some electron transfer on contact to balance Fermi levels which will result in the formation of a double layer of electrical charge at the interface. The electronic theory of adhesion is due primarily to Deryaguin and co-workers [55–57] and they have suggested that the electrostatic forces arising from such contact or junction potentials may contribute significantly to the intrinsic adhesion. The controversy this theory has caused is due to this final statement that such electrostatic forces are an important *cause*, rather than merely a *result*, of high joint strength.

3.4.2 Deryaguin's studies

Deryaguin's theory essentially treats the adhesive/substrate system as a capacitor which is charged due to the contact of the two different materials. Separation of the parts of the capacitor, as during interface rupture, leads to a separation of charge and to a potential difference which increases until a discharge occurs. Adhesion is presumed to be due to the existence of these attractive forces across the electrical double layer and the energy, A_c, that is required to separate the interface is given as:

$$A_c = \frac{h_c \epsilon_d}{8\pi} \cdot \left(\frac{dV_c}{dh_c}\right)^2 \tag{3.8}$$

where V_c is the discharge potential at the discharge gap, h_c, and ϵ_d is the dielectric constant. Now, except at very low pressures, this theory requires a variation of the measured work of adhesion with the pressure of the gas in which the adhesive fracture tests are conducted. Derjaguin *et al.* [56] therefore conducted peeling experiments on plasticized poly(vinyl chloride)/glass, natural rubber/glass and natural rubber/steel interfaces in argon or air at different gas pressures and the results are shown in Table 3.2. As may be seen, except at low pressures, there does appear to be a dependence of the peel energy upon the gas pressure and the theoretical values for the work, A_c, to separate the various interfaces deduced from Equation 3.8 are in excellent agreement with the experimentally determined values. However, several major problems arise with these experiments and calculations. Firstly, Weidner [58] has repeated some of these experiments and found no increase in the peel strength of pressure-sensitive tapes when tested in vacuum as opposed to atmospheric pressure. Secondly, as commented by Wake [59], the technique for actually determining the values of the various parameters in Equation 3.8, and hence the value of A_c, turns out to be a circular argument. Thus, the excellent

Table 3.2 Comparison of the experimental peel energies and the theoretical condenser discharge energy, A_c, for various interfaces

Gas pressure (mm Hg)	Measured peel energy (mJ/m^2)	Charge density (electrostatic units/cm^2)	Condenser discharge energy, A_c (cgs units)
Plasticized poly(vinyl chloride)/glass in argon			
760	1.8×10^4	2.9×10^3	1.8×10^4
175	3.2×10^4	3.0×10^3	3.1×10^4
100	4.0×10^4	2.8×10^3	4.1×10^4
50	5.0×10^4	2.8×10^3	5.1×10^4
Natural rubber/glass in air			
760	7.9×10^4	8.5×10^3	6.1×10^4
100	13.2×10^4	6.3×10^3	13.6×10^4
10	17.5×10^4	5.6×10^3	21.0×10^4
Natural rubber/steel in air			
760	10.5×10^3	2.2×10^3	9.5×10^3
200	15.8×10^3	1.7×10^3	16.9×10^3
100	17.7×10^3	1.8×10^3	22.2×10^3

(1 mJ/m^2 = 1 erg/cm^2 in cgs units)

agreement between theory and experiment is really fortuitous. Thirdly, as remarked earlier, the majority of the measured peel energy represents energy dissipated through viscous and viscoelastic responses of the materials and this energy should not be included in the value equated with the electrical energy. This implies that the theoretical values of A_c are far too large in value.

3.4.3 Weaver's studies

Weaver [60–63] has employed a 'scratch test' to investigate the adhesion of various vacuum-deposited metallic films on a range of insulator polymeric substrates. This test technique necessitates an extremely subjective interpretation and, although as a consequence the experiments have been heavily criticized [64–67], the general trend of Weaver's results are intriguing. Weaver reported that upon aging various metal/polymer interfaces large increases in the scratch resistance were found for copper, silver or gold coatings and smaller increases for aluminium. This appeared to be true irrespective of the nature of the polymeric substrates, although some polymers such as polypropylene and polytetrafluoroethylene exhibited much smaller effects than others like polycarbonate and poly(methyl methacrylate). Previous work [68,69] on metallic coatings on glass had suggested that increases in scratch resistance upon aging metal/glass interfaces were due to increased oxidation of the metal surface with time leading to stronger interfacial bonding, possibly due to an oxide bond between the metal and glass with an oxygen atom acting as a bridge [70].

However, Weaver considered that the inclusion of gold in the above list made any such analogous explanation unlikely for the metal/polymer interfaces.

The key experiment was considered to be the deposition of the metal onto the polymeric surface, aging the specimen to increase the measured interfacial strength, and then to replace the coated polymer in a vacuum chamber and expose it to a glow discharge for a few minutes. The glow discharge was found to produce no changes in the measured strength of copper coated onto glass, where primary chemical bonding via an oxide layer occurs, but for copper on poly(methyl methacrylate) the high interfacial strength which had developed on aging was reduced to the initial level and started to build again on further aging. Weaver concluded that the way in which an ionizing discharge removed the increased adhesion confirmed the electrical nature of the increased adhesion, although the exact mechanism of charge transfer between the film and substrate remained unidentified.

Deryaguin *et al.* [55–57] and Raff and Sharan [71] have suggested that an electron transfer mechanism from the metal to the polymer so as to equalize the Fermi levels is responsible for the intrinsic electronic adhesion, but Weaver considered that this should be a fairly rapid process and, although it might explain the initial adhesion, it could not explain the slow build-up in adhesion that was observed upon aging. Also, if the electrons have to enter the conduction band of the polymeric adhesive, a metal electron would have to acquire extra energy of the order of 2–3 eV and Weaver considered the chances of this happening to be negligible.

Finally, according to Deryaguin's mechanism the magnitude of the charge transferred, and therefore the adhesion, should depend directly on the height of the potential barrier at the metal/polymer interface, but Weaver [60] found no correlation between the work function of the metal and the measured strength. In the light of these and other observations Weaver suggested that the Fermi level in the polymer was initially above that in the metal and thus the charge transfer producing adhesion might be due to positive hole injection, i.e. electron transfer from polymer to metal. Further, under these conditions the charge build-up would be slow since there are very few conduction electrons in any insulator and even these are likely to lie in electron traps which would greatly reduce the electron mobility and increase the time required for charge to build up at the interface. The final charge concentration under these conditions might easily depend on the number of available electrons rather than the difference in energy levels and it would be impossible to trace any correlation between final adhesion and work function.

3.4.4 Criticisms of the electronic theory

Some criticisms of the importance of the electronic theory influencing the intrinsic adhesion have already been raised and other comments on its general applicability will now be considered. Several groups of workers [72–75] have investigated the improvement in the adhesion of thin metal films which is recorded if the insulator substrate is subjected to a low-pressure glow discharge prior to coating. Stoddart *et al.* [73] found no net surface electric charge on the substrate but found that the electron emission was altered, indicating that the surface electronic states of the substrate had been changed. The mechanisms of adhesion, and the relative importance of any electrostatic contribution, remain therefore somewhat vague. Recognizing this, Wake [76] has suggested that, while the nature of the charge-carrying bodies in polymers is obscure, it is known that additives and impurities may drastically affect their ability to form electrical double layers and thus it would be extremely worthwhile to generate data on radiation polymerized material made from a highly purified monomer.

Skinner *et al.* [77] have calculated the tensile strength of a metal/polymer/metal joint due to the existence of electrical double layers at the interface as a function of the volume charge density. For boundary charge densities of 10^{17}, 10^{19} and 10^{21} electrons/cm^3 corresponding values of joint strength of 0.4 kPa, 40 kPa and 4 MPa were deduced. However, the experimental measurements revealed that the maximum charge density was of the order of 10^{19} electrons/cm^3 but the joint strength was of the order of 7 MPa. Roberts [78] has studied the adhesion of crosslinked rubbers to glass, aluminized-glass and steel substrates and directly measured the surface charge densities. He concluded that surface charge densities were apparent when the rubber was separated from the substrate; and these were sufficient to attract dust, resulting in reduced adhesion if subsequent re-contact was attempted. However, their contribution to the intrinsic adhesion was always less than 10% and, for example, for the natural rubber/glass interface the contribution from an electrical double layer was found to be about 10^{-5} mJ/m^2, which is negligible compared to the contribution from van der Waals forces of about 60 mJ/m^2.

von Harrach and Chapman [79] have measured charge densities on a glass insulator substrate coated with a vacuum-deposited layer of gold, silver or copper and from such measurements deduced the electrostatic contributions to the work of adhesion. In Table 3.3 the values calculated are compared with the

Table 3.3 Electrostatic contribution to the adhesion of vacuum-deposited metals on glass substrates [79]

Metal film	Work of adhesion (mJ/m^2) Electrostatic contribution	van der Waals contribution	Adhesive peel energy (mJ/m^2)
Gold	5	950	1400 ± 300
Copper	80	400	800 ± 200
Silver	115	800	1000 ± 200

thermodynamic work of adhesion, W_A, deduced assuming only van der Waals interfacial attractions, and the measured adhesive peel energy for removing the metallic coating. The contribution from electrostatic forces is far lower than that from van der Waals interactions.

3.4.5 Summary

To summarize, in certain specialized interfaces the influence of an electrostatic double layer has been clearly demonstrated, for example, in the case of zirconium-coated gold spheres on cadmium sulphide single crystal substrates [80,81]. The adhesive force was measured by a centrifuge technique as the intensity of the illumination on the adhesive system was varied. This changed the electronic properties of the photoconducting CdS substrates and hence the electrostatic double layer force at the interfaces could be varied leaving the other contributions, e.g. van der Waals, unaffected. Qualitatively the experimental results were in agreement with a simple model of the metal/semiconductor contact. However, it appears that, for typical adhesive/substrate interfaces, any electrical double layer which is generated does not contribute significantly to the intrinsic adhesion. Further, any electrical phenomena observed during the joint fracture process probably arise from the failure event, rather than cause the adhesion between the materials.

3.5 ADSORPTION THEORY

3.5.1 Introduction

The adsorption theory of adhesion is the most widely applicable theory and proposes that, provided sufficiently intimate molecular contact is achieved at the interface, the materials will adhere because of the interatomic and intermolecular forces which are established between the atoms and molecules in the surfaces of the adhesive and substrate. The most common such forces are van der Waals forces and these are referred to as secondary bonds. Also in this category may be included hydrogen bonds. In addition, chemical bonds may sometimes be formed across the interface. This is termed chemisorption and involves ionic, covalent or metallic interfacial bonds being established; these types of bonds are referred to as primary bonds. The terms primary and secondary are in a sense a measure, albeit somewhat arbitary, of the relative strengths of the bonds. This may be appreciated from Table 3.4, where the various types of bonds are shown together with estimates of the range of magnitude of their respective bond energies. Also, it has been proposed that donor–acceptor interactions may occur across an interface and these are typically intermediate in strength between secondary and primary bonds. Finally, it has also been suggested that interfacial molecular complex structures

Table 3.4 Bond types and typical bond energies [82–84]

Type	Bond energy (kJ/mol)
Primary bonds	
Ionic	600–1100
Covalent	60–700
Metallic	110–350
Donor–acceptor bonds	
Bronsted acid–base interactions (i.e. up to a primary ionic bond)	Up to 1000
Lewis acid–base interactions	Up to 80
Secondary bonds	
Hydrogen bonds	
Hydrogen bonds involving fluorine	Up to 40
Hydrogen bonds excluding fluorine	10–25
van der Waals bonds	
Permanent dipole–dipole interactions	4–20
Dipole–induced dipole interactions	Less than 2
Dispersion (London) forces	0.08–40

may, in certain circumstances, be established. These various types of interfacial interactions and their importance in adhesives technology are discussed below.

3.5.2 Secondary force interactions

3.5.2.1 Introduction

Huntsberger [85,86] and others [87,88] have calculated the attractive forces between two planar bulk phases due solely to dispersion forces from a knowledge of the surface free energies. They showed, for example, that even at a separation of one nanometer the attractive force would result in a joint strength in tension of approximately 100 MPa. This is considerably higher than the experimental strength of most adhesive joints. This discrepancy between theoretical and experimental joint strengths is due to air-filled voids, defects or geometric features acting as stress raisers when the joint is loaded, causing rupture of the joint at stresses very much below the theoretical value (see Chapters 6 and 7). However, this calculation does reveal that high joint strengths may, in theory, result from the intrinsic adhesion that arises solely from dispersion forces acting across the interface.

The assertion that secondary forces are sufficient to account for commonly measured joint strengths is supported by much experimental work [e.g. 14,15,27,89–102] which demonstrates that the mechanism of adhesion in many different adhesive joints involves only interfacial secondary forces. Indeed, most of the evidence suggests that to establish stronger primary bonding across the interface requires the use of special techniques; for example, the incorporation of particular side groups along the polymer chain of the adhesive or the use of certain phenolic and organometallic primers or simultaneously bonding and crosslinking elastomeric adhesives against a reactive substrate surface. These techniques are discussed later.

Considering some of the studies reported in the literature, then the adhesion of epoxy adhesives to degreased, chromic acid etched and chromic acid and sulphuric acid anodized aluminium alloys appears [89,90] to involve only secondary forces, although in all cases the initial joint strength is high and, for the etched and anodized aluminium joints, the locus of joint failure is by cohesive fracture in the adhesive.

Of course, at such interfaces hydrogen bonds may well be established and, as may be seen from Table 3.4, the strength of these types of secondary bonds typically lies above that of van der Waals bonds. Indeed, the formation of hydrogen bonds across the interface appears to enhance the intrinsic adhesion and has often been observed, as has been discussed by many authors. For example, Kusaksa and Suetaka [95] have employed attenuated total reflectance infrared spectroscopy to study the interfacial bonding between a cyanoacrylate adhesive and an anodized aluminium. They observed a lowering of the C=O stretching frequency and a shift of the antisymmetric stretching vibration of the C–O–C group to a higher frequency in the infrared spectrum of the cyanoacrylate when it was adsorbed onto the surface of the aluminium. These changes were interpreted as being due to the formation of interfacial hydrogen bonds between the carbonyl groups on the cyanoacrylate adhesive and hydroxyl groups on the surface of the aluminium oxide.

Furukawa *et al.* [96] have used inelastic electron tunnelling (IET) spectroscopy to study the adsorption of aminophenyltrimethoxysilane on aluminium oxide. This technique is based upon the quantum effect tunnelling of electrons through a thin insulating layer sandwiched between two metal films. The insulating layer is typically about 2 to 3 nm thick for optimum resolution but it can be composed of two parts, the oxide of one metal substrate and a superimposed monolayer of the organic material of interest. Small voltages are applied across the insulating layer and the vibrational spectra of its components are determined by the energy lost by tunnelling electrons. Data is generally obtained in terms of the second derivative ($\partial^2 I/\partial V^2$) of the tunnelling current, I, as a function of the applied voltage, V, since this is the analog of an infrared spectrum. Furukawa *et al.* [96] concluded that hydrogen bonding across the silane/oxide interface occurred via the amino group on the silane and surface hydroxyl groups on the oxide.

Considering plastic substrates, Pritchard [97] cites the dipping of nylon cords into a complex adhesive mixture of rubber and resorcinol–formaldehyde, as used in the production of tyres, as an example where hydrogen bonding may

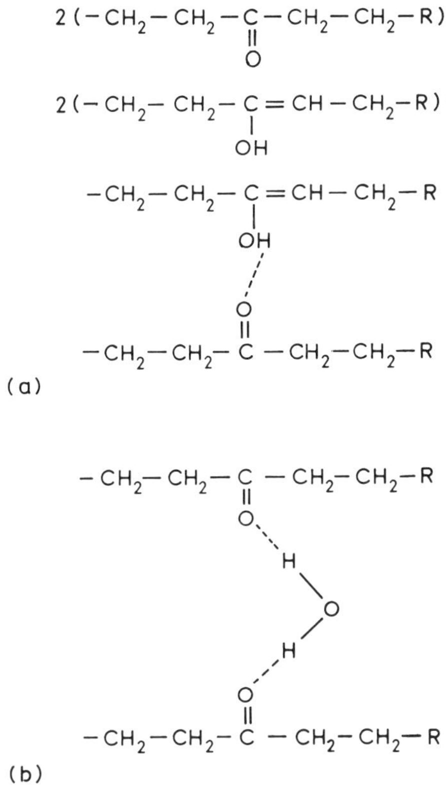

Figure 3.8 Possible hydrogen-bonding mechanisms in the autohesion of oxidized polyethylene. (a) Via tautomerization, (b) via a water molecule.

play an important role. It was considered that the resorcinol–formaldehyde was adsorbed onto the nylon surface via hydrogen bonds through the phenolic groups. Another example where hydrogen bonding has been shown to play an important role is in the autohesion of polymers which have been subjected to surface oxidation by immersion in certain acids [98,99] or exposure to an electrical (corona) discharge [100–102]. In the case of polyethylene, the formation of hydrogen bonds has been suggested to be the result of tautomerization according to a series of transformations shown in Fig. 3.8a. Here the carbonyl group at the surface of the oxidized polyethylene may first undergo the tautomerization of the ketone to the enolic form. The enolic form then exhibits an acidic character and may form a hydrogen bond with a carbonyl group on the adjacent surface. A second possible mechanism of hydrogen bond formation is via a water molecule [98] as illustrated in Fig. 3.8b. Some supporting evidence for these schemes comes from the work of Blythe *et al.*

[102] who have employed X-ray photoelectron spectroscopy (XPS) to analyse the surface of oxidized polyethylene. However, their results also indicated that several other hydrogen bonding mechanisms are conceivable.

The use of XPS, also known as electron spectroscopy for chemical analysis (ESCA), will be discussed many times in the present book since it can provide very good quantitative and qualitative elemental analysis of a surface and reasonably good determination of the chemical bonding. XPS involves the bombardment of the specimen surface with mono-energetic X-rays in ultra-high vacuum ($<10^{-7}$ torr, $<10^{-5}$ Pa). As the photons travel through the material some are absorbed and their energy is transferred to electrons which can be ejected from the specimen. The spectrum, the electron intensity versus the binding energy of the electron to the atom, is obtained by pulse-counting techniques. However, only photoelectrons originating very near the surface (about 1.5 to 9 nm) have a high probability of escaping from the surface without losing energy; and only such electrons may be distinguished; hence the surface sensitivity of this technique. A complementary technique is Auger electron spectroscopy (AES) and here Auger electrons are excited from the atoms in the uppermost surface regions using a small diameter electron beam (0.1 to 50 μm), thereby giving good spatial resolution. Since XPS depends upon X-ray photons to excite photoelectrons, spatial resolution comparable to AES cannot as yet be obtained. The advantage of XPS is, however, that charging effects are minimized and the surfaces of insulators can be readily analysed. With both techniques the specimen surface may be progressively eroded by sputtering with an argon ion gun so that a depth profile through the surface regions of the specimen may be obtained.

3.5.2.2 Thermodynamic work of adhesion

The thermodynamic work of adhesion, W_A, required to separate a unit area of a solid and a liquid phase forming an interface across which secondary forces are acting may be related to the surface and interfacial free energies by the Dupré equation. The reversible work of adhesion, W_A, in an inert medium may be expressed by:

$$W_A = (\text{Sum of the surface free energies of the solid and liquid phases} - \text{the interfacial free energy}) \tag{3.9}$$

i.e.

$$W_A = \gamma_s + \gamma_{lv} - \gamma_{sl} \tag{3.10}$$

Note that the value of surface free energy appropriate to the Dupré equation is γ_s, rather than γ_{sv}. As pointed out by Wake [103], if the latter term is used in the Dupré equation then the separation to which the value of W_A refers is the separation of the liquid from a solid which is still covered with a layer of vapour, i.e. this implies that the adhesion which has failed is the adhesion between the bulk liquid and a layer of molecules of the liquid one molecule thick.

Now, by eliminating the interfacial free energy by using Equation 2.3:

$$W_A = \gamma_{lv}(1+\cos\theta)+\Pi_s \tag{3.11}$$

This equation strictly only applies to a solid/liquid interface but, by assuming that the surface free energy of a liquid does not change significantly when it solidifies isothermally and ignoring any shrinkage stresses, it may be applied to solid-adhesive/substrate interfaces. However, for the case of adhesive/substrate interfaces an often more useful expression for Equation 3.9 may be derived by using Equation 2.17 to eliminate the interfacial free energy:

$$W_A = 2(\gamma_a^D\gamma_s^D)^{1/2}+2(\gamma_a^P\gamma_s^P)^{1/2} \tag{3.12}$$

where the subscripts a and s now represent the adhesive and substrate, respectively. It is of interest to note that to maximize the thermodynamic work of adhesion, W_A, the values of the surface free energy components of *both* the adhesive and substrate should be as large as possible. But this requirement conflicts with the one needed to obtain good interfacial contact where it was shown (Section 2.3.1) that, ignoring Π_s:

$$\gamma_s \geq \gamma_{as}+\gamma_a \tag{3.13}$$

where γ_{as} is the interfacial free energy between the adhesive and substrate.

3.5.2.3 W_A and joint strength

It is a recurrent hypothesis in the literature [104–112] that it should be possible to correlate the measured strengths of adhesive joints with the thermodynamic work of adhesion, W_A. Indeed, shortly after Zisman introduced his concept of the critical surface tension of wetting, γ_c, Levine *et al.* [107] reported a correlation between the strength of joints prepared from an epoxy adhesive bonding various polymeric substrates and the γ_c value of the substrate, as shown in Fig. 3.9. However, whilst some subsequent authors [108–110] have confirmed such correlations, others [111,112] have found none. The main causes for this conflict are that, firstly, the test methods commonly employed for measuring the strengths of adhesive joints are not well suited to theoretical analysis, especially with respect to separating out the contributions from the intrinsic adhesion forces and from viscoelastic and plastic energy losses in the adhesive and substrates. Secondly, such a correlation would only be expected when the locus of joint failure is interfacial.

Now, by adopting a continuum fracture mechanics approach, the work of Andrews and Kinloch [14,15,113] and Gent and Kinloch [27] defined a geometry-independent measure of joint strength, the adhesive fracture energy,

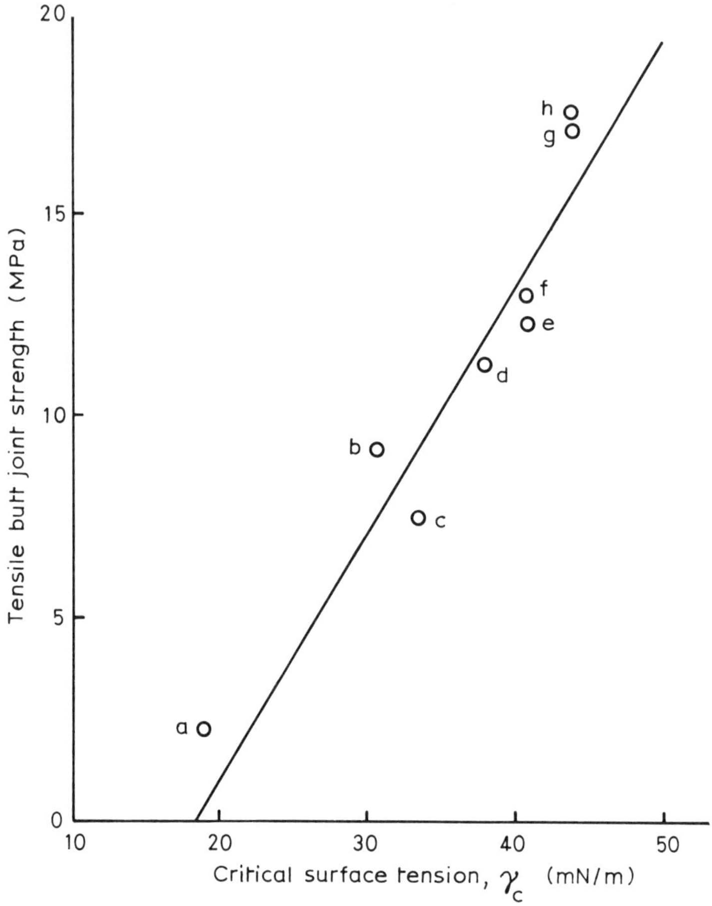

Figure 3.9 Relation between measured joint strength and critical surface tension, γ_c, of the substrate for an epoxy bonding various polymeric substrates [107]. (a) Polytetrafluoroethylene, (b) poly(vinyl fluoride), (c) polystyrene, (d) poly(vinyl alcohol), (e) poly(vinylidene chloride), (f) poly(vinyl chloride), (g) poly(ethylene terephthalate) and (h) poly(1,4-cyclohexylene dimethylene terephthalate).

G_c. These studies are described in detail in Chapter 7, but essentially from experimental and theoretical considerations it was demonstrated that the adhesive fracture energy, G_c, for a crosslinked rubbery adhesive/rigid plastic interface could be divided into two major components:

(a) The energy required to propagate a crack through a unit area of interface in the absence of viscoelastic energy losses, i.e. an intrinsic adhesive fracture energy, G_o, which is a direct measure of the bonding forces.
(b) The energy, ψ, dissipated viscoelastically within the rubbery adhesive at the propagating crack, again referred to unit area of interface.

The energy, ψ, dissipated viscoelastically is usually the major contribution to the measured adhesive fracture energy, G_c. This results in the value of G_c being much greater than that of G_o and the value of G_c being highly dependent upon the rate and temperature, both observations being in accord with experimental results. The relationship between these two terms and the measured adhesive energy, G_c, is discussed in Section 7.2.2.2 and is given by:

$$G_c = G_o + \psi \tag{3.14}$$

but it was shown that:

$$\psi = G_o f(\dot{a}, T, \epsilon) \tag{3.15}$$

where f is a function, the value of which depends upon the crack growth rate, $\dot{a}$, the temperature, T and the strain level ϵ. Now let:

$$\Phi_v(\dot{a}, T, \epsilon) = 1 + f(\dot{a}, T, \epsilon) \tag{3.16}$$

then, combining Equations 3.14 to 3.16 yields:

$$G_c = G_o \Phi_v(\dot{a}, T, \epsilon) \tag{3.17}$$

where Φ_v is a mechanical loss function. The dependence of the energy loss term, ψ, upon the intrinsic adhesive fracture energy, G_o, is particularly intriguing. This dependence arises because material around the crack tip can only be subjected to stress, and hence energy losses, whilst the chemical and physical bonds acting ahead of the crack tip remain unbroken. An important aspect of Equation 3.17 is that when viscoelastic and plastic energy losses are negligible, i.e. $\Phi_v(\dot{a}, T, \epsilon) \rightarrow 1$ and $f(\dot{a}, T, \epsilon) \rightarrow 0$, then the measured value of the adhesive fracture energy is equal to G_o. Thus, a direct measure of the interatomic and intermolecular bonding forces across the fracture plane may be obtained. Since in an adhesive joint the fracture plane may wander between the interface and the materials forming the interface, the intrinsic adhesive fracture energy, G_o, may be expressed as a weighted average of the various possible failure modes:

$$G_o = iG_o\,(\text{interfacial}) + b'\,G_o\,(\text{adhesive}) + sG_o\,(\text{substrate}) \tag{3.18}$$

where G_o (interfacial), G_c (adhesive) and G_o (substrate) are the intrinsic fracture energies for interfacial, cohesive-in-adhesive and cohesive-in-substrate failure, respectively, and i, b' and s are the respective area fractions of interfacial failure, failure in adhesive and failure in substrate: i.e. $i + b' + s = 1$. For those joints which exhibit a solely interfacial locus of failure and in which only secondary bonding is established across the interface then $i = 1$, G_o (interfacial) $= W_A$ and, therefore, the measured value of G_o should be equal to the thermodynamic work of adhesion, W_A.

The values of G_o were ascertained for various joints from the mechanical property studies and, where it was found that interfacial failure had occurred, the values of G_o were compared to corresponding values of W_A calculated from

Table 3.5 Values of the intrinsic fracture energy, G_o, and the thermodynamic work of adhesion, W_A, for a styrene–butadiene rubber adhering to various substrates [14,15,113]

Substrate	G_o (mJ/m^2)	W_A (mJ/m^2)
Fluorinated ethylene–propylene copolymer	22	48
Plasma-treated ethylene–propylene copolymer	69	57
Poly(chlorotrifluoroethylene)	75	63
Nylon 11	71	71
Poly(ethylene terephthalate)	79	72

Equation 3.12. The results are shown in Table 3.5 and, as may be seen, the values of the two parameters are in good agreement. Thus, the nature of the intrinsic adhesion forces acting across these interfaces was secondary interfacial bonds.

An important general point emerges from the above work. Namely, that although the energy, ψ, which is dissipated viscoelastically and plastically and is usually far greater in value than W_A, the value of ψ is dependent upon the value of W_A. Indeed, for the model joints described above the value of ψ is directly proportional to the value of W_A. Hence, designing an interface to maximize the interfacial bonding forces (but bearing in mind the possibly conflicting need to attain good interfacial contact) is not a trivial pursuit. Obviously, this has long been recognized in industry as evidenced by most adhesives being polar in character and often having the capability to establish interfacial hydrogen bonds, as discussed above.

3.5.3 Donor–acceptor interactions

It has been argued, particularly by Fowkes and co-workers [114–119], that the formation of acid–base interactions between the adhesive and substrate may represent a major type of intrinsic adhesion force that operates across the interface. This classification includes hydrogen bonds which are considered to be a subset of acid–base interactions. The acid may be an electron acceptor, in the Lewis sense, or a proton donor in the Bronsted sense. The base may be an electron donor, in the Lewis sense, or a proton acceptor, in the Bronsted sense. Materials may have one of three types of acid–base capability [114,120,121]:

(a) Act as electron acceptors (or proton donors, i.e. acids). For example, this includes partially halogenated molecules, including polymers such as poly(vinyl chloride), chlorinated polyethylenes or polypropylenes, poly(vinylidene fluoride) and ethylene–acrylic acid co-polymers, and other surfaces such as silica, Fe_2O_3 and Fe_3O_4.

(b) Act as electron donors (or proton acceptors, i.e. bases). For example, this includes esters, ketones, ethers or aromatics, and polymers such as poly(methyl methacrylate), polystyrene, ethylene–vinyl acetate copolymers, polycarbonate, polyimides and other surfaces such as calcium carbonate, amorphous Al_2O_3, hydrous Fe_2O_3 and amorphous hydroxides of iron oxide.

(c) Act both as electron acceptors (or proton donors) and electron donors (or proton acceptors). For example, molecules such as amides, amines and alcohols which includes polymers such as polyamides and poly(vinyl alcohol), and other surfaces such as α-Al_2O_3, AlOOH(boehmite) and $Al(OH)_3$ (bayerite).

As discussed in Section 2.4.1.4, Fowkes *et al.* considered that the contributions from acid–base interfacial interactions could not be predicted from a geometric mean relation but proposed that they could be described by a relation of the form:

$$-\Delta H^{AB} = C^A C^B + E^A E^B \tag{2.19}$$

where ΔH^{AB} is the enthalpy of the acid–base interactions, and C^A and E^A are two constants which are needed to characterize the acid; similarly C^B and E^B characterize the base. Fowkes and Mostafa [115] have demonstrated the application of Equation 2.19 in the area of polymer adsorption. For example, they determined the amount of adsorption resulting from stirring silica (an acidic solid) into a solution of poly(methyl methacrylate) (i.e. PMMA, a basic polymer) dissolved in either basic, neutral or acidic solvents. (The basic solvents were characterized by their ΔH^{AB} value with tert-butanol and the acidic liquids by their ΔH^{AB} value with ethyl acetate.) The results are shown in Fig. 3.10. It may be seen that with increasing basicity of the solvent the amount of adsorption of PMMA on the silica decreases markedly. This is because the basic solvent is competing with the basic polymer for the acidic silica SiOH sites of the silica; when the solvent is more basic than the polymer's ester carbonyl oxygen groups little adsorption of the polymer occurs. Similarly, it may also be seen that with increasing solvent acidity the adsorption of PMMA upon the silica decreases. This is because the acidic solvent and acidic solid are both competing for the basic polymer; little polymer adsorption occurs upon the silica when the solvent is more acidic than the surface of the silica. Thus, adsorption is at a maximum when the solvent is neutral.

Fowkes and Mostafa [115] have suggested that the work of adhesion, W_A, is given by:

$$W_A = 2(\gamma^{AD}\gamma^{BD})^{1/2} - k_5(C^A C^B + E^A E^B)n^{AB} + W_A^P \tag{3.19}$$

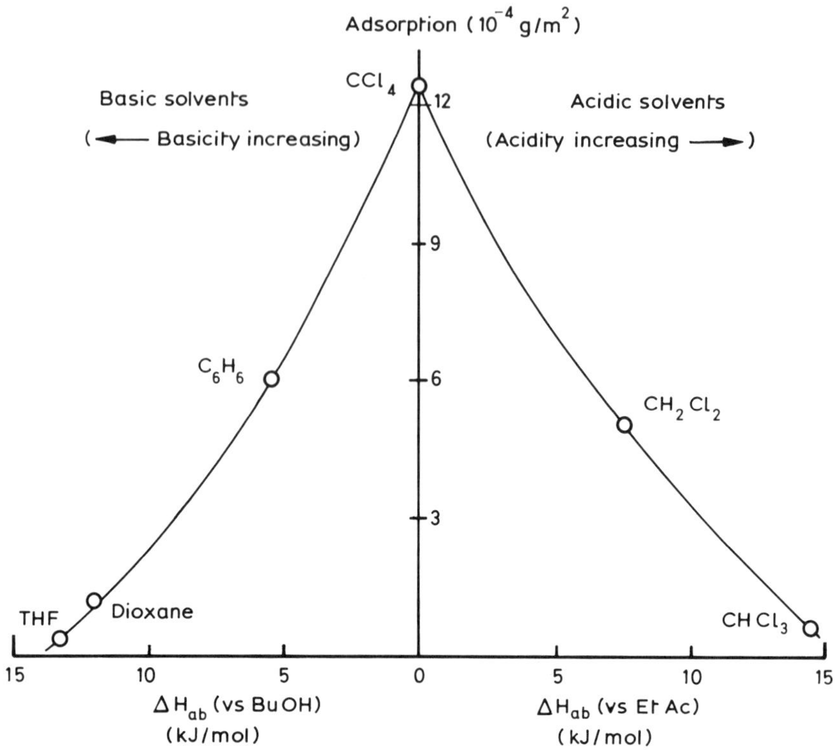

Figure 3.10 Adsorption of poly(methyl methacrylate), a basic polymer, onto the acidic surface of silica from basic, neutral and acidic solvents [115].

where the thermodynamic work of adhesion, W_A^{AB}, arising from acid–base interactions is:

$$W_A^{AB} = -k_5(C^A C^B + E^A E^B) n^{AB} \tag{3.20}$$

and where k_5 is a constant, close to unity, which converts enthalpy per unit area into surface free energy, n^{AB} is the number of moles of acid–base pairs per unit area and W_A^P is the thermodynamic work of adhesion arising from dipole–dipole interfacial interactions, which was usually considered to be negligible. According to these ideas, the intrinsic interfacial adhesion will obviously be increased if W_A^{AB} can be maximized and, indeed, the pretreatment of substrate surfaces often generates chemical groups which have electron accepting or donating capability. However, whether increases in the intrinsic adhesion arise from acid–base interactions or dipole–dipole interactions has not been firmly established. One problem is that the predictive capability of Equations 3.19 and 3.20 is very restricted due to the lack of data on the values of C and E. Further, other work [114,122] indicates that even when values of these constants are available they may not correctly predict interfacial acid–base interactions, and that dipolar interactions cannot be considered negligible.

Schultz *et al.* [149] have also used an acid–base approach but adopted the approach of Gutman [150], rather than that used by Fowkes and colleagues. The Gutman method describes a semi-empirical acid–base scale and, using inverse gas chromatography, Schultz *et al.* were able to derive a semi-quantitative 'specific interaction parameter' for the degree of acid–base interactions acting between various pretreated carbon fibres and an epoxy matrix. It appeared that measured increases in the shear strength of the fibre–matrix interface correlated better with the degree of acid–base interactions, via the calculated 'specific interaction parameter', than with the value of W_A, as deduced from contact angle measurements and Equation 3.12.

3.5.4 Primary force interactions

3.5.4.1 Introduction

Depending upon the chemistry of the interface, various types of primary bonds have been reported to be formed across an interface. For example, in the work of Andrews and Kinloch [14,15,113] and Gent and co-workers [123–126] on bonding and crosslinking elastomers against reactive substrates it appears that interfacial covalent bonds are formed, as discussed in greater detail below. Similarly, Klein *et al.* [127] found infrared evidence of covalent primary bonds between a polyurethane adhesive and epoxy-based primers, and such interfacial interactions gave the highest joint strengths, although it should be noted that such covalent bonds will, of course, usually have a degree of ionic character due to the different electronegativities of the atoms involved in the bond.

Interfacial primary bonds which are highly ionic in character have been

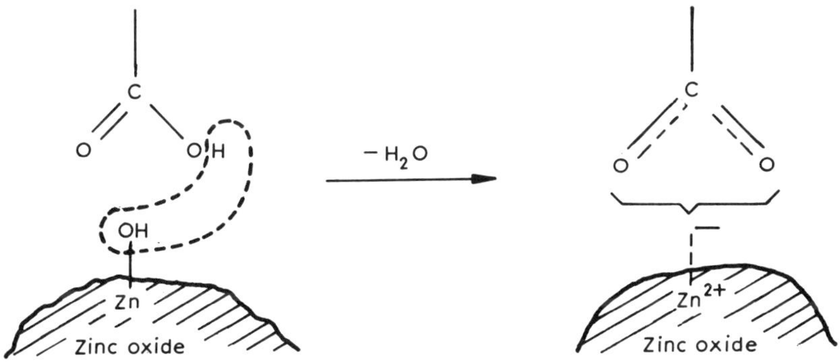

Figure 3.11 Possible reaction scheme for a poly(acrylic acid) bonding to a zinc oxide surface [129].

reported [128–130] between polymeric adhesives such as poly(acrylic acid) and metal oxides. The importance of such studies is that materials based upon poly(acrylic acid) containing certain powdered metal oxides are currently being used as filling materials for restoring teeth and for adhesively bonding inlays and crowns into place. For example, the bonding between the poly(acrylic acid) and a zinc oxide surface has been suggested [129] to be of the form shown in Fig. 3.11. Chu *et al.* [130] have used inelastic electron tunnelling spectroscopy to examine the nature of the interfacial bonding of 4-hydroxybenzoic and 4-aminobenzoic acid to aluminium oxide; these chemicals having been proposed as coupling agents. They concluded that the bonds that were established were primary ionic bonds ($COO^{-}Al^{+}$) with the phenyl ring oriented perpendicular to the oxide surface. However, no results were reported on the effectiveness of these proposed coupling agents and it is likely that, although interfacial primary bonds were formed, the fact that the adsorbates could not polymerize or form multiple surface attachments would lead to them acting as weak boundary layers, unless deposited as monolayers.

3.5.4.2 Organometallic primers

One of the most controversial areas for the role of primary interfacial bonds has been when adhesion promoters, or coupling agents, are employed to enhance the joint strength or, more commonly, to enhance the environmental resistance of the interface to attack by moisture. The various types of adhesion promoters that have been employed are discussed in detail in Chapters 4 and 8, where the aspects of surface pretreatment and environmental attack respectively are considered. It is sufficient to note here that the most common type is the silane-based primers [131,132] which have the general structure $X_3Si(CH_2)_nY$, where $n = 0$ to 3, X is a hydrolysable group on silicon and Y is an organofunctional group usually selected for reactivity with a given resin. They are widely used as a surface finish on glass fibres prior to forming a glass fibre reinforced composite, as an additive in adhesive formulations and as a primer applied to the substrate prior to adhesive bonding.

Until recently there has been little known of the mechanism by which the intrinsic adhesion of the interface is enhanced, especially with respect to environmental attack. The generally, but not universally, accepted mechanism of adhesion has been the formation of strong, primary interfacial bonds, mainly covalent in nature but with some degree of ionic character due to the different electronegativities of the atoms forming the primary bonds. For the silane/substrate interface this arises from the formation of Si–O–substrate bonds. For the adhesive/silane interface this may arise from the reaction of the Y group on the silane with reactive groups in the adhesive; although, at this interface some form of interpenetrating network structure might well be formed instead of simple primary bonds. Other mechanisms by which such primers work have been suggested, and include the idea that the primer may result in better wetting at the interfaces [133] and the possible formation of a boundary layer in the adhesive differing in chemical and physical properties to that of the bulk resin [134].

However, studies by Koenig and Shih [135] employing laser Raman spectroscopy and Chiang *et al.* [136] using Fourier transform infrared spectroscopy have clearly established the presence of Si–O–Si bonding across a silane/glass interface. The work of Chiang *et al.* examined the bonding and structure of γ-aminopropyltriethoxysilane on silica and glass fibre surfaces. Apart from identifying the interfacial primary bonds, they also demonstrated that the silane condensed to form a polysiloxane on the surfaces and that on the silica surface the amino groups orientated towards the surface and formed hydrogen bonds with hydroxyl groups on the silica surface. The proposed interfacial

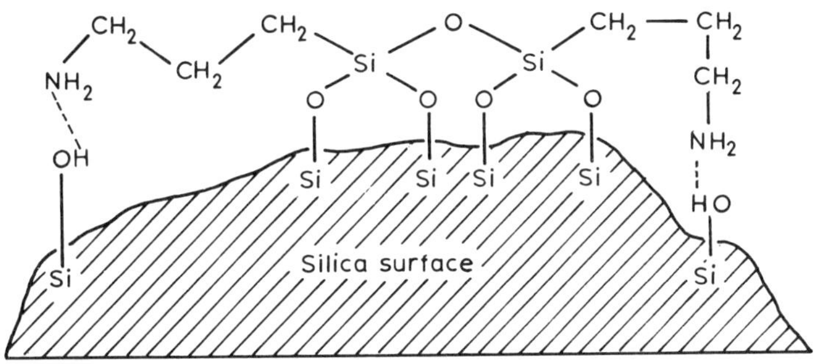

Figure 3.12 The structure and interfacial bonding of γ-aminopropyltriethoxysilane on a silica surface [136].

bonding mechanism of the γ-aminopropyltriethoxysilane on the silica substrate is shown in Fig. 3.12.

Gettings and Kinloch [137] have investigated the nature of the interfacial bonding between γ-glycidoxypropyltrimethoxysilane adsorbed onto a mild steel substrate using X-ray photoelectron and static secondary ion mass (SIMS) spectroscopy. With the SIMS technique, ionized particles ejected from the surface by the action of an argon beam are mass analysed. In 'static' SIMS the current densities used are very low (of the order of 10^{-10} A/cm) and, therefore, the first one or two monolayers of the surface can be investigated. Either atoms or molecules can be ionized and thus details of the chemical state of the atoms in a surface can be inferred. The static SIMS technique proved especially rewarding and the spectrum is shown in Fig. 3.13. By comparison with other spectra, it was concluded that the presence of SiO_2H^-, $SiOH^+$ and SiO_2^- arose from the silane primer polymerizing to give a polysiloxane structure on the metal substrate, thus giving the possibility of many adsorption sites per

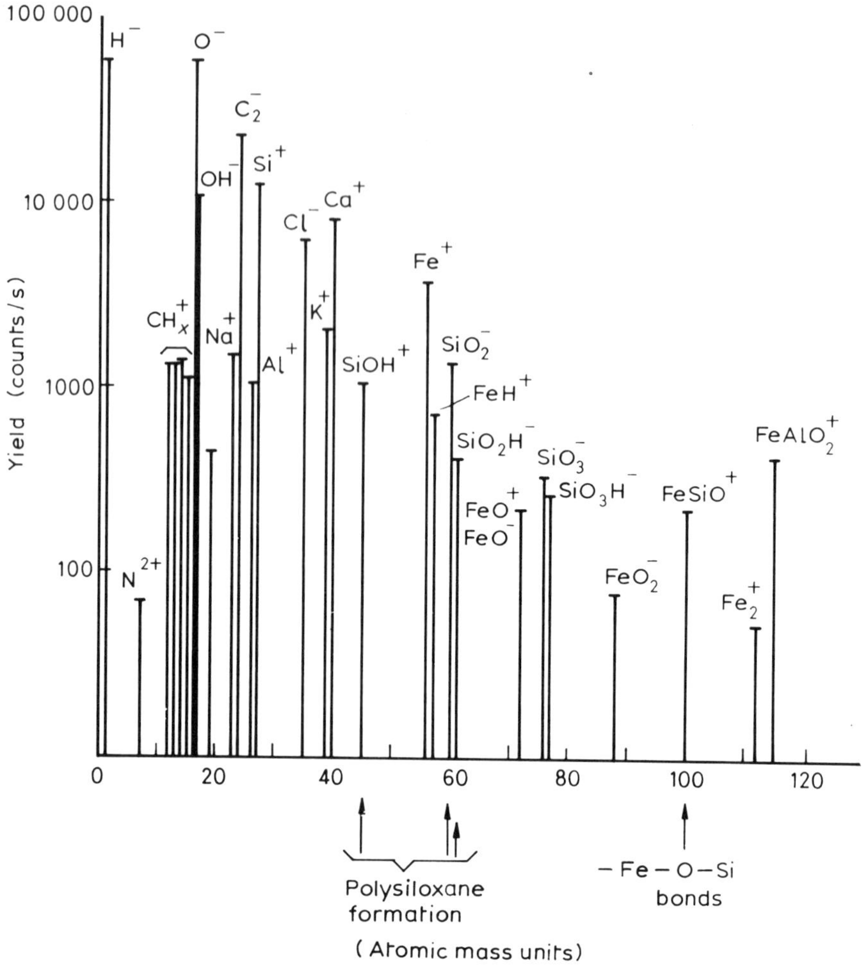

Figure 3.13 Positive and negative static SIMS spectrum from a mild steel substrate coated with a γ-glycidoxypropyltrimethoxysilane primer [137].

molecule and resulting in the primer having a reasonable mechanical strength, and not therefore behaving as a weak boundary layer. Also of interest was the detection of the $FeSiO^+$ ion which provided strong evidence for the formation of a chemical bond, probably Fe–O–Si, between the metal oxide and the siloxane primer. Similar studies [138] have been conducted on stainless steel surfaces and for the adsorption of γ-glycidoxypropyltrimethoxysilane the SIMS spectra revealed the presence of both $FeOSi^+$ and $CrOSi^+$ radicals, indicating the formation of Fe–O–Si and Cr–O–Si bonds across the interface. These results suggest that the reaction mechanism of the silane primer is probably as

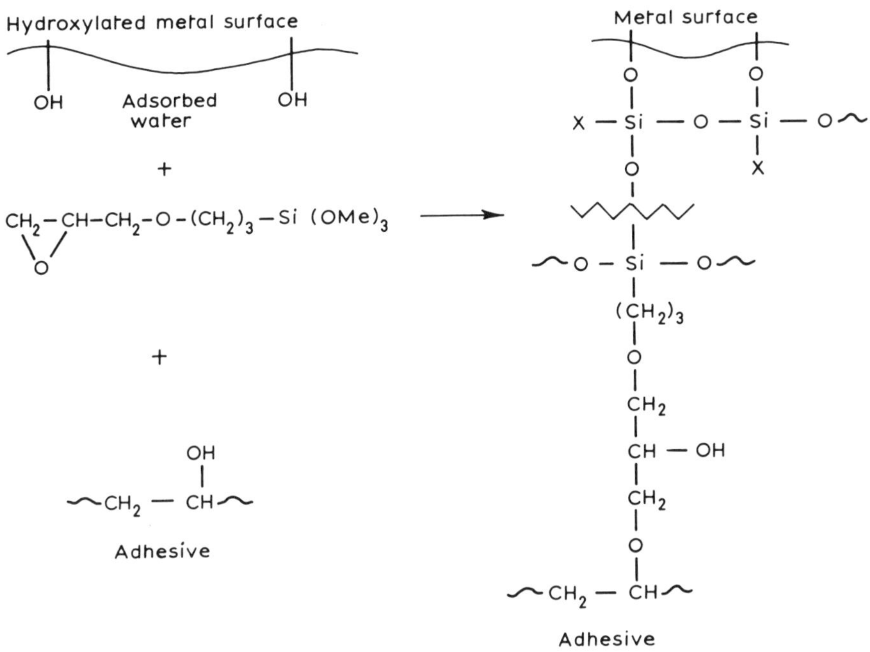

Figure 3.14 Proposed schematic diagram of the mechanism of adhesion for epoxy adhesive/γ-glycidoxy propyltrimethoxysilane primer/ferric oxide interfaces.

illustrated in Fig. 3.14. The technological importance of these results is that only in those cases where there was evidence of primary interfacial bonds being formed between the silane primer and the substrate was there any corresponding increase in the environmental resistance of the adhesive joint.

3.5.4.3 Values of the intrinsic adhesive fracture energy, G_o

Andrews and Kinloch [14,15,113] have extended their studies described in Section 3.5.2.3 to include interfaces where covalent bonds could be established across the interface. This was achieved by introducing C=C double bonds into the surface regions of fluorocarbon polymeric substrates, by using a suitable etching treatment, and then crosslinking the rubbery adhesive, using dicumy peroxide via a free radical reaction, after interfacial contact with the etched substrate had been established. For such interfaces the values of the intrinsic adhesive fracture energy, G_o, were far higher than the corresponding values of the thermodynamic work of adhesion, W_A. The values of W_A assume, of course, that only secondary bonds act at the interface. Two further points of interest also emerged. Firstly, the increases in the value of G_o arising from the interfacial covalent bonding were reflected in very much higher joint strengths, since the energy dissipated viscoelastically in the adhesive was proportional to the value of G_o. Secondly, at a certain level of interfacial covalent bonding no further increase in the value of G_o was recorded, and this was due to the locus of failure moving from the interface into the adhesive and substrate.

Gent *et al.* [123–126] have also demonstrated the influential effect that primary interfacial bonds may exert by conducting some elegant experiments on the adhesion of polybutadiene to glass using silane primers. The surface density of interfacial chemical bonds between the glass substrate and the crosslinked polybutadiene adhesive was varied by first treating the glass with a mixture of vinyl- and ethyl-terminated silanes. The silane based primer is known to form Si–O–Si bonds to the glass substrate (Section 3.5.4.2) and it was reasonably assumed that the vinyl groups of the former silane primer would form covalent bonds to the polybutadiene when the rubber was crosslinked via a free radical reaction after interfacial contact to the primed glass substrate was established. However, it was assumed that the ethyl groups on the ethyl-terminated silane would be quite unreactive towards the polybutadiene. In agreement with these assumptions it was found that the measured fracture energy increased in direct proportion to the amount of vinyl silane used for priming the glass, as shown in Fig. 3.15. Gent and co-workers [124-126] have since demonstrated that increased joint strengths may also be attained when interfacial covalent bonds are formed across the interfaces when bonding rubbers together, as previously observed by Trostyanskaya *et al.* [139] and Kildunovich *et al.* [140].

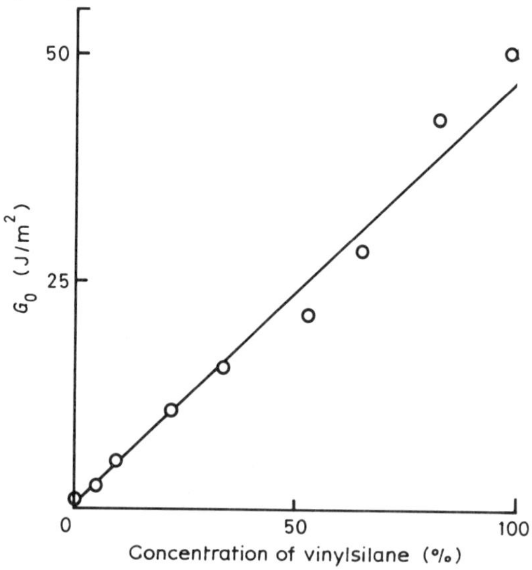

Figure 3.15 Intrinsic adhesive fracture energy, G_o, for polybutadiene/glass interfaces versus the fraction of vinylsilane in the mixture of vinyl and ethyl silanes used to treat the glass substrate surface [123].

3.5.5 Molecular complexes

It is difficult to define the term 'molecular complex' precisely, but in general it is applied to compounds where the number of bonds formed by one of the atoms is greater than expected from the usual valency considerations. Such complexes are usually formed between one molecule which has the capability to accept electrons (i.e. a Lewis acid) and another which can readily donate electrons (i.e. a Lewis base). When both electrons for the bond come from one atom donating a pair of electrons then the bond is commonly referred to as a 'coordinate' bond. However, donor–acceptor interactions are not the only form of bonding which is possible, and in the present text the term complex will not necessarily imply a particular type of bond. The atoms, or ions, which bond to the central atom, invariably a metal atom, are referred to as 'ligands' and ligands are characterized according to how many donor centres they have. Hence, ligands that bond through only one atom (e.g. N in NH_3) are termed 'monodentate' (literally one-toothed). Ethylenediamine ($NH_2CH_2CH_2NH_2$) and diethylenetriamine ($NH_2CH_2CH_2NHCH_2CH_2NH_2$), which have two and three donor nitrogen atoms, respectively, are termed 'bidentate' and 'tridentate'. When both donor atoms of a bidentate ligand bond to the same metal centre, thus giving a ring system, the compound formed is termed a 'chelate'. The name comes from the Greek word *chela*, which refers to the claw of a lobster or crab. The term chelate does not apply, however, if the bidentate ligand forms a bridging system with its two donor atoms bonded to separate acceptor centres.

Considering the formation of complex systems at interfaces of interest to the adhesives technologist, several workers [141–147] have reported that such interfacial bonding does indeed occur and may contribute significantly to the intrinsic adhesion. For example, DeNicola and Bell [142] have studied polyfunctional chelating agents for improving the durability of epoxy/steel interfaces. They employed mercaptoester chelating primers which are capable of establishing intrinsically strong adhesion forces by reacting with metal ions and forming organometallic chelate ring structures. However, the metal surface first had to be treated with an ammonium citrate solution and it was assumed that this was necessary to provide a source of iron cations to form the chelate. This work also emphasized the need for the chelating primer to be able to attach itself to more than one site on the substrate surface, i.e. to act as a polymeric chelate. Complexes may also be formed at the interface in the case of copper substrates [144,145] with both benzotriazole and 2-mercaptoimidazole based primers. Cotton *et al.* [144] have shown that various benzotriazole compounds may promote good adhesion to copper, or silver or cadmium, and alloys. Such primers contain a group which is reactive towards the adhesive and a ligand group which can form a polymeric complex with the metal. Gardiner *et al.* [145] have used the new technique of Raman microscopy to obtain vibrational spectra *in situ* in an aqueous environment from copper substrates coated with various 2-mercaptoimidazole based primers and shown that Cu(I)-ligand coordination complexes are formed.

3.6 CONCLUDING REMARKS

From the discussions in this chapter it is evident that in certain circumstances any of the four main mechanisms of adhesion may be responsible for the intrinsic adhesion which enables an interface to resist separation under the application of an applied stress. Nevertheless, the evidence clearly indicates that for most adhesive/substrate interfaces of interest to the adhesives technologist the main mechanism of adhesion is the adsorption theory. This theory states that, providing intimate molecular contact is achieved at the interface, the materials will adhere because of the interatomic and intermolecular forces which are established between the atoms and molecules in the surfaces of the adhesive and substrate.

However, several other interesting points emerge. Firstly, although the adsorption theory may be the one of most general applicability, there may be distinct advantages in bonding to a rough surface: not necessarily because of any mechanical interlocking mechanism but due to an improved degree of interfacial contact and more extensive energy dissipative deformations being initiated in the adhesive. Secondly, the importance of initiating such deformations was also evident when discussing the influence of secondary and primary interfacial bonds on the measured joint strengths and, in this respect, the conclusions reached clearly indicated that maximizing the intrinsic adhesion forces is not a trivial pursuit. Thirdly, the benefit of establishing strong interfacial bonds was further demonstrated in connection with increasing the environmental resistance of the adhesive/substrate to attack by moisture. For example, the roles of interfacial primary bonds and chelate complexes were briefly referred to, and this aspect will be considered in greater detail in Chapters 4 and 8.

The previous two chapters have emphasized the importance of the chemistry, topography and morphology of the substrate's surface in ensuring both that intimate interfacial contact is achieved and that adequately strong and stable intrinsic adhesion forces are forged across the interface. Therefore, the next chapter will consider the various surface pretreatments that are employed to attain these aims.

REFERENCES

1. Tabor, D. and Winterton, R. H. S. (1969) *Proc. Roy. Soc.* **A312**, 435.
2. Israelachvili, J. N. and Tabor, D. (1972) *Proc. Roy. Soc.* **A331**, 19.
3. Johnson, K. L., Kendall, K. and Roberts, A. D. (1971) *Proc. Roy. Soc.* **A324**, 301.

4. Knock, K. K. and Locke, M. C. (1981) *Proceedings of the 159th Meeting of the American Electrochemical Society*, Minneapolis, May.
5. Kato, K. (1967) *Polymer* **8**, 33.
6. Kato, K. (1968) *Polymer* **9**, 419.
7. Heynmann, K. (1966) *Production Finish* **19**, 38.
8. Matsunaga, M., Haguida, Y. and Ito, K. (1968) *Metal Finish* **66**(11), 80.
9. Rantell, A. (1969) *Trans. Inst. Metal Finish.* **47**, 197.
10. Sykes, J. M. and Hoar, T. P. (1969) *J. Polymer Sci.* **A1**, 1385.
11. Abu-Isa, I. A. (1971) *J. Appl. Polymer Sci.* **15**, 2865.
12. Roberts, R., Ryan, F. W., Schonhorn, H., Sessler, G. M. and West, J. E. (1976) *J. Appl. Polymer Sci.* **20**, 225.
13. Perrins, L. E. and Pettett, K. (1971) *Plastics and Polymers* **39**, 391.
14. Andrews, E. H. and Kinloch, A. J. (1973) *Proc. Roy. Soc.* **A332**, 385.
15. Andrews, E. H. and Kinloch, A. J. (1973) *Proc. Roy. Soc.* **A332**, 401.
16. Gent, A. N. and Schultz, J. (1972) *J. Adhesion* **3**, 281.
17. Wake, W. C. (1982) *Adhesion and the Formulation of Adhesives*, Applied Science Pub., London, p. 70.
18. Jennings, C. W. (1972) *J. Adhesion* **4**, 25.
19. Arrowsmith, D. J. (1970) *Trans. Inst. Metal Finish.* **48**, 88.
20. Evans, J. R. and Packham, D. E. (1977) *Adhesion–1* (ed. K. W. Allen) Applied Science Pub., London, p. 297.
21. Evans, J. R. and Packham, D. E. (1978) *J. Adhesion* **9**, 267.
22. Evans, J. R. and Packham, D. E. (1979) *J. Adhesion* **10**, 39.
23. Evans, J. R. and Packham, D. E. (1979) *J. Adhesion* **10**, 177.
24. Packham, D. E. (1981) *Developments in Adhesives–2* (ed. A. J. Kinloch), Applied Science Pub., London p. 315.
25. Packham, D. E. (1983) *Adhesion Aspects of Polymeric Coatings* (ed. K. L. Mitta), Plenum, New York, p. 19.
26. Hine, P. J., Muddarris, S. El and Packham, D. E. (1984) *J. Adhesion* **17**, 207.
27. Gent, A. N. and Kinloch, A. J. (1971) *J. Polymer Sci.* **A2**, 659.
28. Bair, H. E., Matsuoka, S., Vadimsky, R. G. and Wang, T. T. (1971) *J. Adhesion* **3**, 89.
29. Wang, T. T. and Vazirani, H. N. (1972) *J. Adhesion* **4**, 353.
30. Voyutskii, S. S. (1963) *Autohesion and Adhesion of High Polymers*, Wiley Interscience, New York.
31. Voyutskii, S. S. (1962) *Adhesives Age* **5**(4), 30.
32. Voyutskii, S. S., Markin, Yu I., Gorchakova, V. M. and Gul, V. E. (1965) *Adhesives Age* **8**(11), 24.
33. Vasenin, R. M. (1969) *Adhesion, Fundamentals and Practice*, McLaren and Son, London, p. 29.
34. Vasenin, R. M. (1965) *Adhesives Age* **8**(5), 21.
35. Vasenin, R. M. (1965) *Adhesives Age* **8**(6), 30.
36. Campion, R. P. (1975) *J. Adhesion* **7**, 1.
37. Campion, R. P. (1977) *Adhesion–1* (ed. K. W. Allen), Applied Science Pub., London, p. 63.
38. Beuche, F., Cashin, W. M. and Debye, P. (1952) *J. Chem. Phys.* **20**, 1950.
39. Bresler, S. E., Zakharov, G. M. and Kirillov, S. V. (1961) *Vysok Soed.* **3**, 1072.
40. Voyutskii, S. S. (1971) *J. Adhesion* **3**, 69.
41. Morozova, L. P. and Krotova, N. A. (1957) *Dokl. Akad. Nauk. SSR* **115**, 747.
42. Morozova, L. P. and Krotova, N. A. (1958) *Kolloid. Zh.* **20**, 59.
43. Krotova, N. A. and Morozova, L. P. (1959) *Dokl. Akad. Nauk. SSSR* **127**, 141.

44. Anand, J. N. and Karam, H. J. (1969) *J. Adhesion* **1**, 16.
45. Anand, J. N. and Balwinski, R. Z. (1969) *J. Adhesion* **1**, 24.
46. Anand, J. N. (1969) *J. Adhesion* **1**, 31.
47. Anand, J. N. and Dipzinski, L. (1970) *J. Adhesion* **2**, 16.
48. Anand, J. N. (1970). *J. Adhesion* **2**, 23.
49. Anand, J. N. (1973) *J. Adhesion* **5**, 265.
50. Titow, W. V. (1978) *Adhesion–2* (ed. K. W. Allen), Applied Science Pub., London, p. 181.
51. Gabler, K. and Potente, H. (1980) *J. Adhesion* **11**, 145.
52. Dodin, M. G. (1981) *J. Adhesion* **12**, 99.
53. Currie, J., Depelsenaire, P., Goleau, R. and Sacher, E. (1984) *J. Colloid Interf. Sci.* **97**, 410.
54. Yasuda, H. K., Sharma, A. K., Hale, E. B. and James, W. J. (1982) *J. Adhesion* **13**, 269.
55. Deryaguin, B. V. (1955) *Research* **8**, 70.
56. Deryaguin, B. V., Krotova, N. A., Karassev, V. V., Kirillova, Y. M. and Aleinikova, I. N. (1957) *Proceedings of the 2nd International Congress on Surface Activity–III*, Butterworths, London, p. 417.
57. Deryaguin, B. V. and Smilga, V. P. (1969) *Adhesion, Fundamentals and Practice*, McLaren and Son, London, p. 152.
58. Weidner, C. L. (1963) *Adhesives Age* **6**(7), 30.
59. Wake, W. C. (1982) *Adhesion and the Formulation of Adhesives*, Applied Science Pub., London, p. 89.
60. Weaver, C. (1969) *Adhesion, Fundamentals and Practice*, McLaren and Son, London, p. 46.
61. Weaver, C. (1969) *Aspects of Adhesion–5*, (ed. D. J. Alner), University of London Press, London, p. 202.
62. Weaver, C. (1972) *Faraday Special Discussions* **2**, 18.
63. Weaver, C. (1975) *J. Vac. Sci. Technol.* **12**, 18.
64. Chapman, B. N. (1970) *Aspects of Adhesion–6*, (ed. D. J. Alner), University of London Press, London, p. 43.
65. Butler, D. W., Stoddart, C. T. H. and Stuart, P. R. (1970) *Aspects of Adhesion–6*, (ed. D. J. Alner), University of London Press, London, p. 55.
66. Butler, D. W., Stoddart, C. T. H. and Stuart, P. R. (1970) *J. Phys. D.* **3**, 877.
67. Oroshnik, J. and Croll, W. K. (1977) *Adhesion Measurements of Thin Films, Thick Films and Bulk Coating* (ed. K. L. Mittal), American Society for Testing Materials, Special Technical Publication **640**, p. 158.
68. Benjamin, P. and Weaver, C. (1960) *Proc. Roy. Soc.* **A254**, 177.
69. Benjamin, P. and Weaver, C. (1961) *Proc. Roy. Soc.* **A261**, 515.
70. Bateson, S. (1952) *Vacuum* **2**, 365.
71. Raff, R. A. V. and Sharan, A. M. (1969) *J. Appl. Polymer Sci.* **13**, 1129.
72. Mattox, D. M. (1973) *J. Vac. Sci. Technol.* **10**, 47.
73. Stoddart, C. T. H., Clarke, D. R. and Robbie, C. J. (1970), *J. Adhesion* **2**, 241.
74. Terr, D. G., Delcea, B. L. and Kirkham, A. J. (1976) *J. Adhesion* **8**, 171.
75. Teer, D. G. (1977) *J. Adhesion* **8**, 289.
76. Wake, W. C. (1978) *Polymer* **19**, 291.
77. Skinner, S. M., Savage, R. L. and Rutzler, J. E. (1953) *J. Appl. Phys.* **24**, 439.
78. Roberts, A. D. (1977) *Adhesion–1* (ed. K. W. Allen), Applied Science Publishers, London, p. 207.
79. Harrach, Graf von H. and Chapman, B. N. (1972) *Thin Solid Films* **13**, 157.
80. Krupp, J. and Schnabel, W. (1973) *J. Adhesion* **5**, 296.

81. Schnabel, W. (1969) *Physics of Adhesion* (ed. R. Polke and P. Hohn), Universitat Karlsruhe, Karlsruhe, p. 102.
82. Pauling, L. (1960) *The Nature of the Chemical Bond*, Cornell University Press, New York.
83. Good, R. J. (1967) *Treatise on Adhesion and Adhesives*, Vol. 1 (ed. R. L. Patrick), Marcel Dekker, New York, p. 15.
84. Fowkes, F. M. (1983) *Physicochemical Aspects of Polymer Surfaces*, Vol. 2 (ed. K. L. Mittal), Plenum, New York, p. 583.
85. Huntsberger, J. R. (1967) *Treatise on Adhesion and Adhesives*, Vol. 1, (ed. R. L. Patrick), Marcel Dekker, New York, p. 119.
86. Huntsberger, J. R. (1970) *Adhesives Age* **13**(11), 43.
87. Orowan, E. (1970) *J. Franklin Inst.* **290**, 493.
88. Tabor, D. (1951) *Rep. Prog. Appl. Chem.* **36**, 621.
89. Bell, J. P. and McCarvill, W. T. (1974) *J. Appl. Polymer Sci.* **18**, 2243.
90. Yaniv, A. E., Klein, I. E., Sharon, J. and Dodiuk, H. (1983) *Surf. Interface Anal.* **5**, 93.
91. Gledhill, R. A. and Kinloch, A. J. (1974) *J. Adhesion* **6**, 315.
92. Kinloch, A. J. (1979) *J. Adhesion* **10**, 193.
93. Owens, D. K. (1974) *J. Appl. Polymer Sci.* **18**, 1869.
94. Baldwin, W. B., Milun, A. J. and Wittcoff, H. A. (1969) *Prepr. Amer. Chem. Soc., Div. Org. Coat. Plast. Chem.* **29**, 30.
95. Kusaka, I. and W. Suetaka, (1980) *Spectrochim. Acta* **36A**, 647.
96. Furukawa, T., Eib, N. K., Mittal, K. L. and Anderson, H. R. (1982) *Surf. Interf. Anal.* **4**, 240.
97. Pritchard, W. H. (1970) *Aspects of Adhesion–6* (ed. D. J. Alner), University of London Press, London p. 11.
98. Baszkin, A. and Ter-Minassian-Saraga, L. (1978) *Polymer* **19**, 1083.
99. Leclerq, B., Sotton, M., Baszkin, A. and Ter-Minassian-Saraga, L. (1977) *Polymer* **18**, 675.
100. Owens, D. K. (1975) *J. Appl. Polymer Sci.* **19**, 265.
101. Owens, D. K. (1979) *J. Appl. Polymer Sci.* **19**, 3315.
102. Blythe, A. R., Briggs, D., Kendall, C. R., Rance, D. G. and Zichy, V. J. I. (1978) *Polymer* **19**, 1273.
103. Wake, W. C. (1982) *Adhesion and the Formulation of Adhesives*, Applied Science Pub., London, p. 54.
104. Dukes, W. A. and Kinloch, A. J. (1977) *Developments in Adhesives–1* (ed. W. C. Wake), Applied Science Pub., London, p. 251.
105. Mittal, K. L. (1975) *Adhesion Science and Technology* (ed. L. H. Lee), Plenum Press, New York, p. 129.
106. Mittal, K. L. (1977) *Polymer Eng. Sci.* **17**, 467.
107. Levine, M., Ilkka, G. and Weiss, P. (1964) *J. Polymer Sci.* **B2**, 215.
108. Dahlquist, C. A. (1964) *ASTM, STP* **360**, 46.
109. Dahlquist, C. A. (1969) *Aspects of Adhesion–5* (ed. D. J. Alner), University of London Press, London, p. 183.
110. Scholberg, H. M. and Hatfield, M. R. (1952) *Adhesion and Adhesives, Fundamentals and Practice*, MacKay, London, p. 34.
111. Kaelble, D. H. (1969) *J. Adhesion* **1**, 102.
112. Cherry, B. W. and Muddarris, S. L. (1970) *J. Adhesion* **2**, 42.
113. Andrews, E. H. and Kinloch, A. J. (1974) *J. Polymer Sci. Symp.* **46**, 1.
114. Fowkes, F. M. and Maruchi, S. (1977) *Org. Coatings Plastics Chem.* **37**, 605.
115. Fowkes, F. M. and Mostafa, M. A. (1978) *Ind. Eng. Chem. Prod. Res. Dev.* **17**, 3.

116. Fowkes, F. M. (1983) *Physicochemical Aspects of Polymer Surfaces*, Vol. 2 (ed. K. L. Mittal), Plenum, New York, p. 583.
117. Fowkes, F. M. (1984) *Rub. Chem. Technol.* **57**, 328.
118. Fowkes, F. M., Tischler, D. O., Wolfe, J. A., Lannigan, L. A., Ademu-John, C. M. and Halliwell, M. J. (1984) *J. Polymer Sci., Polymer Chem. Ed.* **22**, 547.
119. Fowkes, F. M., Sun, C. Y. and Joslin, S. T. (1981) *Corrosion Control Organic Coatings* (ed. H. Leidheiser), NACE, Houston, Texas, p. 1.
120. Bolger, J. C. and Michaels, A. S. (1968) *Interface Conversions for Polymer Coatings* (ed. P. Weiss and G. D. Cheever), Elsevier, New York, p. 3.
121. Rance, D. G. (1985) *Industrial Adhesion Problems* (eds D. M. Brewis and D. Briggs), Orbital Press, Oxford, p. 48.
122. Huntsberger, J. R. (1981) *J. Adhesion* **12**, 3.
123. Gent, A. N. and Ahagon, A. (1975) *J. Polymer Sci., Polymer Phys. Ed.* **13**, 1285.
124. Chang, R. J., Gent, A. N., Hsu, C. C. and Sehgal, K. C. (1980) *J. Appl. Polymer Sci.* **25**, 163.
125. Chang, R. J. and Gent, A. N. (1981) *J. Polymer Sci., Polymer Phys. Ed.* **19**, 1619.
126. Gent, A. N. (1981) *Int. J. Adhesion Adhesives* **1**, 175.
127. Klein, I. E., Sharon, J., Yaniv, A. E., Dodiuk, H. and Katz, D. (1983) *Int. J. Adhesion Adhesives* **3**, 159.
128. Crisp, S., Prosser, H. J. and Wilson, A. D. (1976) *J. Mater. Sci.* **11**, 36.
129. Sugama, T., Kukacka, L. E. and Carciello, N. (1984) *J. Mater. Sci.* **19**, 4045.
130. Chu, H. T., Eib, N. K., Gent, A. N. and Henriksen, P. N. (1979) *Advances in Chemistry Series 174* (ed. J. L. Koenig), American Chemical Society, Washington DC, p. 87.
131. Plueddemann, E. P. (1982) *Silane Coupling Agents*, Plenum Press, New York.
132. Comyn, J. (1986) *Structural Adhesives: Developments in Resins and Primers* (ed. A. J. Kinloch), Applied Science Pub., London, p. 269.
133. Zisman, W. A. (1971) *Ind. Eng. Chem. Prod. Res. Devel.* **11**, 170.
134. Bascom, W. D. and Patrick, R. L. (1974) *Adhesives Age* **17**(10), 25.
135. Koenig, J. L. and Shih, P. T. K. (1971) *J. Colloid Interf. Sci.* **36**, 247.
136. Chiang, C. H., Ishida, H. and Koenig, J. L. (1980) *J. Colloid Interf. Sci.* **74**, 396.
137. Gettings, M. and Kinloch, A. J. (1977) *J. Mater. Sci.* **12**, 2049.
138. Gettings, M. and Kinloch, A. J. (1980) *Surface Interf. Anal.* **1**, 189.
139. Trostyanaskaya, E. B., Golovkin, G. S. and Komarov, G. V. (1966) *Soviet Rub. Technol.* **25**, 13.
140. Koldunovich, G. E., Epshtein, V. G. and Chekanova, A. A. (1970) *Soviet Rub. Technol.* **29**, 22.
141. Barraud, A., Rosilio, C. and Ruaudel-Teixier, A. (1980) *Thin Solid Films* **68**, 7.
142. DeNicola, A. J. and Bell, J. P. (1983) *Epoxy Resin Chemistry*, American Chemical Society, Washington DC, p. 119.
143. Kollek, H. (1985) *Int. J. Adhesion Adhesives* **5**, 75.
144. Cotton, J. B., Coldfield, S., Mack, P. A. and Price, R. US Patent 3,837,964, 1974.
145. Gardiner, D. J., Gorvin, A. C., Gutteridge, C., Jackson, A. R. and Raper, E. S. (1985) *Corrosion Sci.* **25**, 1019.
146. Kummert, R. and Stumm, W. (1980) *J. Colloid Interf. Sci.* **75**, 373.
147. Yang, C. Q. (1985) *Polymer Mater. Sci. Eng.* **53**, 579.
148. Jud, K., Kausch, H. H. and Williams, J. G. (1981) *J. Materials Sci.* **16**, 204.
149. Schultz, J., Lavelle, L. and Martin, C. (1987) *J. Adhesion*, to be published.
150. Gutman, V. (1978) *The Donor–Acceptor Approach to Molecular Interactions*, Plenum Press, New York.

CHAPTER FOUR

Surface pretreatments

4.1 INTRODUCTION

Although in many applications no form of pretreatment is employed for the substrate materials prior to adhesive bonding, to attain the maximum in joint performance some form of surface pretreatment for the substrate materials being joined is almost always necessary. Adhesives are available which will join substrates which are covered in light machine or protective oils or low molecular weight plasticizers but some control over the state of the surface is still usually necessary. For example, the type of protective oil needs to be 'matched' to the adhesive and its thickness needs to be controlled, so that the adhesive can dissolve and displace the oil. Further, the reproducibility and service life of such adhesive joints is still invariably greatest when some form of surface pretreatment is employed.

Various reviews [1–6] have given details of the common surface pretreatments that have been developed, mainly empirically, both for non-metals and metals to give adhesive joints which possess the maximum strength, a low coefficient of variation on the strength values (i.e. are reproducible) and have long service lives in hostile environments. It is not the purpose of the present chapter to repeat the many recipes that are listed in these reviews but rather to indicate the types of pretreatment that are employed for different substrates and identify the scientific principles involved.

The purpose of any particular surface pretreatment may be manifold but the main aims are usually one or more of the following:

(a) To remove, or prevent the subsequent formation of, any weak surface boundary layer on the substrate. Obviously, if any weak layer initially on the substrate is not removed then the locus of joint failure is likely to be through this region and a relatively low joint strength will be observed. Examples of such weak boundary layers include low molecular weight species (e.g. plasticizers which have migrated to the surfaces of polymers), weak oxide scale on metallic substrates, air-borne contamination and layers of protective oils or greases. It should be noted that often processing or protective oils can become oxidized and very firmly attached and ingrained on the surface of a metal oxide. Such layers may require several degreasing treatments or even acid etching to remove them completely. Weak boundary layers may also form after the adhesive joint has been prepared. One such example is the problem of processing additives, e.g. stearates in polymeric films, diffusing to the interface after the film has been laminated. Another example is the hydration of aluminium oxide by water that has diffused through the adhesive; such an oxide is mechanically weak and leads to a significant loss of joint strength. To combat the formation of such *in situ* weak boundary layers pretreatments will be described which stabilize the oxide and so prevent this mechanism of environmental attack.

(b) To maximize the degree of intimate molecular contact that is attained between the adhesive (or primer) and the substrate during the bonding operation. From the discussions in Chapter 2 it is evident that in the case of many low-energy surfaces, such as fluorocarbon and olefin polymers, this requires a chemical modification to increase the surface free energy and/or to introduce surface groups which have the potential for acid–base interactions. Further, for all substrates, some degree of surface roughening may often assist in attaining good interfacial contact.

(c) To ensure that the level of intrinsic adhesion forces which are established across the interface(s) are sufficient for obtaining both the initial joint strength and subsequent service life that is required. In the case of low-energy polymeric surfaces these aims may again be achieved by selecting a pretreatment which increases the surface free energy of the substrate. It may also be necessary to introduce chemical groups into the surface regions which can form acid–base or stronger primary bonds with the adhesive. In the case of high-energy metallic substrates the main problem that is commonly encountered is not one of attaining sufficient intrinsic adhesion for high initial joint strengths, but one of establishing intrinsic adhesion forces which are resistant to environmental attack by moisture. This is frequently achieved by employing especially formulated primers, e.g. organometallic primers, which can establish interfacial bonds which are more stable than van der Waals bonds.

(d) To generate a specific surface topography on the substrate. As discussed in the previous chapter, only in a few instances does it appear that mechanical interlocking is a major mechanism of intrinsic adhesion. However, certain fibrous-like topographies may increase the amount of energy dissipation that occurs in the adhesive during fracture of the joint, thereby increasing joint strength.

(e) To assist in the hardening of the adhesive. This is not a common aim of a surface pretreatment but for a few adhesives, such as the cyanoacrylates, primers are available which are applied to the substrate(s) and catalyse the polymerization of the liquid adhesive monomer.

(f) To protect the surface of the substrate prior to the bonding operation. This is frequently necessary in the case of high-energy substrates such as metals. This arises because, having pretreated the surface by a degreasing, abrasion or chemical method, the surface is highly active not only towards the adhesive but also to atmospheric contamination. Indeed, workshop schedules usually require the freshly cleaned surface to be bonded or coated within a few hours of the pretreatment having been conducted. To meet this requirement and retain manufacturing flexibility it is a common practice to apply a primer, which is compatible with the adhesive, to the surface soon after pretreatment. The primed surface can then be left, often for several months, before application of the adhesive. The primer may also be formulated to meet some of the other requirements listed previously.

4.2 LOW-ENERGY SURFACES

4.2.1 Introduction

As may be seen from Tables 2.3 and 2.5, the surfaces of plastics, rubbers and fibre composites have low surface free energies (typically less than about 50 mJ/m^2), in distinction from those of metals, with energies generally greater than about 500 mJ/m^2. With many plastics, plastic laminates and rubbers it is possible to obtain reasonable or even good adhesive joint strengths with little or no pretreatment. This is shown by the results in Table 4.1.

However, some polymeric substrates, such as polytetrafluoroethylene, polyethylene, polypropylene and cured butyl rubber, are particularly difficult to bond and some form of pretreatment is nearly always necessary. Also, even for

Table 4.1 Tensile strength of various plastic substrates bonded with an epoxy–polyamide adhesive [7]

Plastic substrate	Failure strength (MPa)	Locus of failure
Polytetrafluoroethylene	1.2	Interfacial
Poly(vinylidene fluoride)	26.4	Interfacial/substrate
Poly(vinyl fluoride)	11.9	Interfacial/substrate
Low-density polyethylene	0.7	Interfacial
Nylon 11	16.7	Interfacial/substrate
Poly(ethylene terephthalate)	17.2	Interfacial/substrate
Nylon 6	11.2	Interfacial/substrate
Nylon 6.6	18.6	Interfacial/substrate

those polymers which possess relatively high surface free energies, and may be joined successfully without any treatment, it is often found that the highest joint strengths are observed when some form of pretreatment is employed. The mechanisms whereby a pretreatment can lead to an increase in the joint strength are:

(i) The removal of any weak boundary layer, such as may occur from low molecular weight contamination, antioxidants, processing aids or the 'tail' of the polymer's molecular weight distribution. Also, the prevention of a weak boundary layer developing once the adhesive joint has been formed.
(ii) The introduction of specific groups on the surface to aid the formation of interfacial polar interactions, hydrogen bonds, acid–base interactions or primary bonds with the adhesive; thereby increasing the ease of attaining, and the extent of, interfacial contact and the degree of intrinsic adhesion. This chemical modification of the polymer surface is usually reflected in higher values for the surface free energy, thermodynamic work of adhesion (Equations 3.12 and 3.19) and intrinsic adhesive fracture energy (Equation 3.18).
(iii) The creation of a surface topography suitable for mechanical interlocking to occur, so increasing the intrinsic adhesion.

The following discussions will give examples to demonstrate that each of the above may be the main reason that a particular pretreatment is successful in a given instance. However, the mechanism that may operate in any situation is also dependent upon the adhesive and bonding conditions employed. For example, some types of adhesive may readily dissolve/displace a weak boundary layer and therefore not require a pretreatment to carry out this process. Under different bonding conditions, e.g. at a lower cure temperature, or when using a different chemical type of adhesive, this cleaning action may not occur and a surface pretreatment would now be necessary. Another example is where some types of adhesive may be able to graft onto chemical groups introduced onto the substrate's surface, but obviously other adhesives might not possess the appropriate reactive species to enable interfacial grafting to occur. Thus, the pretreatment and the adhesive, or primer, need to be considered as an 'adhesive system'.

Finally, the results discussed in Section 2.4.3 on the effects of orientation of surface groups on polymeric substrates should be recalled, namely, that at a given temperature polymer chains may become sufficiently mobile such that any polar surface groups may redistribute into the bulk of the material, and the extent of this effect also depends upon the nature of the liquid adhesive in contact with the surface. This decrease in concentration of polar groups at the interface is mirrored in the subsequently measured joint strengths. It is also possible that reorganization of surface groups may occur in the vacuum chamber of modern surface analytical equipment, especially under the action of bombarding electrons, X-rays or ions, and this possibility makes the data from such techniques even more difficult to interpret!

In the following sections the various types of polymers and the effect of different pretreatments will be discussed. However, the use of plasma treatments for treating a wide range of polymers will be considered separately, since the principles involved are general ones and are conveniently reviewed in a separate section.

4.2.2 Fluorocarbon polymers

The fluorocarbon family includes polytetrafluoroethylene (PTFE), poly(chlorotrifluoroethylene) (PCTFE), fluorinated ethylene–propylene copolymer (FEP), poly(vinyl fluoride) and poly(vinylidene fluoride). They all possess low surface free energies, especially the fully fluorinated polymers, and γ_s values typically lie between about 15 and 35 mJ/m^2 as may be seen from Table 2.3. They are also highly inert polymers. Thus, some form of pretreatment is virtually always necessary if high joint strengths are required when joining these substrates. Increasing the surface free energy is an obvious route to increasing the joint strength and this is a major effect of all the surface pretreatments. However, the surface region of PTFE shears easily and molecules are pulled out and transferred to the adhesive at relatively low applied loads. Therefore, increasing the cohesive strength of the surface regions may also be advantageous, although whether surface treatments actually do promote stronger surface regions has yet to be resolved.

4.2.2.1 Abrasion

Simple abrasion is usually a very poor surface treatment for fluorocarbon polymers. This may be seen from the results shown in Table 4.2 where joint strengths for PTFE and PCTFE, bonded using an epoxy–polyamide adhesive, are given and the fluorocarbons have been pretreated by either abrasion or a

Table 4.2 Effect of various surface pretreatments on the joint strength of PTFE and PCTFE [11]

	Butt joint strength (MPa)			
Fluorocarbon	No surface treatment	Abraded	Etched	Abraded and etched
PTFE	Fell apart	Fell apart	10.8	13.2
PCTFE	2.6	7.7	19.4	20.7

Table 4.3 Abrasion surface pretreatments for PTFE

	Lap shear strength (MPa)	
Joint	Abraded in air	Abraded beneath adhesive (epoxy–vinyl acetate mixture)
PTFE/aluminium	0.98	5.71
PTFE/PTFE	0.98	6.67

chemical etch (see below) pretreatment. However, Lerchenthal and co-workers [8–10] have claimed that abrading substrates beneath 'reactive' adhesives may dramatically increase joint strengths compared to normal abrasion methods in air. Results for PTFE bonded to itself and to aluminium with an epoxy–polyamide adhesive are shown in Table 4.3. Lerchenthal *et al.* suggested that the increases from abrading beneath a reactive adhesive arise from the abrasion process rupturing polymer chains in the substrate and these free-radical terminated molecules then reacting with the adhesive to give interfacial primary bonds. However, it is equally possible that the higher joint strengths observed when the abrasion pretreatment is conducted with the adhesive present are due to improved wetting, as the input of mechanical energy will override the predictions from the wetting equilibria and force the liquid adhesive to spread over the surface and attain good molecular contact with the substrate.

4.2.2.2 Etching treatments

The treatment for fluorocarbon polymers is usually based upon metallic sodium dissolved in a solution of naphthalene in a polar aprotic solvent such as tetrahydrofuran, although other etch treatments such as a solution of alkali metal in liquid ammonia [12] have been reported [13]. Details for the preparation of the treatment solutions have been published [14] and solutions of sodium naphthalenide are commercially available, e.g. under the trade name 'Tetraetch' from W. L. Gore Associates Inc. These commercial pretreatments are relatively easy and safe to operate and widely used in industry. Also, it is possible to purchase fluorocarbon polymers already treated by such chemical processes.

The sodium naphthalenide pretreatment requires immersion of the polymer in fresh treating solution, which typically contains about 15–23 g of sodium per litre, at room temperature for a few seconds, then rinsing in water, then in acetone, followed by thorough washing in water again and finally drying. The etching process produces a dull brown-black surface layer and this coloration intensifies as the etching time is increased. It has been suggested [15] that this colour change may provide a useful quality control check for procurement of uniformly and reproducibly etched material.

Nelson *et al.* [16] have shown that the etching solution attacks only the surface regions of the fluorocarbon polymer. They examined a cut section of etched PTFE and, from optically microscopy, determined that the depth of the coloured surface region was about 1 μm. However, they reported that the bulk electrical properties of the treated and untreated polymer were virtually identical and that their electron diffraction patterns were indistinguishable, indicating that no changes had been induced in the bulk polymer's crystallinity. These observations may be attributed to the thinness of the treated surface layer and thus the properties of the layer are not measured when analytical techniques are employed which monitor the bulk properties of the material. Indeed, Benderley [17] has since reported that the treated layer does have a substantially lower surface resistivity than the untreated material.

Purvis and Beck [18] have ascribed the brown coloration to a carbonaceous surface being formed as a result of the extraction of fluorine atoms by the etching solution. Their detection of fluoride ions in the expended etching solution provided evidence for this proposed mechanism. Further support comes from the work of Miller *et al.* [19] and Andrews and Kinloch [20]. These latter authors exposed untreated and etched FEP to a 1% w/w solution of osmium tetroxide for six days. Osmium tetroxide reacts with carbon–carbon double bonds to produce a dark appearance. In the case of the untreated material no visible colour change occurred after immersion. However, the surface of the etched FEP was darkened even further; by sectioning perpendicular to the surface and examination in the optical microscope the depth of the darkened surface layer was found to be about 1 μm, in good agreement with work of previous investigators. It is the presence of these conjugated carbon–carbon double bonds in the surface regions that are responsible for the brown-black coloration.

Several workers [21–23] have more recently employed X-ray photoelectron spectroscopy (XPS) to identify the chemical composition of the surface regions of fluorocarbon polymers etched in sodium naphthalenide solutions. These studies have confirmed that the defluorinated surface consists mainly of carbon and oxygen and that unsaturation is introduced by the etching pretreatment. The oxygen is present as C=O, COH and CO_2H groups, and these groups are produced after exposure of the etched surface to air. This latter conclusion was reached from studies which revealed that if the treatment was conducted in an inert atmosphere, and the polymer was handled immediately after treatment in the inert environment, then no oxygen was detected in the surface regions; the surface regions were composed almost entirely of carbon.

As might be expected, the chemical changes induced by etching the fluorocarbon polymer in air cause a large increase in the surface free energy of

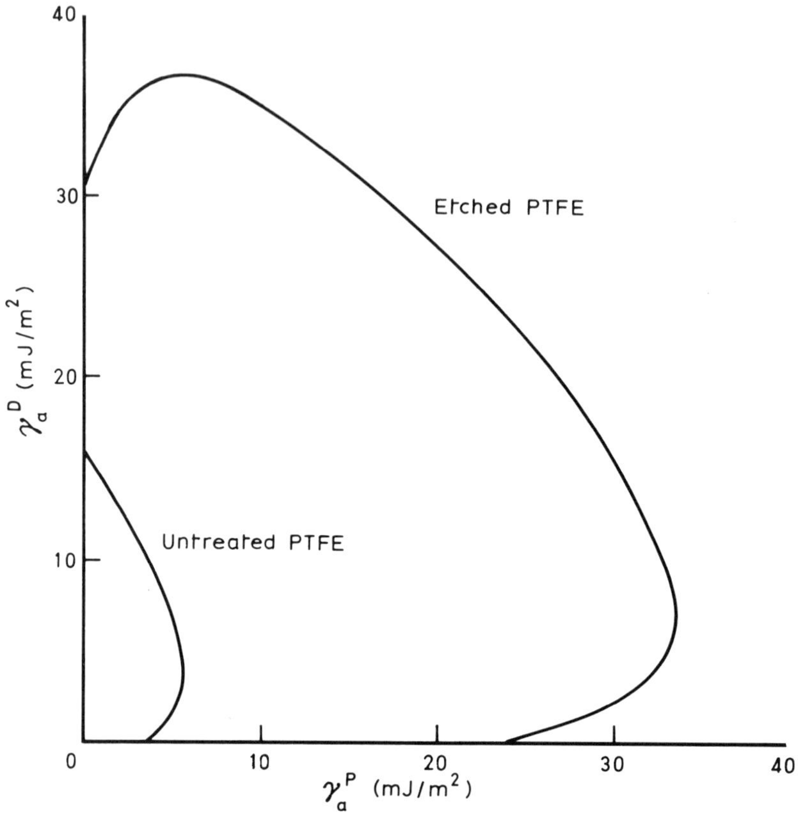

Figure 4.1 Wettability envelopes of polytetrafluoroethylene and etched PTFE [24].

the material [20,24,25]. This is clearly indicated by the wettability envelopes which are shown in Fig. 4.1 [24]. These envelopes have been constructed as described in Section 2.6.2 using Equation 2.34 and it will be recalled that any adhesive with values of γ_a^D and γ_a^P which lie inside the envelope will, from thermodynamic considerations, spread spontaneously and completely wet the substrate. No common adhesives lie inside the envelope in the case of the untreated PTFE but many lie inside the envelope for the etched PTFE. Also, of course, from the discussions in Section 3.5.2.2 and Equation 3.12 the higher values of surface free energy for the etched PTFE will increase the value of the thermodynamic work of adhesion, W_A. Thus, better interfacial contact and stronger intrinsic adhesion forces are achieved with the etched fluorocarbon polymers and may result in dramatic improvements in joint strengths, as illustrated in Table 4.2. Similar increases in peel strength have been recorded

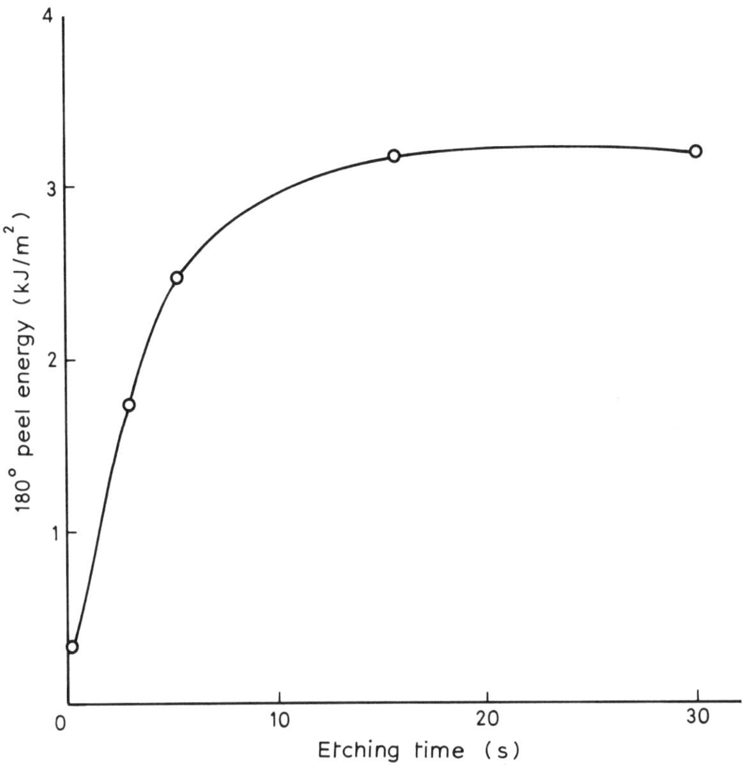

Figure 4.2 Effect of a sodium naphthalenide etching treatment on the peel strength of PTFE/polychloroprene adhesive/PTFE joints.

when PTFE is bonded to itself with a polychloroprene rubbery adhesive, as shown in Fig. 4.2.

The studies of Andrews and Kinloch [20,26] illustrate two further points of interest concerning the etching of fluorocarbon polymers. Firstly, the presence of unsaturated groups in the treated polymer enables interfacial primary bonds to be established if, for example, a rubbery adhesive containing dicumyl peroxide (a free radical initiator) is crosslinked in contact with the pretreated substrate. Secondly, however, if a relatively long etching time is used then the surface regions are degraded and become mechanically weak, i.e. a weak boundary layer is formed which provides a path for joint fracture at low applied stress levels. These points are illustrated in Fig. 4.3 which shows the intrinsic adhesive fracture energy, G_o, for crosslinked styrene–butadiene–rubbery adhesive and etched FEP substrates as a function of etching time in a solution of sodium naphthalenide. The value of G_o is only equal to that of W_A for the

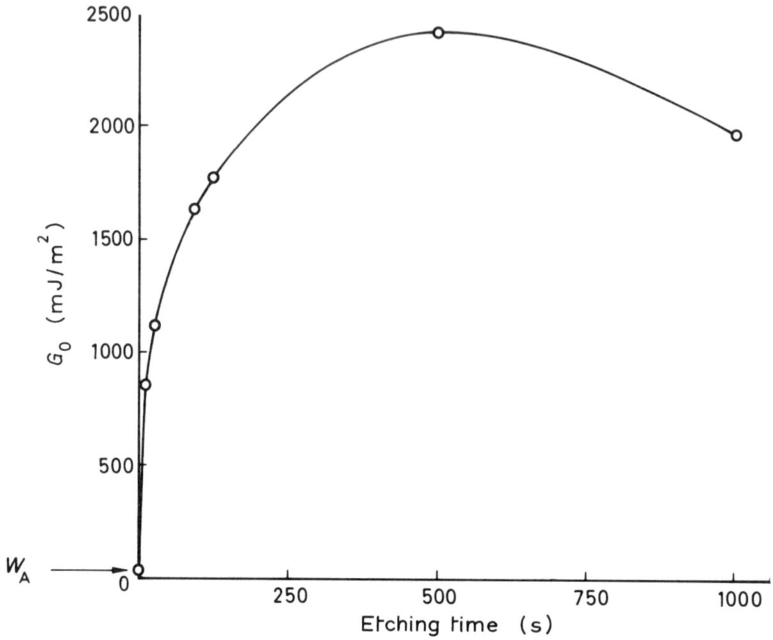

Figure 4.3 Intrinsic adhesive fracture energy, G_o, versus etching time for rubber/FEP joints [20].

untreated FEP, otherwise the value of G_o is far greater than W_A due to the formation of interfacial primary bonds and increases initially with etching time as the concentration of carbon–carbon double bonds, and hence concentration of interfacial primary bonds, increases. However, at longer etching times degradation of the etched surface regions occurs and the locus of joint failure increasingly becomes via fracture through this relatively weak region. This is reflected in the value of G_o falling in accord with Equation 3.18. These two effects combine to give an optimum etching time in order to obtain the maximum joint strength.

Applications in which proprietary sodium naphthalenide pretreatments have played an important role include: bonding of PTFE bushes to metals, bonding PTFE insulation to electrical components, preparing seals to PTFE and PCTFE insulated connectors, preparing seals between polyethylene and PTFE insulated wire and connector terminals, and improving potting seals. However, it should be noted that one problem that may occur is that the unsaturated and oxidized layer on the fluorocarbon is more reactive not only to the adhesive but also to ultraviolet (u.v.) radiation. Thus, if u.v. light can reach the interface in a sufficient concentration, the treated regions of the substrate may be attacked and become mechanically weak, so leading to premature joint failure. The incorporation of u.v. absorbers into the substrate and/or adhesive may significantly delay such problems.

4.2.2.3 Electrochemical treatments

Dahm and colleagues [27–30] have described pretreatments which improve the adhesion of PTFE based upon electrochemical treatment methods. They showed that the surface of PTFE, when held in contact with a cathode at a potential more cathodic than −1.5 V versus the saturated calomel electrode in an aprotic solvent containing a tetra-alkylammonium support electrode, is reduced to yield a black surface layer which grows outward from the point of contact. The appearance and nature of the treated surface layer resembled that of the chemically etched PTFE and exhibited similar wetting and adhesion characteristics. Indeed, as for the etched PTFE, Dahm *et al.* [29] attributed the improved joint strengths that were recorded for the electrochemically treated PTFE to the greatly increased surface free energy of the material rather than to the removal of any weak boundary layer that was initially present on the untreated surface. However, one advantage of the electrochemical treatment, compared to the etching methods, is that since the process involves essentially a surface reaction it is self-limiting and overtreatment, leading to very thick and weak surface layers, is not a problem. A second advantage is that only one side of the PTFE film or sheet is treated at a time and hence there is no need for masking.

4.2.2.4 Melting against high-energy surfaces

Schonhorn and Ryan [31] have reported that heterogeneous nucleating and crystallization of FEP melt against high-energy surfaces (e.g. gold, aluminium) produces a transcrystalline surface on the fluorocarbon of high mechanical strength and increased surface free energy. Indeed, the critical surface tension may rise from 18.8 to 40.4 mJ/m^2. Other workers [22] have used XPS and suggested that this increase in the value of γ_c is due to the formation of a thin layer of oxygen-containing hydrocarbon material rather than a highly crystalline layer of FEP. Whatever the reason, a surface is produced which is amenable to conventional adhesive bonding. The lap-shear strengths of aluminium/epoxy/FEP joints where the FEP has been subjected to this and other surface pretreatments is shown in Table 4.4.

From the results shown in Table 4.4, Schonhorn and Ryan concluded that the gold films which were deposited electrochemically are not effective, perhaps because they are contaminated or complexed with a variety of organic compounds (e.g. cyanide, citrate) normally used in plating baths. Evaporated films of gold were considered to be far more effective in generating a transcrystalline region of high strength and surface free energy. Further, gold

Table 4.4 Various surface pretreatments for FEP

Surface pretreatment of FEP	Lap-shear strength (MPa)
None	0.69
Nucleated and crystallized against electrochemically deposited gold	4.82
Nucleated and crystallized against evaporated aluminium	12.41
Nucleated and crystallized against evaporated gold	19.30
Etched in sodium naphthalenide solution for 40 s	20.00

appears to be more effective than aluminium oxide in this respect. The basic difference may be the density of nucleating sites on the surface. If there are too few, ordered transcrystalline growth may not take place.

4.2.2.5 Aluminium–alkali treatment

Roberts and co-workers [32] have reported a method for increasing the interfacial strength of gold bonded to FEP which yields a high joint strength without affecting the electrical properties of the polymer. The method consists of depositing a 100 nm layer of aluminium onto the FEP film by evaporation, removing it by washing the sample in dilute sodium hydroxide, followed by deposition of gold by evaporation. XPS indicated that significant chemical modification of the polymer's surface had occurred as a result of the aluminium deposition, apparently consisting of the generation of a thin and possibly continuous surface layer of oxygen-containing hydrocarbon material. In contrast scanning electron microscopy revealed that the surface topography was relatively unaffected. It appears therefore that the increased joint strength may be attributed to the increased surface free energy of the fluorocarbon.

4.2.3 Polyolefins

Like the fluorocarbon polymers discussed above, polyolefins such as low- and high-density polyethylene, polypropylene and poly(4-methyl 1-pentene) cannot usually be joined by adhesives to give reproducible high strengths unless some form of surface pretreatment is first employed.

4.2.3.1 Solvent cleaning

The simplest treatment is solvent cleaning [33], and the effect of dipping a polypropylene substrate into a trichloroethylene solvent vapour prior to bonding to aluminium using an epoxy adhesive is shown in Fig. 4.4. The effect of a short exposure time on the joint strength is spectacular but overtreatment

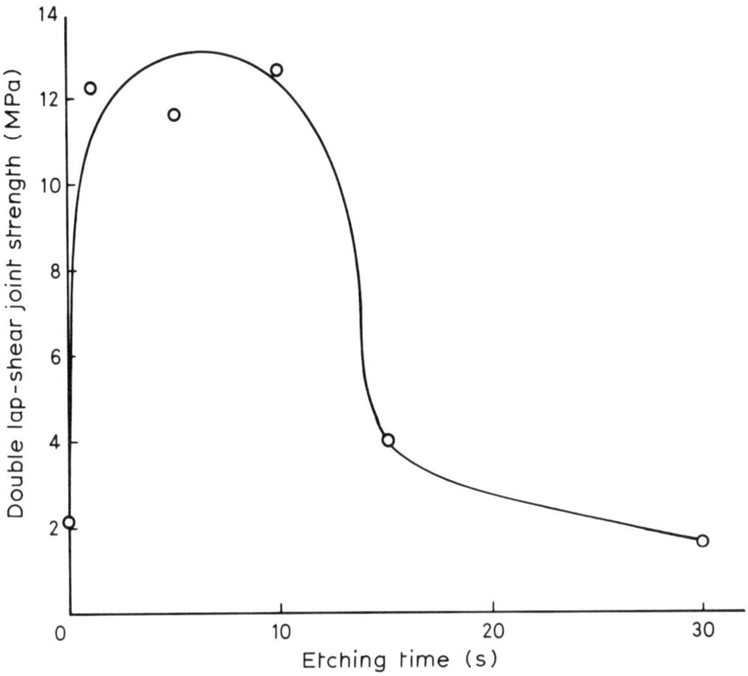

Figure 4.4 Effect of a trichloroethylene vapour surface pretreatment on the strength of polypropylene joints [33].

may readily occur causing a rapid decline in joint strength. The trichloroethylene vapour imparted a dull white finish to the smooth polymer surface, and from scanning electron microscopy studies the authors concluded that the untreated polymer possessed a weak surface boundary layer which was removed by the solvent treatment. This treatment also resulted in a highly porous surface (possibly caused by the solvent preferentially attacking the amorphous regions of the surface). Thus, there is also the possibility that a semi-interpenetrating network structure was produced as the liquid adhesive penetrated the porous surface regions of the substrate and then cured. It appears that the improved adhesion cannot be attributed to an increase in the

Table 4.5 Wettability of untreated and solvent–vapour cleaned polypropylene

Pretreatment	Contact angle of water	Contact angle of epoxy adhesive
None	90°	26°
Solvent vapour	98°	34°

surface free energy of the polypropylene since, from contact angle measurements shown in Table 4.5, the solvent-treated surfaces have a wettability inferior to that of the untreated polymer. Finally, it is noteworthy that a rapid decline in joint strength is observed at etching times longer than about 10 s and this is probably due to degradation and weakening of the surface layer. This feature makes this treatment difficult to operate in industry. Finally, it has also been reported [34] that this treatment results in large improvements in the adhesive strength when joining high-density polyethylene.

4.2.3.2 Flame treatments

As discussed in detail by Briggs [35], although flame treatment has been applied to polyolefin films, it is now mainly used to pretreat thicker section components such as blow-moulded bottles. The increased adhesion that results enables the bottles to be successfully ink-printed. The component to be treated is passed over one or more burners, each burner consisting of a large number of closely spaced jets; the burners are fed with an air–gas mixture in controlled proportions and the gas can be from the mains (largely methane) or bottled (propane or butane). Typically the component spends about 1 s, or less, in the flame. The most important variables are the position of the component relative to the flame; the air:gas ratio; the air/gas flow rate; the nature of the gas; the burner–surface separation; and the exposure time.

Only a few authors [36–39] have considered the flame treatment of polyolefins. It is generally agreed that for optimum treatment an oxidizing flame should be employed, which requires a slightly higher concentration of oxygen over that required for complete combustion; for methane and propane excesses of 10 and 15%, respectively, are recommended. However, as might be expected, the flame treatment can readily lead to degradation of the polymer's surface regions and some of the optima in process parameters are quite sharp.

The most extensive examination of the effect of flame treatment and the mechanism of surface modification has been the work of Briggs and co-workers [35,39]. They employed XPS to study flame-treated low-density polyethylene. As commonly assumed, the treatment was shown to cause surface oxidation and the depth of the oxidized layer was between 4 and 9 nm. Indeed, during its passage through the flame the polyolefin surface can experience temperatures of around 2000 °C; as a result thermal oxidation, which is a chain-reaction free radical process, may take place. The results of the surface analysis are shown in Table 4.6 where the increased oxidation is evident by the O:C and N:C ratios increasing and this chemical modification of the polyolefin's surface regions is reflected in higher lap-shear strengths. The data also demonstrates that antioxidant in the polyethylene does not affect the efficiency of the flame treatment. Since the oxidation is a free radical process it should be prevented by the antioxidant. However, the lack of effectiveness of the antioxidant may

Table 4.6 XPS and joint strength data for flame-treated polyethylene [39]

Polymer	Treatment time (s)	Natural gas flow (cm^3/s)	Air flow (cm^3/s)	O:C ratio (at. %)	N:C ratio (at. %)	Lap-shear strength (MPa)
Alkathene 47†	0	0	0	0.25	0	0.55
Alkathene 47	1.2	37	150	16.9	0.94	6.6
Alkathene 47	4.8	37	150	31.0	3.2	7.2
Alkathene 47	1.2	74	317	15.3	2.2	6.8
Alkathene 47	1.2	18.5	75	6.8	0	5.1
Alkathene 11‡	0	0	0	<0.25	0	0.36
Alkathene 11	1.2	37	150	20.5	1.5	5.6
Alkathane 11	4.8	37	150	33.4	3.2	7.2
Alkathene 11	1.2	74	317	13.7	2.5	6.4
Alkathene 11	1.2	18.5	75	5.1	0	5.7

† Contains no additives. Alkathene is an ICI trade name.
‡ Contains 0.02% 2,6-di-*t*-butyl-*p*-cresol as an antioxidant. Note that it does not affect flame treatment efficiency.

arise from its volatilization, so that within the surface layers there is an insufficient concentration of the antioxidant to be effective.

4.2.3.3 Chemical treatments

Many chemical pretreatments have been reported for polyolefins but the only one of commercial importance is etching in chromic acid. It is typically used for the treatment of irregularly shaped or large objects when it would be difficult to use either flame or corona-discharge (Section 4.2.7) treatments. The chromic acid etches the polyolefin surfaces and results in loss of material into solution.

Blais *et al.* [40] have reported the use of chromic acid solutions to etch low- and high-density polyethylene and isotactic polypropylene. Apart from measuring joint strengths they also characterized the etched surfaces by attenuated total reflectance (ATR) infrared spectroscopy, contact angle measurements and electron microscopy. Spectroscopic examination revealed extensive chemical changes only in the case of low-density polyethylene, with an increasing concentration of –OH, >C=O and possibly $-S(=O)_2OH$ groups in the surface layer. However, ATR infrared spectroscopy is not a true surface technique since it has a relatively large sampling depth compared to the likely etch depth. Indeed, this problem has been clearly demonstrated by the work of Briggs and co-workers [41,42]. They used XPS to study similarly treated low-density polyethylene and polypropylene and showed that oxidation and sulphonation occurred in the outermost surface regions of both materials with evidence for –COH, >C=O, –COOH (carboxylic acid) and $-S(=O)_2OH$

groups. They also detected [42] residual chrominium in the surface regions, although this element appeared to have little influence on the joint strength with epoxy adhesives but may be important in improving the adhesion in the case of electroless plating [43,44]. More recent work [45] also using XPS has indicated that for chromic acid treated polyethylene the main groups introduced are carboxylic acid functions and, in particular, no residual chromium-, nitrogen- or sulphur-containing species were evident. The absence of such species, compared to the earlier work of Briggs *et al.*, described above, was attributed to the very dilute concentration of acid used in this later study. The density of surface carboxylic acid groups was estimated to be 16×10^{14} cm^{-2} and from studying the acid–base behaviour it was concluded that these surface carboxylic acid groups were less acidic than acetic or poly(acrylic acid). The initial ionization of the surface carboxylic acid groups occurred when the pH of the contacting solution was approximately at a pH of 6, compared to pH values of about 3.5 and 4 for acrylic acid and poly(acrylic acid), respectively.

The work of Briggs *et al.* [41] has shown that the essential difference between etching polypropylene for 1 min at 20 °C or 6 h at 70 °C was in the depth of attack achieved, rather than any increase in the degree of surface oxidation. Further, for all the polyolefins they examined, only in the case of polypropylene etched under the former condition was the depth of oxidation less than 30 nm. In the case of polyethylene, however, both the etching depth and degree of surface oxidation did increase with etching time. Nevertheless, for both polyolefins considerable improvement in joint strengths are recorded after etching for only 1 min at 20 °C but increased etching times did not lead to increased joint strengths, possibly due to the limiting strength of the polyolefin substrate. Blais *et al.* [40] have also studied the effect of etching time in chromic acid solution. They recognized that the enhancement of adhesion by a pretreatment would eventually lead to the situation where joint failure occurs by cohesive fracture of the polyolefin substrate, as observed by Briggs and co-workers. Since measurements under these conditions obviously do not reflect the effectiveness of the surface treatment they therefore used (i) an adhesive which would give an extremely low joint strength to an untreated polyolefin surface, (ii) an adhesive with reasonable cohesive strength and (iii) a joint geometry sensitive to variations in interfacial adhesion. They chose a poly(acrylic acid) adhesive, plasticized with a little ethylene glycol, and a modified 90° peel test geometry. The peel energy is shown as a function of etching time in Fig. 4.5. The polypropylene shows an approximately 170-fold increase in peel energy after a 100 s etch but little improvement for a more prolonged attack, while both polyethylenes exhibit a rapid increase in joint strength with longer etching times.

The work of both Blais *et al.* and Briggs *et al.* indicates that there is little correlation between the contact angles of water on the etched surface and resulting joint strength. One reason for this might be indicated from the work

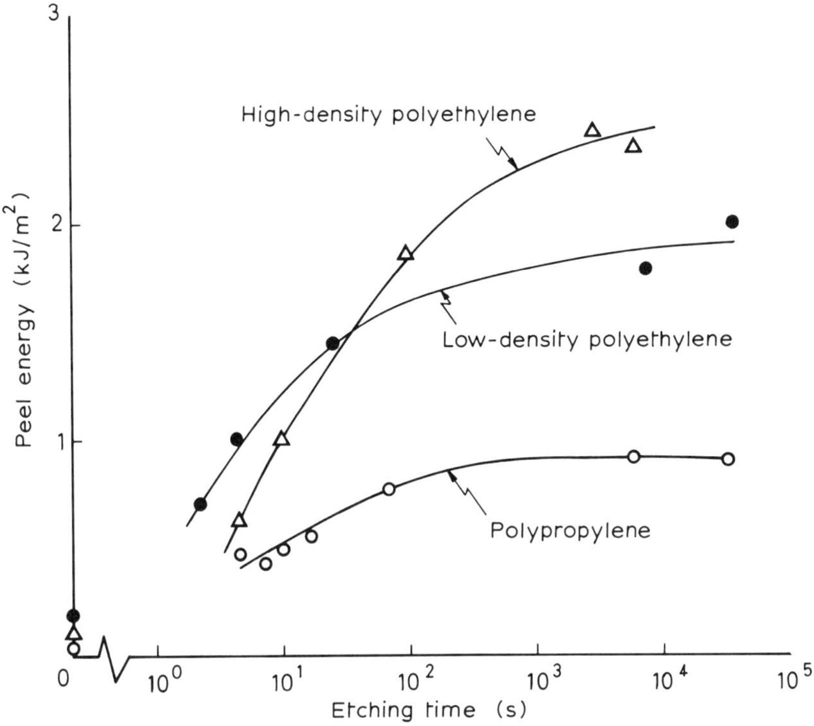

Figure 4.5 Peel energies for etched polyolefin substrates bonded with a poly(acrylic acid) adhesive as a function of etch time in chromic acid solution [40].

of Shields [46] who has shown that chromic acid etching may produce a very rough surface, and the discussions in Section 2.3.3 revealed that surface roughness alone may result in contact angle changes. Thus, when both the surface chemistry and roughness are being affected simultaneously any interpretation of contact angle data is difficult. As might be expected, it has been observed [40] that chromic acid often selectively etches the regions of low and zero crystallinity.

With the chemical and physical changes induced by chromic acid etching elucidated, it might be hoped that a unified mechanism for improvement in adhesion would be identified. Shields [46] concluded that the effectiveness of chromic acid treatments was mainly due to their ability to produce a roughened surface which increased the surface area and enabled mechanical interlocking to occur. However, Blais *et al.* observed only very slight surface roughening and the considerable increases in joint strength that they recorded for their relatively smooth etched polyolefin surfaces suggests that mechanical interlocking cannot be the major mechanism operating. Briggs *et al.* [41,42]

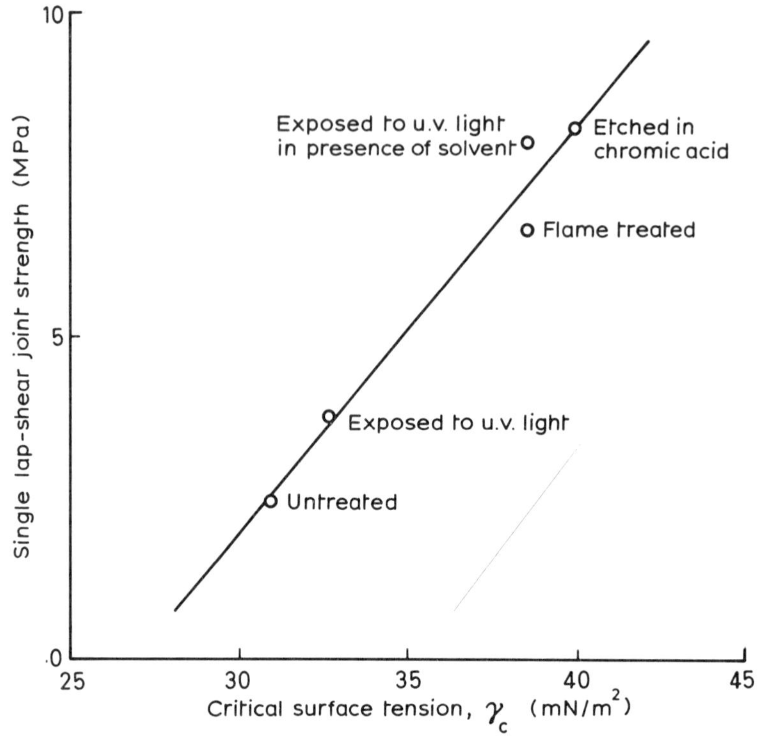

Figure 4.6 Single-lap shear joint strength versus critical surface tension, γ_c, for low-density polyethylene pretreated by various methods [47].

ascribed the increases in adhesion to increased surface polarity and reported that there is a good correlation between joint strength and degree of surface oxidation (expressed in terms of the number of carbon–oxygen functions with an assumed constant average stoichiometry), but there is no correlation between the increased adhesion and $-S(=O_2)OH$ group concentration. Whether the former correlation with surface oxidation arises because of improved wetting or a higher intrinsic adhesion, or of course a combination of these effects, has yet to be established. Similarly, whether such effects arise because of increased van der Waals attractions or acid–base interactions is unclear. Some, albeit limited, support for the former comes from the work of Rauhut [47]. He has found a correlation between increased surface polarity, as reflected by the critical surface tension, γ_c, and joint strength, as shown in Fig. 4.6. However, more recently Chew *et al.* [48] have suggested that the interfacial interactions are more meaningfully assessed in terms of acid–base concepts.

Others favour a weak boundary layer mechanism. Blais *et al.* [40] feel that, while increased oxidation may play a minor role in the case of polyethylene, the major mechanism is the removal by the etchant of a weak boundary layer covering the polyolefin's surface; Garnish and Haskins [33] also proposed this mechanism, as discussed previously. However, XPS examination by Briggs and co-workers [41] of a fractured joint which had consisted of untreated polyethylene and epoxy adhesive gave no evidence of transfer of any significant amounts of polyethylene substrate to the epoxy; this is obviously evidence against the weak boundary layer concept. This conclusion has recently been confirmed by Briggs [227] using the more sensitive surface analytical technique of static secondary ion mass spectroscopy (SIMS). (Although it is noteworthy that the SIMS method has shown [49] that poor adhesion of a particular grade of untreated low-density polyethylene to poly(ethylene terephthalate) for a film laminate was due to the transfer of low molecular weight hydrocarbon material from the polyethylene to the interface during contact in the peeled laminate, thereby creating a weak boundary layer at the interface.) As a final point on this subject, it has been reported [48] that if the oxidized layer on the etched polyethylene is removed then the adhesion to the surface so formed, with any original weak boundary layer removed, is poor – yet again showing that the improved adhesion that is realized from using these treatments does not generally arise from the removal of a weak boundary layer but rather the chemical modifications that are induced in the surface regions of the substrate.

Considering other chemical treatments, Chew *et al.* [48] have demonstrated that low- and high-density polyethylene films may be photobrominated to produce surfaces with high levels of adhesion to an epoxy adhesive. The polyethylene was brominated by exposure to saturated bromine vapour in a nitrogen atmosphere with simultaneous irradiation by light from tungsten filaments.

Morris [50,51] has proposed another chemical pretreatment based upon the use of an aqueous ammonium peroxydisulphate solution for treating polyethylene and polypropylene; the effect of immersion time and temperature is illustrated in Fig. 4.7. No differences were apparent between the ATR infrared spectra of untreated and treated specimens; hence any oxidation of the polymer by the peroxydisulphate was confined to the immediate surface regions. This fact was confirmed by the critical surface tension of the treated material being about 5 mJ/m^2 higher in value. Also, an insoluble gel fraction was present in the treated polymer, about 1% by weight of the original polymer, but was absent in the untreated specimens. Morris concluded that treatment of polyethylene and polypropylene with a peroxydisulphate solution led to the formation of a surface layer of crosslinked, entangled molecules and that this increased the mechanical strength of the initially weak surface regions and resulted in higher joint strengths. Again no direct evidence was furnished

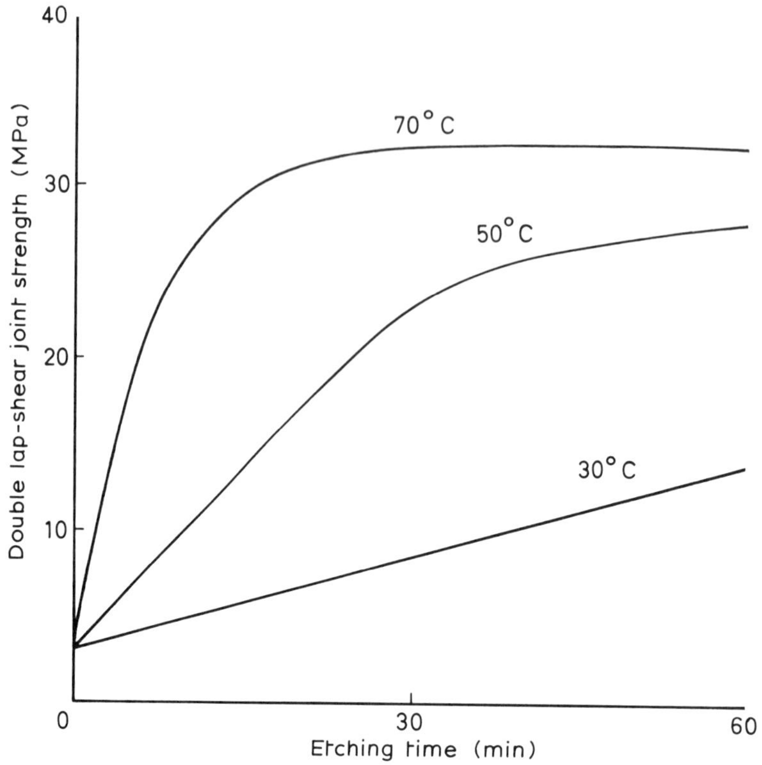

Figure 4.7 Effect of pretreatment time and temperature on the joint strength of low-density polyethylene treated with 0.26 M solution of ammonium peroxydisulphate [50,51].

for the presence on the original untreated polymer surface of a weak boundary layer, which needed strengthening.

To try and summarize these discussions, then there is no single readily identifiable mechanism. It is obvious that any weak boundary layer must be removed prior to adhesive bonding but it is not well established that such layers are commonly found on most grades of polyolefins. Further, even if such layers are present and then removed, good adhesion to the polyolefin will not automatically follow unless the surface has been chemically modified, typically by oxidizing the surface regions. Hence, the essential basic mechanism appears to be one of chemical modification, but with some polyolefins the removal of a weak boundary layer may be a necessary first step. As mentioned earlier, the type of adhesive employed will play a role in influencing the exact mechanisms which will operate, and hence the exact details of the pretreatment which will be required.

4.2.3.4 Melting against high-energy surfaces

Briggs *et al.* [52,53] have melted polyethylene against an aluminium foil and then either bonded the laminate to another substrate and fractured this joint or removed the aluminium with sodium hydroxide and then bonded the exposed polyethylene surface. Using XPS they found that melting the polyethylene against the aluminium surface had led to an oxidized thin surface region, a similar conclusion reached by Dwight and Riggs [22] in the case of FEP (Section 4.2.2.4). The measured joint strengths increased as the level of oxidation increased and the improved strengths may therefore be ascribed to the better wetting and higher intrinsic adhesion arising from oxidation of the polymer's surface.

It is interesting to consider the origin of the oxidizing species, and they noted that melting polyethylenes against aluminium gave levels of surface oxidation 12 to 15 times greater than when polyethylenes were pressed against poly(ethylene terephthalate) film; further these latter polyethylenes gave low joint strengths as expected. The very much higher levels of oxidation when melted against aluminium might be due to a specific catalytic effect or due to trapped air in the aluminium surface, which is much rougher than in the poly(ethylene terephthalate).

Finally, it has also been found that antioxidants in the polyethylene reduced the level of oxidation induced by this treatment technique. However, exposure of the polyethylene to an atmosphere rich in ozone just prior to melting against the aluminium foil permitted substantial oxidation of the polymer and increased the subsequent joint strength. They concluded that ozone reacts with the antioxidant at the surface nullifying its suppressive effect on the oxidation, rather than by directly oxidizing the polymer via a different mechanism.

4.2.4 Other plastic substrates

As mentioned earlier, other plastics such as poly(vinyl chloride), poly(ethylene terephthalate), polyacetals, nylons and polyimides do not present such a severe problem to the adhesives technologist as do fluorocarbon polymers or polyolefins. Nevertheless, in order to obtain very high joint strengths approaching the cohesive strength of the substrate, some form of surface pretreatment is often necessary.

Again plasma treatments, which are common pretreatments for these plastics, will be discussed in a later section; however, Table 4.7 shows the effect of various surface treatments on a nylon 6.6. For the epoxy adhesive, solvent cleaning and abrading is as effective as the chromic acid etching treatments and only a potassium permanganate etch followed by application of a resorcinol–formaldehyde primer results in the highest joint strengths (i.e. when the cohesive strength of the nylon 6.6 is reached). The interface between the nylon

Table 4.7 Effect of surface treatment on the lap-shear strength of nylon 6.6/aluminium joints [46]

Surface treatment	Adhesive	Lap-shear strength (MPa)
Solvent cleaning	Epoxy	2.92
	Cyanoacrylate	3.95
Solvent cleaning + abrasion	Epoxy	3.58
	Cyanoacrylate	3.37
15% chromic acid; etched for 7 min at 20 °C	Epoxy	4.31
	Cyanoacrylate	4.76
50% chromic acid; etched for 30 sec at 50 °C	Epoxy	3.21
	Cyanoacrylate	6.42
50% chromic acid; etched for 120 s at 50 °C	Epoxy	2.77
	Cyanoacrylate	5.72
Acid potassium permanganate; 5 min at 90 °C	Epoxy	3.76
	Cyanoacrylate	6.45
As above + resorcinol–formaldehyde primer	Epoxy	6.05
	Cyanoacrylate	5.56

and resorcinol–formaldehyde primer probably has intermolecular hydrogen, and possibly covalent, bonds acting across it and this is reflected by the high joint strengths. In the case of the cyanoacrylate adhesive a chromic acid etch, particularly at elevated temperatures, results in high joint strengths. This may possibly arise from (i) the greater penetration of the cyanoacrylate, due to its much lower viscosity, into the etch pits on the nylon's surface and (ii) the cyanoacrylate's greater reactivity towards the more polar etched surface. These results illustrate that the total interface, i.e. adhesive as well as substrate, must be considered when selecting an effective treatment.

Abu-Isa [54] has studied the iodine treatment of nylon materials. This pretreatment essentially consists of dipping the nylon in iodine potassium iodide solution and quickly rinsing the adsorbed iodine out of the nylon surfaces using glycol–water–thiosulphate as a wash solution; this considerably increased the strength of a metal film subsequently plated onto the metal surface. It was found that the strength of the plating was dependent upon the nature of the nylon surface (crystallinity, molecular weight, etc.), the duration of the iodine treatment and the nature and conditions of the wash solution. Scanning electron micrographs of the nylon surface demonstrated that only a shallow etch pattern was introduced and thus mechanical interlocking between the polymer and the plating was not considered to be of prime importance. However, under certain circumstances the overall degree of crystallinity was decreased and the type of crystalline structure present changed. Further, the polymer was softened, low molecular weight fractions and monomers extracted and the level of surface polarity raised. These last changes were considered to be responsible for the enhancement of adhesion to the treated nylon plastic.

More recently Cope [55] has examined a range of nylon plastics and, as might be expected, the shorter the hydrocarbon chain length between amide links, the easier the material is to bond; thus nylon 6 and nylon 6.6 were found to give higher joint strengths than nylon 11 or nylon 12. Also, he reported that the joint strength to nylons could be improved by using a flame treatment and the mechanism of improved adhesion appeared to be via oxidation of the surface region leading to an increase in the surface free energy.

Heat pretreatment has been reported to increase the adhesion to polyester films [56–58] but gains from such a treatment are comparatively small. As commented by Briggs *et al.* [56] the increases have been variously ascribed to removal of low molecular weight species, such as oligomers or diethylene glycol, from the surface or by relaxing surface molecular orientation, but it is also possible that it increases surface crystallinity and hence the surface free energy. Apparently various attempts to develop primer systems for polyester film to give improved adhesion have, up to the present, been largely unsuccessful. However, it has been reported [59] that silane-based primers, such as the γ-glycidoxypropyltrimethoxysilane in hexa(methoxymethyl)melamine, may be applied to biaxially drawn poly(ethylene terephthalate) films prior to their lamination with an ethylene–vinyl acetate terpolymer or an elastomeric polyester hot-melt adhesive. The use of the primer greatly increased the peel strength of the laminated film.

An example where a weak boundary layer can most definitely represent a major problem in attaining good joint strength is in the case of plasticized poly(vinyl chloride). Adhesive bonding to rigid poly(vinyl chloride) presents little problem but in plasticized poly(vinyl chloride), containing low molecular weight liquids to soften the plastic, these plasticizers migrate to the surface prior to bonding and to the interface after bonding. Such a plasticizer layer acts as a weak boundary layer. Typically, solvent cleaning followed by application of a primer is used to prevent this problem.

4.2.5 Plastic laminate materials

Plastic composites based upon thermosetting resins containing glass fibres or carbon fibres are being increasingly used in many diverse applications. These materials are usually based upon epoxy or polyester resins and are, therefore, more polar than the fluorocarbons or polyolefins which makes them more receptive to adhesive bonding. Also they do not, of course, form oxides or corrode in moist environments. This combination of properties results in surface pretreatments being required which simply remove contaminants such as oils, mould lubricants or general dirt. There are two main techniques used to achieve this: (i) the peel-ply method and (ii) abrasion and solvent cleaning, often conducted after a peel-ply surface has been exposed.

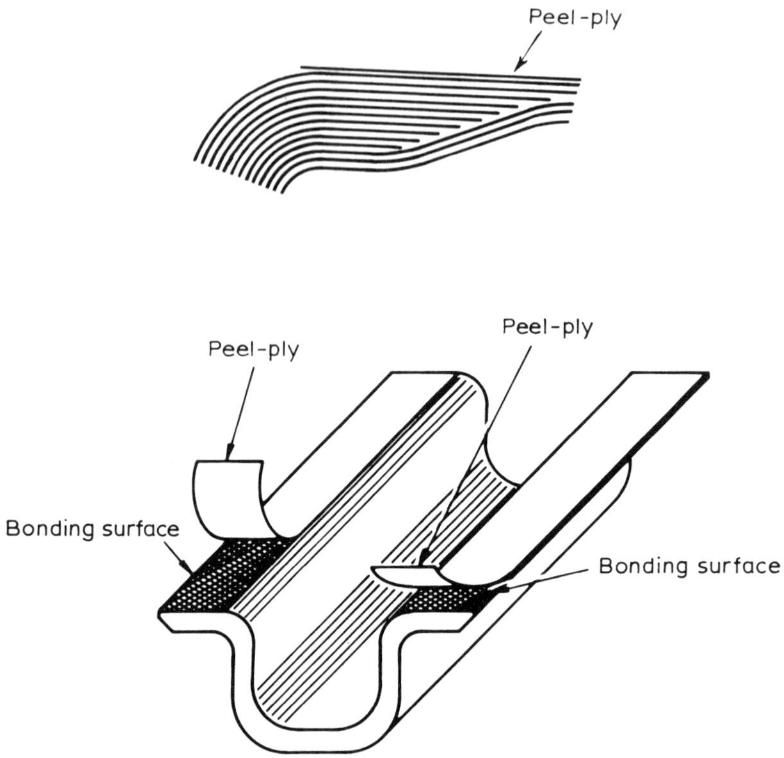

Figure 4.8 Schematic of a reinforced plastic-laminate with a peel-ply surface.

The peel-ply method is illustrated in Fig. 4.8. In designing the laminate one ply of fabric, such as Dacron or an equivalent, should be installed at the bonding surfaces, as shown, and just prior to bonding the peel-ply is removed. In principle a clean bondable surface is exposed. However, many workers [60–64] have commented that it is extremely difficult to ensure that the peel-ply has not left behind sufficient contamination, in the form of release agents used in the manufacture of the peel-ply, to reduce both the strength of the adhesively bonded composite and greatly increase the coefficient of variation. If joints are prepared from the 'as-moulded' composite surface then the joint strength is still dependent upon the extent of surface contamination; i.e. is a function of the nature of the surface in contact with the composite during moulding. Parker and Waghorne [61] have used XPS to study the effect of the concentration of contamination on the resulting joint strength. Silicone- and fluorine-containing contaminates were the main types detected, reflecting the use of these elements in release coatings on the peel-plies. The effect of concentration of contamination on the composite surface, for the fluorinated

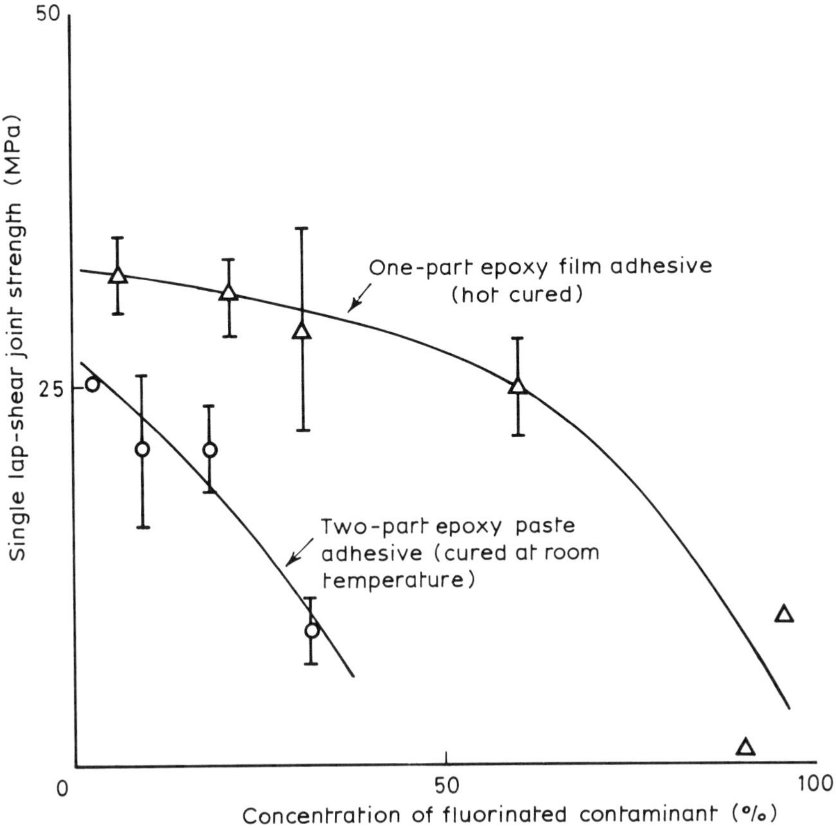

Figure 4.9 Effect of residual fluorinated contamination remaining on the surface of a carbon fibre composite after removal of the peel-ply on the subsequent joint strength [61].

contaminants, on the subsequent joint strengths is shown in Fig. 4.9 and, as may be seen, the joint strength decreases rapidly as the concentration increases but the rate of loss of joint strength is somewhat dependent upon the adhesive employed. Also, as might be expected, the locus of joint failure was increasingly at the adhesive/composite interface as the degree of contamination increased. The better performance of the hot-cured one-part epoxy adhesive shown in Fig. 4.9 clearly illustrates the superior ability of this type of adhesive to displace and dissolve surface contamination on the substrate compared to the room-temperature, or at least low-temperature, cured two-part epoxy adhesive.

To overcome the above problems it is either necessary to use a peel-ply which does not leave contaminants behind, which often means not using release agents and so makes the peel-ply difficult to remove [62], or to

Table 4.8 Effect of surface preparation on the strength of carbon fibre reinforced epoxy/epoxy adhesive joints [63]

Surface preparation	Single-lap shear strength (MPa)	Coefficient of variation (%)
Untreated	27.5	14
Peel-ply, polyester based	23.7	18
Peel-ply, nylon based†	7.9	17
Peel-ply, nylon based	24.6	15
Sandpaper	31.2	3.3
Grit-blast	29.2	8.5
Abrasive disc	29.8	5.1

† A very high level of fluorine contamination detected by XPS.

eliminate the residual contamination. This latter route has been the one successfully followed in industry and an abrasion treatment followed by a solvent wipe to remove the abrasion products, has proved to be most effective [59–65]. A light grit-blast with alumina particles has been found to be very efficient in removing the residual contamination but methodical hand-abrasion, especially using commercial abrasive pads, may be equally effective. This is demonstrated from the data given in Table 4.8.

Matienzo *et al.* [64] have suggested that the relatively high level of surface roughness induced by some of the treatment methods may lead to mechanical interlocking increasing the intrinsic adhesion. However, other work [63] discounts this mechanism, and even the studies by former workers lead to the conclusion that ensuring a contamination-free surface is the factor of overwhelming importance. Thus, the main function of a typical peel-ply is to protect the surface until just prior to bonding, at which time an abrasion/solvent cleaning pretreatment is then necessary to remove any residual contamination transferred upon stripping the peel-ply away. Other treatments have been reported [66] for composites, including treatment in acid solutions, but none has been as effective in giving high and reproducible joint strengths as abrasion and solvent cleaning techniques.

Another problem that may arise upon bonding fibre-laminates is that absorbed moisture in the composite may be evolved during the adhesive bonding cycle and lead to poor adhesion and air-voids in the adhesive layer [67,68]. The obvious method of overcoming this problem is to dry the composite prior to bonding, and this is the usual route followed although it can be difficult to achieve in practice when undertaking repair work. Also, when undertaking adhesive repairs, it may be necessary to remove absorbed hydraulic oils, etc. from the composite before satisfactory adhesion can be achieved.

More recently fibre composites have been commercially introduced which are based upon carbon fibres in a thermoplastic matrix, namely, poly(ether ether ketone). Such composites may be formed using thermoplastic processing techniques and have superior impact properties to those based upon current thermosetting resins. However, recent work [69] has revealed that bonding thermoplastic composites using abrasion and solvent-cleaning pretreatment techniques and epoxy adhesives leads to significantly lower joint strengths, compared to conventional composites. The reasons for this appear to be lack of sufficient intrinsic adhesion either of the adhesive to the composite substrate or, in some instances, of the outermost layer of the matrix to the underlying fibres. Whether the answer to these observations lies in changing the manufacturing technology or using a more active pretreatment method, e.g. a corona discharge, has yet to be resolved. On the other hand, if it is possible to use a very high-temperature hot-melt adhesive, work by Stein *et al.* [70] has shown that a film of poly(ether ether ketone) may be used successfully to bond fibre composites based upon this polymer.

4.2.6 Rubbers

Rubbers are most easily bonded to other materials during vulcanization, when they are crosslinked. Proprietary one-component and two-component adhesive systems for such rubber/rubber and rubber/metal bonding are available [71–73]. The uncured rubber surface should be clean and free from contaminants such as mould-release agents and the metal substrate should be vapour-degreased followed by grit- or shot-blasting. Degreasing by means of aqueous alkaline solutions followed by acid etching can also be used; special pretreatments for specific metals are usually laid down by suppliers of the adhesives where they are considered advisable [74].

In some instances, however, it may be necessary to bond rubbers after they have been vulcanized. This method of bonding is referred to as post-vulcanization bonding and a main advantage is that complicated rubber shapes may be bonded. Indeed, it is often not technically or economically possible to bond such components by the bonding/vulcanizing one-step method. For post-vulcanization bonding one- or two-component primer/adhesive systems are commercially available. Wake [75] has suggested that, if the joint has to bear only a light load, cleaning the cured rubber surface with solvent, thereby swelling it somewhat, followed by application of the adhesive may be sufficient. If the joint has really to be load-bearing then it is necessary to remove the existing surface of the vulcanized rubber by abrasion immediately before applying the adhesive to be used. The effects of different surface treatments for the vulcanisate prior to joining has been considered by Spearman and Hutchinson [76] and some of their results are given in Table 4.9. Natural rubber is sensitive to the type of surface treatment, with abrasion being the most effective. Polychloroprene is relatively insensitive. For all three rubbers

Table 4.9 Butt joint strengths (MPa) of rubber/metal joints for various vulcanized rubbers and surface treatments [76]

Vulcanized rubber	Solvent wipe	Ultrasonic degreasing	Hot alkaline rinse	Abrasion	Chlorination	Chemical modification by cyclization
Natural	4.5	5.4	6.5	9.4	5.6	7.0
SBR	1.8	6.1	6.2	7.9	6.9	4.5
Polychloroprene	5.1	4.9	6.7	6.8	6.9	7.0

abrasion is a good pretreatment and results in high joint strengths and cohesive failure of the rubber substrate. However, other cured rubbers such as butyl and ethylene–propylene–diene rubber are far more difficult to bond than the above, more polar and more unsaturated natural, styrene–butadiene and polychloroprene rubbers.

Considerable work [77] has also been conducted on pretreatments for thermoplastic block copolymers based upon SBS, namely, polystyrene–polybutadiene–polystyrene, and one of the widest uses of the SBS copolymer is in the footwear industry where it is used as a soling material. To obtain good adhesion to the urethane- or polychloroprene-based adhesives various halogenation pretreatments have been developed. There appeared to be no evidence of a weak boundary layer on the original SBS surface and Greenwood and Wake [78] have suggested that a mechanism for the greatly improved adhesion of the treated surface with a urethane adhesive may be due to hydrogen bonding, as shown schematically in Fig. 4.10.

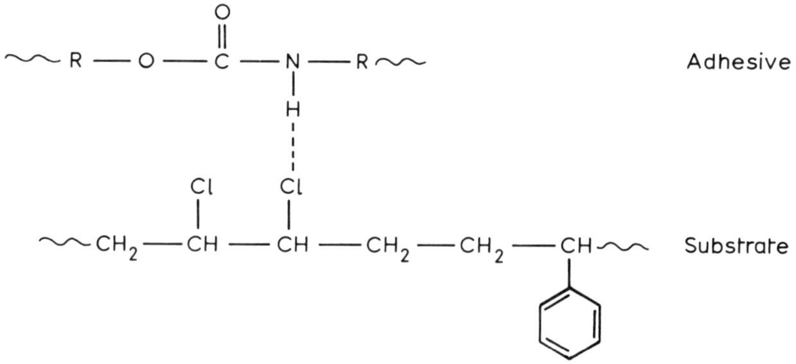

Figure 4.10 Proposed mechanism for hydrogen bonding across the interface of a SBS thermoplastic block copolymer substrate after being subjected to an halogenation pretreatment and using a polyurethane adhesive [78].

4.2.7 Plasma treatments

4.2.7.1 Introduction

A plasma is an excited gas which consists of atoms, molecules, ions, free radicals, free electrons and metastable species and the use of plasmas to pretreat polymers (and metals, as discussed later) has been known for about 20 years. The plasma may typically be generated in air, termed a 'corona' discharge, or under reduced pressure, termed a 'glow discharge'. Also, the gas may be non-polymerizable or be able to polymerize and deposit onto the substrate surface. A schematic representation [79,80] of the numerous interactions which are possible in a gas plasma impinging upon a substrate is illustrated in Fig. 4.11.

The major effects of a plasma pretreatment can include the following and, of course, several may occur simultaneously:

(i) Cleaning – the excited species may have sufficient energy to displace low molecular weight contamination, e.g. processing additives, which are present on the surface. (The displacement process may be preceded by the plasma degrading, and reducing the molecular weight of, the contamination.)

(ii) Degradation and ablation – the plasma may cause degradation of the surface regions of the polymeric substrate and lead to removal of material from the surface. Plasmas of air or oxygen have a marked tendency to cause degradation and ablation compared to plasmas of inert gases such as helium and argon [81–83]. Further, some polymers, such as poly(ethylene terephthalate), are more prone to these effects than others.

(iii) Crosslinking – the surface region of the polymer may become crosslinked, although Yasuda *et al.* [82] have reported that in most polymers this effect is marginal. However, polyethylene and unsaturated polymeric substrates do tend to crosslink. This may be a disadvantage, e.g. lack of heat sealing, or an advantage, e.g. hinder the development of a weak boundary layer at the interface of a joint by decreasing the diffusion of low molecular weight species.

(iv) Oxidation – the plasma often leads to the introduction of polar groups, such as carbonyl groups, into the surface regions of the substrate, as discussed in detail below. XPS has clearly confirmed that this may occur even when inert gases [84] or fluorine [85] are employed. In such cases the oxygen is suggested to arise from traces of oxygen being present in the treatment chamber (e.g. in the gas, on the chamber walls or in the polymeric substrate) or from post-treatment reaction occurring when the polymer is removed from the treatment equipment.

(v) Polymerization and grafting onto the substrate's surface – the plasma may induce the polymerization of a constituent of the gas phase and so cause

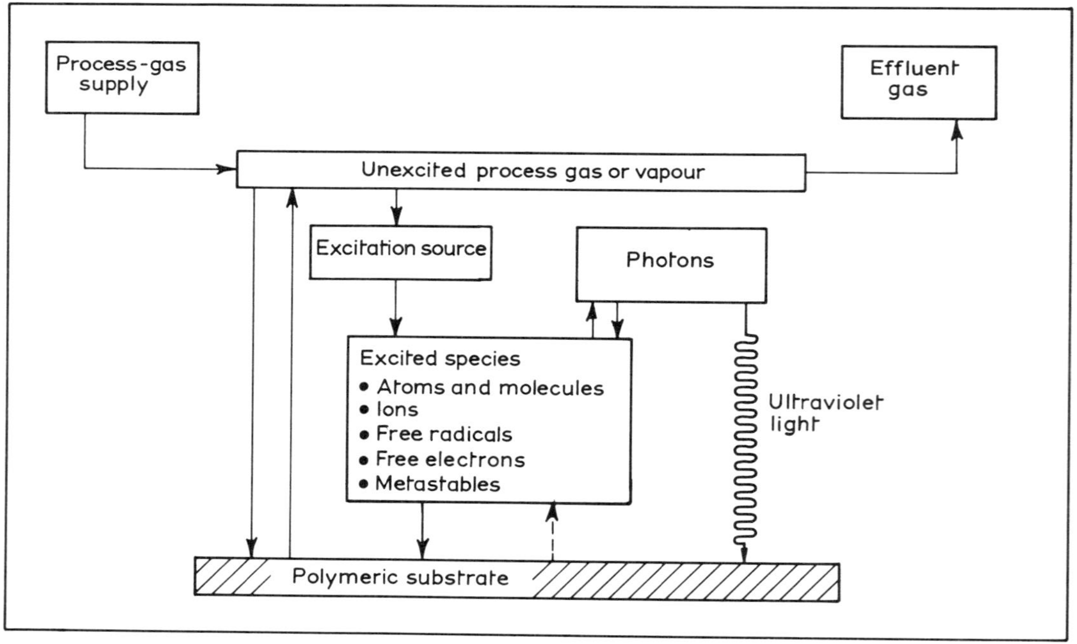

Figure 4.11 A schematic representation of the numerous interactions which are possible in a gas plasma impinging upon a substrate. The process gas or vapour is excited into an energetic state containing many different species and metastables that react with and modify the polymeric substrate. The dashed line indicates that the nature of the excited species will usually be changed after impinging upon the substrate. As the excited species return to lower energy states they may emit light (photons) that gives rise to the 'corona' or 'glow-discharge' often observed and may also further modify the substrate. The effects of such interactions, including the possibility of a constituent of the process gas being selected so that it polymerizes on the substrate, are discussed in Section 4.2.7.

the deposition of a thin polymeric film onto the substrate. Such a film may be grafted to the underlying substrate. Most of the organic, organosilicone and organometallic vapours tend to form polymeric films on the substrate when subjected to a glow discharge, and the deposition of such films is the main factor under these circumstances. However, it is also possible that some of the gaseous species from cleaning and ablation of the surface may subsequently polymerize and redeposit as a polymeric film on the substrate.

(vi) Ion implantation – the plasma may cause foreign atoms to be implanted in the surface regions of the substrate and these atoms may either aid adhesion, e.g. via the adhesive grafting to them, or adversely affect the joint strength and durability. Ion implantation may also be used to try and achieve an interphase region of properties intermediate between those of the substrate and adhesive.

4.2.7.2 Corona-discharge treatments

As discussed in detail by Briggs [35], the corona-discharge treatment is a long established method for treating polymers prior to adhesive bonding, printing with inks, lamination to other films, vacuum metallization with metals, etc. Indeed, it is the most widely used method for pretreating polyolefin film and is also employed for conical or cone-shaped containers (e.g. tubs). The basis for the technique is shown in Fig.4.12 and it typically involves passing the polymer film over an earthed metal electrode which is covered with an insulating dielectric material, such as chlorosulphonated polyethylene rubber, silicone rubber, epoxy, epoxy–glass or polyester. The metal electrode, usually made from aluminium, is separated from the polymer film surface by approximately 1 to 2 mm and a high-frequency (typically 10 to 20 kHz) generator and a step-up transformer provides a high voltage to the electrode. In each half-cycle the applied voltage, about 20 kV maximum, increases until it exceeds the threshold value for electrical breakdown of the air gap when the air is ionized and, hence a plasma is generated. The atmospheric pressure plasma is called a corona discharge, hence the name for this type of treatment. As commented by Briggs [35], in practice discrete intense sparks can also be seen originating from localized regions of the electrode, so corona is not a truly exact description for this type of treatment. Apart from using air as the medium in which the plasma is generated, liquids or oxygen-containing gases are often injected into the air gap in an attempt to increase treatment efficiency. Finally, .on the process aspects, many other factors, such as air flow rate and film temperature, may influence the efficiency of the pretreatment but the two major factors [35] are the power output into the discharge and the speed of throughput of the polymer being treated.

Turning to the mechanisms whereby corona-discharge treatment can

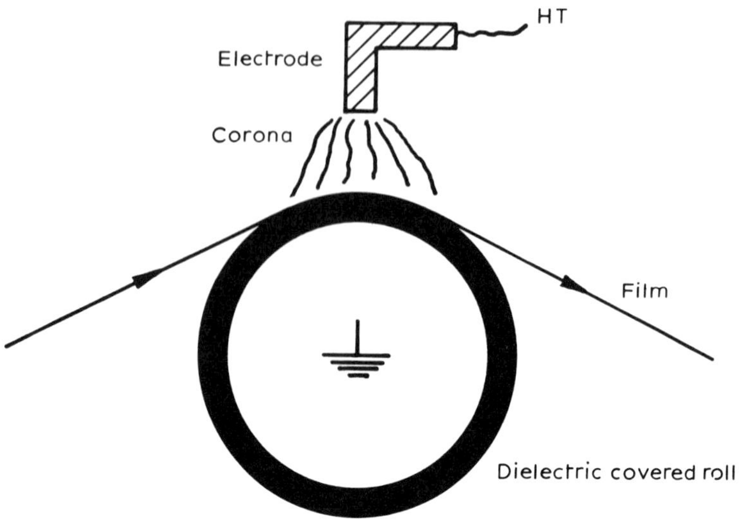

Figure 4.12 Schematic diagram of the electrical ('corona') discharge treatment process [35].

increase the adhesion to polymeric substrates, Rossman [86] demonstrated many years ago that in the case of polyethylene this treatment causes the formation of unsaturated (C=C) and carbonyl (>C=O) groups and ascribed the improved joining properties to the increased surface free energy of the polymer's surface (see Fig. 4.13). This hypothesis is supported by the work of Carley and Kitze [87] who have shown that the value of the dispersion force component, γ_s^D, does in fact decrease somewhat but that the polar force component, γ_s^P, increases significantly with increasing degree of treatment. It was the increasing values of γ_s^P, reflecting the increased oxidation of the surface, that could be directly correlated with increasing joint strength.

Kim *et al.* [88] have advanced a different theory. They have proposed that the mechanism of improved adhesion involves some type of electret formation induced in the polymer by the corona discharge. Electrets are long-lived electrical charges induced when dielectric materials are subjected to strong d.c. fields. Although 60 Hz was used by these workers they claimed that charge effects produced by one cycle are not completely neutralized by the subsequent cycle and a discrete charge is built up on the surface by application of an a.c. field. They demonstrated that a charge pattern did exist on the corona-treated polyethylene by spreading a light dielectric powder on the surface of the polymer and observing a clear 'Lichtenburg figure' pattern. An important feature of this work is that the surface charge due to electret formation is known to decay slowly with time and, indeed, they found the adhesion of the

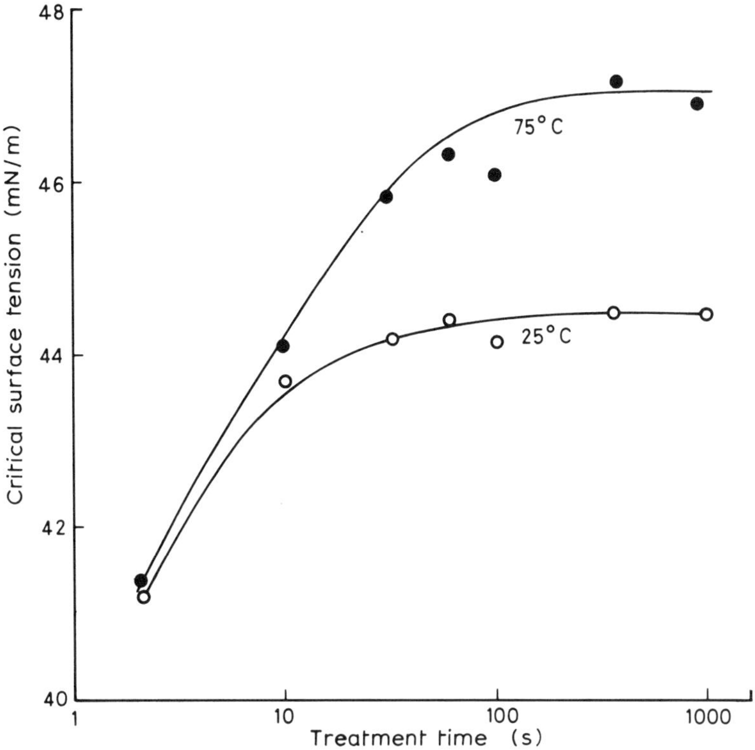

Figure 4.13 Critical surface tension, γ_c, versus time of treatment for low-density polyethylene pretreated in an air corona at different temperatures [88].

treated polyethylene decayed as the time between treating and bonding was increased. However, this effect is well known and can be explained by surface chemistry considerations; i.e. the oxidized species formed on the polymer's surface adsorb and react with airborne (hydrocarbon) contamination and, indeed the value of γ_s^P does fall upon storage of the treated polymer [87].

This mechanism of electret formation is considered to be vague and physically unacceptable by Owens [89]. His work clearly showed that the high autoadhesion between two films of corona-treated low-density polyethylene could be completely destroyed by the application of hydrogen-bonding liquids to the joint and that the effects of such liquids is completely reversible; joints allowed to dry regain their full, original strength. As discussed in Section 3.5.2.1 (Fig. 3.8a), he proposed that the corona treatment formed keto groups on the polyethylene chain; these groups enolize and the enolic hydrogens exhibit an acidic character and form a hydrogen bond with carbonyl groups on the adjacent substrate when the two layers of polyethylene are heated together

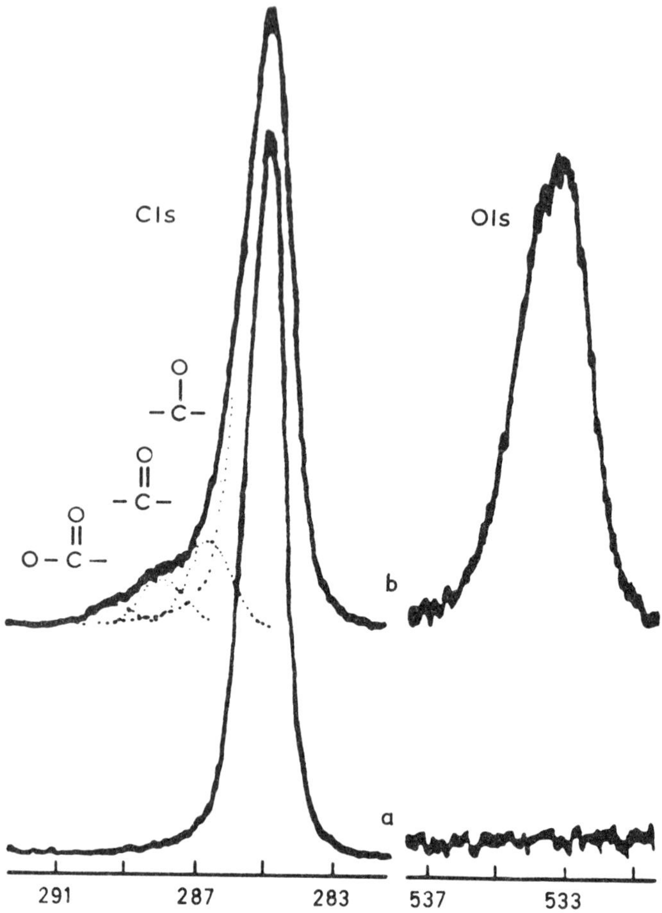

Figure 4.14 High resolution C1s and O1s spectra for low-density polyethylene from X-ray photoelectron spectroscopy [91]. (a) Untreated, (b) treated in an air corona (13.7 kV peak voltage at 50 Hz) for 30 s.

under pressure. Another similar mechanism [90] would be the formation of a hydrogen bond via a water molecule (see Fig. 3.8b). Studies [91–93] using XPS have confirmed the increased oxidation induced by corona discharge, as may be clearly seen in the typical XPS spectrum for untreated and treated low-density polyethylene shown in Fig. 4.14. Such oxidation is suggested to arise via a free radical mechanism. Further, these studies support the mechanism proposed by Owens [89], as illustrated in Fig. 3.8a. It is noteworthy, following the discussions in Chapters 2 and 3, that unless the liquids in any contact angle measurements are especially selected the values of γ_s^D and γ_s^P deduced might well not reflect the potential of the treated surfaces to form such hydrogen bonds. Thus, a direct correlation between values of surface free energy, deduced from contact angle measurements, and joint strength might not always be readily established.

In a few studies the importance of considering the role of weak boundary layers has been mentioned. For example, where the polymer contains low molecular weight additives, such as slip additives in film formulations, it has been found necessary to pretreat for longer times to achieve effective results [87]. A second example is when overtreatment occurs, e.g. very long treatment times, and degradation of the surface regions of the polymer results, so generating a classic weak boundary layer and giving low joint strengths [87].

Other polymers have also been pretreated using corona-discharge methods, although invariably they are more difficult to treat as effectively as low-density polyethylene. Polypropylene [94], poly(ethylene terephthalate) [95–97], a copolymer of ethylene and vinyl acetate [86] and polytetrafluoroethylene [98] are amongst the polymers which have been subjected to corona treatments and where the subsequent adhesion has been improved. The mechanisms of adhesion have only been investigated in depth for poly(ethylene terephthalate) and the principal mechanism was found to be the same as for polyethylene; i.e. increased oxidation of the surface regions and hence the possibility of improved interfacial contact and higher intrinsic adhesion. Again, the role of interfacial hydrogen bonds has been emphasized. However, it appears that the problem of reorientation (Sections 2.4.3 and 4.2.1) is particularly pronounced with treated poly(ethylene terephthalate) and the polar groups at the surface redistribute to form internal hydrogen bonds, so reducing the subsequent joint strength.

4.2.7.3 Glow-discharge treatments (non-polymerizable gases)

In 1966 Schonhorn and Hansen [99] reported a highly effective pretreatment for the surface preparation of low surface energy polymers for adhesive bonding. Essentially their technique consisted of exposing the polymer surface to an inert gas plasma at reduced pressure generated by an electrodeless glow discharge (i.e. radiofrequency field). For polyethylene only very short treatment times were necessary (Fig. 4.15) while longer contact times were required for other polymers such as polytetrafluoroethylene. Most subsequent work has also employed an electrodeless technique for generating the plasma but it may also be generated between metal electrodes. Hall *et al.* [100,101] have examined the joint strength of many polymers as a function of length of exposure to excited helium or oxygen. Polymers studied include low- and high-density polyethylene, poly(4-methyl 1-pentene), poly(vinyl fluoride), poly(vinylidene fluoride), fluorinated ethylene–propylene copolymer, polyacetals, polyamides, poly(ethylene terephthalate) and polystyrene. Generally, the joint strength increased rapidly initially and then remained nearly constant,

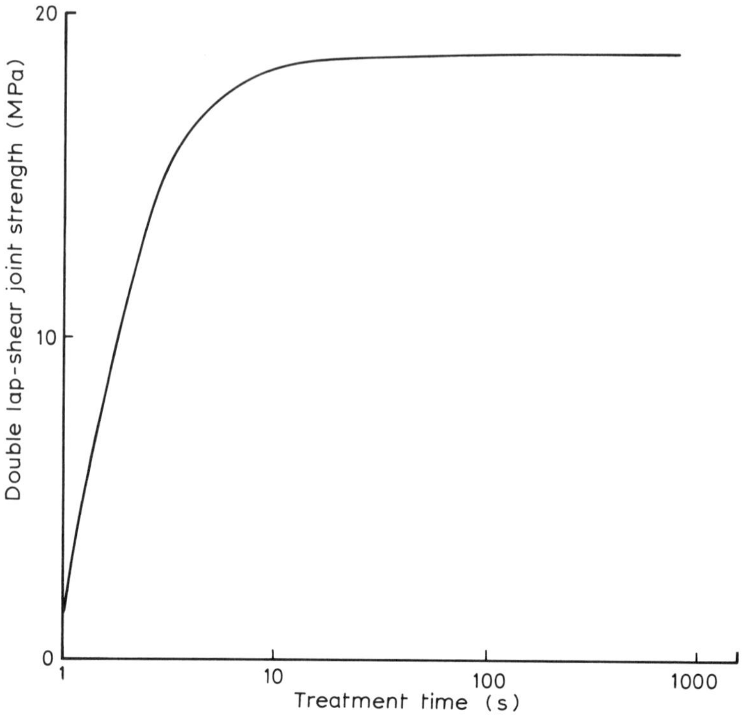

Figure 4.15 Lap-shear strength of aluminium/epoxy/LDPE joints as a function of the time of pretreatment of the LDPE in a glow-discharge plasma of helium [106].

perhaps decreasing in some cases at long exposure times. Further work has dealt with the effect of gas employed [80,82,85,101–103], gas pressure [102], electrode separation [102] and discharge current [82,102]. More recently commercial equipment [80] has become available based upon forming the plasma by a radiofrequency generator and a very wide range of polymers have been examined, including the newer engineering thermoplastics such as polycarbonate, poly(ether ether ketone), poly(phenylene oxide), polyacetal, poly(ether sulphone) and poly(phenylene sulphide) and difficult to bond rubbers such as silicone rubbers. The effect of various plasmas on the subsequent joint strength of poly(ether sulphone) substrates is shown in Table 4.10 and, as may be seen, the plasma pretreatments can be most effective.

Turning to the mechanisms whereby the glow-discharge treatments may enhance joint strengths, from their pioneering work in this field Schonhorn and co-workers [99,103–106] reported that such treatments did not change the wettability of the polyethylenes they pretreated, as measured by the critical surface tension, γ_c. However, gel permeation chromatography showed that some highly crosslinked polymer was formed after treatment and the weight average molecular weight increased from 86 000 to 135 000. Also, ATR infrared spectroscopy revealed only the presence of transethylenic unsaturation after treatment and a thin intractable skin was readily observed by melting the polymer. Schonhorn concluded that wettability was not the sole criterion for the formation of an adhesive joint but what may be of greater importance is to enhance the cohesive strength of a weak surface region. He and co-workers suggested [99,103–107] that many melt crystallized polymers possess an inherent weak boundary layer, about 0.02 to 0.1 μm thick, which is composed of amorphous low molecular weight materials rejected by the crystallization process which starts in the bulk and proceeds outwards. It was therefore concluded that the increases in joint strength that were achieved by this technique were primarily due to increasing the cohesive strength of the polymer in the surface regions by the introduction of crosslinks. The surface

Table 4.10 The effect of various glow-discharge treatments on the joint strength of poly(ether sulphone) substrates bonded with an epoxy adhesive [80]

Plasma gas	Single-lap-shear strength (MPa)	Locus of failure
None (solvent wipe, control)	0.90	Interfacial
Argon	21.7	Substrate
Air	21.4	Substrate
Ammonia	12.9	Substrate
Nitrogen	19.5	Substrate
Oxygen	18.3	Substrate

treatment was therefore christened CASING (Crosslinking by Activated Species of Inert Gases).

Unfortunately, as discussed previously, ATR infrared spectroscopy is not sensitive to the chemistry of the uppermost surface layers and techniques such as XPS were not available when these studies were conducted. Thus, Schonhorn and colleagues largely had to infer the basic point of their proposed mechanism, namely, that a weak amorphous surface is produced when polymeric substrates are melt crystallized in air. Considering all the evidence now available from studies on these types of pretreatment, as well as the other treatments discussed above, it would appear that the hypothesis that most polymers are covered with a weak layer of amorphous low molecular weight material is untenable. As described previously, in those instances where some form of weak boundary is present then its removal by a pretreatment may be a necessary step to attaining good joint strengths but with many polymers chemical modification of the surface is still essential. Indeed, many workers [20,80,84,102,108,109] have established that the surface free energy of the polymeric substrate may be increased by glow-discharge treatments due to oxidation of the surface region.

Thus, it appears that the general mechanisms involved are: (i) the cleaning action of the plasma in removing surface contamination, e.g. mould-release agents and additives that have migrated to the surface; (ii) introduction of chemical groups which increase the surface free energy and can lead to acid–base interactions, so increasing the degree of interfacial contact and intrinsic adhesion; and (iii) the introduction of chemically active species into the substrate's surface regions which can greatly increase the intrinsic adhesion via grafting of the adhesive, or a primer.

4.2.7.4 Plasma-induced polymerization treatments

Polymer films produced by a glow discharge often possess chemical and physical properties that are superior to conventionally polymerized materials [110,111]. They are prepared by a totally dry vapour phase deposition process and may be conformally coated onto virtually any substrate, thereby completely changing the surface properties of the substrate. Further, the substrate may be cleaned by the plasma prior to deposition and, under the correct conditions, it is often possible to graft the deposited polymeric film to the underlying substrate.

The techniques of plasma-induced polymerization to modify the surfaces of a polymeric substrate for subsequent adhesive bonding have not been widely employed so far. Inagaki *et al.* [112] have pretreated several different plastics, such as nylon, acrylonitrile–butadiene–styrene copolymer, polycarbonate, poly(vinyl chloride), polyethylene and polypropylene, by first cleaning the substrate by exposing it to an argon plasma for a few minutes, then injecting the monomer gas into the treatment chamber. The monomer gases used were either trimethylsilyldimethylamine or hexamethyldisilazane. Polymers were deposited from both the monomers and the former adhered to the polymeric substrates more readily, but the adhesion was greatly influenced by the nature of the substrate, as would be expected.

Using a somewhat different approach Yamakawa and co-workers [113–115] have modified polyethylene and poly(chlorotrifluoroethylene) substrates by the vapour phase polymerization and grafting of methylacrylate onto the substrate's surface, using γ-irradiation followed by hydrolysis. The surface regions therefore consist of an outer carboxyl-containing homopolymer layer of hydrolysed poly(methyl acrylate) and an inner graft copolymer layer. The treated substrates were bonded, using an epoxy adhesive, and the resulting joint strengths were far greater than for the untreated substrates, or from using other pretreatment methods, as illustrated in Table 4.11.

Turning to the mechanisms of adhesion, the graft copolymer layer between the polyethylene substrate and hydrolysed poly(methyl acrylate) surface layer provides one intrinsically strong interfacial region. In the case of the apparently very good wetting and adhesion of the epoxy adhesive to the hydrolysed poly(methyl acrylate) outer layer, the increased surface free energy afforded by the hydrolysed poly(methyl acrylate) was considered to be a minor factor. Instead, Yamakawa *et al.* suggested that this layer had a porous structure and that the epoxy adhesive penetrated this layer and hardened to form a semi-interpenetrating network structure, thus, leading to a second intrinsically strong and tough interfacial region.

Table 4.11 T-peel strength (kJ/m^2) of treated polyethylene substrates bonded using an epoxy adhesive [115]

Treatment	LDPE†	MDPE‡
None	<0.04	<0.04
Chromic acid etch	0.82	1.56
O_2 plasma (5–30 min; 25–50 W)	2.34	1.95
He plasma (5–30 min; 75–100 W)	0.16	1.13
Hydrolysed poly(methyl acrylate) graft	>11.7§	>15.6§

† LDPE, low-density polyethylene.
‡ MDPE, medium-density polyethylene.
§ Substrate failure.

4.3 HIGH-ENERGY SURFACES

4.3.1 Introduction

Metals, metal oxides and ceramics typically have surface free energies greater than about 500 mJ/m^2 and are therefore termed high-energy solid surfaces. Thus, from the discussions in Chapter 2, one would think that such a surface would be readily wetted, giving a high degree of intimate interfacial contact, by organic adhesives with values of surface free energies typically less than about 50 mJ/m^2. Further, the high surface free energy of the substrate will result in relatively strong interfacial secondary force interactions. This is reflected in high values for the thermodynamic work of adhesion, W_A, for such interfaces. For example, from the values given in Tables 2.3 and 2.5 and using Equation 3.12, an epoxy/steel interface has a value of W_A of about 300 mJ/m^2 compared to 50–90 mJ/m^2 for an epoxy/polymeric substrate interface.

For attaining good initial joint strengths the above intuitive arguments are essentially sound, but several factors may complicate the bonding operation, to greater or lesser extents, depending upon the exact circumstances. These factors may all give rise to a weak boundary layer being present on the surface of the high-energy substrate, and so prevent the attainment of the good joint strengths expected from the above arguments. These factors are:

(i) Many substrates may often be coated by protective oils, greases, waxes, etc. and may also be covered by mill scale and corrosion products. If such layers are not removed prior to adhesive bonding then they may act as weak boundary layers in the joint. However, in the case of protective oils, greases, waxes, etc., many modern adhesives, particularly acrylics, hot-cured epoxies and solvent-based formulations, will dissolve or displace protective oils, etc. However, the protective layer employed needs to be 'matched' to the adhesive type and its thickness should be controlled. Indeed, from the author's own experience, some protective greases become so ingrained into the substrate's surface that a chemical etch treatment has to be used to ensure their removal!

(ii) It should be recalled (Section 2.6.2) that truly clean high-energy surfaces will instantly adsorb atmospheric contamination. If such multilayer contamination is not dissolved or displaced by the adhesive it will act as a weak boundary layer. From XPS and Auger spectroscopy it has been found [116] that this contamination contains adsorbed hydrocarbons. This will assist the displacement process since most common adhesives are usually relatively polar in character and thus will have a thermodynamic tendency to displace such non-polar contamination, especially if only physisorbed.

(iii) In the case of a very few metals, notably copper and magnesium and their alloys, the oxide layer produced naturally, or even by an abrasion pretreatment, appears to be very unreceptive to the adhesive, and mechanically weak. Hence, a chemical etch and/or primer treatment is often necessary to ensure adequate initial joint strength.

The technological implications of the above points are basically that, to attain good initial joint strengths, in many instances no surface treatment is necessary; and even when the state of the incoming substrate material is such that relatively thick layers of protective greases, mill scale, etc. need to be removed then a simple degreasing and, possibly, mechanical abrasion, pretreatment are all that is required.

However, there is one overriding reason why more complex, often chemical pretreatments, are employed to treat high-energy surfaces prior to bonding; namely, to ensure an adequate service life of the joint when it is exposed to aqueous environments. Unfortunately, one of the most hostile environments for adhesive joints consisting of high-energy substrates is water and this, of course, is also one of the most commonly encountered. Water will always gain entry to a joint by diffusing through the organic adhesive layer, and possibly via other routes as well, e.g. diffusing along the interface. The fundamental reasons and mechanisms underlying the problem of environmental attack will be reviewed in detail in Chapter 8 and it is sufficient for the present discussions to note that a recurring theme when considering joint durability is the importance of the stability of the adhesive (or primer)/substrate interfacial regions. This, in turn, will be greatly influenced by the pretreatment which had been selected for the substrate prior to bonding. Basically, the pretreatment must:

(i) Ensure that the intrinsic adhesion forces established between the adhesive and substrate are stable in the presence of water.
(ii) Ensure that the surface regions of the substrate in the joint are stable and will not weaken and form an *in situ* weak boundary layer. For example, in the case of metallic substrates the pretreatment must generate an oxide which will not undergo hydration and so become mechanically weak.

Thus, the choice of a particular pretreatment, or often a combination of pretreatments, largely depends upon the nature of the substrate and its initial surface condition, the type of adhesive to be employed and the service environment likely to be experienced. Again this emphasizes a 'systems' approach to the problem of selecting an adequate, and yet economically viable, pretreatment.

4.3.2 Solvent cleaning

Substrate surfaces are frequently contaminated with oils, greases, etc. and a common pretreatment is degreasing either with steam, or by wiping the surface with solvent-dipped clean cloths, or most effectively in liquid and vapour degreasing baths, possibly using an ultrasonic agitator. Commonly used solvents are 1,1,1-trichloroethane or perchloroethylene, and they should be checked periodically for the formation of corrosive acid.

Following the removal of greases and oils, the residual contaminants are usually removed with an alkaline cleaner which emulsifies the surface contaminants and keeps them in suspension. Uninhibited (etching type) strong alkaline cleaners are used for ferrous metals, titanium and certain copper alloys, but aluminium generally requires inhibited solutions if etching of the material is to be avoided. Alkaline cleaners are available commercially or may be prepared from the appropriate ingredients to existing formulations [3,6]. The alkaline cleaning process is often followed by a mechanical or chemical treatment since it leaves the surface unreceptive to many adhesives.

4.3.3 Mechanical abrasion

The methods available include wire brushes, sand and emery papers, abrasive pads and grit- or shot-blasting. The techniques of grit- or shot-blasting give the most reproducible results and, wherever practicable, are preferred for industrial processes. The equipment basically consists of an air pressure-fed grit-blasting machine in which abrasive is stored in a pressure vessel, and is introduced through a small feed orifice into the blast line. The abrasive is typically angular chilled iron abrasive of size G04 to British Standard 254 or angular alumina abrasive of 180/220 mesh. It is ultimately propelled through a nozzle of convergent/divergent cross-section onto the work surface. It has been shown [117] that variations in pressure in the grit-blasting machine, angle of incidence of the jet or type of abrasive have little influence on the resultant joint strength, and these results are shown in Table 4.12. In these experiments, as in normal practice, a final solvent liquid and vapour rinse was used to remove abrasive dust which would otherwise remain at the adhesive/substrate interface and act as a weak boundary layer.

Table 4.12 Mechanical abrasion as a surface pretreatment for epoxy/stainless steel joints

Surface pretreatment of stainless steel	Joint strength (MPa)	Locus of failure
Degreased only	6.14	35% interfacial
Degreased and abraded:		
(i) At 90° incidence and (a) 280 kPa	6.89	Cohesive in adhesive
(b) 550 kPa	6.79	Cohesive in adhesive
(ii) At 45° incidence and (a) 280 kPa	6.88	Cohesive in adhesive
(b) 550 kPa	7.03	Cohesive in adhesive
Degreased and etched in chromic acid	5.72	45% interfacial

Jennings [118] has conducted extensive and detailed comparisons between bonding to relatively smooth (polished) surfaces of aluminium alloy and stainless steel substrates and rough (machined or abraded) surfaces. Some typical results from room-temperature tests are shown in Table 4.13, where an epoxy–polyamide adhesive was employed. It may be seen that all the joint strengths are of a reasonable value but the rougher the substrate surface the stronger the joint, with sand-blasting giving the strongest joints. Interestingly, as the test temperature was raised, making the adhesive more ductile, the differences between the strengths made from the smooth and rough surfaces disappeared. Jennings considered that the better wetting and increased surface area afforded by the roughened substrates probably made some contribution to the higher joint strengths, but commented that these factors could not explain the observed effect of test temperature. He suggested that the effect of the surface topography on the local stress distribution was possibly the main factor. Essentially, if the macroscopic surface roughness is random, as for sand-blasting (see Fig. 3.1), it can be effective [119] in preventing any small cracks, flaws, voids or other points of stress concentration from aligning and rapidly

Table 4.13 Joint strength as a function of surface roughness [118]

Surface condition	Butt joint strength (MPa)	Coefficient of variation (%)
Aluminium alloy (6061)		
Polished 1 μm diamond paste	28.8	24.4
Abraded through 600 SiC paper	30.9	24.9
Abraded through 280 SiC paper	39.0	17.5
Abraded through 180 SiC paper	36.7	20.4
Sandblasted with 40 to 50 mesh SiO_2 grit	48.5	14.4
Stainless steel (304)		
Polished 1 μm diamond paste	27.8	20.8
Regular machined grooves	35.2	20.0
Sandblasted with 40 to 50 mesh SiO_2 grit	53.4	10.8

propagating along any line of interfacial weakness in the joint. Obviously, such alignment and propagation are more likely for a polished and smooth substrate surface, which results in a very planar interface, than a substrate with an irregular surface profile, which gives a very tortuous interface. Also, this effect of surface roughness on the ease of crack propagation would be expected to diminish in importance as the adhesive became less brittle and more ductile, in accordance with the experimental observations.

Thus, as discussed in Section 3.2, it appears that roughening the surface of the substrate may lead to increased joint strengths but that such improvements do not generally arise from simple mechanical interlocking. They may arise from the very effective cleaning action associated with the abrasion process, the increased surface area for bonding and the often improved kinetics of wetting. They may also arise from more subtle effects due to changes in the local stress distribution. For example, an increase in roughness may increase the localized energy dissipation in the adhesive near the interface and prevent any cracks at, or close to, the interface from aligning and then readily propagating.

4.3.4 Chemical treatments

4.3.4.1 Introduction

Whilst solvent degreasing and abrasion, or even no pretreatment, is often sufficient to attain the required initial design joint strength there are a few high-energy substrates which require more complex treatments to attain a reasonable level of initial joint strength. Further, with nearly all high-energy substrates the maximum durability to aqueous (and solvent) environments is achieved when a chemical pretreatment, and/or a primer, is employed. Thus, joints involving these materials which are to experience wet/humid weathering conditions or exposure to certain solvents will need particular attention to be paid to the selection of surface pretreatments, with the economic implications also being assessed.

Apart from the direct economic implications, chemical pretreatments also require careful monitoring of the various baths and may present a waste disposal problem. Further, the treatment must be selected so as not to affect adversely the properties of the material being pretreated. For example, during the treatment of certain alloys, such as certain titanium alloys and martensitic steels, hydrogen must not be generated since this could lead to hydrogen embrittlement.

The present section will discuss chemical etching and anodizing treatments and the following Section will discuss the use of primers. As before, the discussions will concentrate on the general principles and the reader is referred elsewhere [3,5,6] for detailed recipes.

4.3.4.2 'Difficult to bond' metals

Two metals which usually need some form of chemical pretreatment simply to achieve reasonable initial joint strengths are copper and magnesium and their alloys. It is not entirely clear why it is usually so difficult to attain high joint strengths to the natural oxides of these metals, or to oxides generated by a mechanical abrasion pretreatment. From the author's own experience it seems to be a combination of these oxides being difficult for the adhesive to wet readily and the oxides being mechanically weak, and so acting as a weak boundary layer in the bonded joint. Indeed, in the case of copper, recent work by Park and Bell [120], using energy-dispersive X-ray analysis and the deposition of silver onto any exposed copper oxide, has confirmed that the oxide layer formed in air is relatively weak and is commonly the locus for joint failure.

The chemical treatment of copper to give good joint strengths was discussed in Sections 3.2.4 and 3.2.5 and many of the treatments apparently result in a relatively strong oxide layer which has a microfibrous topography (Figs 3.2a, b and 3.3a). This topography reinforces the interfacial region, promotes localized plastic energy losses in the adhesive and, with specific topographies (Table 3.1), provides sites for true mechanical interlocking to occur. The proprietary oxidizing solution used in some of these studies has also been reported [121] to be an effective pretreatment for copper alloys such as beryllium–copper, phosphor–bronze, but not for yellow-brass (65% copper) since a porous and weak oxide was formed. However, these workers postulated that the good interfacial adhesion observed with all but the last alloy was due to the oxide consisting largely of cupric oxide which oxidized the adhesive, a hot-melt polyethylene, and was itself partially reduced to cuprous oxide. The polyolefin surface would then possess a large concentration of polar groups which would significantly increase the intrinsic adhesion arising from interfacial secondary bonds. Later work by Evans and Packham [122] has indicated that this mechanism does not operate and that the mechanisms leading to high joint strengths are a result of the formation of the relatively strong and microfibrous oxide structure, as discussed above.

Finally, in any discussion on bonding copper, mention must be made of the strong bonding that occurs between sulphur-containing natural rubber and copper alloys, e.g. brass, which develops during vulcanization. This has been known for over a century and extensive use is made of this technology in steel-belted radial tyres. The reinforcing steel cords are coated with a layer of brass, 0.1 to 0.2 μm thick, and the bond is formed during curing of the tyre. Our understanding of the mechanisms involved has increased greatly in recent years due to the elegant work by van Ooij and colleagues [123–126]. Essentially, the mechanism is unique in that during the crosslinking of the polymer by the sulphur an interphase of reaction products is formed which act as the 'adhesive' to the metal. This interphase region consists of ZnO, ZnS and non-stoichiometric copper sulphide. The exact type of bonding between the copper sulphide surface and the cured rubber is still uncertain and chemical bonds of the type rubber–S–Cu have been suggested, but no definitive evidence has been produced to support the proposal. van Ooij believes that the bonding is probably due to an 'interpenetrating network' type of structure, with the rubber molecules penetrating the porous sulphide layer and then crosslinking. The intrinsic adhesion established is extremely strong and the interfacial strength far exceeds the tear strength of the rubber, hence this technique has not so far been displaced by the use of conventional adhesives or primers.

4.3.4.3 Steels

Steels are frequently bonded with no, or a minimum of, pretreatment being employed and good, reproducible initial joint strengths are readily obtained using a combination of degreasing/mechanical abrasion techniques, as shown above. Also, this combination of pretreatments may even give a reasonable service life [127] when the joint is exposed to normal weathering conditions, especially if the applied stress levels are low and the bonded area relatively large. However, to obtain the longest service life some form of chemical pretreatment and/or primer (Section 4.3.5) is inevitably necessary.

In the case of mild and alloy steels the use of phosphate coatings [126,128–131] to impart corrosion protection has been employed since about the beginning of the present century. The function of the phosphated layer is to reduce the corrosion of the steel by providing a mechanical layer against aggressive species and such layers do not have any electrochemical or inhibiting properties. From the author's experience the use of phosphate coatings as a pretreatment for steels appears to offer no significant advantages compared to degreasing/abrasion, both with respect to initial joint strength and subsequent joint durability. Recent work by Brockmann [127] confirms these observations. Indeed, the author's work suggested that, unless the phosphating process was controlled, a thick phosphate layer would develop. Whilst this is possibly good for corrosion protection of the underlying steel, it was mechanically weak and fracture of the joint occurred at low loads through this layer.

A similar problem may occur when bonding to galvanized steel. Here the corrosion protection for the exposed steel is afforded by a layer of zinc. The enhanced corrosion protection comes from the metallic zinc being more resistant to natural weathering than the steel and this feature is related to the nature of the corrosion products formed by the zinc and steel. In such conditions, zinc forms a dense layer of hydroxide salts which adhere well to the underlying zinc and have low solubilities in water, and therefore essentially passivate the metal: on the other hand iron or steel corrode because the corrosion products formed are porous, brittle and non-adherent. Various types of galvanizing processes are employed in industry and some result in relatively thick layers which are mechanically weak and so, upon adhesive bonding with modern tough and strong adhesives, again provide a path for joint fracture at low fracture loads.

Considering stainless steels, the ultimate performance of joints from these substrates is observed when the stainless steel is chemically pretreated [132–135]. A typical pretreatment consists of degreasing and water rinsing, followed by etching in sulphuric acid at 60 °C, water rinsing, 'de-smutting' in chromic acid at 60 °C, water rinsing and drying. The de-smutting stage is essential if high joint strengths are to be realized since the sulphuric acid etch, and similar acid etches, leave behind a deposit of carbon on the surface, known as 'smut', which will act as a weak boundary layer if not removed. In the case of stainless steel very little is known about the detailed mechanisms whereby a treatment schedule, such as the typical one given above, leads to improved joint performance; especially as to why the environmental resistance of the interfacial regions is far greater compared to that observed when an abrasion pretreatment is employed. One of the reasons for this lack of understanding of the mechanisms involved is that the effectiveness of the pretreatment appears to be very sensitive to the manufacturing path used for the steel [133–135], as may be seen from the data shown in Fig. 4.16. The basic cause for this observation possibly arises from the level of martensitic structure present in the surface regions compared to the level of austenitic structure. The relative amounts of these structures is influenced by the manufacturing path and appears to control the rate of etching in the acid bath, and hence the topography of the etched substrate. The presence of martensitic structure leads to a faster etching rate and rougher surface on the etched substrate and, also

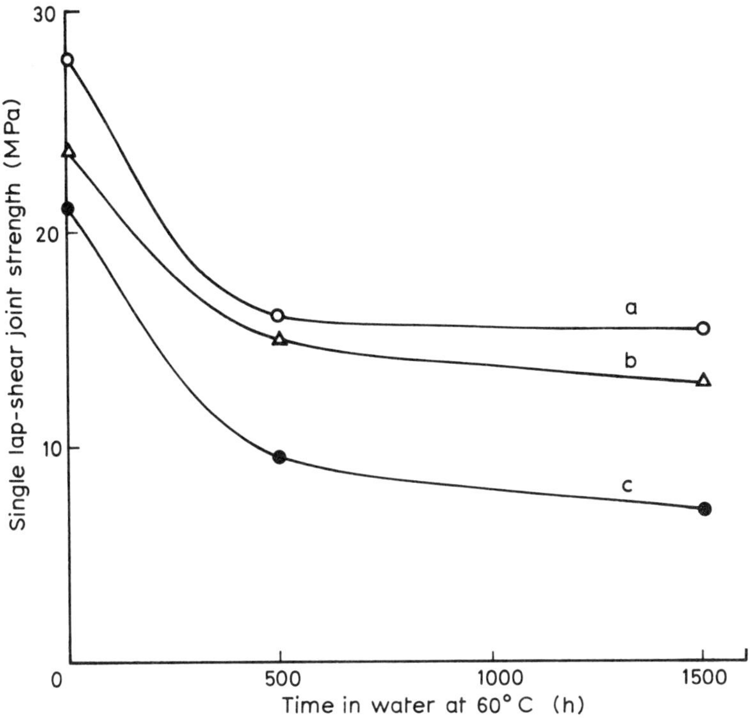

Figure 4.16 Single-lap-shear strength of joints prepared from stainless steels as a function of immersion time in water. The stainless steels had nominally the same bulk chemical composition but had followed somewhat different manufacturing paths, as shown below. The steels were etched in sulphuric acid and de-smutted in chromic acid prior to bonding [133].

	Melting	*Rolling*	*Flattening*	*Heat treatment*	
				1st	*2nd (in air)*
(a)	In air	Yes	Roller level	Air anneal + pickle	240–260 °C
(b)	In vacuum	No	None	Bright anneal (NH_3)	340–360 °C
(c)	In air	Yes	Stretch flatten	Bright anneal (NH_3)	340–360 °C

one which has a somewhat different chemical composition. In Fig. 4.16 the steel which had been manufactured by route (a) possessed a mixed austenitic/martensitic structure, with steels manufactured by routes (b) and (c) possessing only austensitic structures. However, whether the presence of the martensitic structure, and associated surface characteristics, were truly responsible for the different durabilities which were observed was not conclusively established.

4.3.4.4 Titanium and its alloys

The adhesive bonding of titanium and its alloys has primarily been developed and employed by the aerospace community. An abrasion/degreasing pretreatment, typically based upon grit-blasting with alumina particles, often results in both a very good initial strength (see Table 4.14) and an adequate joint durability [136]. However, for the longest service life upon exposure to moisture some form of chemical pretreatment is invariably necessary [137,138].

Various chemical treatments have been proposed for pure titanium and the 6Al–4V–Ti alloy and they are summarized in Table 4.15.

Table 4.14 Effect of surface pretreatment on the initial strength of 6Al–4V–Ti alloy bonded with a hot-cured epoxy adhesive [136]

Pretreatment	Lap-shear-strength (MPa)	90° Peel energy (kJ/m^2)
Grit blasting	54.7	5.1
Chemical:		
Phosphate fluoride etch	49.7	5.4
Alkaline peroxide etch	49.4	5.5

As usual, the above chemical treatments have been largely developed empirically and our understanding of why different treatments lead to such different durabilities upon exposure of the joint to moisture is limited, but will be reviewed in Chapter 8.

4.3.4.5 Aluminium and its alloys

Many applications involving the adhesive bonding of aluminium and its alloys employ no pretreatment or just simply a degreasing, and possibly an abrasion, treatment for the aluminium substrate. However, whilst such approaches to pretreatment will often yield adequate values for the initial joint strength, the highest joint strengths with the lowest coefficients of variation are usually observed when a chemical pretreatment is employed [2,147]. This is demonstrated by the results shown in Table 4.16 which reveal that grit-blasting is markedly inferior to chemical pretreatments for obtaining high and reproducible initial strengths. This should be contrasted to the bonding of steel and titanium substrates where, for initial joint strength, a grit-blasting pretreatment is one of the best treatments. Thus, it appears that for ease of obtaining good initial joint strengths, aluminium substrates lie between materials like copper and magnesium (Section 4.3.4.2) at one end of the spectrum and the steels (Section 4.3.4.3) and titanium (Section 4.3.4.4) at the other end. Whether this

Table 4.15 Surface pretreatments commonly used for bonding titanium and its alloys

Pretreatment	Initial joint strength	Joint durability	References
Mechanical abrasion			
Hand abrasion	Adequate	Poor	137
Alumina grit-blasting	Good	Adequate	136
Chemical			
Alkaline cleaners	Poor	Poor	137
Commercial alkaline etch ('TURCO' 5578)	Adequate	Adequate	3
Nitric-hydrofluoric acid etch	Adequate	Poor	137
Phosphate–fluoride etch	Adequate	Poor	139,140
'Modified' phosphate fluoride etch	Adequate	Poor (but a slight improvement on the above)	136,140
Commercial nitric–chromic acid mixture with fluorides ('Pasa Jell' 107)	Adequate	Adequate	137
Alkaline–peroxide	Good	Good	136,141
Mechano-chemical			
VAST process†	Good	Poor	137,142, 143
Electrochemical			
Anodizing in chromic acid–fluoride mixture	Good	Good	144
Anodizing in alkaline peroxide solutions	Good	Good	145
Cathodic deposition from non-aqueous solutions of aluminium nitrate	Adequate	Adequate	137,146

† VAST stands for Vought Abrasive Surface Treatment and in this process the substrate is blasted in a specially designed chamber with a slurry of fine abrasive containing hydrofluorosilic acid.

Table 4.16 Effect of pretreatment on the initial single lap-shear strength of clad aluminium alloy bonded with various two-part adhesives [147]

Treatment	Cold-cured epoxy		Cold-cured acrylic	
	Strength (MPa)	Coefficient of variation (%)	Strength (MPa)	Coefficient of variation (%)
Solvent wipe	11.7	44.7	11.3	30.1
Grit blasting	12.0	14.1	25.5	16.4
Chromic acid etching	24.8	13.0	31.0	3.2
Chromic acid anodizing	24.1	9.7	31.0	2.4

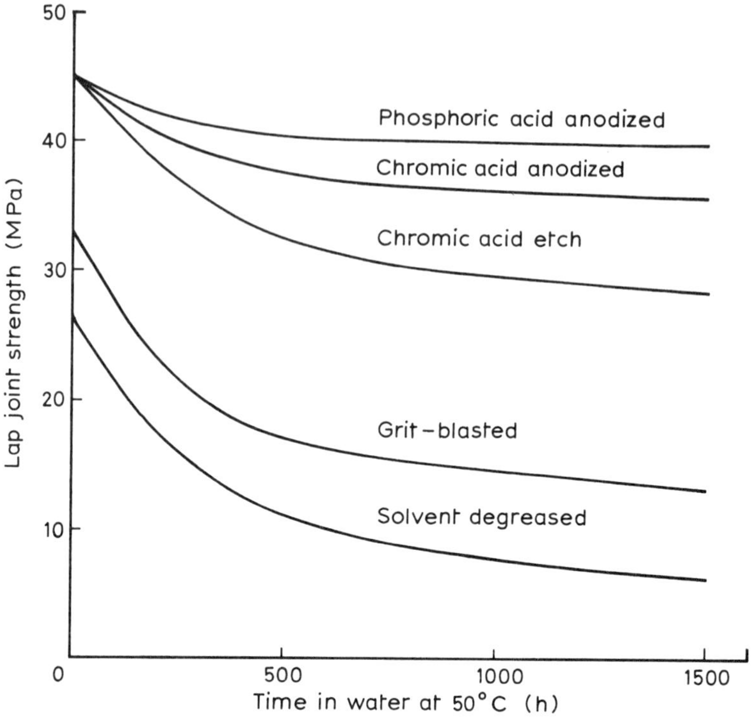

Figure 4.17 Effect of surface pretreatment on the durability of aluminium alloy/toughened epoxy joints subjected to accelerated aging in water at 50 °C.

arises from increases in the wettability and intrinsic adhesion to the chemically generated oxides or to such oxides being inherently stronger has not been conclusively resolved.

Considering the aspect of joint durability, then as might be expected from the above observations, aluminium components which are bonded are particularly prone to environmental attack upon outdoor weathering unless some chemical pretreatment has been employed. Such a pretreatment is often followed by the application of a specially formulated corrosion-inhibiting primer, designed to enhance further the environmental resistance of the interfacial regions. This feature is emphasized by the results shown in Fig. 4.17 where the effect of various pretreatments upon the initial and long-term strength of epoxy/aluminium alloy joints is given; note that grit-blasting is a particularly poor pretreatment for ensuring a long service life from bonded aluminium and its alloys.

The pretreatments shown in the above examples are all in common use in the aerospace industries and the chemical treatments actually involve a number of steps, with the treatment being identified by the distinguishing chemical etch or anodizing step. Again, the reader is referred elsewhere [3,6,148–151] for complete details of the actual processes but they basically consist of:

(a) *Chromic acid etch:* following degreasing, the substrate is etched for 15 to 30 min in chromic acid at 60 to 68 °C;
(b) *Chromic acid anodizing:* the above chromic acid etch is carried out and the substrate is then anodized in a chromic acid solution (4–5% CrO_3) at 40 °C. Eight 5 V steps in voltage are applied within 10 min; the substrate is then held at 40 V for 20 min; the voltage is then increased to 50 V within 5 min; and this level is maintained for 5 min.
(c) *Phosphoric acid anodizing:* the above chromic acid etch is carried out and the substrate is then anodized at 10 V for 25 min at 20 to 25 °C in an aqueous solution containing 10% by weight of 85% orthophosphoric acid in distilled water.

In all cases the final steps are washing in cold running tap water for 20 min and drying in warm air. Note that in all cases the anodizing processes do not involve 'sealing' the oxide, i.e. exposing the anodized oxide to hot water. This would seal the pores of the oxide and prevent penetration of the adhesive or primer. Also, the exact details for the above common types of chemical treatments do vary somewhat according to which agency issued the specification, but they have all been developed via empirical testing and apparently minor changes may greatly reduce the effectiveness of the treatment, so the inexperienced technologist diverts from the specification at his peril. For example, in the case of chromic acid etching, using distilled water can lead to a thicker and weaker oxide on the substrate which then acts as a weak boundary layer and leads to poor joint strengths [152–155]. The presence of soluble cations in tap water apparently prevents the growth of such a layer. However, too high a concentration of foreign ions in the final rinse water may also adversely affect the treated surface and the optimum value is in the range 50 to 200 ppm of inorganic salts [155].

The use of adhesives for joining aluminium and its alloys in general engineering is increasing rapidly and such applications often require a long service life from a joint which will be exposed to aqueous environments and, at present, this leads to the requirement for some form of surface pretreatment to be employed. Many general engineering companies that wish to use adhesives technology are often reluctant to undertake such complex chemical pretreatments and there is, therefore, an urgent need for alternative pretreatment procedures to be developed. Two such lines of development are the use of primers, which will be discussed in the following sections, and the supply of grades of material which have already been pretreated for adhesive bonding, with the specific aim of ensuring good joint durability. This latter approach is the subject of a recent patent application [156]. In this patent the use of either a conversion-coating process or the deposition of fine silica particles coated with chrominium trioxide for pretreating various aluminium alloys is described. However, a key aspect is that the pretreated surface may then be covered with a protective press-lubricant, chosen so that it may be displaced or dissolved by the adhesive during the bonding operation. The protected, pretreated material may now be stored, transported, etc. and is still 'active' for up to about six months from the time it was pretreated. A second important feature is that such a treated aluminium alloy may be cut and formed into complex shapes without significantly damaging the pretreated surface and reducing the subsequent joint performance. Thirdly, the oxide layer generated by this pretreatment is sufficiently thin still to enable spot welding of the material; this can be a useful method oi jigging the component during an assembly operation. However, the thinness of the oxide layer, compared to those generated by traditional aerospace treatments, may well result in inferior corrosion protection for the aluminium. Also, from the data given in the patent, whilst the pretreatments described do impart quite good joint durability they are inferior in this respect to the best of the aerospace pretreatments, as may be seen later in Section 8.3.2.3. (It is noteworthy that in the patent application the results using the phosphoric acid anodize pretreatment, which are included for comparative purposes, are significantly inferior to what one might expect: possibly because the adhesives employed are highly viscous and could not penetrate and completely wet the deep pores of the anodized surface. In aerospace applications a primer is always employed with the phosphoric acid anodized pretreatment which overcomes this problem.)

Again, the underlying reasons for the environment resistance of adhesive joints consisting of aluminium and its alloys being so highly dependent upon the choice of surface pretreatment are more conveniently discussed in Chapter 8.

4.3.5 Primers

4.3.5.1 Introduction

The use of primers to pretreat high surface energy substrates prior to adhesive bonding is becoming of increasing importance in industrial applications. The main reasons why a primer is applied as either the only form of pretreatment or, more typically, as the final stage in a multistage pretreatment schedule are:

(i) To improve the performance of the bonded component. Primers that have several different types of function fall into this category:

 (a) Some adhesives, e.g. high-temperature polyimides, often possess poor wetting characteristics, due to their viscosity never reaching a relatively low value during the bonding operation. Applying the adhesive in a solvent is undesirable since complete solvent removal from the adhesive in the joint is virtually impossible and this leads to the mechanical properties being adversely affected. Therefore, a primer is formulated which is basically a solvented version of the adhesive but which will now completely wet the substrate and can then be cured to the required extent and the solvent removed prior to the (undiluted) adhesive being applied. The opportunity is often taken to add ingredients to the primer to improve such aspects as thermal stability and environmental resistance (see below).

(b) When using certain adhesives or working with 'difficult to bond' substrates the role of the primer is to establish strong interfacial forces to both the adhesive and substrate. Examples would be the use of silane-based primers to ensure adequate adhesion between silicone rubber adhesives and a variety of substrates and the use of nylon- or benzotriazole-based primers to obtain good bonding to copper substrates.

(c) As might be expected from the above discussions, a common use of primers is to increase the service life of the joint, rather than simply increase the initial strength. This is typically achieved by establishing strong and moisture-resistant interfacial bonds and protecting the substrate's surface regions from hydration and corrosion, and so forming an *in situ* weak boundary layer. Phenolic- and silane-based primers are frequently employed to attain this goal and are often formulated together with corrosion-inhibiting species such as chromate compounds.

(d) With the older more brittle adhesives a lower-modulus tough primer was sometimes used to increase the peel strength of the joint. However, with the advent of the newer tough adhesives the need for this type of primer has largely disappeared.

(ii) To increase production flexibility in the bonding operation. After pretreating a high-energy substrate the active surface will readily adsorb atmospheric contamination, and after a certain 'surface exposure time' this may lead to inferior joint performance especially with respect to durability [157–160]. The allowable surface exposure time may be comparatively short, often only a matter of a few hours in the case of chemically pretreated substrates. To overcome the problems this causes in an industrial production line it is a common practice to apply a primer to the pretreated surface within a few hours, usually within three or four hours. Such primers are air- and/or oven-dried and usually form non-tacky or only slightly tacky films so that the substrate parts can be handled and stored conveniently. The actual application of the adhesive may then be delayed for up to several months. The primers are typically based upon epoxy polyurethane or phenolic materials and they are often formulated so that they also assist in ensuring good environmental resistance to the bonded component.

4.3.5.2 Silane-based primers

Silane-based primers were first developed in the 1940s and 1950s to pretreat glass fibres for use in fibre composites in order to increase the environmental stability of the glass/polymeric matrix interface and so improve the water resistance of these composite materials. The structure of the silanes may be represented by the general structure $Y(CH_2)_nSiX_3$, where $n = 0$ to 3, Y is an organofunctional group usually selected for reactivity with a given matrix or adhesive and X is a hydrolysable group on silicon. Silane-based primers have been used to increase the performance of interfaces in applications as diverse as fibre composites [e.g. 161–163], paints [e.g. 164–169], adhesives [e.g. 161,162], mineral-filled polymers [e.g. 161], coatings [e.g. 161,162] and microelectronics [e.g. 170–172]. Some representative silane-based primers which are commercially available are shown in Table 4.17.

Table 4.17 Representative commercial silane-based primers [161,162]

Organofunctional group	Chemical structure
Vinyl	$CH_2{=}CHSi(OCH_3)_3$
Chloropropyl	$ClCH_2CH_2CH_2Si(OCH_3)_3$
Glycidoxy	$\underset{\diagdown O \diagup}{CH_2CH}CH_2OCH_2CH_2CH_2Si(OCH_3)_3$
Methacrylate	$CH_2{=}C(CH_3)COOCH_2CH_2CH_2Si(OCH_3)_3$
Primary amine	$H_2NCH_2CH_2CH_2Si(OC_2H_5)_3$
Diamine	$H_2NCH_2CH_2NHCH_2CH_2CH_2Si(OCH_3)_3$
Mercapto	$HSCH_2CH_2CH_2Si(OCH_3)_3$
Cationic styryl	$CH_2{=}CHC_6H_4CH_2NHCH_2CH_2NH(CH_2)_3Si(OCH_3)_3 \cdot HCl$

As mentioned above and in Section 3.5.4.2, the most common reason for employing silane-based primers is to increase the durability of adhesive joints. Such primers have been used to achieve this goal on substrates as diverse as glass, silica, silicon, copper, steels, aluminium alloys and titanium alloys, and their role in increasing the service life of adhesive joints is discussed in detail in Chapter 8. However, an example of their use to increase significantly the initial strength of an adhesive joint comes from the work of Sung *et al.* [173]. These workers used γ-aminopropyltriethoxysilane (see Table 4.17) to increase the peel strength of a model hot-melt adhesive, polyethylene, to single-crystal alumina (α-Al_2O_3) plates. In Fig. 4.18 the peel strength is shown as a function of concentration of the γ-aminopropyltriethoxysilane solution, applied from an aqueous solution at a pH of 10.4 and dried at 25 °C under vacuum.

Obviously to understand the role of the silane primer in this application, and in those to be discussed in Chapter 8, requires the type of interfacial bonding and the structure of the silane to be identified. This is an extremely complex area; for example, the solvent [174], the pH [175,176] and the drying conditions [177,178], as well as the type of silane and substrate, are known to affect the type of silane/substrate bonding and the structure of the silane primer film that is formed. The reader is referred elsewhere [161,162] for a detailed review of this subject. These reviews reveal that the full range of modern surface analytical techniques, including laser–Raman, Fourier transform infrared, inelastic electron tunnelling and carbon-13 nuclear magnetic resonance spectroscopy, have been employed to gain a firmer understanding of how silane-based primers function.

The silane primer is usually applied from dilute solution, typically from either water or a water/ethanol mixture. It is generally considered that for the optimum results in terms of joint performance the silane primer should first hydrolyse, preferably prior to being applied to the substrate surface [179], to form the trihydroxy silanol. Once on the substrate it should undergo a condensation reaction to polymerize to form a crosslinked polysiloxane, and therefore not act as a low molecular weight weak boundary layer, and should chemisorb onto the substrate surface, and so establish good intrinsic adhesion. This is shown schematically in Fig. 3.14. When primary interfacial bonds are established they are generally considered to be mainly covalent in nature, but with a degree of ionic character, or involve the formation of a molecular complex at the interface [180–185]. This type of surface reaction has led to the term 'coupling agent' for these primers, and this term is also used for primers which are thought to react in a similar manner. The silane is generally adsorbed via the formation of one or two $-Si-O-$substrate bonds (see Figs 3.12 and 3.14) via condensation reactions between the $-SiOH$ groups and hydroxy groups on the substrate, and the silane molecule may be orientated parallel or perpendicular to the substrate surface (cf. Figs 3.12 and 3.14). The silane molecule is also sometimes adsorbed onto the substrate tripod-like [186], via the three alkoxy groups, and may even be adsorbed via the organic functional group. The detailed reasons for these different types of configuration of the adsorbed silane are not completely understood but depend upon the manner in which the primer is deposited. It has been suggested [181,187,188] that for some silanes these polymerization and interfacial reactions may proceed more rapidly in the presence of catalysts, e.g. acetic acid or *n*-propylamine.

Experimental evidence [116,163,174] has shown that typically the silane primer layer is not a monolayer but is about 2 to 100 nm thick, and that the thinner layers are deposited from the solutions containing lower concentrations of silane and those using the more polar solvents. Further, the polysiloxane films are not always continuous but may form agglomerates [174,189]. This feature may be responsible for the initial increase in joint strength with increasing concentration of silane solution shown in Fig. 4.18; a complete coverage of silane primer only being formed at about 2% v/v concentration. Further, whilst there is often a thin, chemisorbed layer of crosslinked poly-

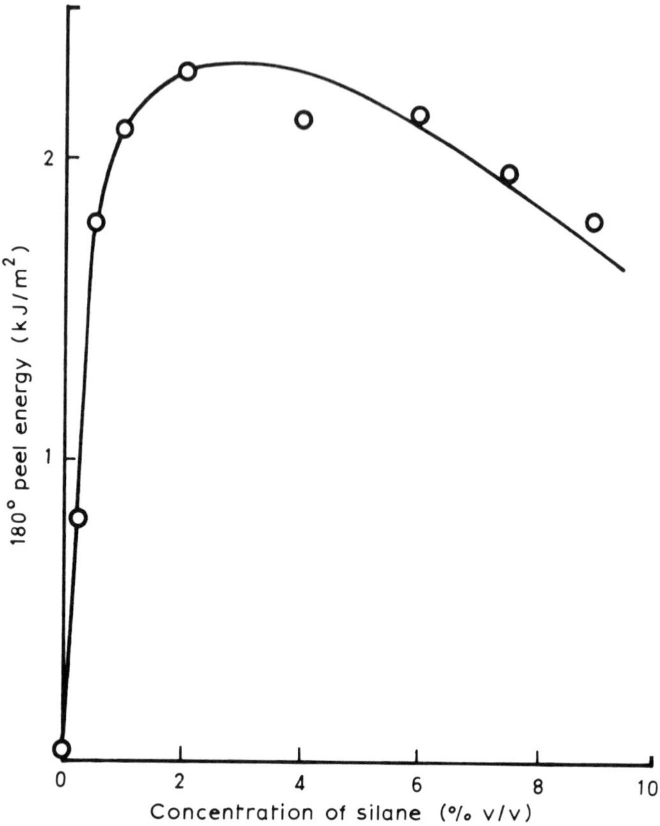

Figure 4.18 Effect of concentration of γ-aminopropyltriethoxysilane in an aqueous solution used as a primer to pretreat sapphire on the 180° peel energy of an α-Al_2O_3/polyethylene laminate [173].

siloxane network near the substrate surface, the outer regions of the polysiloxane layer appear to consist of oligomers which are mainly physisorbed and have an open porous structure [174,190,191]. The formation of a relatively thick, low molecular weight polysiloxane, through which joint failure occurs, is probably responsible for the decrease in peel strength with the higher silane concentration shown in Fig. 4.18.

Turning to the silane primer/adhesive interface, three mechanisms of adhesion have been suggested. Firstly, the classical explanation is, of course, that the functional group on the silane reacts with the adhesive and there is evidence [192,193] that this can indeed occur. For example, in the use of γ-aminopropyltriethoxysilane as a primer for germanium crystals prior to coating with a polyimide it has been found that the amino group on the silane reacts with the polyamic acid group in the coating (a precursor of the final polyimide) to give an interfacial amide linkage, which on further heating yields an interfacial imide linkage. Secondly, the polysiloxane layer has an open porous structure and the liquid adhesive may penetrate this and then harden to form an interpenetrating network interphase region. Thirdly, the work of Sung *et al.* [173], shown in Fig. 4.18, uses a polyethylene adhesive with little capability for reacting with the polysiloxane primer layer and the available evidence suggested that limited intermolecular diffusion had occurred across the polyethylene/polysiloxane interface. This was based on a slight depression of the melting point of the polyethylene because of silane miscibility, and measurement of silicone distribution across the interface using energy dispersive X-ray analysis. Apart from the mechanisms of adhesion given above, the organic functional group on the silane primer might influence the effectiveness of the primer via two other mechanisms. Namely, the importance of the organic functionality might be (i) to impart a structure that prevents the silane primers from preferential cyclizing and forming intramolecular hydrogen bonds, as opposed to polymerizing and chemisorbing on the substrate, and (ii) to act as a catalyst for the formation of the crosslinked siloxane polymers on the substrate and for the chemisorption reactions. Nevertheless, whatever the underlying mechanisms, the technological importance of selecting the 'right' organic functionality for giving the best joint performance is well established.

4.3.5.3 Other primers

Silane-based primers are by far the most widely used organometallic primers, but others based upon organotitanates [162,194–197], organozirconates [197, 198], organozircoaluminates [199] and chrome complexes [200,201] have also been reported and find some application in adhesives technology. The organotitanates have the general formula $YOTi(OX)_3$ where Y is typically an isopropyl group and X is a larger group such as a stearate; X may contain a functional organic group. It is generally considered that they form interfacial primary bonds to the substrate via a reaction with surface protons:

$$YOTi(OX)_3 + \text{substrate–OH} \rightarrow \text{substrate–O–Ti}(OX)_3 + YOH \qquad (4.1)$$

The organotitanates appear to be very effective in increasing the ease of dispersing particulate fillers in resins and polymers, by preventing agglomerates from forming and reducing the viscosity and in increasing the subsequent adhesion of the filler to the matrix. They have not, however, found wide use in adhesives technology but are used in the manufacture of some film laminates. The zirconate-based primers are thought to react with the substrate in a similar manner and are used commercially as primers for polyolefin films prior to printing. The zircoaluminates, like the titanates, currently find their main application as primers for coating fillers. Turning to the chrome complexes, the most common commercially available type is the aqueous solution of chromium (III) fumarato coordination compound, in which Cr^{3+} ions are linked by fumaric groups and hydroxy bridges to form inorganic polymers which have complex structures. It is widely used as a primer for aluminium foil to increase the strength and durability of aluminium/polyethylene interfaces. The polyethylene is typically extrusion coated as a hot-melt onto the primed aluminium for applications as diverse as packaging and communication cable shielding. Using scanning Auger and X-ray photoelectron spectroscopy Yang [201] has shown that the chromium (III) fumarato primer is not present simply as a monolayer but penetrates into the oxide on the aluminium where it establishes primary bonds to the aluminium oxide. As in the work of Sung *et al.* [173] described above, it was sugggested that the mechanism of adhesion at the primer/polyethylene interface was one of interdiffusion of compatible segments of the primer with segments of polyethylene chains.

Other types of primers that have been employed, and which were referred to in Section 3.5.5, are those based upon mercaptoesters and benzotriazole compounds [202–204]. The former type are polyfunctional chelating agents, e.g. pentaerythritol tetrathioglycolate and pentaerythritol tetramercapto-propionate, and have been used on steel substrates. After first pretreating the steel by abrasion and immersion in ammonium citrate, such primers lead to strong adhesion forces by reacting with metal ions in the oxide to form organometallic chelate ring structures. However, as might be expected, they generally offer little improvement in initial strength but may significantly increase the durability of the joint upon exposure to moisture, as discussed in more detail in Chapter 8. Various benzotriazole compounds, containing chemical groups reactive towards potential adhesives, e.g. ϵ-aminobenzotriazole, have been described for pretreating copper, silver and cadmium surfaces. They form a polymeric complex with the metal surface via the benzotriazole group and it is assumed that the amino group, for example, reacts with the adhesive. With these ‘difficult to bond’ metals the use of such primers may not only increase the joint durability, again as discussed in Chapter 8, but may also significantly increase the initial strength of the joints, as shown in Table 4.18.

The primers given in Table 4.18 were also shown to be effective when bonding copper with other adhesives and to polymeric films, such as fluorocarbon polymers, polyolefins, polyester and polyamides, to produce laminates.

Another type of primer which has already been mentioned are those especially formulated to be applied to chemically treated aluminium alloys in particular, with the main aim of further improving the durability of joints prepared from these pretreated substrates [e.g. 205–209]. They are commonly referred to as ‘corrosion inhibiting primers’ and are proprietary formulations which typically contain phenolic, epoxy and chromate species. They are used extensively in modern aircraft structures. They will be discussed in greater

Table 4.18 Effect of benzotriazole primers on the strength of epoxy adhesive/copper joints [204]

Primer	Peel energy (kJ/m^2)
None (abraded and solvent cleaned)	0.74
5-[2,4-*bis*-[di-(β-hydroxyethyl)amino]-*s*-triazine-6-ylamino]-benzotriazole	1.4
None (dendritic surface†)	1.5
As above plus:	
5-[2,4-*bis*-[di-(β-hydroxyethyl)amino]-*s*-triazine-6-ylamino]-benzotriazole	1.8
5-(ϵ-aminocaproylamine)-benzotriazole	1.9

† Typical electrolytic pretreatment used for copper when producing rigid and flexible laminates for the electronics industry, see Section 3.2.4.

detail in Chapter 8 but essentially they must establish stable interfacial bonds to the oxide and contain enough corrosion-inhibiting pigment to provide adequate corrosion protection for the underlying substrate. They are usually partially cured at elevated temperatures to give a tack-free surface and the primed substrate can then be handled and stored until the adhesive bonding operation.

Finally, mention should be made of the concept of a 'sacrificial' primer which enables metals to be readily bonded underwater [210,211]. From the comments in Section 2.6.2 it is obvious that when trying to apply an adhesive to a substrate underwater it will be difficult for the adhesive to completely displace the water adsorbed on the surface and establish interfacial contact with the substrate; a layer of water would remain adsorbed onto the substrate and act as a weak boundary layer. The idea of the sacrificial primer is that the substrate is firstly cleaned, typically by an abrasion pretreatment, but simultaneously a primer is applied which is formulated so that it displaces any water and is more readily adsorbed onto the substrate surface. This primer, although water repellant, is also formulated so that it is compatible with the adhesive to the extent that it, in turn, is displaced and dissolved by the adhesive when it is applied. Thus, the concept of a sacrificial primer enables the adhesive to wet the substrate completely without a weak boundary layer of water being retained at the interface. The adhesive is also especially formulated to aid these processes and contains silanes which will diffuse to the interface and establish strong and stable interfacial bonds to ensure a long service life for the joint in its underwater environment [212].

4.3.6 Plasma treatments

As discussed in Section 4.2.7 a plasma is an excited gas which consists of atoms, molecules, free electrons and metastable species and, whilst the use of plasmas to pretreat polymeric substrates is well established, very little has been reported on the use of plasmas in the treatment of glasses, ceramics and metals prior to adhesive bonding.

In the case of high-energy substrates the major effects of the plasma can include the following and, of course, several may occur simultaneously.

4.3.6.1 Cleaning and etching

The excited species may have sufficient energy to displace low molecular weight contamination which is present on the surface. The displacement process may be aided by the plasma first degrading, and then reducing the molecule weight of, the typical organic contamination. The plasma may also cause etching of the surface, which will aid the cleaning process and increase the surface rugosity. O'Kane and Mittal [213] have studied the surface cleaning efficiency of glow-discharge plasmas of argon and helium–oxygen with respect to rhodium and iron–cobalt alloys. They reported that a plasma pretreatment was more efficient than solvent cleaning, especially for the removal of sulphur and carbon contamination. This observation is confirmed from experiments where aluminium alloy has been subjected to a corona-discharge pretreatment [214] and then bonded with an epoxy primer/adhesive combination. The joint strengths from corona-treated alloys were significantly higher than those using just a solvent wipe, but were of approximately the same level as those which employed an abrasion treatment. Further, it was reported that corona-treated surfaces gave joints with poorer durability to moisture. From previous discussions these results imply that corona-treated surfaces lead to initial joint strengths significantly lower than those from chemical pretreatments (Section 4.3.4.5) and, obviously, would lead to a very short service life of the joint upon exposure to moisture. In the case of stainless steel substrates, Baker [215] has shown that plasma cleaned surfaces give lower joint strengths than a standard solvent-degreasing and grit-blasting pretreatment.

4.3.6.2 Polymerization of a 'primer' onto the substrate's surface

The plasma may induce the polymerization of a constituent of the gas phase and so cause deposition of a thin polymeric film, or primer, onto the substrate. Several workers have explored this route for depositing a primer, to which another polymer is subsequently adhered, or for depositing a protective coating directly onto a metal substrate [216–223]. For example, Nichols *et al.* [217] have deposited polymeric primer films, about 30 nm thick, from glow-discharge polymerization of methane tetrafluoroethylene or tetramethyldisiloxane onto platinum wires. These primers were then coated with a few-micron-thick layer of either poly(*p*-xylylene) or poly(chloro-*p*-xylylene), which are good electrical insulators in thin-film applications provided this coating

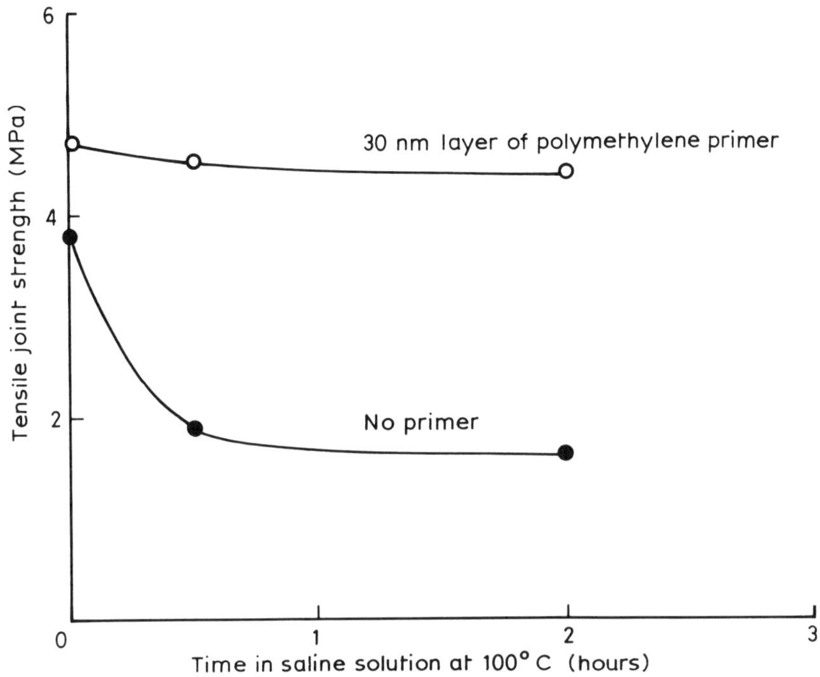

Figure 4.19 Effect of depositing a plasma polymerized primer of polymethylene onto a platinum substrate prior to coating with poly(chloro-*p*-xylylene) [217].

does not become debonded from the underlying metal substrate. The insulated platinum wires require a long service life in a saline environment as they are intended for implanting into the human body. It was found that the presence of the plasma-polymerized primer greatly increased the environmental resistance of the coated platinum and so greatly extended the service life, as shown in Fig. 4.19. Schreiber *et al.* [222,223] have studied the microwave plasma polymerization of organosilicones onto metallic substrates. They found that for optimum environmental stability of the coated interface the substrate should be heated above a certain critical temperature during the deposition and polymerization processes. The reasons for this were not completely clear but are related to the structure of the resulting plasma-polymerized coating and the nature of the interfacial reactions. Whether a suitable organometallic primer could have been simply painted or sprayed onto the substrates in the above experiments to achieve similar results is unclear, but the plasma polymerization of the primer can offer some advantages such as better precleaning of the substrate, good control of the thickness of the primer layer and, possibly, increasing the likelihood of interfacial reactions occurring. However, it is obviously more expensive and usually less convenient in a production environment and therefore definite advantages need to be established before the adhesives technologist will adopt such a pretreatment method.

4.3.6.3 Ion implantation

The plasma may cause foreign atoms to be implanted in the surface regions of the substrate [224,225]. The ion to be implanted may be directly implanted into the surface. Alternatively, a thin film may be coated onto the substrate surface and inert ions of argon or nitrogen, for example fired at the surface and used to drive atoms in the coating into the substrate. This latter technique is known as recoil implantation or ion beam mixing. Yasuda *et al.* [224] have described these concepts as 'atomic interfacial mixing', and this was briefly referred to in Section 3.3.6. For example, if an atomic species such as C^+ is implanted into the surface regions of a metallic substrate, then such C^+ species should be capable of establishing primary covalent bonds to a deposited film of a plasma-polymerized organic polymer, thereby ensuring good intrinsic adhesion which may also be resistant to rupture by ingressing moisture. Indeed, in the small number of experiments so far conducted this concept appears to be valid. When a plasma polymer of methane was deposited onto platinum both the initial adhesion and the environmental resistance of the polymer/platinum interface were increased if C^+ ions had first been implanted into the metal's surface regions; implanting Ar^+, Kr^+ or Si^+ did not show any improvements over platinum substrates which had not been subjected to ion implantation.

4.3.6.4 Plasma spraying

This is a very well-established process for coating metals with metals and ceramics to attain specific surface properties [226]. Typically powder (particle size 20–100 μm) of the material to be sprayed is injected into a plasma jet, e.g. of Ar/H_2, and is rapidly melted and accelerated to produce a stream of molten particles which are projected onto the substrate. Large temperature and velocity gradients exist in the jet with peak values ranging up to 10000 K and 1000 m/s, respectively. The velocity of the particles impinging on the substrate ranges from 100 to 400 m/s and the thermal conditions are therefore such that refractory materials like ZrO_2, Al_2O_3, molybdenum or tungsten may be deposited successfully. For example, tungsten carbide may be plasma sprayed onto steel components to give a much more wear-resistant surface for application temperatures up to about 540 °C. In the context of the present discussions the interest centres upon whether a particular coating could be selected to attain better initial joint strengths and improved joint durability when bonding high-energy substrates. The question remains unanswered. However, some exploratory work by the author has indicated that under the correct spraying conditions the coating may adhere firmly enough to the underlying substrate (mild steel or aluminium alloy) but that the common spray coatings which were employed did not offer any better durability for epoxy bonded joints than when a simple degreasing/grit-blasting pretreatment was used for the substrate.

4.4 CONCLUDING REMARKS

This chapter has clearly demonstrated how complex is the subject of surface pretreatment for the substrate materials prior to adhesive bonding. In many applications little or no surface pretreatment is necessary prior to application of the adhesive but usually to obtain the maximum properties from the bonded component some form of pretreatment is necessary. Obviously, the performance required from the bonded component must be weighed against any economic penalties which will be incurred by selecting a given pretreatment procedure.

The treatments used in industry have nearly all been developed empirically and our fundamental understanding of the mechanisms by which they function has developed slowly. Although, more recently, considerable advances have been made in this area, largely due to advances in sensitive surface analytical techniques which have enabled the substrate surfaces to be characterized in detail. The current understanding has been reviewed in this chapter and as our knowledge grows it is to be hoped that improved pretreatments will be developed. Such pretreatments may lead to better joint performance or, equally importantly, to more economic routes to the performance that currently is only achievable via very complex and costly pretreatments. In this respect, the potential of still relatively novel pretreatments, such as organometallic primers and primers applied via plasma processes, must be considerable. Similarly, there is great potential for developing the concept of the materials supplier producing grades of metals, etc. which have already been pretreated and covered with a suitable protective layer. Thus, the adhesives user has no need to undertake any form of pretreatment and can also store the substrate materials until needed. In the case of high-energy substrates, the necessity to pretreat the substrate to attain good environmental resistance for the bonded component has been referred to on many occasions. The mechanisms by which certain surface pretreatments can greatly increase the service life will be discussed in Chapter 8, where the mechanisms of such environmental attack are reviewed.

We have so far considered the establishment of interfacial contact between the adhesive and substrate, the nature of the intrinsic adhesion forces which hold the interface together, and the role of surface pretreatments to assist in the above processes. The next stage in the formation of an adhesive joint is usually, but not always, the hardening of the adhesive so that it is capable of transmitting the required mechanical loads from one substrate to the other. This aspect is considered in Chapter 5.

REFERENCES

1. Martin, J. J. (1967) *Adhesion and Adhesives–2* (eds R. Houwink and G. Salomon) Elsevier, New York.
2. Dukes W. A. and Kinloch, A. J. (1976) *Sira Technical Review* No. 3, Sira,London.
3. Snogren, R. C. (1975) *Handbook of Surface Preparation*, Palmerton, New York.
4. ASTM (1980) *Recommended Practice for the Preparation of Surfaces of Plastics Prior to Adhesive Bonding*, ASTM D2093-69.
5. Brewis, D. M. (ed.) (1982) *Surface Analysis and Pretreatment of Plastics and Metals*, Applied Science Pub., London.
6. Shields, J. (1985) *Adhesives Handbook*, Butterworths, London.
7. Brewis, D. M. (1985) *Industrial Adhesion Problems* (eds. D. M. Brewis and D. Briggs) Orbital Press, Oxford, p. 128.
8. Lerchenthal, C. H., Brenman, M. and Yits'haq, N. (1975) *J. Polymer Sci., Polymer Chem. Ed.* **13**, 737.
9. Lerchenthal, C. H. and Brenman, M. (1976) *Polymer Eng. Sci.* **16**, 747.
10. Lerchenthal, C. H. and Brenman, M. (1976) *Polymer Eng. Sci.* **16**, 760.
11. Delollis, N. J. and Montoya, O. (1963) *Adhesives Age* **6**(1), 32.
12. Minnesota Mining and Manufacturing Co. (1957) British Patent 765,284.
13. Dahm, R. H. (1982) *Surface Analysis and Pretreatment of Plastics and Metals* (ed. D. M. Brewis) Applied Science Pub., London, p. 227.
14. Goldie, W. (1969) *Metallic Coating of Plastics–2*, Electrochemical Publications Ltd, London, p. 314.
15. Meier, J. F. and Petrie, E. M. (1973) *J. Appl. Polymer Sci.* **17**, 1007.
16. Nelson, E. R., Kilduff, T. J. and Benderley, A. A. (1958) *Ind. Eng. Chem.* **50**, 329.
17. Benderley, A. A. (1962) *Symposium on Adhesives for Structural Applications*, Interscience, New York, p. 107.
18. Purvis, R. J. and Beck, W. R. (1957) US Patent 2,789,063.
19. Miller, M. L., Postal, R. H., Sawyer, P. N., Martin, J. G. and Kaplit, M. (1970) *J. Appl. Polymer Sci.* **14**, 257.
20. Andrews, E. H. and Kinloch, A. J. (1973) *Proc. Roy. Soc.* **A332**, 385.
21. Brecht, H., Mayer, F. and Binder, H. (1973) *Makromol. Chem.* **33**, 89.
22. Dwight, D. W. and Riggs, W. M. (1974) *J. Colloid Interf. Sci.* **47**, 650.
23. Dwight, D. W. (1977) *J. Colloid Interf. Sci.* **59**, 447.
24. Cirlin, E. H. and Kaelble, D. H. (1973) *J. Polymer Sci., Polymer Chem. Ed.* **11**, 785.
25. Allan, A. J. G. (1957) *J. Polymer Sci.* **24**, 461.
26. Andrews, E. H. and Kinloch, A. J. (1973) *Proc. Roy. Soc.* **A332**, 401.
27. Brewis, D. M., Barker, D. J., Dahm, R. H. and Hoy, L. R. J. (1978) *Electrochim. Acta.* **23**, 1107.
28. Brewis, D. M., Barker, D. J., Dahm, R. H. and Hoy, L. R. J. (1979) *J. Mater. Sci.* **14**, 749.
29. Dahm, R. H., Barker, D. J., Brewis, D. M. and Hoy, L. R. J. (1979) *Adhesion–4* (ed. K. W. Allen) Applied Science Pub., London, p. 215.
30. Brewis, D. M., Dahm, R. H. and Konieckzo, M. B. (1975) *Makromol. Chem.* **43**, 191.

31. Schonhorn, H. and Ryan, F. W. (1969) *J. Polymer Sci.*, **7** (A2), 105.
32. Roberts, R. F., Ryan, F. W., Schohorn, H., Sessler, G. M. and West, J. E. (1976) *J. Appl. Polymer Sci.* **20**, 255.
33. Garnish, E. W. and Haskins, G. G. *Aspects of Adhesion–5* (ed. D. J. Alner), University of London Press, London, p. 259.
34. Brewis, D. M. (1985) *Prog. Rub. Plastics Technol.* **1**, 1.
35. Briggs, D. (1982) *Surface Analysis and Pretreatment of Plastics and Metals* (ed. D. M. Brewis), Applied Science Pub., London, p. 199.
36. Hurst, C. and Schanzle, R. E. (1966) *Mod. Packaging* **40**(2), 163.
37. Leeds, S. (1961) *TAPPI*, **44**(4), 244.
38. Ayres, R. L. and Shofner, D. L. (1972) *SPE J.* **28**(12), 51.
39. Briggs, D., Brewis, D. M. and Konieczko, M. B. (1979) *J. Mater. Sci.* **14**, 1344.
40. Blais, P., Carlsson, J., Csullog, C. W. and Wiles, D. M. (1974) *J. Colloid Interf. Sci.* **47**, 636.
41. Briggs, D., Brewis, D. M. and Konieczko, M. B. (1976) *J. Mater. Sci.* **11**, 1270.
42. Briggs, D., Vichy, V. J., Brewis, D. M., Comyn, J., Dahm, R. H., Green, M. A. and Konieczko, M. B. (1980) *Surf. Interface Anal.* **2**, 107.
43. Ghorashi, H. M. (1977) *Plating Surface Finish.* April, 42.
44. Briggs, D. (1982) *J. Adhesion* **13**, 287.
45. Farley, S., Reamey, R. H., McCarthy, T. J., Deutch, J. and Whitesides, G. M. (1985) *Langmuir* **1**, 725.
46. Shields, J. (1972) SIRA Report (Kent, UK) No. R500.
47. Rauhut, H. W. (1969) *Adhesives Age* **12**(12), 28.
48. Chew, A., Dahm, R. H., Brewis, D. M., Briggs, D. and Rance, D. G. (1986) *J. Colloid Interf. Sci.* **110**, 88.
49. Briggs, D. (1984) *Polymer* **25**, 1379.
50. Morris, C. E. M. (1970) *J. Appl. Polymer Sci.* **14**, 2171.
51. Morris, C. E. M. (1971) *J. Appl. Polymer Sci.* **15**, 501.
52. Briggs, D., Brewis, D. M. and Konieczko, M. B. (1977) *J. Mater. Sci.* **12**, 429.
53. Briggs, D., Brewis, D. M. and Konieczko, M. B. (1978) *Eur. Polymer J.* **14**, 1.
54. Abu-Isa, I. A. (1971) *J. Appl. Polymer Sci.* **15**, 2865.
55. Cope, B. C. (1985) *Adhesives, Sealants and Encapsulants Conference '85*, Conference Proceedings Network Events, Bucks, p. 355.
56. Briggs, D., Ashley, R. J., Ford, K. S. and Kelly, R. S. A. (1985) *Industrial Adhesion Problems* (eds D. M. Brewis and D. Briggs), Orbital Press, Oxford, p. 199.
57. ICI, UK Patent 1,370,893.
58. Du Pont, UK Patent 743,281.
59. Plueddemann, E. P. (1985) *Industrial Adhesion Problems* (eds D. M. Brewis and D. Briggs), Orbital Press, Oxford, p. 148.
60. Stone, M. H. (1981) *Int. J. Adhesion Adhesives* **1**, 271.
61. Parker, B. M. and Waghorne, R. M. (1982) *Composites* **13**, 280.
62. Hart-Smith, L. J., Ochsner, R. W. and Radecky, R. L. (1984) Douglas Service Report, First Quarter.
63. Pocius, A. V. and Wenz, R. P. (1985) *SAMPE J.* Sept./Oct. 50.
64. Matienzo, L. J., Venables, J. D., Fudge, J. D., and Velton, J. J. (1985) *30th National SAMPE Symp.* March, p. 302.
65. Bowditch, M. R. and Stannard, K. J. (1982) *Composites* **13**, 298.
66. Crane, L. W., Hammermesh, C. L. and Maus, L. (1976) *SAMPE J.* **12**, 6.
67. Myhre, S. H., Labor, J. D. and Aker, S. C. (1982) *Composites* **13**, 289.
68. Parker, B. M. (1983) *Composites* **14**, 226.
69. Kinloch, A. J. and Taig, C. M. (1987) *J. Adhesion* **21**, to be published.

70. Stein, B. A., Hodges, W. T. and Tyeryar, J. R. (1985) *26th Structures, Structural Dynamics and Materials Conference*, AIAA Paper 85-0750-CP.
71. Buchan, S. (1959) *Rubber-to-Metal Bonding*, Crosby-Lockwood, London.
72. Anon (1973) *Rub. J.* **155**(1), 14.
73. Peterson, C. H. (1965) *Adhesives Age* **8**(7), 22.
74. Cox, D. R. (1969) *Rub. J.* **151**(4), 49.
75. Wake, W. C. (1967) *Adhesion and Adhesives*, Vol. 1 (eds R. Houwink and G. Salomon), Elsevier, New York, p. 395.
76. Spearman, B. P. and Hutchinson, J. D. (1974) *Adhesives Age* **17**(4), 30.
77. Carter, A. R. and Petit, D. (1977) *Developments in Adhesives–1* (ed. W. C. Wake), Applied Science Publishers, London, p. 223.
78. Greenwood, L. and Wake, W. C. (1970) *Adhesion des Elastomers*, Association Francaise Ing. du Caout. Plastiques, Paris, p. 19.
79. Yasuda, H. K., Sharma, A. K., Hale, E. B. and James, W. J. (1982) *J. Adhesion* **13**, 269.
80. Rose, P. W. and Liston, E. M. (1985) *Plastics Eng.* **43** (10), 41.
81. Yasuda, T., Gazicki, M. and Yasuda, H. (1984) *J. Appl. Polymer Sci., Appl. Polymer Symp.* **38**, 201.
82. Yasuda, H., Lamaze, C. E. and Sakaoku, K. (1973) *J. Appl. Polymer Sci.* **17**, 137.
83. Gazicki, M., and Yasuda, H. (1983) *Proceedings of the 184th ACS Meeting*, American Chemical Society, Washington DC, p. 308.
84. Collins, G. C. S., Lowe, A. C. and Nicholas, D. (1973) *Eur. Polymer J.* **9**, 1173.
85. Anand, M., Cohen, R. E. and Baddour, R. F. (1981) *Polymer* **22**, 361.
86. Rossman, K. (1956) *J. Polymer Sci.* **19**, 141.
87. Carley, J. F. and Kitze, P. T. (1978) *Polymer Eng. Sci.* **18**, 326.
88. Kim, C. Y., Evans, J. and Goring, D. A. I. (1971) *J. Appl. Polymer Sci.* **15**, 1365.
89. Owens, D. K. (1975) *J. Appl. Polymer Sci.* **19**, 265.
90. Baszkin, A. and Ter-Minassian-Saraga, L. (1978) *Polymer* **19**, 1083.
91. Blythe, A. R., Briggs, D., Kendall, C. R., Rance, D. G. and Zichy, V. J. I. (1978) *Polymer* **19**, 1273.
92. Briggs, D. and Kendall, C. R. (1979) *Polymer* **20**, 1053.
93. Briggs, D. and Kendall, C. R. (1982) *Int. J. Adhesion Adhesives* **2**, 13.
94. Briggs, D., Kendall, C. R., Blythe, A. R. and Wootton, A. B. (1983) *Polymer* **24**, 47.
95. Briggs, D., Rance, D. G., Kendall, C. R. and Blythe, A. R. (1980) *Polymer* **21**, 895.
96. Leclercq, B., Sotton, M., Baszkin, A. and Ter-Minassian-Saraga, L. (1977) *Polymer* **18**, 675.
97. Owens, D. K. (1975) *J. Appl. Polymer Sci.* **19**, 3315.
98. Ryan, D. L. (1962) BP 890,446.
99. Schonhorn, H. and Hansen, R. H. (1966) *J. Polymer Sci.* **B4**, 203.
100. Hall, J. R., Westerdahl, C. A., Devine, A. T. and Bodnar, M. J. (1969) *J. Appl. Polymer Sci.* **13**, 2085.
101. Hall, J. R., Westerdahl, C. A., Bodnar, M. J. and Levi, D. W. (1972) *J. Appl. Polymer Sci.* **16**, 1465.
102. Malpass, B. W. and Bright, K. (1969) *Aspects of Adhesion–5* (ed. D. J. Alner), University London Press, London, p. 214.
103. Schonhorn, H., Ryan, F. W. and Hansen, R. H. (1970) *J. Adhesion* **2**, 93.
104. Schonhorn, H. and Hansen, R. H. (1967) *J. Appl. Polymer Sci.* **11**, 1461.
105. Schonhorn, H. (1969) *Adhesion, Fundamentals and Practice*, McLaren and Son, London, p. 12.

106. Schonhorn, H. and Hansen, R. H. (1969) *Adhesion, Fundamentals and Practice*, McLaren and Son, London, p. 22.
107. Schonhorn, H. (1978) *Polymer Surfaces* (eds D. T. Clark and W. J. Feast), John Wiley, New York, p. 213.
108. Sowell, R. R., De Lollis, N. J., Gregory, J. N. and Montoya, O. (1972) *J. Adhesion* **4**, 15.
109. De Lollis, N. J. (1973) *Rub. Chem. Technol.* **46**, 549.
110. Kaplan, S. and Dilks, A. (1984) *J. Appl. Polymer Sci., Appl. Polymer Symp.* **38**, 105.
111. Yasuda, H. (1981) *J. Polymer Sci. Macromol. Rev.* **16**, 199.
112. Inagaki, N., Kishi, A. and Katsuura, K. (1982) *Int. J. Adhesion Adhesives* **2**, 233.
113. Yamakawa, S. (1979) *Macromol.* **12**, 1222.
114. Yamakawa, S. and Yamamoto, F. (1980) *J. Appl. Polymer Sci.* **25**, 25.
115. Yamakawa, S. and Yamamoto, F. (1980) *J. Appl. Polymer Sci.* **25**, 41.
116. Gettings, M., Baker, F. S. and Kinloch, A. J. (1977) *J. Appl. Polymer Sci.* **21**, 2375.
117. Atkins, R. W., Greenwood, L. and Punter, T. J. (1973). Unpublished work.
118. Jennings, C. W. (1972) *J. Adhesion* **4**, 25.
119. Zisman, W. A. (1964) *Contact Angle, Wettability and Adhesion* (ed. R. F. Gould), American Chemical Society, Washington DC, p. 44.
120. Park, J. M. and Bell, J. P. (1983) *Adhesion Aspects of Polymeric Coatings* (ed. K. L. Mittal), Plenum Press, New York, p. 205.
121. Baker, R. G. and Spencer, A. T. (1960) *Indust. Eng. Chem.* **52**, 1015.
122. Evans, J. R. G. and Packham, D. E. (1978) *J. Adhesion* **9**, 267.
123. van Ooij, W. J. (1979) *Rub. Chem. Technol.* **52**, 605.
124. van Ooij, W. J. (1984) *Rub. Chem. Technol.* **57**, 421.
125. van Ooij, W. J. and Biemond, M. E. F. (1984) *Rub. Chem. Technol.* **57**, 686.
126. van Ooij, W. J. (1985) *Industrial Adhesion Problems* (eds D. M. Brewis and D.Briggs), Orbital Press, Oxford, p. 87.
127. Brookman, W. (1983) *Durability of Structural Adhesives* (ed. A. J. Kinloch), Applied Science Pub., London, p. 281.
128. Andrew, J. F., Clarke, S. G. and Longhurst, E. E. (1954) *J. Appl. Chem.* **4**, 581.
129. Andrew, J. F. and Donovan, P. D. (1970) *Trans. Inst. Met. Finish.* **48**, 152.
130. Andrew, J. F. and Donovan, P. D. (1971) *Trans. Inst. Met. Finish.* **49**, 162.
131. Bender, H. S., Cheever, G. D. and Wojtkowiak, J. J. (1980) *Prog. Organ. Coat.* **8**, 241.
132. Allen, K. W. and Alsalim,H. S. (1977) *J. Adhesion* **8**, 183.
133. Gettings, M. and Kinloch, A. J. (1979) *Surf. Interf. Anal.* **1**, 165.
134. Gettings, M. and Kinloch, A. J. (1979) *Surf. Interf. Anal.* **1**, 189.
135. Pocius, A. V., Almer, C. J., Waid, R. D., Wilson, T. H. and Davidian, B. E. (1984) *SAMPE J.*, Nov./Dec., 11.
136. Mahoon, A. and Cotter, J. L. (1978) *SAMPE Tech. Conf. Series* **10**, 425.
137. Mahoon, A. (1983) *Durability of Structural Adhesives* (ed. A. J. Kinloch), Applied Science Pub., London, p. 255.
138. Venables, J. D. (1984) *J. Mater. Sci.* **19**, 2431.
139. Miller, P. D., Jeffreys, R. A. and Pray, H. A. (1974) *Met. Prog.* May, 61.
140. Wegman, R. F. and Bodnar, M. J. (1973) *SAMPE Q.* **5**, 28.
141. Wightman, J. P. (1981) *SAMPE Q* **13**, 1.
142. Allen, K. W., Alsalim, H. S. and Wake, W. C. (1974) *J. Adhesion* **6**, 153.
143. Hohman, A. E. and Lively, G. W. (1975) US Patent 3,891,456.
144. Boeing Airplane Company, US Patent 3,959,091.

145. British Aerospace (1981) UK Patent 2,074,608A.
146. Northrop Corporation (1978) US Patent 409,475.
147. Allen, K. W., Chan, S. Y. T. and Armstrong, K. B. (1981) *Int. J. Adhesion Adhesives* **2**, 239.
148. Moloney, A. C. (1982) *Surface Analysis and Pretreatment of Plastics and Metals* (ed. D. M. Brewis), Applied Science Pub., London, p. 175.
149. Ministry of Defence (UK) (1965) Defence Specification 151.
150. Ministry of Defence (UK) (1970) Defence Standard 03–2/1.
151. Boeing Aeroplane Company Specification BAC 5555.
152. Wegman, R. F. (1967) *Adhesives Age*, **10**(1), 20.
153. McCarvill, W. T. and Bell, J. P. (1974) *J. Appl. Polymer Sci.* **18**, 335.
154. McCarvill, W. T. and Bell, J. P. (1974) *J. Appl. Polymer Sci.* **18**, 343.
155. McGarel, W. and Farnham, H. A. (1985) *Adhesives, Sealants and Encapsulants, Vol. 3,* Conference Proceedings, Network Events, Buckingham, p. 34.
156. Selwood, P. G., Maddison, A. and Sheasby, P. G. (1984) UK Patent Application 2,139,540A.
157. Ross, M. C., Wegman, R. F., Bondnar, M. J. and Tanner, W. C. (1974) *SAMPE J.* **6**(2), 12.
158. Ross, M. C., Wegman, R. F., Bondnar, M. J. and Tanner, W. C. (1974) *SAMPE J.* **6**(4), 14.
159. Ross, M. C., Wegman, R. F., Bondnar, M. J. and Tanner (1974) *SAMPE J* **6**(3), 20.
160. Kaelble, D. H. and Dynes, P. J. (1975) *J. Colloid Interf. Sci.* **52**, 562.
161. Plueddmann, E. P. (1982) *Silane Coupling Agents*, Plenum Press, New York.
162. Comyn, J. (1986) *Structural Adhesives: Developments in Resins and Primers* (ed. A. J. Kinloch), Applied Science Pub., London, 269.
163. Ishida, H. and Kumar, G. (eds) (1985) *Molecular Characterisation of Composite Interfaces*, Plenum Press, New York.
164. Walker, P. (1980) *J. Coatings Technol.* **52**, 49.
165. Walker, P. (1982) *J. Oil Colour Chem. Assoc.* **65**, 415.
166. Walker, P. (1982) *J. Oil Colour Chem. Assoc.* **65**, 436.
167. Walker, P. (1983) *J. Oil Colour Chem. Assoc.* **66**, 188.
168. Walker, P. (1984) *J. Oil Colour Chem. Assoc.* **67**, 108.
169. Walker, P. (1984) *J. Oil Colour Chem. Assoc.* **67**, 126.
170. Belton, D. J., van Pelt, P. and Morgan, A. E. (1973) *ASTM STP* **804**, 273.
171. Garton, A. (1984) *J. Polymer Sci., Polymer Chem. Ed.* **22**, 1495.
172. Surayanarana, D. and Mittal, K. L. (1984) *J. Applied Polymer Sci.* **29**, 2039.
173. Sung, N. H., Kaul, A., Chin,, I. and Sung, C. S. P. (1982) *Polymer Eng. Sci.* **22**, 637.
174. Bascom, W. D. (1972) *Macromolecules* **5**, 792.
175. Naviroj, S., Culler, S. R., Koenig, J. L. and Ishida, H. (1984) *J. Colloid Interf. Sci.* **97**, 308.
176. Boerio, F. J. and Williams, J. W. (1981) *Applic. Surf. Sci.* **7**, 19.
177. Chiang, C. H. and Koenig, J. L. (1981) *J. Colloid Interf. Sci.* **83**, 361.
178. Boerio, F. J., Dillingham, R. G., Snow, A. R. and Gosselin, C. A. (1984) *Am. Chem. Soc. Preprints* **50**, 444.
179. Koenig, J. L. and Shih, P. T. K. (1975) *Mater. Sci. Eng.* **20**, 136.
180. Kinloch, A. J. and Gettings, M. (1977) *J. Mater. Sci.* **12**, 2511.
181. Koenig, J. L. and Shih, P. T. K. (1971) *J. Colloid Interf. Sci.* **36**, 247.
182. Chiang, C. H., Ishida, H. and Koenig, J. L. (1980) *J. Colloid Interf. Sci.* **74**, 396.
183. Gettings, M. and Kinloch, A. J. (1980) *Surf. Interf. Anal.* **1**, 189.
184. Naviroj, S., Koenig, J. L., Ishida, H. (1985) *J. Adhesion* **18**, 93.

185. Boerio, F.J., Williams, J. W. and Burkstrand, J. M. (1983) *J. Colloid Interf. Sci.* **91**, 485.
186. Brewis, D. M., Comyn, J., Oxley, R. G., Reynolds, S., Werrett, C. R. and Kinloch, A.J. (1984) *Surf. Interf. Anal.* **6**, 40.
187. Kaas, R. L. and Kardos, J. L. (1971) *Polymer Eng. Sci.* **11**, 11.
188. Gent, A. N. and Hsu, E. C. (1974) *Macromolecules* **7**, 933.
189. Marsden, J. G. and Sterman, S. (1966) *Indust. Eng. Chem.* **58**(3), 33.
190. Schrader, M. E., Lerner, I. and D'Oria, F. J. (1967) *Modern Plastics* **45**, 195.
191. Ishida, I. (1985) *Molecular Characterisation of Composite Interfaces* (eds H. Ishida and G. Kumar), Plenum Press, New York, p. 25.
192. Chiang, C. H. and Koenig, J. L. (1981) *Polymer Composites* **2**, 192.
193. Linde, H. G. (1982) *J. Polymer Sci., Polymer Chem. Ed.* **20**, 1031.
194. Monte, S. J. and Sugerman, G. (1983) *Elastomerics* **115**, 30.
195. Monte, S. J. and Sugerman, G. (1985) *Polymer Eng. Sci.* **24**, 1369.
196. Monte, S. J. and Sugerman, G. (1983) *Adhesion Aspects of Polymeric Coatings* (ed. K. L.Mittal), Plenum Press, New York, p. 421.
197. Calvert, P. D., Lalanandham, R. P. and Walton, D. R. M. (1983) *Adhesion Aspects of Polymeric Coatings* (ed. K. L. Mittal), Plenum Press, New York, p. 457.
198. Moles, P. J. (1981) *Polymer Paint Colour* **173**, 391.
199. Cohen, L. B. (1983) *Plastics Eng.* **39**(11), 29.
200. Deyrup, A. J. (1974) US Patent 3,787,326.
201. Yang, C. Q. (1985) *Polymeric Mater. Sci. Eng.* **53**, 579.
202. DeNicola, A. J. and Bell, J. P. (1983) *Epoxy Resin Chemistry*, American Chemical Society, Washington DC, p. 119.
203. Park, J. M. and Bell, J. P.(1983) *Adhesion Aspects of Polymeric Coatings* (ed. K. L. Mittal), Plenum Press, New York, p. 205.
204. Cotton, J. B., Coldfield, S., Mack, P. A. and Price, R. (1974) US Patent 3,837,964.
205. Scardino, W. and Marceau, J. A. (1977) *J. Appl. Polymer Sci., Appl. Polymer Symp.* **32**, 51.
206. Albericci, P. (1983) *Durability of Structural Adhesives* (ed. A. J. Kinloch), Applied Science Pub., London, p. 317.
207. Pocius, A. V.(1981) *SAMPE Meeting*, Munich, Nov. p. 1.
208. Pike, R.A. (1986) *Int. J. Adhesion Adhesives* **6**, 21.
209. McMillan, J. C.(1981) *Developments in Adhesives–2* (ed. A. J. Kinloch), Applied Science Pub., London, p. 243.
210. Bowditch, M. R. and Stannard, K. J. (1985) *Adhesives Sealants and Encapsulants–3*, Conference Proceedings, Network Events,Bucks, p. 66.
211. Bowditch, M. R.(1982) UK Patent 2,083,377B.
212. Bowditch, M. R. (1981) UK Patent 1,598,431.
213. O'Kane, D. F.and Mittal, K. L. (1974) *J. Vac. Sci. Technol.* **11**, 567.
214. Beevers, A. and Njegic, A. (1985) *Adhesives Sealants and Encapsulants–2*, Conference Proceedings, Network Events, Bucks, p. 349.
215. Baker, F. S. (1979) *J. Adhesion* **10**, 107.
216. O'Kane, D. F. and Mittal, K. L. (1976) *J. Adhesion* **8**, 93.
217. Nichols, M. J., Hahn, A. W., James, W. J., Sharma, A. K. and Yasuda, H. K. (1984) *J. Appl. Polymer Sci., Appl. Polymer Symp.* **38**, 21.
218. Sahdir, R. K., James, W. J. and Auerbach, R. A. ((1984) *J. Appl. Polymer Sci. Appl. Polymer Symp.* **38**, 99.
219. Charlson, E. J., Charlson, E. M., Sharma, A. K. and Yasuda, H. K. (1984) *J. Appl. Polymer Sci., Appl. Polymer Symp.* **38**, 137.
220. Tyczkowski, J. and Kryszewski, M. (1984) *J. Appl. Polymer Sci., Appl. Polymer Symp.* **38**, 149.

221. Freitag, W.O., Yasuda, H. K. and Sharma, A. K. (1984) *J. Appl. Polymer Sci., Appl. Polymer Symp.* **38**, 185.
222. Schreiber, H. P., Wertheimer, M. R. and Wrobel, A. M. (1980) *Thin Solid Films* **72**, 487.
223. Wertheimer, M. R., Klemberg-Sapieha, J. E. and Schreiber, H. P. (1984) *Thin Solid Films* **115**, 109.
224. Yasuda, H. K., Sharma, A. K., Hale, E. B. and James, W. J. (1982) *J. Adhesion* **13**, 269.
225. Surendran, G., James, W. J., Brealey, W. and Hale, E.B. (1984) *J. Appl. Polymer Sci., Appl. Polymer Symp.* **38**, 75.
226. Zaat, J. H. (1983) *Ann. Rev. Mater. Sci.* **13**, 9.
227. Briggs, D. (1987) *J. Adhesion*, to be published.

CHAPTER FIVE

Hardening of the adhesive

5.1 INTRODUCTION

As discussed in Chapter 2, for the formation of intrinsic adhesion forces across the adhesive/substrate interface it is a necessary, though sometimes insufficient, requirement that intimate interfacial contact occurs between the adhesive and substrate. This means that at some time in the bonding operation the adhesive must be in a liquid form, ideally with a relatively low viscosity. However, in service most adhesive joints carry some mechanical loads, even if only those resulting from the weight of the substrates, and to be able to transfer mechanical loads efficiently from one substrate to the other requires that the adhesive is a solid with an adequate stiffness and strength. These conflicting requirements mean that once the adhesive has established interfacial contact, i.e. completely wet the substrate, it must then 'harden' or 'set'. In this chapter the methods by which the adhesive may be transformed from a liquid state to a solid state will be considered and this classification will also be used to introduce briefly some of the main chemical types of adhesive to those unfamiliar with this aspect of adhesives technology.

This area of adhesives technology is obviously where the polymer chemist plays a crucial role but a review of the basic chemistry of adhesives and of the formulation of adhesives is well beyond the scope of the present book. For such information the reader should consult references, such as Skeist's excellent *Handbook of Adhesives* [1] where these aspects are reviewed in detail for every imaginable chemical type of adhesive. There are also many specialized articles and chapters devoted to particular adhesive types such as those available on natural materials [2], solutions and latices [3], hot-melts [4], polyurethanes [5,6], phenolics [7], epoxies [8,9], acrylics [10,11], imide resins [12,13] and pressure-sensitive adhesives [14,15]. To translate from a particular chemical type of adhesive into a commercially available product reference may be made to the technical data sheets provided by the adhesives companies and to adhesive product directories, such as that produced by Shields [16].

It should be noted that the properties of any commercial adhesive will be greatly influenced not only by the chemical type of adhesive but also by the many additives typically introduced by the formulator. Additives include plasticizers, tackifiers, toughening agents, coupling agents, antioxidants, catalysts, fillers, etc. and they will often play a crucial role in influencing such factors as the rate of hardening and viscosity characteristics of the adhesive and the mechanical performance of the adhesive joint. Indeed, the technical skill and art of the formulator largely lies in developing particular grades of adhesive from a common chemical family of polymers or monomers which possess the optimum compromise of chemical and physical properties for a given application.

Further, in the present chapter the selection of adhesives for specific applications will not be considered in detail. The main reason being that this aspect is also well covered by product directories such as the one by Shields [16] and, in any event, it is even more satisfactorily dealt with by employing a microcomputer program. 'User friendly' programs are now commercially available [17] and enable the adhesives user and designer to sort through literally millions of permutations created by interactions between the nature of the substrates to be bonded, component design, production conditions and the required ultimate performance and operating environment of the bonded joint. Such programs typically recommend adhesives, by chemical type or trade name, and place the recommendations in an order of preference; with advisory notes as to the reasons for the final order of selection and as to why certain adhesives were rejected *en route* to the recommended list.

Therefore, as mentioned earlier, the aim of this chapter is to consider briefly the ways in which the adhesive may be transformed from the liquid to the solid state and so provide a general background for the reader unfamiliar with the chemistry of adhesives. The hardening of the liquid adhesive may be accomplished in several ways: by solvent or dispersing medium removal, by cooling and by chemical reaction. As will be shown, more than one of these mechanisms of hardening may operate in any given adhesive formulation. Also, there are those adhesives which are an exception to this classification and do not undergo hardening. These different ways of hardening the adhesive, and the non-hardening adhesives, will now be briefly discussed and the principles involved illustrated by reference to some common examples found in the home and in industry.

5.2 HARDENING BY SOLVENT OR DISPERSING MEDIUM REMOVAL

5.2.1 Introduction

Solutions and latices usually contain the already polymerized adhesive dissolved or dispersed in organic solvents or water. In the case of a solution the adhesive is dissolved in the liquid which therefore acts as a solvent, whilst a latex consists of a dispersion or emulsion of the adhesive in a liquid medium, normally water. In a solution the concentration of 'solids', i.e. material which will remain in the final adhesive layer, is usually under 30% for organic solvents but may be up to 35–55% with water as the carrier, especially with latices. The presence of these liquids lowers the viscosity sufficiently to enable the adhesive to wet the substrate readily but once this has been achieved the solvent or dispersing medium must be removed. With porous substrates such as paper or wood they may be drawn away from the adhesive layer by capillary action. However, if both substrates are non-porous it is necessary to evaporate the water or organic liquid before bringing the two adhesively coated substrate surfaces into contact to form the joint.

The need to use and then remove a liquid carrier for the adhesive introduces several disadvantages. Firstly, the liquid may attack the substrate, e.g. water may corrode metallic substrates and organic solvents may dissolve or craze plastic substrates. Secondly, when the substrates are both non-porous the need to evaporate the liquid prior to forming the joint greatly slows down the production rate and this is one of the main disadvantages of solutions and latices in adhesive applications. Thirdly, in such bonding operations the mechanism of adhesion which operates when the dried adhesively coated substrate surfaces are brought together is interdiffusion. From the discussions in Section 3.3 this limits the type of adhesive that may be used to a rubber. However, a thermoplastic adhesive may be used but an additional thermal-welding step is then required. Fourthly, the use of organic solvents often pose major problems because of cost and possible toxicity or flammability. Obviously water-based adhesives overcome these problems but drying times are longer for such adhesives and the performance of the bonded joints is often inferior. This poorer performance is often attributable to the lack of 'cleaning action' of water in removing contaminants on the substrate surface, compared to that of organic solvents, again emphasizing the interaction between the choice of substrate surface pretreatment and that of the adhesive which was noted in the previous chapter.

5.2.2 Examples

The main polymers used as solutions and latices are shown in Table 5.1, but it should be recalled that any commercial adhesive will be a complex formulation.

For example, starch is the principal water-dispersible natural polymer used industrially as an adhesive and in its many forms it is widely employed in the paper and paper products industry. Poly(vinyl acetate) and its copolymers are synthetic materials and are usually used as aqueous latices. They have found widespread used in the paper industry, where their often superior properties offset their higher price, and are used in bookbinding, milk cartons, envelopes, etc. and are also very familiar as the household 'white glue' for woodworking.

Table 5.1 Main types of polymers used in adhesive formulations based upon hardening via removal of a solvent or dispersing medium [18]

Water-based	*Water- or solvent-based*
Starch	Natural rubber
Gums	Styrene–butadiene rubbers
Albumen	Butyl rubber
Sodium silicate	Poly(chloroprene) rubbers
Casein	Nitrile rubbers
Sodium carboxymethylcellulose	Reclaim rubbers
Lignin	Polyurethane rubbers
Poly(vinyl alcohol)	Poly(vinyl acetate) and copolymers
	Poly(vinyl ether)
Solvent-based	Poly(vinyl chloride) and copolymers
Nitrocellulose	Polyacrylate and poly(methacrylate)
Cellulose acetate butyrate	Polyamides (nylons)
Cyclized rubber	Asphalt
Polyisobutylene	Urea formaldehyde
	Phenol formaldehyde
	Resorcinol formaldehyde
	Rosin esters

The 'contact' adhesives sold for domestic use are mainly based upon polychloroprene dissolved in a suitable solvent or mixture of solvents; typical solvents include toluene, methylethyl ketone, hexane and acetone. Polychloroprene is popular for this application particularly because of its rapid development of joint strength coupled with its relatively high ultimate strength. These properties are associated with the ability of the polychloroprene macromolecules to crystallize as the solvent evaporates and deposits the polymer from solution. Nitrile rubbers are also normally used as solvent adhesives and, being relatively strongly polar in nature, give good joints to both rigid and plasticized poly(vinyl chloride). To achieve improved physical properties, especially higher strengths at somewhat elevated temperatures, solvent-based rubbery adhesives, such as polychloroprene and nitrile rubbers, may also be crosslinked after the joint has been formed. This additional mechanism of hardening may be achieved via the incorporation of a curing agent into the adhesive which initiates crosslinking at elevated temperatures or at room temperature. The former would typically be used as a one-part adhesive formulation and the latter would be a two-part formulation, using, for example, a polydiisocyanate which is mixed into the adhesive just prior to it being applied to the substrate. Obviously, the rubbery adhesive must not significantly crosslink, and so cease to flow, before it has wetted the substrate, the solvent has been removed and the joint prepared.

5.3 HARDENING BY COOLING

5.3.1 Introduction

The basic concept with the adhesives that harden by cooling is that a thermoplastic polymer will soften and melt when heated and, if its viscosity falls to a sufficiently low value, the molten polymer may successfully wet the substrate materials to be bonded. Upon cooling, the thermoplastic polymeric adhesive will resolidify, and possibly recrystallize, and therefore develop high cohesive strength. For obvious reasons, these adhesives which harden by cooling are referred to as 'hot-melt' adhesives. Therefore, as usually supplied they contain no water or volatile solvents, i.e. they possess a 100% solids level.

Thus, the adhesive is solid at room temperature but becomes liquid when heated, normally to temperatures in the range 65 to 180 °C, although reference was made in Section 4.2.5 to the use of poly(ether ether ketone) as a hot-melt adhesive for composite and metal bonding in the aerospace industry where a temperature of 400 °C is employed! The polymers from which the adhesives are formulated have a low heat content so that when the heating source is removed the adhesive rapidly loses heat to the substrates and cools sufficiently to solidify in seconds. Since solvent, dispersion medium removal and chemical reaction are not involved, hot-melt adhesives provide a most rapid method of obtaining an adequate joint strength so that assemblies can be bonded without complex jigs being necessary to keep the substrates in place whilst the adhesive slowly hardens. Nevertheless, pressure must still usually be applied to the contacting surfaces immediately the adhesive is applied, and must be maintained until the adhesive has cooled and hardened sufficiently; although this is often only a matter of a few seconds at most. If, however, the cooling of the molten adhesive is too rapid then inadequate wetting may occur. This situation is encountered if the substrate has a high thermal conductivity since the adhesive in immediate contact with the substrate surface may be prematurely chilled before it has completely wetted the substrate. Thus, as commented by Wake [4], edge veneering of block board by strips of various plastic materials using a hot-melt adhesive is very successful but bonding the parts of metal frames together would not be regarded as a viable procedure, although it has been attempted and achieved by first warming the metal substrates.

5.3.1 Examples

The earliest types of hot-melt adhesives were based on natural materials such as beeswax, i.e. sealing wax, but virtually all modern hot-melt adhesives are based upon synthetic thermoplastic polymers. They are to be found in a variety of forms such as free-films, coated onto tapes, powders, rods, pellets, blocks and are usually applied with automatic equipment, although hand-held 'guns' are used in domestic and some industrial applications. The use of automatic equipment leads, for both technical and economic reasons, to very fast speeds of bonding and this is a most important feature of this type of adhesive, as commented above. However, this means that the polymeric adhesive needs to possess a low surface free energy and low viscosity when molten, so that it may very readily wet completely a wide range of substrates. Also, it should possess a high temperature coefficient of viscosity so that it hardens rapidly upon cooling. The hardened adhesive is usually required to exhibit a reasonable strength, high toughness and often to be relatively flexible. Obviously, the properties of a hot-melt adhesive at elevated temperatures are going to be limited, or alternatively a very high softening point polymer must be chosen which leads to a high application temperature, as noted above in the case of a hot-melt adhesive based upon poly(ether ether ketone).

One of the most widely used polymers in the formulation of hot-melt adhesives is the copolymer of ethylene and vinyl acetate (EVA). For example, such copolymers containing abut 28 to 33% vinyl acetate are widely used as the basis of hot-melt adhesives for bookbinding. They not only bond the individual pages of a paperback book together but they also attach the cover. For such an application the adhesive must have excellent flexibility, toughness and fatigue resistance, and these properties need to be maintained from about −20 to 60 °C. EVA hot-melts are also used for carton and case sealing, carpet applications and in many other industries.

The other main polymers used as hot-melts are polyethylene homopolymers and polyamide homopolymers, copolymers and blends. A particular attribute of the hot-melt adhesives based upon polyamides is that just above their melting point the viscosity of the adhesive is low enough so that it can be readily applied to form the joint, but, once formed, much of the strength of the bonded joint will still be retained until just below the melting point. This arises from the highly polar groups in the polyamide chains which, below the melting temperature, give interchain hydrogen bonding resulting in high polymer strength at relatively low molecular weights. However, once the polymer has melted the low molecular weight means that the melt has a low viscosity. Polyamide-based hot-melt adhesives are finding increasing use in the shoe, automotive, packaging, electronic, electrical, wood working and general product assembly industries.

5.4 HARDENING BY CHEMICAL REACTION

5.4.1 Introduction

The adhesives which harden primarily by chemical reaction are usually applied to the substrates as monomers or oligomers (sometimes referred to as 'resins') and are then polymerized *in situ* in the joint between the substrates. The starting products may already be in the form of liquids which have a low viscosity at room temperature. Alternatively, in order to achieve the requirement of a low viscosity they may be applied from a solution or latex, where the liquid is removed prior to chemical reaction (and in the case of non-porous substrates prior to joint formation) or heat may be applied. Thus, this class of adhesives has a very wide range of forms at room temperature: low-viscosity liquids (either with a solvent present or as 100% solids), thixotropic pastes, powders, rods and tacky or tack-free films. The polymerization reaction may lead to a crosslinked rubber, a thermoplastic or a thermosetting polymeric adhesive.

It is convenient to divide the chemical reactions by which this class of adhesives harden into two groups:

(i) Those which harden by a step-growth polymerization where reactions take place between multifunctional monomer molecules. Functionality is the number of sites at which a monomer molecule is able to react. In order to form linear macromolecules it is necessary that the monomers are bifunctional. If the monomers have a functionality of greater than two it is sometimes possible to form crosslinked polymers, i.e. rubbers or thermosetting plastics, rather than linear molecules, i.e. thermoplastics. Step-growth polymerization is characterized by the gradual formation of polymer chains through successive reactions coupling monomers to each other to form dimers which can also then react with other dimers or unreacted monomer molecules. This type of polymerization is often also called condensation polymerization since it is frequently concerned with condensation reactions taking place; such reactions involve the elimination of a small molecule, usually water but sometimes alcohols, acetic acid, etc., as a by-product of the chemical reaction. This group includes adhesives which are based upon room-temperature vulcanizing (RTV) silicone rubbers, thermosetting phenolic and amino resins, which all involve condensation reactions, and urethanes and many epoxy formulations which typically do not involve condensation reactions so that the elimination of a small volatile molecule does not accompany the polymerization of these adhesives.

(ii) Those which harden by an addition polymerization where the monomer molecules add onto the growing chain one at a time and which does not involve the formation of small molecules such as water. Addition polymerization takes place in three distinct steps: initiation, propagation and termination. Monomers which polymerize by addition polymerization often contain double bonds and are of the general formula $CH_2{=}CR_1R_2$. The double bond is susceptible to attack by either free radical or ionic initiators to form a species known as an active centre. This active centre then propagates a growing chain by the addition of monomer molecules until the active centre is eventually neutralized by a termination reaction. Note that with addition polymerization the reaction only occurs at the reactive end of the growing polymer chain. Amongst this group are many of the newer adhesives, such as those based upon cyanoacrylates, anaerobics and radiation-curable polymers, as well as several epoxy adhesive formulations.

In some cases the above chemical reactions will proceed at room temperature and may be caused by the addition of a hardener, such as a free radical initiator or a crosslinking (curing) agent. These are two-part adhesive formulations and, obviously, the hardener cannot be added until just before the adhesive is to be applied, otherwise the adhesive will polymerize and become too viscous to wet the substrate readily. As might be imagined, the formulator is often developing grades which represent a range of compromises between having a reasonably long working life for the mixed adhesive against the typical request for a rapid rate of hardening. Two-part adhesives have often been disliked on the factory floor because of the need to mix the two parts reasonably consistently, especially as some of the hardeners particularly are somewhat obnoxious. However, the increasing use of automatic mixing and dispensing equipment, often linked with robotic application of the mixed adhesive, is removing this concern. The chemical reaction may also proceed at room temperature when the reaction is catalysed by an external source, such as ultraviolet light, mechanical rupture of an encapsulated hardener, the presence of moisture (e.g. cyanoacrylates, RTV silicones or urethanes) and the absence of oxygen (e.g. the anaerobics). These types of adhesives are usually one-part formulations and 'protected' from the external source until needed for use. For example, depending upon the external source, they may be stored in a waterproof container or in a container with a large volume of air, i.e. a good supply of oxygen, above the adhesive. Finally, a general aspect of the hardening of adhesives at room temperature which is frequently overlooked is that, whilst sufficient strength for handling the joint may be developed within anything from a few minutes to a few hours, the chemical reactions often continue, albeit at an ever slower rate, for a week or so. It may therefore take a relatively long time for the full joint strength to be developed.

In other formulations the chemical reactions will only proceed at elevated temperatures. Most of the chemical types of adhesive mentioned above may be formulated to harden in this manner and are usually one-part formulations containing an hardener which is only activated when heat is applied. However, some chemical rection may occur slowly at room temperature and hence the limited storage life of this type of adhesive and the common practice of storage in a freezer to prolong the shelf-life.

The relative slowness of hardening the adhesive by chemical reaction, coupled with the fact that it occurs *in situ*, means that some mechanical design feature of the joint or some form of jigging is usually necessary to prevent the substrates moving relative to one another and rupturing the joint prematurely before the adhesive has developed sufficient strength. For most modern adhesives only contact pressure needs to be maintained and this enables small clamps, or the occasional spot-weld or rivet to be successfully employed to meet this requirement. However, for some of the high-strength adhesives which harden by a condensation reaction, such as phenolic resins, and where the volatile by-products will give a large concentration of voids in the adhesive layer, significantly reducing the mechanical properties, then a relatively high bonding pressure must be applied to prevent this sequence of events; for example, a press or autoclave must be employed.

Compared to the adhesives which harden either by solvent or dispersing medium removal or by cooling, the present class which primarily harden by chemical reaction offers many advantages. Firstly, there is the fact that most are used as 100% solids, and so there is no water or organic liquid present which eliminates the problems described in Section 5.2.1. Also, with water- or organic liquid-containing adhesives some liquid inevitably remains in the adhesive layer and, with high strength adhesives, this may significantly reduce their mechanical performance. Secondly, unlike the hot-melts, many formulations are available which may be hardened at room temperature, so removing the need for heating equipment, and, also, they are able to wet readily and bond substrates with a high thermal conductivity, e.g. metals. Thirdly, their temperature resistance is usually far superior to both the other two methods for hardening the adhesive. The main disadvantages of the adhesives which harden by chemical reaction are (i) their slow rate of hardening compared to the hot-melts and the economic implications of this, although some chemical types can now achieve a handling strength within seconds as discussed below, (ii) the need to sometimes mix two parts together, although automatic mixers and dispensers are becoming increasingly available as noted previously, and (iii) their often greater cost.

5.4.2 Examples

One of the main applications of adhesives which harden by chemical reaction is in bonding structural components in engineering industries such as aerospace, road and rail transport, manufactured household goods, building and electrical. There is no universally agreed definition of a *structural adhesive* but one might suggest that such adhesives are those which when hardened possess a relatively high modulus and high strength and may be used to join relatively rigid substrates so that a truly load-bearing joint is constructed. Obviously, some adhesives which harden by liquid removal or by cooling may also be termed structural but, for the reasons discussed above, the main ones are those which harden by chemical reaction.

5.4.2.1 Phenolic/resorcinolic adhesives

These are usually medium- or high-viscosity liquids, but powdered and film forms are available, and they harden by a step-growth (condensation) polymerization. They were the first structural adhesives to be developed for metallic substrates and the original formulation consisted of a hot-cured phenolic, modified with poly(vinyl formal) to reduce the brittleness of the highly crosslinked phenolic that results upon curing. This adhesive formulation is still widely used in the European aircraft industries for bonding aluminium alloys, mainly because of its excellent durability, when used in conjunction with the appropriate surface pretreatment for the aluminium alloy (see Chapter 8). However, except in this application and for laminating and bonding wood this family of adhesives has not proved particularly attractive in mass-production industries where the problems of high temperatures for the hardening reaction and the associated need to maintain a high bonding pressure to overcome the problem of volatiles are not easily accommodated.

5.4.2.2 Epoxy adhesives

Epoxy adhesives represent the most common structural adhesives and have gained wide acceptance in many diverse industries. They essentially consist of an epoxy resin, often based upon the diglycidyl ether of bisphenol A, and harden to give a thermosetting polymer by step-growth polymerization or addition polymerization.

In step-growth polymerization a curing agent possessing active chemical groups, such as mercaptan, amine, amide or acid anhydride groups, is added and reacts with the epoxy groups. This is illustrated by a typical reaction scheme shown in Fig. 5.1. Note that the polymerization does not usually involve a condensation reaction, so *no* volatiles are evolved. Such reactions may proceed either at room or elevated temperatures. The curing agent is added by the user in the case of a room-temperature hardening reaction, i.e. a two-part formulation, and the multifunctional hardeners that are commonly used are based upon polysulphides, aliphatic polyamines (e.g. tetraethylenepentamine, TEPA) and polyamine-amides. When the hardening reaction proceeds only at an elevated temperature, such as with the aromatic amine hardeners, e.g. diaminodiphenyl methane or diaminodiphenyl sulphone, the supplier often adds the hardener and the adhesives user purchases a one-part formulation.

The addition polymerization reaction involves polymerizing through the intermolecular reaction of the epoxy groups driven by an ionic reaction initiated by one of several types of catalyst which are present in the adhesive formulation, such as dicyandiamide, certain aromatic amines, BF_3-amine complexes or imidazole complexes. The polymerization reaction typically only

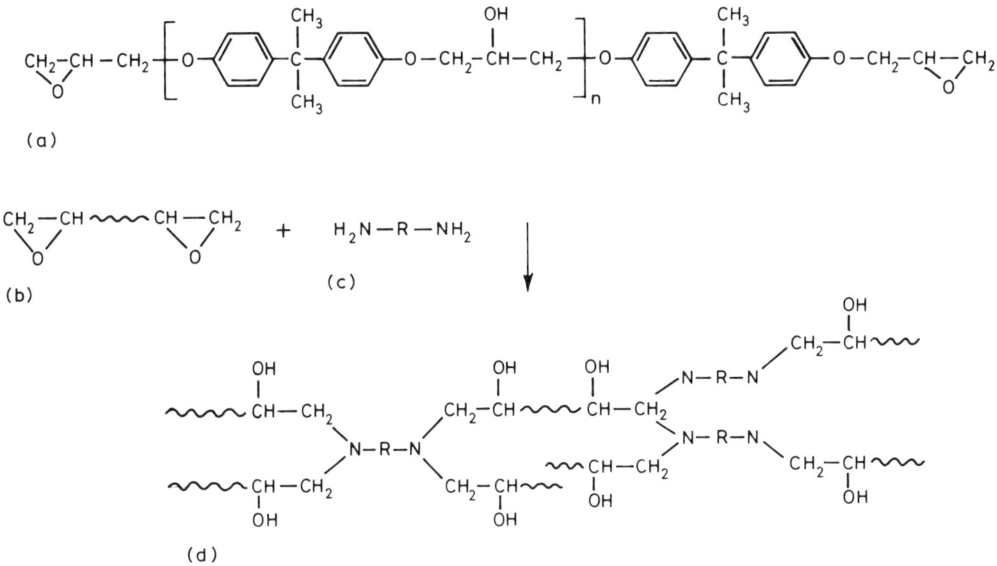

Figure 5.1 Typical epoxy resin and hardening reaction using a primary diamine hardener. (a) A bisphenol A-epichlorohydrin based epoxy resin. The most important type is when $n \approx 0.2$ and the liquid resin is essentially (b) a simple diglycidyl ether of bisphenol A (DGEBA). (c) A primary diamine hardener, (d) a crosslinked, thermosetting epoxy adhesive.

proceeds at an elevated temperature and formulations employing the latent hardeners listed above are the basis for many of the current generation of one-part hot-cured epoxy formulations, especially those toughened by the inclusion of about 10% w/w of a rubber in the formulation.

The modern principle of adding the rubber, often a low molecular weight (liquid) carboxyl-terminated copolymer of butadiene and acrylonitrile, is that it is designed to react and phase separate before the epoxy hardening reaction has proceeded very far. This results in a microstructure of small well-bonded rubber particles in a matrix of epoxy adhesive which has subsequently hardened. This microstructure, shown in Fig. 5.2, may lead [19] to significant increases in the toughness of the adhesive without major losses in other important properties such as stiffness, creep resistance and elevated temperature properties. The micromechanical reasons for this are discussed in more detail in Section 7.6.3.2. Both room- and hot-curing epoxy formulations may be modified in this manner and this technology has also been applied to many other adhesive types, as discussed below.

The wide variety of epoxy resins and hardeners which are commercially available, together with the ability to toughen these strong and temperature

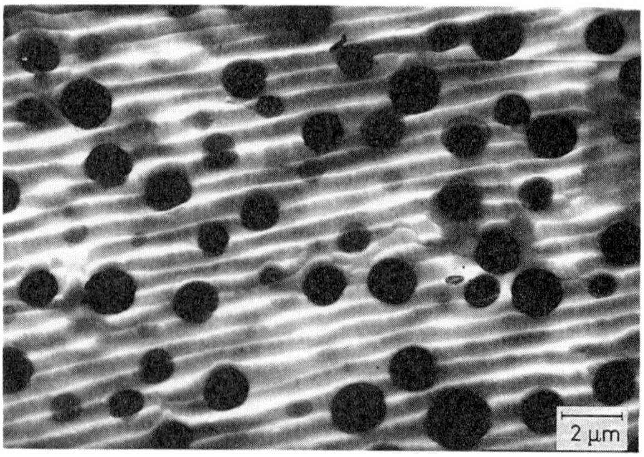

Figure 5.2 Electron micrograph of the typical microstructure of a rubber-toughened epoxy adhesive showing the dispersed rubbery phase of particles in the matrix of a thermosetting epoxy polymer. (The wavy lines are artefacts due to the cutting of a very thin section of material.)

resistant thermoset polymers, has enabled the formulator to produce an extremely large number of grades of epoxy adhesives; each possesses its own particular advantages and disadvantages and each finds a niche in a specific application.

5.4.2.3 Acrylics

Acrylics is the family of structural adhesives based upon acrylic resins; it is also diverse, and often confusing even to the specialist. The family is illustrated in Fig. 5.3 and includes a wide range of different acrylic resins and hardening reactions. The hardening of acrylics typically involves an ionic or free-radical initiated addition polymerization and is frequently a very fast reaction leading to rapid development of joint strength, so removing the necessity for complex jigs, etc; although making final adjustments to the relative position of the substrates extremely difficult.

For example, one chemical type of acrylics are the cyanoacrylates and they typically possess a very low viscosity and undergo an anionic polymerization of the double bond of the acrylic ester to yield a thermoplastic adhesive. The polymerization is base (alkali) catalysed and the catalyst normally takes the form of traces of surface moisture, which are in fact usually slightly alkaline in character; acids strongly inhibit the hardening reaction. Hence, the observation that the cyanoacrylates polymerize far more rapidly on high-energy substrates, which more readily adsorb atmospheric moisture, than polymeric substrates. Indeed, when used to bond close-fitting metallic components a joint which has sufficient strength to be handled can be produced in two or three seconds. They

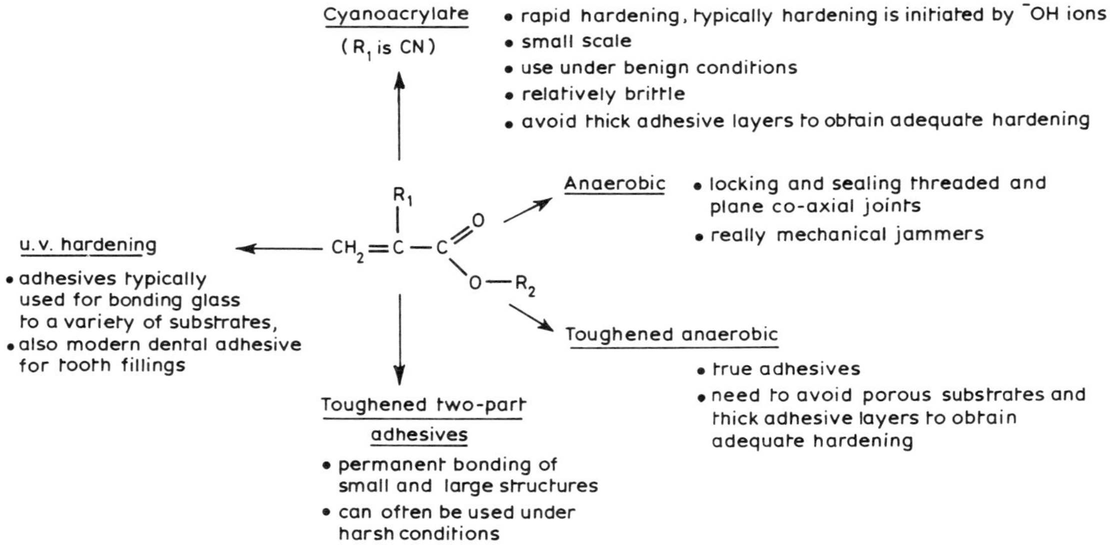

Figure 5.3 The family of acrylic adhesives used for structural and semistructural applications; R_1 and R_2 represent various organic groups [10].

are commonly used to assemble all kinds of small rubber, plastic and metal components. However, although cyanoacrylate adhesives are nominally quite strong, the thermoplastic nature of most cyanoacrylates coupled with their low softening point and sensitivity to humidity on metallic surfaces generally restricts their use to lightly loaded components in benign conditions.

Another member of the acrylic family is known as the toughened acrylic and they are based upon a variety of acrylic monomers with viscosities which range from thin liquids to thixotropic pastes. Like the toughened epoxies, they contain an initially soluble rubber additive which phase separates during the polymerization reaction to give small rubber particles, a micron or so in diameter, which greatly increases the toughness of the adhesive. Some are truly one-part adhesives hardening at room temperature, i.e. the anaerobic versions, whilst others will still harden at room temperature but require some form of hardener to be mixed either directly into the acrylic monomer or painted as a primer onto the substrate(s) just prior to bonding.

5.4.2.4 Polyurethanes

These are characterized by the presence of urethane linkages which are typically formed by step-growth polymerization of a hydroxyl-containing compound, often derived from a polyether, polyester or hydrocarbon, to an isocyanate group such as in a polyisocyanate. A condensation reaction is not involved so no volatiles are emitted during the hardening reaction. It should be noted that polyurethane adhesives are widely used for non-structural applications. For example, the chemistry may be designed to yield an elastomeric adhesive which hardens at room temperature and this may be used in the form of a two-part, 100% solids adhesive or a one-part, 100% solids adhesive whose hardening reaction is initiated by moisture. The latter type of material is widely used for bonding windscreens into motor cars. (Alternatively, the polyurethane elastomer may be formed and then dissolved in a solvent to give a solvent adhesive which hardens by solvent evaporation, as discussed above.) As structural adhesives they are typically used as 100% solids which are either one-part (heat cured) or two-part (room-temperature cured) and such adhesives are being increasingly used very successfully in the automobile industry for joining thermoplastics and glass fibre reinforced plastic substrates. However, whilst they possess excellent low-temperature performance, their use as structural adhesives has been limited by their poor elevated temperature performance (see Fig. 6.20, page 230). Further, some chemical types, for example, those derived from polyesters, are prone to hydrolytic degradation and when polyurethanes are employed to bond metallic components the resistance of the joint to attack by moisture appears to be particularly poor; although the use of chemical pretreatments and primers may improve the service life of the joint, as discussed later in Chapter 8.

5.4.2.5 Polyimides

Polyimides are the most recent structural adhesives which polymerize, either via a condensation or an addition reaction, to give a polyimide adhesive. They do not possess the ease of processing or toughness of the epoxy or acrylic adhesives but have a far superior resistance to high temperatures. They are therefore of considerable interest to the aerospace community and it is also predicted that they will be used widely in electronic applications. Again there is a family of polyimides, with the ones possessing the best temperature resistance tending to be those which are more difficult to employ in an industrial environment and which are the more brittle. Some, for example, retain most of their room-temperature mechanical properties up to a test temperature of about 350 °C and have service life expectations of about 30 000 h at about 220 °C, or even somewhat higher. The latest ones are expected to perform satisfactorily during continuous exposure to temperatures of the order of 300 °C. This exciting area of adhesives technology has recently been the subject of several good reviews [12,13,20].

5.5 NON-HARDENING ADHESIVES

Unlike all the other classes described above these adhesives do not typically undergo hardening after they have been applied to the substrate surfaces and the joint formed. They are generally in the form of an already polymerized adhesive which is coated onto one or both sides of a backing material such as cellulose, polyester, foamed polyurethane, poly(vinyl chloride), aluminium or lead. The adhesive is usually permanently tacky and based upon natural rubber, styrene–butadiene rubber (random and, more recently, block copolymers), polyisobutylene or an acrylic polymer, but as usual is a complex formulation containing many additives. The adhesive is formulated so that it flows sufficiently, when hand pressure is applied to the joint for a short period of time, to wet the substrate adequately in order to attain a certain, albeit minimal, level of joint strength. Thus, the term 'pressure-sensitive' is often applied to this class of adhesives.

In fact one of their main attractions for some applications is that they wet the substrate so poorly that they can be removed from the substrate surface without damaging the substrate or leaving any adhesive residue, i.e. can be guaranteed to fail by 100% interfacial failure at a low peel strength. Thus, their widespread use as masking tapes prior to painting an article and the development of adhesive-backed notelets which enable a message note to be adhered to a page of a book, letter, etc. but removed cleanly later. They are also widely used of course for producing surgical tape, office sticky tape, self-adhesive labels and for bonding decorative trim to motor-cars. The common forms of pressure-sensitive tapes carry nominal loads, although there are some formulations which can be crosslinked by heat after the joint has been formed and these may be lightly loaded.

It is interesting to note that pressure-sensitive tapes and labels are usually interlined with a release-coated backing to prevent them from sticking together prior to use. This *abhesive* requirement leads to the selection of a release coating with a significantly lower surface free energy than the pressure-sensitive adhesive (see Section 2.3.1). Silicone coatings meet this criterion and are commonly employed for this purpose.

5.6 CONCLUDING REMARKS

It is hoped that this brief overview will give the reader who is unfamiliar with the science and technology of the chemistry and formulation of common types of adhesives a general appreciation of the ways in which the adhesive may be transformed from the liquid to the solid state. Also, the aims of this chapter have been to give a broad appreciation of why there is such a myriad of adhesives commercially available and to have introduced the non-specialist to some of the major chemical types.

As stated at the beginning of the chapter, it is beyond the scope of the present book to review in detail the chemistry and formulation of adhesives and the reader is referred elsewhere for such information. Also, the selection of a particular adhesive for a given application has not been covered in any detail. If involved in the selection of adhesives for industrial applications, the use of a computer-aided selection program by an 'informed' designer or materials scientist or engineer is to be highly recommended. This is a most powerful tool for reducing the choice from an evergrowing multitude of commercially available adhesives formulations down to a more sensible number.

The present chapter and the three previous chapters have all been concerned with various aspects of the formation of adhesive joints. The following three chapters are concerned with the mechanics and mechanisms of failure of adhesive joints.

REFERENCES

1. Skeist, I. (ed.) (1977) *Handbook of Adhesives*, 2nd edn, van Nostrand Reinhold Co., New York. 2nd Edition.
2. Bolker, H. (1974) *Natural and Synthetic Products*, Marcel Dekker, New York.
3. Welch, M. J. and Keating, W. F. (1982) *Adhesion–6* (ed. K. W. Allen), Applied Science Pub., London, p. 149.
4. Wake, W. C. (1982) *Adhesion and the Formulation of Adhesives*, Applied Science Pub., London, p. 267.
5. Dollhausen, M. and Warrach, W. (1982) *Adhesives Age* **25**(6), 28.
6. Bluestein, C. (1981) *Adhesives Age* **24**(6), 31.

7. Knop, A. and Scheib, W. (1979) *Chemistry and Applications of Phenolic Resins*, Springer-Verlag, West Berlin.
8. Lee, H. and Neville, K. (1967) *Epoxy Resins*, McGraw-Hill, London.
9. Garnish, E. W. (1986) *Structural Adhesives: Developments in Resins and Primers* (ed. A. J. Kinloch), Applied Science Pub., London, p. 57.
10. Lees, W. A. (1979) *British Polymer J.* **11**, 64.
11. Martin, F. R. (1986) *Structural Adhesives: Developments in Resins and Primers* (ed. A. J. Kinloch), Applied Science Pub., London, p. 29.
12. Stenzenberger, H. (1986) *Structural Adhesives: Developments in Resins and Primers* (ed. A. J. Kinloch), Applied Science Pub., London, p. 77.
13. St Clair, T. L. (1985) *Polyimides* (ed. K. L. Mittal), Plenum Press, New York, p. 977.
14. Hodgson, M. F. (1979) *Adhesion–3* (ed. K. W. Allen), Applied Science Pub., London, p. 207.
15. Krenceski, M. A., Johnson, J. F. and Temin, S. C. (1986) *J. Macromol. Sci., Rev. Macromol. Chem. Phys.* **C26**, 143.
16. Shields, J. (1984) *Adhesives Handbook*, 3rd edn, Butterworths, Guildford.
17. Lees, W. A. (1984) *Adhesives in Engineering Design*, Design Council, London. (A companion computer-aided adhesives selection program for an Apple, BBC and IBM microcomputer is also available.)
18. Lichman, J. (1977) *Handbook of Adhesives* (ed. I. Skeist), van Nostrand Reinhold Co., New York, p. 679.
19. Kinloch, A. J. (1986) *Structural Adhesives: Developments in Resins and Primers* (ed. A. J. Kinloch), Applied Science Pub., London, p. 127.
20. Subrahmanian, K. P. (1986) *Structural Adhesives: Chemistry and Technology* (ed. S. R. Hartshorn), Plenum Press, New York, p. 309.

CHAPTER SIX

Mechanical behaviour of adhesive joints

6.1 INTRODUCTION

The previous four chapters have discussed various aspects in the formation of adhesive joints. This chapter, and the following two, will consider various aspects of the mechanical behaviour and the mechanics and mechanisms of failure of adhesive joints.

First, common joint designs to be found in many industrial applications will be discussed and some basic design considerations reviewed.

Secondly, the standard test methods typically used to assess the mechanical behaviour of adhesive joints in the laboratory will be considered. However, as will be seen, even when interfacial failure does occur the standard test methods do not enable the magnitude of intrinsic adhesion forces acting across the interface to be determined. Also, even in the more commonly observed instances when the locus of joint failure is via cohesive fracture in the adhesive layer, most test methods do not permit the basic engineering properties of the adhesive to be deduced, although such properties are obviously essential if design calculations are to be undertaken. Indeed, it is most important to appreciate that the measured strengths of almost all the common joint designs and standard test methods depend not only upon the degree of intrinsic adhesion achieved and the mechanical properties of the adhesive employed but also upon those of the substrates and upon the specific geometry of the joint. For example, a single-lap joint loaded in tension is widely used as a simple and practical joint design in many industrial applications and is also one of the most commonly used standard test methods. However, the stiffness, strength and resistance to creep under an applied load of this type of joint are strong functions of such parameters as the modulus and thickness of the substrates and the length of bonded overlap. This general observation arises from the fact that the geometries employed in these apparently 'simple' joint designs possess complex stress distributions.

Thirdly, therefore, the present chapter will consider the nature and magnitude of the stresses in the common types of joint designs and test methods. This information should permit quantitative joint design studies and failure predictions to be undertaken and the effects of parameters such as adhesive properties, joint geometry and test rate and temperature on the measured joint strength to be understood. However, to achieve these objectives also requires a knowledge of the basic engineering properties of the adhesive and substrates, e.g. tensile and shear moduli, yield and fracture stresses and strains. Methods for measuring the basic properties of the adhesive and for analysing the stresses in joints are now reasonably well established. Indeed, to conduct the stress analysis, programs for micro- or macrocomputers, depending upon the complexity of the analysis, are commercially available. However, to predict the actual failure load further requires a suitable failure criterion to be proposed and verified, and the various criteria which have been employed are discussed. It will be seen that the above approach does enable many aspects of the mechanical behaviour of adhesive joints to be readily understood and even predicted. However, there is another complementary approach, namely the use of continuum fracture mechanics. This approach is discussed in detail in Chapter 7 and is generally more useful for problems concerning crack growth and fracture, than for aspects of joint design. Nevertheless, in some aspects of joint design, and the associated topic of interpreting the effects of joint geometry on the measured mechanical behaviour, continuum fracture mechanics can be a most valuable tool. Where this is the case, the use of a continuum fracture mechanics is included in the present chapter in the interests of clarity and to avoid unnecessary repetition.

Finally, the above are concerned with the design and destructive testing of adhesive joints but non-destructive testing is also an important aspect of adhesives technology and will be discussed in the present chapter.

6.2 COMMON JOINT DESIGNS

There are four types of stresses which are commonly referred to when considering adhesive bonded joints. These are normal (or direct) stresses, shear stresses, cleavage stresses and peel stresses and they are illustrated in Fig. 6.1. As the name suggests, normal or direct stresses are normal to the plane on which they act and may be tensile or compressive whilst shear stresses are parallel to the plane on which they act. These two types of stresses represent the two sorts of components into which the total stress on any arbitrary plane can be divided. However, it is convenient to identify two further types of stress: cleavage stresses which typically arise as the result of an offset tensile force or bending moment and peel stresses which arise if one or both of the substrates are flexible. (These last two terms are frequently used interchangeably by authors.)

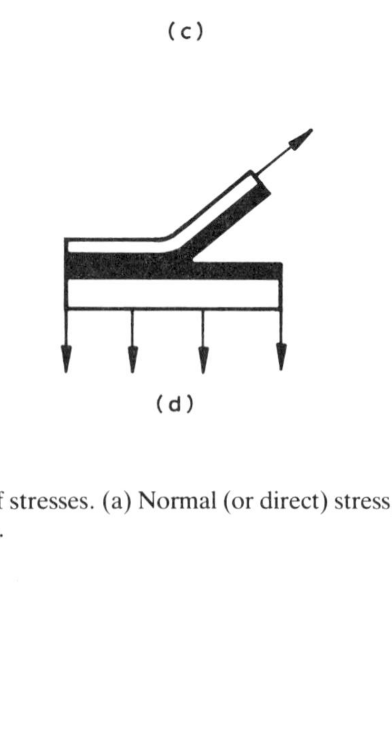

Figure 6.1 Types of stresses. (a) Normal (or direct) stress, (b) shear stress, (c) cleavage stress, (d) peel stress.

There are several important general aspects to bear in mind when designing adhesive joints. As shown later, the modulus and the strength of the polymeric materials used as adhesives are far lower than those of metals, ceramics or continuous-fibre composites. Thus, when employed to join such materials their main application is in bonding relatively thin sheets or panels. Also, the strain capability and toughness of adhesives is invariably higher when measured in compression or shear as opposed to tension, cleavage or peel and this fact greatly influences the practical design of adhesive joints. For example, the effect of an applied peel load is to place a very high stress and strain concentration on the boundary line of the crack front, thus unless the joint is wide or the applied peel load is low, fracture will occur relatively easily. The new generation of 'toughened' adhesives, mentioned in Chapter 5 and also discussed in Chapter 7, do possess a greatly improved resistance to peel loads but the designer is still well advised to keep such stresses to a minimum. A second example is the problem of designing joints such as the butt joint where tensile stresses are the main stresses encountered. The strength of adhesives in tension is usually quite good, although inferior to that in compression. However, a major problem in designing adhesive joints to operate in tension is that unless one of the substrates is highly compliant then even a small misalignment of the substrates will result in cleavage stresses. Again the multiphase-toughened adhesives have a far better resistance to such stresses than the older, more brittle, adhesives and should be selected when such stresses are likely to occur. Notwithstanding, if high cleavage or peel stresses must be borne then some means of distributing the load must be found, for example, by restraining movement at the end of a flat, flexible joint with a rivet or spot-weld. If such remedies are not practicable, then complete redesign may be required.

Therefore, from the aspect of joint design, the above general comments lead to the conclusion that the designer should not only attempt to keep stress concentrations to a minimum but also attempt to distribute the imposed loads within the adhesive layer as a combination of compressive and shear stresses; avoiding tensile, cleavage and peel stresses as much as possible. Some common practical joint designs are sketched in Fig. 6.2 and illustrate these various points by showing 'good' and 'poor' designs. The fundamental reasons why different design features so affect the strength of joints, and how such effects may be calculated, will be discussed later when the nature and magnitude of stresses in joints are considered in detail.

Also, as discussed by Lees [1], practical joint designs must take other factors into consideration. They must allow cost-effective assembly and the design must allow the adhesive to be applied in an appropriate pattern: stripes and crosses are preferred so that upon placing the substrates together air entrapment does not occur, which is possible with a closed-loop pattern. Liquid or paste form adhesives may be more accurately placed if they are deposited in a shallow groove, or if there is a convenient ledge nearby that can be used to guide the application gun. Further, to prevent air entrapment one substrate should be lowered slowly over the other, preferably starting from one end. It is also usually better to design joints so that the parts are pressed down upon the adhesive rather than slid into position; 'placement on' rather than 'sliding over' ensures that the adhesive is not pushed out of place. Large bonded areas may require considerable pressure to be maintained whilst the adhesive hardens.

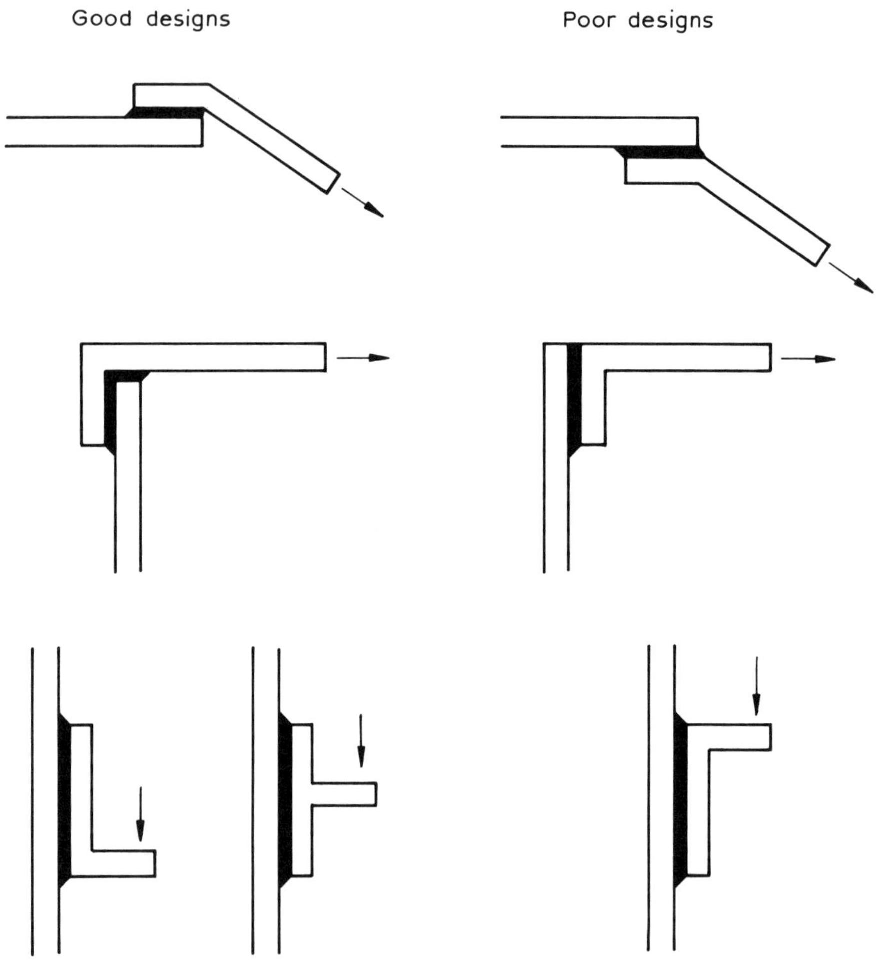

Figure 6.2 Schematic diagram of some 'good' and 'poor' joint designs. The good designs attempt to distribute the imposed loads within the adhesive layer as a combination of shear and compressive stresses; avoiding tensile, cleavage and peel stresses as far as possible.

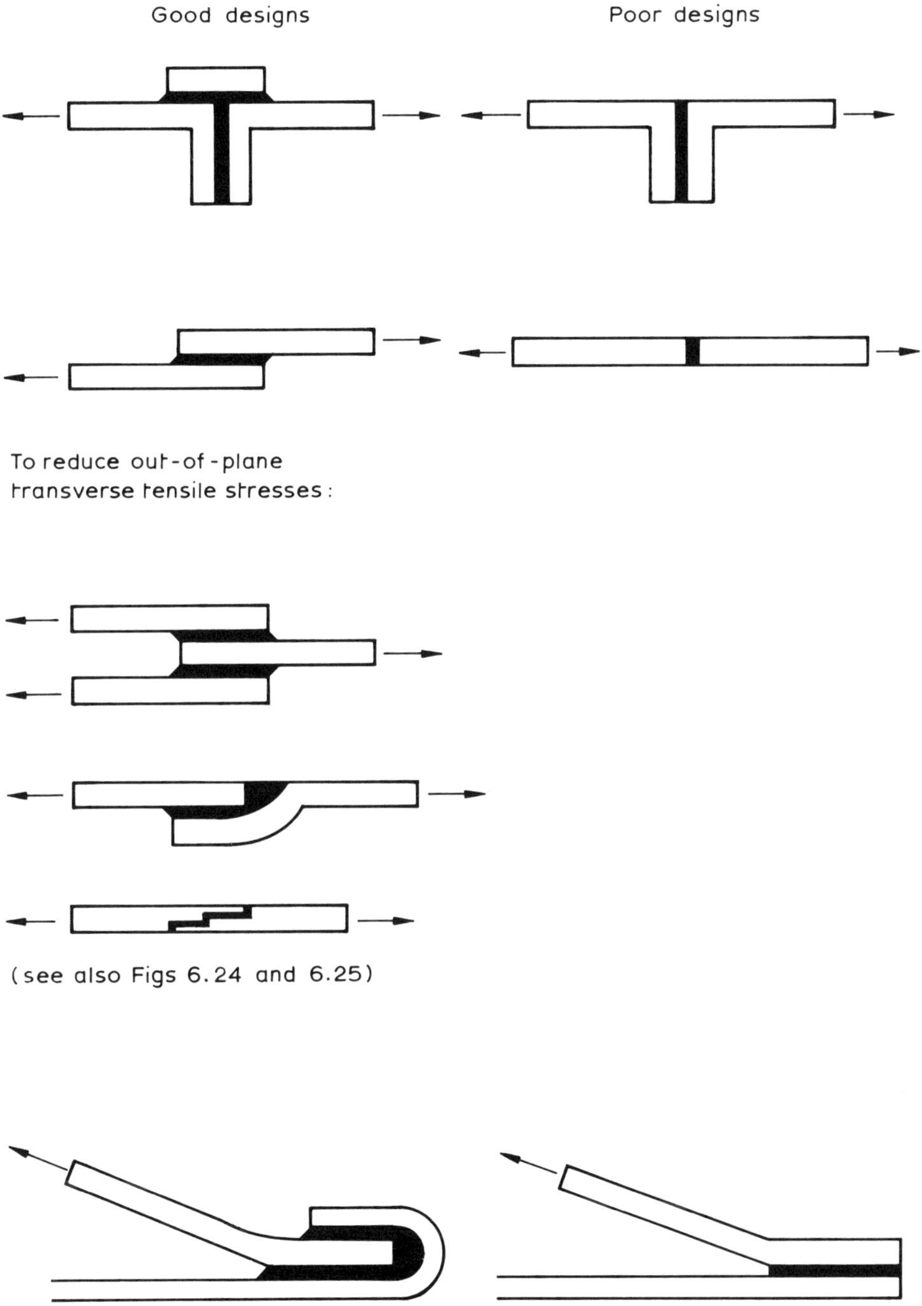

Fig. 6.2—continued

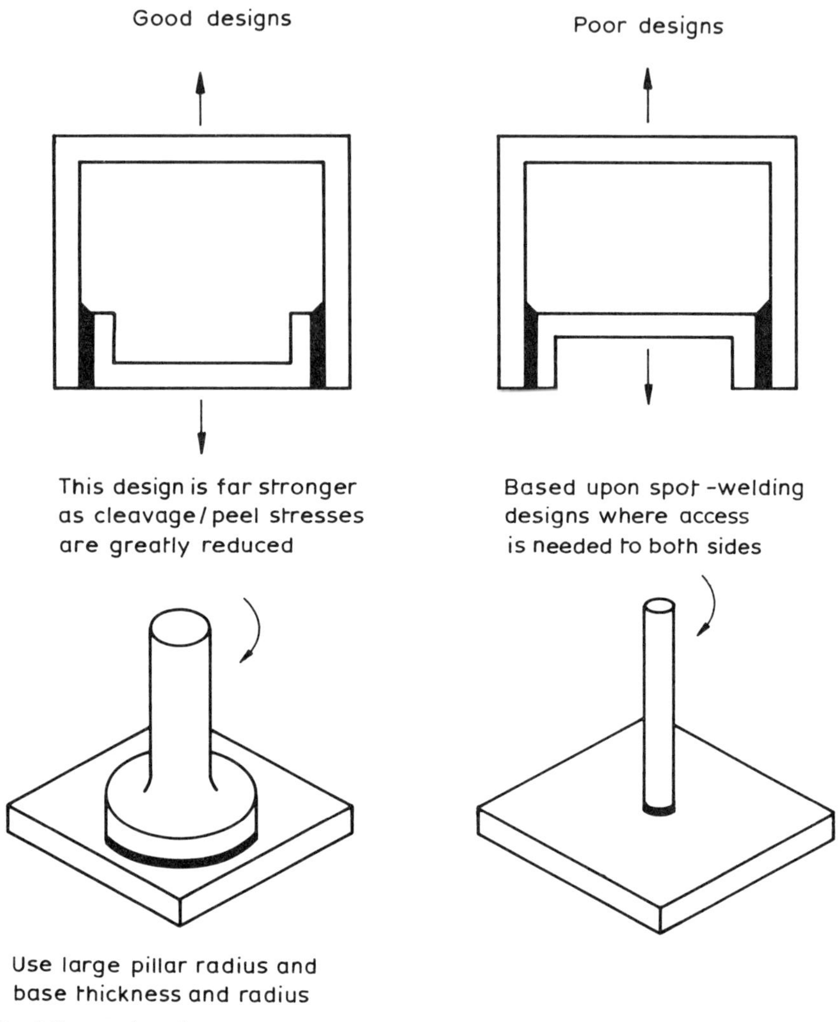

Fig. 6.2—continued

Finally, the excess adhesive squeezed from the joint, commonly termed the 'spew fillet', may contribute to its overall strength by reducing the stress concentration at the edge, so the fillets should only be removed for overriding practical or aesthetic reasons.

6.3 STANDARD TEST METHODS

The most commonly used standard test methods for assessing the performance of an adhesive joint are those issued by the American Society for Testing and

Materials (ASTM series). The ASTM issue a very extensive list of standard tests, and some are also covered in the far smaller series issued by the British Standards Institution (BS series). Other countries and organizations also issue specifications but a discussion or even listing of all the available specifications is beyond the scope of the present text. The ASTM standard tests, together with some of the BS standards, are listed in Table 6.1. The joint geometries are also shown schematically, together with relevant notes concerning the various tests. Since, as mentioned above and discussed in detail below, the measured strength will usually vary considerably if the joint geometry and test conditions specified in the standard are not followed exactly, it is obviously important to adopt the test procedure exactly as described.

With the exception of those few test methods which do measure basic engineering properties of the adhesive, the usefulness of most of the test methods is that they (i) enable comparative tests to be conducted which are 'standard' from one laboratory to another, and (ii) enable the performance of practical joint designs to be estimated by simulating the geometries and stresses likely to be encountered. Several examples will serve to illustrate these points. The single-lap joint test (ASTM D 1002-72) and the peel test (e.g. ASTM D 3167-76) are relatively easy and inexpensive test methods and data from these two standard tests will give the experienced adhesives technologist a good first impression of the initial capability of a new adhesive formulation. Another example is the widespread use of single-lap joints (ASTM D 1002-72) and double-lap joints (ASTM D 3528-76) to identify any decrease in the strength of joints upon exposure to a hostile environment, e.g. moisture, salt-spray, elevated temperatures, etc. Again no basic design data is obtained but the rate of loss of joint strength of any new adhesive system, coupled with experience from how current adhesive systems have performed in both standard tests and in-service, enables a judgement to be made as to the suitability of that adhesive system for that environment. In some applications such tests might represent only a first screening to identify those adhesive systems which warrant further, more detailed examination, whilst in other applications such tests might form the basis for the final selection of the adhesive, primer, substrate pretreatment, etc., i.e. the 'adhesive system'. A third example, would be the use of a 180° peel test (ASTM D 903-49) to determine the peel strength of a flexible substrate bonded to a rigid material. In many practical designs this may be a common loading mode and this test may therefore be used to examine the suitability of the full joint design, e.g. choice of pretreatment and adhesive and thickness of flexible substrate. However, the designer must appreciate that the measured peel energy will also be a function of such parameters as the rate of test, test temperature and service environment. A final example comes from those standards which are directed at particular industrial applications and ASTM D 2558-69 is one such standard which is concerned with evaluating the peel strength of shoe sole-attaching adhesives.

Table 6.1 Standard test methods for adhesive joints

Joint geometry	Standard	Comments
Axially loaded (tensile) butt joints		
	ASTM D 897-78	For substrates in 'block' form
	ASTM D 2094-69 and D 2095-72	Specifically for bar- and rod-shaped substrates
	BS 5350: Part C3: 1979	UK version of above
	ASTM D 429-73	Rubber-to-metal bonding
	ASTM D 816-82	Specifically for rubbery adhesives
	ASTM D 1344-78	Cross-lap specimen specifically for glass substrates
Lap joints loaded in tension		
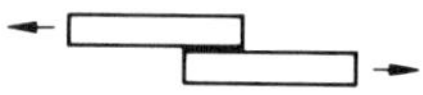	ASTM D 1002-72	Basic metal-to-metal single-lap joint test giving the 'single-lap-shear strength'
	BS 5350: Part C5: 1976	UK version of above test
	ASTM D 2295-72	Single-lap joint test for metal-to-metal joints at elevated test temperatures
	ASTM D 2557-72	As above but at low temperatures
	ASTM D 3163-73	Single-lap joint for rigid plastic substrates
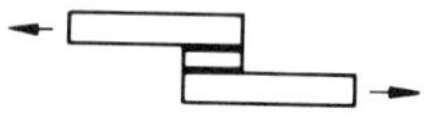	ASTM D 3164-73	Single-lap joint where a small rectangle of plastic substrate is sandwiched between metallic substrates
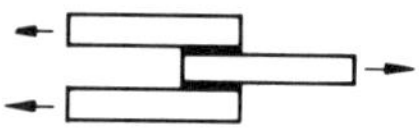	ASTM D 3528-76	Basic metal-to-metal double-lap joint test giving the 'double-lap-shear strength'
	BS 5350: Part C5: 1976	UK version of test
	ASTM D 3165-73	Metal-to-metal laminate test for large bonded areas but may also be used with plastic substrates
	ASTM D 906-82	Specifically for adhesives used in plywood laminate constructions
	ASTM D 2339-82	Specifically for wooden (two-ply) laminate substrates
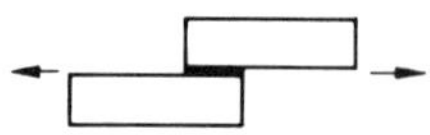	ASTM D 3983-81	Thick substrates used in a single-lap joint; shear modulus and strength of adhesive determined

Table 6.1 continued

Joint geometry	Standard	Comments
Lap joints loaded in compression		
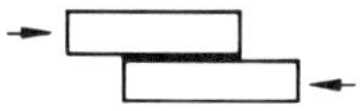	ASTM D 905-49	Mainly intended for wooden substrates
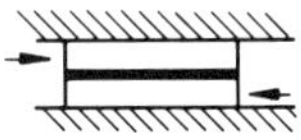	ASTM D 4027-81	For measuring shear modulus and strength of adhesive; uses 'rails' to maintain only shear load
Torsional shear		
	ASTM D 3658-78	Specifically for ultraviolet light-cured glass-to-metal joints
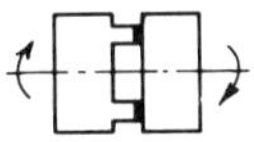	ASTM E 229-70	Uses a 'napkin ring' test for determining shear modulus and shear strength of structural adhesives
Cleavage		
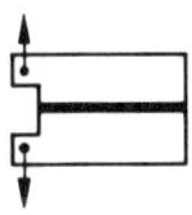	ASTM D 1062-78	Metal-to-metal joints
	ASTM D 3807-79	For engineering plastics substrates
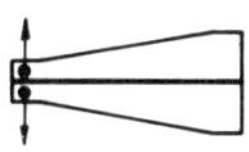	ASTM D 3433-75	Flat and contoured cantilever-beam specimens for determining the adhesive fracture energy, G_{Ic}
Peel joints		
	ASTM D 3167-76, BS 5350: Part C9: 1978	Floating roller test
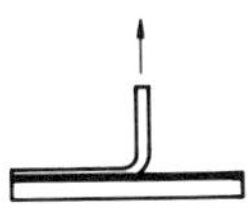	BS 5350: Part C10: 1976	90° peel test for flexible-to-rigid joints
	BS 5350: Part C14: 1979	90° peel test for rigid-to-rigid joints
	ASTM D 903-49, BS 5350: Part C11: 1979	180° peel test

Table 6.1 continued

Joint geometry	Standard	Comments
	ASTM D 1876-72, BS 5350: Part C12: 1979	'T' peel test
	ASTM D 1781-76	Climbing drum peel test for skin-sandwich assemblies
	ASTM D 429-73	For rubber-to-metal bonding
	ASTM D 2558-69	For shoe-soling materials
Impact resistance		
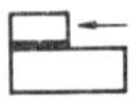	ASTM D 950-82	Uses a block shear joint
Disc shear strength in compression		
	ASTM D 2182-78	Determines adhesive shear strength in compression
Creep resistance		
	ASTM D 1780-72, ASTM D 2294-69	Single-lap joint loaded in tension employed
	ASTM D 2293-69	Single-lap joint, having a long overlap and between 'rails', loaded in compression
	BS 5350: Part C7: 1976	Various test geometries permitted
Fatigue		
	ASTM D 3166-73	Single-lap joint loaded in tension employed
Environmental resistance		
	ASTM D 896-84	General method for assessing resistance of joints to chemicals; may use any ASTM standard test geometry
	ASTM D 904-57	General method for assessing resistance of joints to artificial and natural light; may use any ASTM standard test geometry
	ASTM D 1151-84	General method for assessing resistance of joints to moisture and temperature; may use any ASTM standard test geometry

Table 6.1 continued

Joint geometry	Standard	Comments
	ASTM D 1183-70	General method for assessing resistance of joints to a cyclic laboratory aging environment (moisture and temperature); may use various ASTM standard test geometries
	ASTM D 1828-70	General method for assessing resistance of joints to natural outdoor aging; may use any ASTM standard test geometry
	ASTM D 1879-70	General method for assessing resistance of joints to high-energy irradiation; may use any ASTM standard test geometry
	ASTM D 2918-71	Method for assessing the effect of *stress* and moisture and temperature; uses a peel joint test
	ASTM D 2919-71	As above but uses single-lap joints loaded in tension
	ASTM D 3762-79	As above but uses a wedge (cleavage) test
	ASTM D 3632-77	Effect of a high-pressure oxygen environment on wood-to-wood and wood-to-metal joints; uses a lap joint test
Pressure-sensitive tack		
	ASTM D 2979-71	Inverted probe test
	ASTM D 3121-73	Rolling-ball test

6.4 STRESSES IN ADHESIVE JOINTS

6.4.1 Introduction

To deduce the nature and magnitude of the stresses in an adhesive joint requires the basic mechanical properties of the adhesive and substrate to be known and a mathematical analysis of the joint geometry to be available.

The engineering properties of the substrate materials may usually be found from handbooks of mechanical properties, although in the case of the fibre-composite materials, which possess of course anisotropic properties, some of the data may have to be determined by the adhesives technologist;

ASTM standards should be consulted for details of such test methods. For the adhesive, with the exception of the structural adhesives used in aerospace applications, very little such data is available, although the need for this information is increasingly being recognized. However, several suitable test methods are to be found in the literature and are described below.

Two mathematical methods have been commonly employed to determine the stress distributions in adhesive joints. The first is an analytical approach which defines differential equations which characterize the stresses and strains in the joint and, by using stress functions or other mathematical techniques, closed-form analytical solutions may be obtained. The early analyses of Volkersen [2] and Goland and Reissner [3] used this method, but simply assumed all the materials forming the joint behaved elastically. The later studies of Hart-Smith [4] have also employed this method but, more realistically, allow the adhesive to undergo plastic, as well as, elastic deformation. The second method which has been widely used is that of finite-element analysis which is a numerical method. With finite element analysis the joint is divided into discrete parts, each of which has to be compatible in the force and displacement continuum with adjacent elements. By applying suitable boundary conditions any form of joint may be analysed but the use of a digital computer is essential in order to solve the equations of state numerically. Both linear and non-linear elastic and elastic–plastic finite element analyses have been performed [5,6]. As might be expected, these two different mathematical methods have their respective advantages and disadvantages. For example, the closed-form analytical analyses do not require the computing power that the finite-element method requires, and can often therefore be performed on a microcomputer. Also, once a closed-form solution is known a parametric study of the basic design and material property variables may readily be conducted. The finite-element method usually requires a new computer run for each change in value of any parameter, which may take a considerable time and be expensive. However, for complex geometries a closed-form analytical solution may not always be found, but a finite-element numerical method will always yield a result, and will also far more readily take into account geometric effects such as tapered substrates and adhesive spew fillets. In the discussions below the results of the mathematical methods used to analyse stress distributions in adhesive joints are given, together with appropriate comments on the approximations and assumptions made, but the reader should refer to the original papers for the details of the mathematical method employed.

6.4.2 Engineering properties of the adhesive

To analyse the stresses in adhesive joints requires a knowledge of the basic engineering properties of the adhesive. Typically, the main properties required are the tensile, or Young's modulus (E_a), the shear modulus (G_a) and the yield stresses and the fracture stresses and strains in uniaxial tension and in pure shear. As mentioned above and shown in detail below, most of the common standard joint tests do not enable these properties to be deduced due to the complex state of stress induced in the adhesive layer by the specimen geometry. Two different approaches have been adopted in an effort to overcome these problems: one is to measure these properties by preparing bulk specimens of the adhesive and the second is to measure these properties by using especially designed joint geometries.

The preparation of bulk specimens in order to ascertain the mechanical properties of the adhesive has been an approach adopted by many workers [e.g. 7,8]. Figure 6.3, for example, shows stress versus strain curves for a hot-cured rubber-toughened epoxy tested in uniaxial tension and pure shear [8]. The tensile stress versus strain properties of the adhesive were measured by testing specimens which had been cast and machined into cylindrical 'hour-glass' shaped test pieces. The parallel gauge length was carefully polished in order to avoid premature failure due to surface scratches and a sensitive extensometer was used to measure the displacements. Such uniaxial–tensile tests on bulk specimens of the adhesive are often performed since no adhesive joint geometry results in this simple stress state. However, one problem with many relatively brittle structural adhesives is that the tensile fracture strain particularly is sensitive to how the specimen is prepared, due to the damaging effects of surface marks, scratches, air-bubbles, etc. The shear properties shown in Fig. 6.3 were determined in pure shear by loading bulk cylindrical specimens of the adhesive in torsion. From the measured torque versus twist curve a shear stress versus shear strain curve was derived. This method of test is not widely employed, however, since various joint geometries do enable the shear properties to be measured, as discussed below, and they are usually easier tests to conduct.

Considering the data shown in Fig. 6.3, it is of interest to compare the stress versus strain curves under uniaxial tension and pure shear. Note the very much higher fracture strain and strain energy density at fracture of the epoxy adhesive under a state of pure shear. Such typical observations lead to the statements made earlier concerning the advantages of designing joints to operate under shear rather than tensile loads. This arises, of course, from the tendency for crack growth and fracture to occur prior to gross plastic yielding under the action of tensile stresses in such hard, thermosetting polymers [9]. Indeed, the presence of the rubbery particles in the toughened adhesive is intended to increase the fracture strains observed in both tension and shear compared to the simple, very brittle, structural adhesives, as well as to increase greatly the peel, cleavage and impact performance.

The second approach to determining basic engineering properties of the adhesive has been to employ specific joint geometries, where ideally (i) the stress state is simple, (ii) no stress concentrations are present and (iii) the

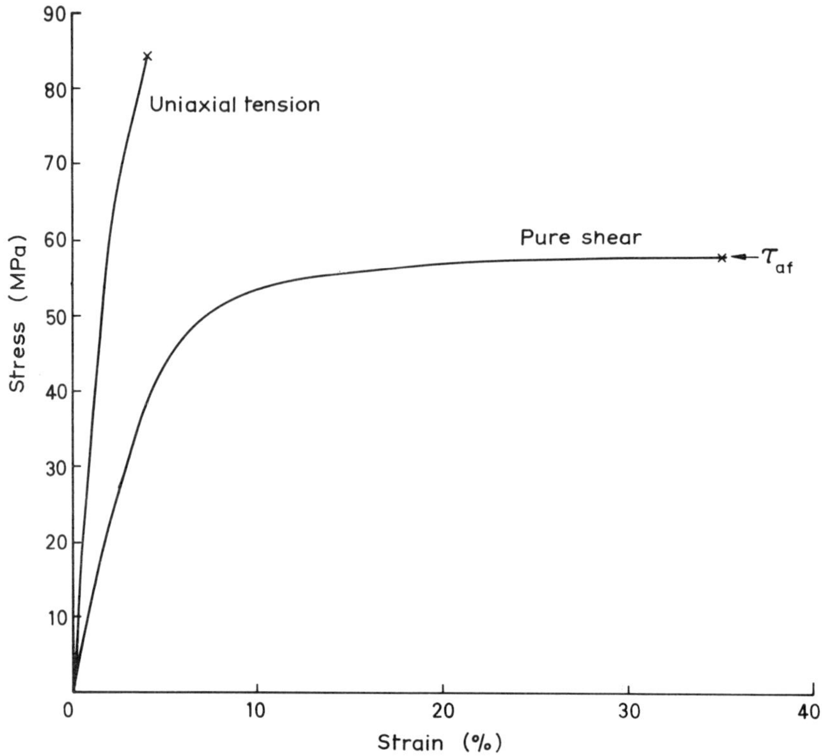

Figure 6.3 Stress versus strain curves for a hot-cured rubber-toughened epoxy adhesive (at 23 °C and an average strain rate of $7.5 \times 10^{-4} s^{-1}$) in uniaxial tension and pure shear [8].

surface pretreatment is such to ensure that the locus of joint failure is cohesive in the adhesive layer. In the case of uniaxial tension no joint design achieves these objectives, as discussed below in Section 6.4.3, although for ascertaining the Young's modulus of the adhesive the apparent modulus obtained from axially loaded butt joints may be corrected, via equations derived from stress analyses, to give the value of E_a (Section 6.4.3.2). This is not usually possible, however, with the fracture properties and, therefore, tensile data is usually obtained from testing bulk adhesive specimens. In the case of the shear properties of the adhesive, several tests have been proposed: ASTM tests D 3983, D 4027 and E 229 (the 'napkin-ring' test) and the 'thick-adherend lap-shear' test. Of these, the two which are most frequently used are the napkin-ring specimen loaded by applying equal and opposite torques [10] and the thick-adherend lap-shear specimen loaded in tension [11–16].

The napkin-ring test (see Table 6.1 and Section 6.4.8) more nearly approaches the ideal of the adhesive being in a uniform state of pure shear; especially if any spew fillets are machined off, since these increase the apparent stiffness of the joint and increase the local shear stress concentrations at the adhesive/substrate interface. However, the manufacture and testing of napkin-ring specimens is somewhat difficult and time consuming. This has led to the modification of this test by using solid circular butt joints [17], and to the introduction of a different test method, namely, the thick-adherend lap-shear specimen which is loaded in tension. This latter specimen is shown in Fig. 6.4 and is somewhat easier and more convenient to construct and test, but does not result in the adhesive layer being in a state of pure shear, as might at first be thought. The use of thick, relatively rigid substrates, is designed to remove both the shear stress concentrations near the ends of the overlap and the cleavage stresses at the ends of the overlap (Section 6.4.4). However, some shear stress concentrations and cleavage stresses do in fact still exist [13,18] and these will affect the measured properties. The limited amount of work [14,17,18] that has been published on comparing results from these two different test methods indicates a reasonable agreement between the tests, as shown in Table 6.2. Indeed, the agreement is surprisingly good, as will be readily appreciated by any reader who has conducted either napkin-ring or thick-adherend tests and dealt with such problems as controlling and measuring

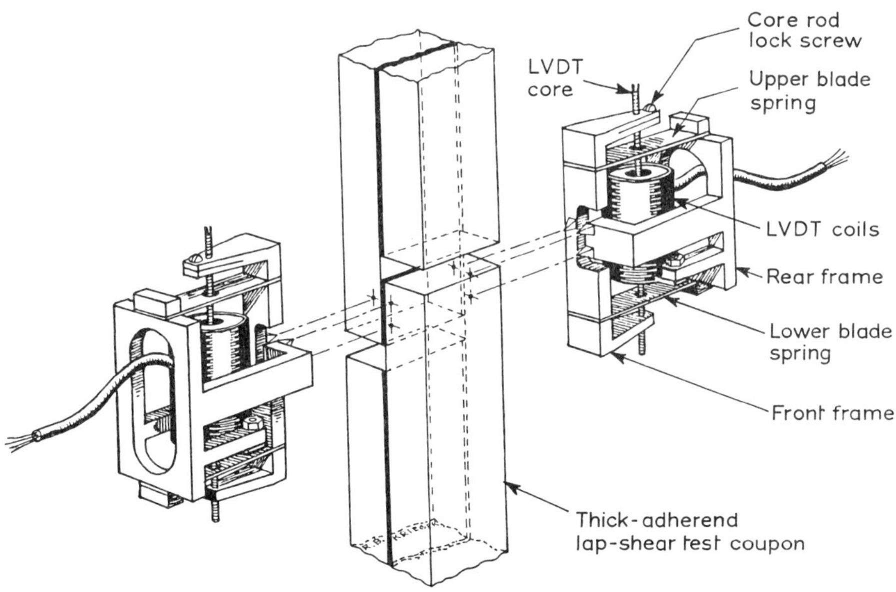

Figure 6.4 Schematic drawing of the 'thick-adherend' specimen, with extensometers for measuring the deformation [13,15].

Table 6.2 Shear properties of some structural adhesives at room temperature and a moderate rate of test

Adhesive†	Modulus††, G_a (GPa)	Yield stress, τ_{ay} (MPa)	Fracture stress, τ_{af} (MPa)	Fracture strain, γ_{af} (%)
FM-73[1]‡	0.58	31	37	61
FM-73‡	0.56	30	38	140
REDUX 775RN§	1.23	50	55	120
BSL 312¶	0.67	38	42	75
Hot-cured (single-part) toughened epoxy	1.13	54	58	35
Cold-cured (two-part) toughened epoxy	0.51	28	35	105
Cold-cured (two-part) toughened acrylic	0.16	8	34	~150

† All properties measured using the thick-adherend shear test specimen, except where superscript 1 shown when the napkin-ring shear specimen was used.
‡ FM-73: a toughened epoxy film adhesive (hot-cured, 60 min/125 °C) from American Cyanamid.
§ REDUX 775RN: a vinyl phenolic film adhesive (hot-cured, 30 min/150 °C) from Ciba-Geigy.
¶ BSL 312: a modified epoxy film adhesive (hot-cured, 30 min/120 °C) from Ciba-Geigy.
†† The subscript 'a' is used to indicate a property of the adhesive.

the thickness of the adhesive layer, aligning the specimens in the test machine, designing and using extensometers to measure the small deformations involved, correcting for deformations in the adherend, etc.

The point to consider concerning the two different approaches to determining the engineering properties of the adhesive is whether the adhesive is the 'same material' in bulk form as when present as a thin adhesive layer between substrates. For example, the presence of the substrates might alter the kinetics of any chemical reaction by which the adhesive hardens, either (i) by changing the local temperature or (ii) by removing curing agent or filler by preferential adsorption. Also, residual internal stresses are more likely to be present in the adhesive when it is cured between substrates. These factors might obviously influence the mechanical properties of the bulk adhesive compared to the *in situ* cured adhesive. Also, since polymeric adhesives are viscoeleastic materials, the dependence of the properties upon test rate must be taken into account when comparing measurements from different sources. Little work has been reported on this topic but recent studies by Knollman *et al.* [19,20] indicate that a variation in modulus might well occur through the thickness of a layer of epoxy-based adhesive bonding aluminium alloy substrates. They have used a method based upon an ultrasonic Rayleigh wave technique which enables the shear modulus of the adhesive to be sampled through the joint in steps of about 50 μm. They found that the adhesive near the adhesive/substrate interface possessed a lower shear modulus than that of the central regions, where it was similar in value to that of the bulk adhesive material. A typical result is shown in Fig. 6.5 and they tentatively ascribed these effects to residual stresses in the adhesive layer, which would decrease in intensity as one examines regions of the adhesive further away from the interface. On the other hand, more recent work by Post *et al.* [21], using high-sensitivity Moire interferometry, has indicated that the strain distribution across the adhesive layer in the

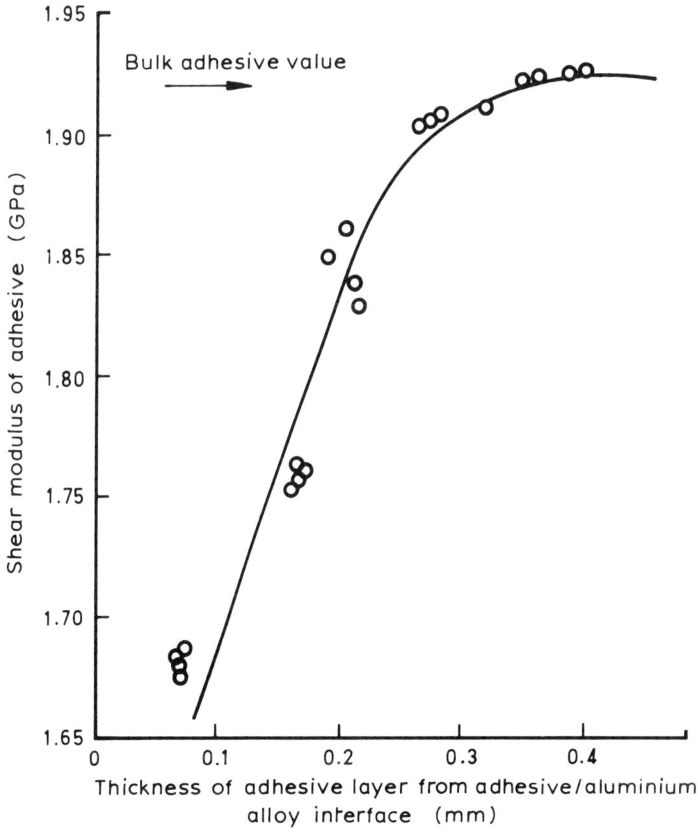

Figure 6.5 Shear modulus, G_a, of an epoxy adhesive, as determined by an ultrasonic Rayleigh wave technique, as a function of the thickness of the adhesive layer from the adhesive/aluminium alloy interface [20].

thick-adherend lap-shear joints that they examined was virtually uniform; suggesting that the adhesive's modulus was uniform across the thickness of the adhesive layer. These aspects are undoubtedly worthy of further investigation. Indeed, the soundness of the mechanical property data which is used in joint design and failure prediction studies is becoming more critical as the stress analyses are rapidly improving and giving a truer picture of the stress distribution in a joint.

The other property of the adhesive which is often required is Poisson's ratio, ν_a, and this is usually deduced from a knowledge of the Young's and shear moduli via the relationship:

$$E_a = 2G_a(1+\nu_a) \tag{6.1}$$

Finally, it should be remembered that polymeric adhesives are viscoelastic materials and that their properties will therefore be dependent upon the test temperature and rate. Also, since polymers absorb many liquids that they may be exposed to in-service, such as moisture, their mechanical behaviour will also depend upon the extent of any chemical attack or plasticization by ingressing liquids. The appropriate test conditions for determining the properties of the adhesive must therefore be carefully selected. The mechanical properties of joints will reflect changes in the adhesive's behaviour arising from the test conditions, although the joints often behave in a more complex manner due to the added effect of stress concentrations arising from geometrical features. Obviously, to discuss in detail all the possible effects of test conditions on the basic engineering properties of adhesives and the properties of joints is beyond the scope of the present text. But representative examples will be given and discussed in the following sections after the stress distribution in various types of joints are considered.

6.4.3 Axially loaded butt or poker-chip joints

6.4.3.1 Stress distribution

This type of joint geometry is shown schematically in Fig. 6.6 and it might be supposed that the tensile stress, σ_z, like the applied normal strain, ϵ_z, is the same everywhere in the adhesive layer. However, this situation would only arise if, under the action of the normal applied stress, σ_o, neither the substrates nor the adhesive tended to deform laterally or did so by the same amount, i.e. if the ratio of Poisson's ratio, ν, to the Young's or tensile modulus, E, were the same for both materials. This condition is seldom, if ever, fulfilled so that the tendency for lateral (Poisson's) deformation will be unequal in the two materials. Usually, the ratio $\nu{:}E$ is higher for the adhesive. The tendency for lateral strain in the adhesive will therefore be far greater than in the substrates but, assuming the interface remains intact, the regions of adhesive near the

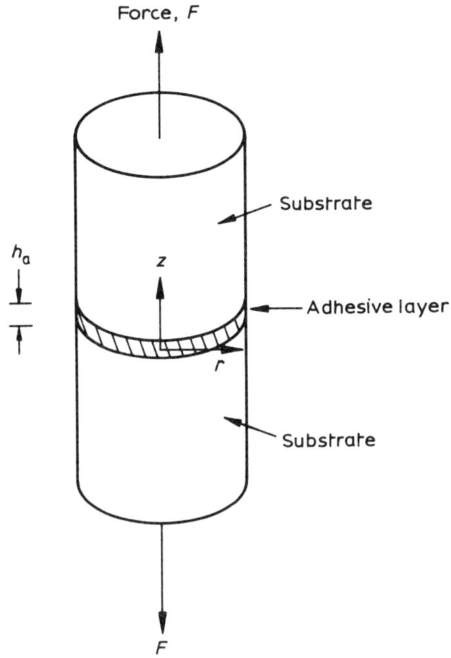

Figure 6.6 Axially loaded butt or 'poker-chip' joint.

interface will be restrained by the far more rigid substrates. This will give rise to a complex distribution of stresses in the adhesive layer.

Alwar and Nagaraja [22] and Adams *et al.* [10] have used an elastic finite-element method to analyse the stress distribution in butt joints loaded in tension and a typical stress distribution is shown in Fig. 6.7 for an epoxy adhesive bonding aluminium alloy substrate. The bonded area comprises two different regions.

Firstly, in the central region of the adhesive layer, the tensile stresses are uniform and the shear stress, τ_{rz}, is zero. The radial, σ_r, and the circumferential, σ_θ, stresses in the adhesive layer were found to be essentially the same as those predicted by Kuenzi and Stevens [23]. These workers used an analytical approach and assumed that the radial and circumferential strains in the adhesive layer were equal to the Poisson's ratio strain in the substrates. Thus:

$$\frac{1}{E_a}(\sigma_\theta - \nu_a[\sigma_z + \sigma_\theta]) = -\left(\frac{\nu_s}{E_s}\right)\sigma_z \tag{6.2a}$$

hence

$$\sigma_\theta = \sigma_r = \left[\nu_a - \frac{E_a \nu_s}{E_s}\right]\left[\frac{\sigma_z}{(1-\nu_a)}\right] \tag{6.2b}$$

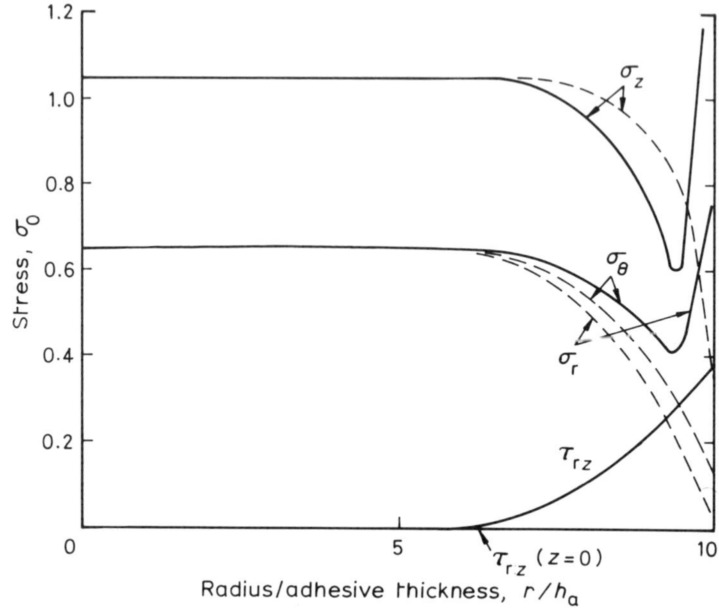

Figure 6.7 Stress distribution in a butt joint loaded in tension [10]. σ_o = average applied axial stress; $\nu_a = 0.4$; $E_a = 2.5$ GPa; $E_s = 69$ GPa; aspect ratio = 20. (Solid lines for $z = \pm h_a/2$ (interface); dashed lines for $z = 0$ (mid-plane).)

where the subscripts 'a' and 's' refer to adhesive and substrate, respectively. Hence, in the central regions of an axially loaded butt joint the adhesive layer is subjected to triaxial tensile stresses. It is noteworthy that if a rubbery adhesive, typically with a Poisson's ratio, ν_a, approaching 0.5, is used to bond relatively rigid substrate (i.e. $E_s \gg E_a$) then, approximately:

$$\sigma_r = \sigma_\theta = \sigma_z \tag{6.3}$$

Thus, in the central region of the joint, the rubbery adhesive is subjected to a hydrostatic tensile stress, a feature which has been recognized for many years [24–28] and which has important consequences, as discussed below.

Secondly, around the periphery of the joint, there is a region where a shear stress, τ_{rz}, may be acting and where both shear and tensile stresses are dependent upon the ratio of the radius of the joint to the thickness of the adhesive layer, i.e. r_0/h_a. These stresses also vary across the thickness of the adhesive layer. On the mid-plane of the adhesive the tensile stresses decrease to low values at the free surface and the shear stress, τ_{rz} ($z = 0$), is always zero. However, at the adhesive/substrate interfaces ($z = \pm h_a/2$) and there are significant stress concentrations at the edge of the joint. The largest stress is the tensile stress, σ_z, and, because of the high stress gradients, the value of this stress concentration is difficult to determine accurately but it has been estimated to be about 2.5. Adams *et al.* [10] have also considered the effect of the excess adhesive around the outside of the joint, i.e. the adhesive 'spew fillet', being present, since this is the usual practical situation. As the fillet transmits some of the applied axial load the level of the stresses in the adhesive layer decreases somewhat. For example, the σ_z stress concentration at the joint edge is reduced to about 1.8 in value, and the size of the fillet has little effect on this value. The important effects of such stress concentrations upon the measured properties of axially loaded butt joints will be discussed next.

6.4.3.2 Mechanical properties

One consequence of the stress distributions outlined above is that, when the substrates are more rigid than the adhesive, the measured, or apparent, Young's modulus, E'_a, (defined as the ratio of the applied tensile stress over the strain across the adhesive layer, i.e. σ_0/ϵ_z) is greater than the true Young's modulus, E_a, of the adhesive. This arises from the adhesive layer being constrained by the more rigid substrates, as discussed above. If it is assumed that the adhesive is perfectly constrained by the substrates, and if the latter are also assumed to be infinitely rigid, then the radial and hoop strains in the adhesive and substrate are zero and the apparent Young's modulus, E'_a, of the adhesive layer is given by:

$$E'_a = \left[\frac{(1-\nu_a)}{(1+\nu_a)(1-2\nu_a)}\right] \cdot E_a \tag{6.4}$$

For a structural adhesive, such as an epoxy, with a Poisson's ratio, ν_a, of about 0.35 the above equation reveals that the modulus measured via a butt joint test may be greater by about 50% or so; and will be even higher if the spew fillet is not removed since this will transmit some of the load. Work by Rutherford *et al.* [29] in 1968, using capacitance methods to measure the small deformations involved, demonstrated that the value of the Young's modulus ascertained from butt joints is indeed greater than the true Young's modulus of the adhesive by approximately this amount and a more recent study by Adams and Coppendale [30], using a resonant bar technique, has confirmed the validity of Equation 6.4. In the case of a rubbery adhesive, where the value of Poisson's ratio tends to 0.5, then the modulus of the adhesive as measured by the butt joint test may be orders of magnitude greater, and will be extremely dependent upon the value of the ratio of the radius of the substrates to the thickness of the adhesive layer. When $\nu_a \rightarrow 0.5$, and assuming the initially vertical cylindrical surfaces of the adhesive take up a parabolic shape, i.e. relaxing the unrealistic condition of perfect constraint, then the apparent Young's modulus, E'_a, of the rubbery adhesive is given by [25]:

$$E'_a = E_a(1 + r_0^2/2h_a^2) \tag{6.5}$$

Hence, for example, for a rubbery adhesive bonding two metal substrates with a ratio of $r_0:h_a$ of 20 the measured value, E'_a, is over two orders of magnitude greater than the true Young's modulus, E_a.

Bryant and Dukes [31] have measured the fracture stress, σ_f, of axially loaded butt joints employing either a rigid epoxy or rubbery silicone adhesive. They reported that for a wide range of radii of the joints, but with relatively thin adhesive layers, the fracture stress, σ_f, is generally independent of the radius of the substrate. However, in accord with other workers [32–38] they reported [31,39] that the fracture stress increased as the thickness of the adhesive layer decreased, as shown in Fig. 6.8. Until recently, the favoured explanations for this effect were based either upon the presence of higher internal stresses [40,41] or upon the statistical probability of larger flaws [32,38,39] in the joints employing relatively thick adhesive layers. However, recent elegant work by Gent [42,43] and Hilton and Gupta [44] has demonstrated that a continuum fracture mechanics approach to this problem provides a simple, and even quantitative, mechanism for this effect.

Gent [43] has examined the strength of butt joints consisting of a rubbery adhesive bonding relatively rigid substrates and measured the strength over a wide range of temperatures and rates. The results were found to yield a single master relation in terms of the effective rate of deformation, $\dot{\epsilon}a_T$ ($\dot{\epsilon}$ is strain rate and a_T is a shift factor), by means of the Williams, Landel and Ferry (WLF) rate–temperature equivalence [45] for viscous materials, as shown in Fig. 6.9. The WLF relation will be used several times in the present and the following chapters and it is, therefore, instructive to consider briefly the basis for this concept.

The WLF equation may be derived from the Doolittle equation [46] which relates the viscosity of a liquid to the free volume, i.e. the volume available for segmental motion. The amount of free volume increases as temperature increases and thus the segmental mobility of molecular chains increases and the degree of internal friction and viscosity, η, decreases. The value of η for a rubber therefore decreases as temperature increases and at $T > T_g$ the value $\eta(T)$ at any temperature may be related to that at the glass transition temperature, $\eta(T_g)$, by the WLF equation for simple viscoelastic materials:

$$\log\left[\frac{\eta(T)}{\eta(T_g)}\right] = \frac{-C_1(T-T_g)}{(C_2+T-T_g)} \tag{6.6}$$

In the WLF equation the constants C_1 and C_2 were originally thought to be universal constants, having values of 17.4 and 51.6, respectively, but they have

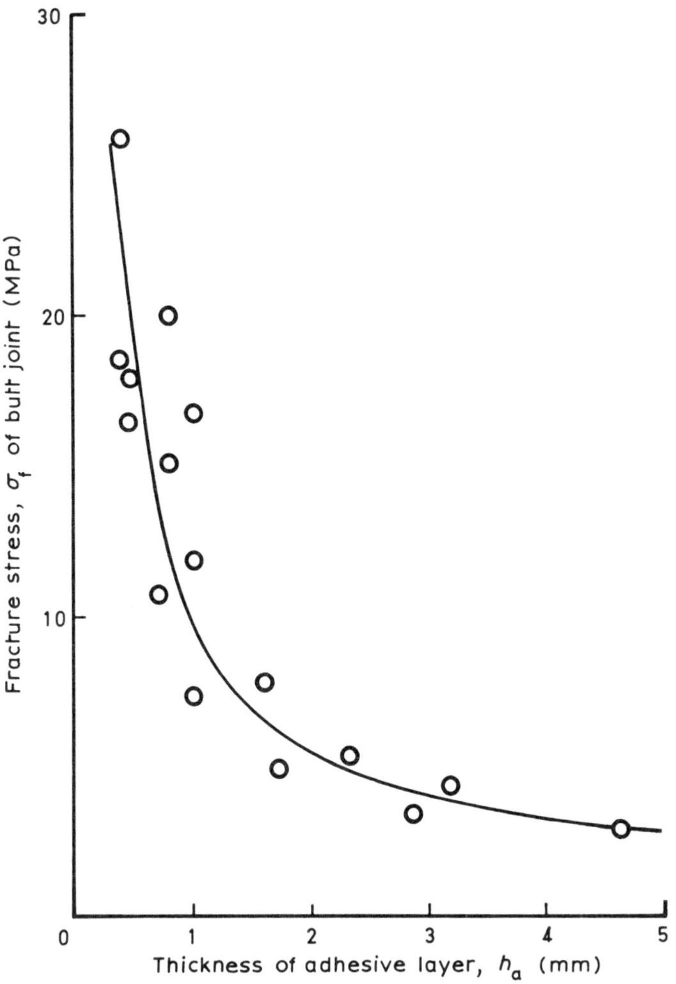

Figure 6.8 Butt joint strength, σ_f, as a function of the thickness of the adhesive layer, h_a, for a poly(methyl methacrylate) adhesive bonding steel substrates [34].

subsequently been found to vary somewhat from polymer to polymer [47]. Now if the effect of test rate and temperature on the value of a measured property largely arises from an internal friction or viscosity mechanism then, at test temperatures $T > T_g$, the WLF equation may be used to produce a single relation between the measured property and rate and temperature. Such a master curve is constructed by multiplying the rate by a shift factor, a_T, where:

$$a_T = \frac{\eta(T)}{\eta(T_g)} \tag{6.7}$$

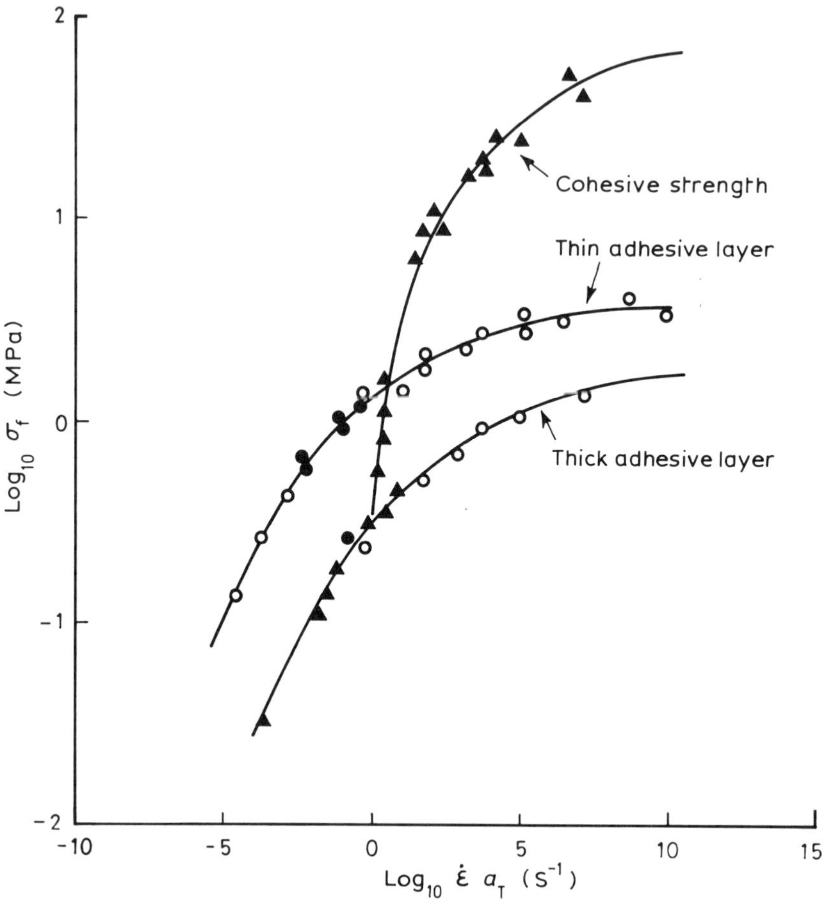

Figure 6.9 Cohesive and adhesive joint tensile fracture stresses plotted against effective rate of extension, $\dot{\varepsilon}a_T$, at 23 °C (styrene–butadiene–rubbery adhesive bonding poly-(ethylene terephthalate substrates) [43]. ▲ Cohesive tensile strength of adhesive. ● Cohesive-in-adhesive failure of butt joints. ○ Interfacial failure of butt joints.

This essentially reduces the rate and temperature to an equivalent rate, with the reference state being the T_g of the polymer; although the equations may be simply manipulated to give an equivalent rate with respect to any desired reference temperature. From Equation 6.6 the value of a_T for a reference temperature of T_g is given by:

$$\log a_T = \frac{-C_1(T - T_g)}{(C_2 + T - T_g)} \tag{6.8}$$

where T is the temperature of test.

Returning to the data of Gent, the fracture stress, σ_f, of the butt joints is shown as a function of the equivalent, or effective, strain rate, $\dot{\epsilon}a_T$, in Fig. 6.9 for a styrene–butadiene–rubber adhesive bonding poly(ethylene terephthalate) substrate. As may be seen the value for the joints containing a thin adhesive layer is always greater than that for those having a thick layer, but their relationship to the cohesive tensile strength of the rubbery adhesive is complex, emphasizing the complex stress distribution in such axially loaded butt joints. However, avoiding the need for detailed stress analyses, Gent interpreted these results using a fracture mechanics approach and proposed that the total strain energy density (i.e. strain energy per unit volume of adhesive), U_d, in the edge regions, where failure initiated due to the high stress concentrations, is given by the sum of the tensile and shear contributions:

$$U_d = \frac{1}{2} E_a \epsilon_z^2 (1 + 3\eta_c^2 r_o^2 / h_a^2) \tag{6.9}$$

where ϵ_z is the tensile strain and η_c is the stress concentration of the edge geometry. The corresponding average tensile stress, σ_o, is obtained from the effective value E'_a of the Young's modulus for the rubbery adhesive layer, so employing Equation 6.5:

$$\sigma_o = E'_a \epsilon_z = E_a \epsilon_z (1 + r_o^2 / 2h_a^2) \tag{6.10}$$

Now, from fracture mechanics principles and for a constant flaw size (see Chapter 7), a suitable failure criterion for joint fracture is when a critical value of U_d is attained. Hence, substituting for ϵ_z from Equation 6.10 into Equation 6.9, gives the adhesive fracture stress, σ_f, necessary to develop a critical strain energy, U_{df}, in the edge regions:

$$\sigma_f = (2E_a U_{df})^{1/2} (1 + r_o^2/2h_a^2)(1 + 3\eta_c^2 r_o^2/h_a^2)^{-1/2} \tag{6.11}$$

Thus, when the radius, r_o, is much larger than the thickness, h_a, of the adhesive layer, Equation 6.11 yields an inverse proportionality between the fracture stress, σ_f, and h_a. When the radius, r_o, is much smaller than h_a the value of σ_f is predicted to be independent of h_a. This general form of dependence is indeed commonly observed. Further, the above analysis shows that it arises from two competing effects: the greater stiffness of the test pieces, necessitating a greater average applied tensile stress to achieve the same level of strain energy, and the increased concentration of shear strain, and hence strain energy, in the edge regions of thinner adhesive layers causing failure at lower average stresses than would otherwise be required. These features also assist in explaining the relationship between adhesive joint strength and cohesive strength of the rubber. At relatively high rates of test and low temperatures (i.e. high $\dot{\epsilon}a_T$ values) the rubber becomes harder and stronger and eventually responds like a glassy solid. Under these conditions the cohesive strength is much greater than the joint strength and this may be partly ascribed to high stress concentrations in the joint. At low rates of extension and high temperatures the material deforms in a ductile manner and under these conditions the adhesive strength is similar or even slightly greater than the cohesive strength.

Hilton and Gupta [44] were concerned with structural adhesive joints consisting of a brittle epoxy adhesive bonding aluminium alloy substrates. They employed a continuum fracture mechanics method using the stress intensity factor approach (Section 7.2.3) and noted that Arin and Erdogan [48] had shown that the geometry factor, Q_{Ii} is a function of the ratio r_o/h_a for cracks in the mid-plane of the adhesive layer. As the adhesive layer decreases in thickness then the value of Q_{Ii} falls. Hence, from Equation 7.22a, for a given applied stress, σ_o, and flaw size, a, the value of the applied stress intensity factor, K_{Ii}, decreases as h_a decreases and is lower for the epoxy as a thin layer than as a bulk material, albeit only marginally for very small flaw sizes. Now a fracture mechanics criterion for joint failure may be taken to be when $K_{Ii} \geq K_{Ic}$. Therefore, assuming that the fracture toughness, K_{Ic}, of the adhesive itself is independent of h_a, this analysis correctly predicts that the adhesive joint strength will increase as h_a decreases.

However, the results of Adams and Coppendale [49] suggest that a more complex situation exists when both brittle and ductile structural adhesives are considered. They examined, both theoretically and experimentally, the joint strength for a range of epoxy adhesives bonding aluminium alloys. They showed that the measured stress–strain behaviour of a butt joint is dependent upon the triaxial stress state induced in the adhesive layer by the restraint of the substrate, as discussed above in Section 6.4.3.1. This may not only cause the apparent modulus of the adhesive layer to be greater than that of the bulk adhesive but may also affect the value of the yield stress of the adhesive relative to its brittle fracture strength. The triaxial stress state causes the adhesive in a butt joint to yield at a stress greater than the uniaxial yield stress of the bulk adhesive. Hence, for a ductile adhesive butt joints may be stronger than the bulk materials; this has been observed, for example, by Jennings [50] for epoxy adhesives tested at somewhat elevated temperatures. Conversely, the presence of the stress concentration can cause a butt joint containing a relatively brittle adhesive to fail at a lower applied tensile stress, σ_f, than the fracture stress, σ_{af}, of a bulk specimen tested in uniaxial tension, and this has also been experimentally observed. Therefore, Adams and Coppendale concluded that the relationship between the strength of a butt joint and that of a bulk specimen of the adhesive depended upon the ductility of the adhesive.

6.4.4 Single lap joints

6.4.4.1 Introduction

Of all the various geometries and methods of loading, the single lap joint loaded in tension has undoubtedly received the most attention from the stress analysts. As mentioned above there are two main reasons for this interest. Firstly, a lap joint which consists of two sheets of the substrates joined by a simple overlay is one of the most common joint designs employed in industry and, secondly, it is a simple and convenient test geometry for evaluating adhesive joints, as may be seen from its frequent use in the standard test methods listed in Table 6.1. However, the stresses in the adhesive layer are not, in practice uniform, and the stress concentrations arise from the differential straining of the bonded substrates and from the eccentricity of the loading path. The early work of Volkersen, who considered only the former of these factors, and Goland and Reissner, who were the first workers to consider both factors, is discussed below, before considering more recent work which considers the more realistic practical situation when the adhesive is modelled as an elastic–plastic material.

6.4.4.2 *Volkersen's analysis*

The simplest useful analysis considers the adhesive to behave only as a linear elastic solid and to deform only in shear. The single lap joint is shown schematically in Fig. 6.10a and the effects of loading the joint in tension with both inextensible and elastic substrates are illustrated in Figs 6.10b and c, respectively. In both cases the substrate bears the full load, F, just before the joint overlap and transmits it gradually to the other substrate through the adhesive. Thus, the stress in Substrate 1 will be highest at A and gradually diminish towards B where it will be zero and the converse will hold for Substrate 2. Now, in the case of the loaded joint with inextensible substrates (see Fig. 6.10b), the non-deformable substrates will move as solid blocks and the adhesive layer will possess a uniform distribution of longitudinal shear strain, γ_{12}, and hence uniform stresses. The variation in stress in the substrates is not, therefore, important if the substrates are inextensible but in practice, of course, they behave at least as elastic materials. For elastic substrates the tensile strain in Substrate 1 at A is larger than at B, with the converse for Substrate 2. This strain must progressively reduce over the bonded overlap length, l_a. Therefore, the picture of the deformation will be as shown in Fig. 6.10c and the distribution of shear stresses in the adhesive layer will be as shown schematically in Fig. 6.10d, with the largest shear strains, and hence stresses, in the adhesive occurring at the very ends of the overlap, at points C and D. (This cannot occur in practice since it implies a complementary shear stress on a free surface and no shear stress can exist on a free surface. The maximum shear stresses really occur near, but not actually at, the ends of the overlap.) Note, that assuming continuity of the adhesive/substrate interface, the uniformly sheared parallelogram of adhesive shown in Fig. 6.10b becomes distorted to the shape illustrated in Fig. 6.10c, and this is known as differential shear.

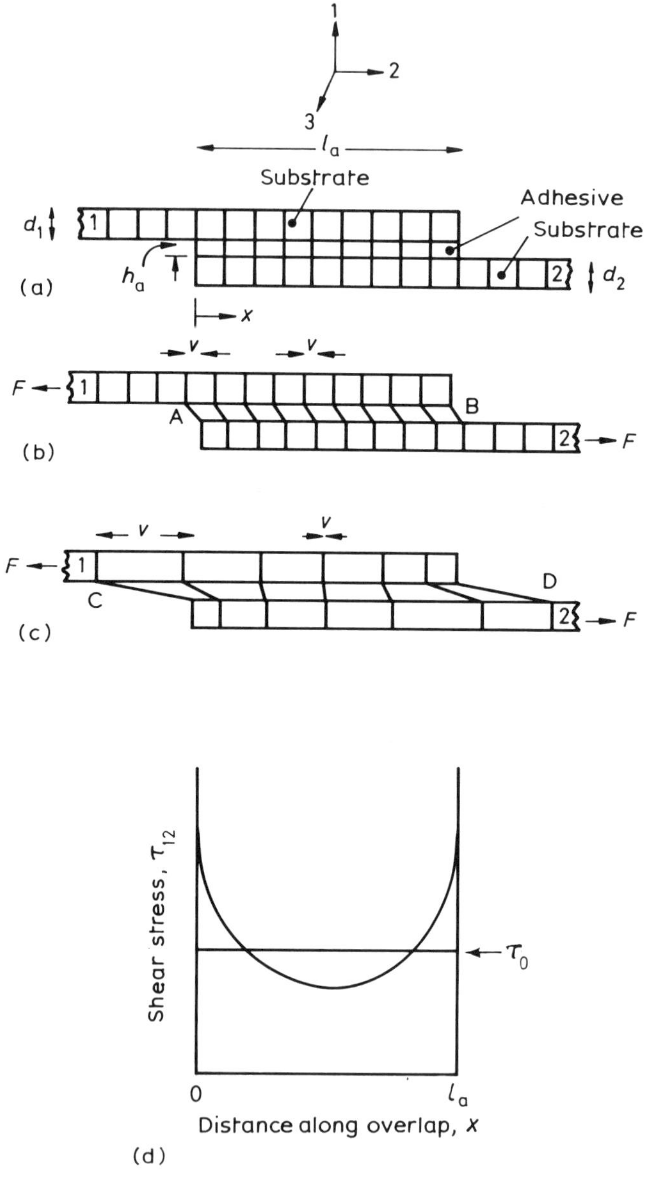

Figure 6.10 Schematic representation of single lap joint. (a) Unloaded, (b) loaded in tension, inextensible substrates, (c) loaded in tension, elastic substrates, and (d) distribution of elastic shear stress in the adhesive layer.

The stresses in the adhesive arising from the differential shear strains were first analysed by Volkersen [2] in 1938. The maximum shear stress, $\tau_{12}(\max)$, in the adhesive is related to the applied shear stress, τ_0, by:

$$\eta_c(\max) = \frac{\tau_{12}(\max)}{\tau_o} \tag{6.12}$$

where η_c is the stress concentration and the value of τ_o is given by:

$$\tau_o = \frac{F}{(\text{bonded area})} = \frac{F}{bl_a} \tag{6.13}$$

where F is the applied tensile load and b is the width of the joint. Assuming that the materials are Hookean solids (i.e. they are elastic materials and stress is proportional to strain) then:

$$\eta_c(\max) = \left(\frac{\Delta}{W}\right)^{1/2} \left[\frac{W - 1 + \cosh(\Delta W)}{\sinh(\Delta W)^{1/2}}\right] \tag{6.14}$$

where Δ is a dimensionless coefficient:

$$\Delta = \frac{G_a l_a^2}{E_{s2} d_2 h_a} \tag{6.15}$$

and W is defined by:

$$W = \frac{(E_{s1} d_1 + E_{s2} d_2)}{E_{s1} d_1} \tag{6.16}$$

and $E_{s1} d_1 > E_{s2} d_2$, E_{s1} is the tensile modulus of Substrate 1, E_{s2} is the tensile modulus of Substrate 2, d_1 is the thickness of Substrate 1, d_2 is the thickness of Substrate 2, G_a is the shear modulus of the adhesive and h_a is the thickness of the adhesive layer.

For substrates for which $E_{s1} d_1$ and $E_{s2} d_2$ are equal, then W reduces to a value of 2 and Equation 6.14 becomes [51]:

$$\eta_c = (\Delta/2)^{1/2} \coth(\Delta/2)^{1/2} \tag{6.17}$$

Thus, the stress concentration factor, η_c, becomes simply a function of a single dimensionless coefficient, Δ. It can, therefore, be readily appreciated that the theory of Volkersen predicts that decreasing the overlap length, l_a, or shear modulus of the adhesive, G_a, or increasing the stiffness of the substrate or the thickness of the adhesive layer, will decrease the shear stress concentrations in the adhesive layer. However, whether these changes result in stronger joints

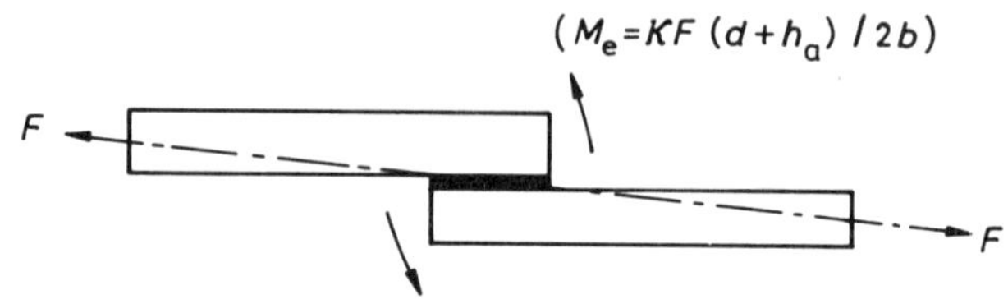

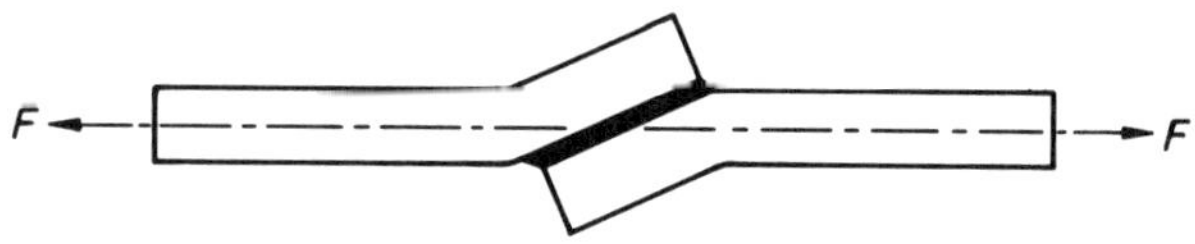

Figure 6.11 Schematic representation of how the eccentricity of the loading path in a single lap joint gives rise to bending moments and how deformation of the substrates upon loading may reduce these bending moments.

will often depend upon other factors. For example, a decrease in G_a, possibly by plasticizing the adhesive, is usually accompanied by a decrease in the shear fracture stress, τ_{af}, of the adhesive. So, if the criterion of joint failure is taken to be when $\tau_{12}(\max) = \tau_{af}$, then little improvement may result from such a change in adhesive formulation; since, although the decrease in G_a will reduce the stress concentration, η_c, in the adhesive, the accompanying decrease in the value of τ_{af} may well offset this advantage. Also, it is notworthy that an increase in the stiffness of the substrates will decrease the shear stress concentrations. Thus, for example, a change in the lay-up sequence of a fibre composite substrate, which will alter its longitudinal modulus, will affect the local stress concentrations and subsequent joint strength. These aspects are considered in more detail later, after further developments in the stress analysis of single lap joints are considered.

6.4.4.3 Goland and Reissner's analysis

The situation examined by Volkersen, whilst it indicates the importance of various parameters on the potential strength of a single lap joint, is incomplete in that it takes no account of the tensile stresses generated in the adhesive as a result of the eccentricity of the loading of the joint. This is shown schematically in Fig. 6.11a and this additional feature was first considered by Goland and Reissner [3] in 1944 who recognized from laboratory experiments with lap joints that the bending of the substrates outside the joint region had a pronounced effect upon the stress distribution in the joint itself. They again assumed that the adhesive and substrates behaved as linear elastic, Hookean materials and expressed the effect of the bending of the substrates in their bending moment factor, κ, and associated rotation factor, κ^*. These parameters, κ and κ^*, are not independent of one another but κ is usually the dominant term; κ is dimensionless and is the ratio of the existing bending moment just before the bonded overlap to the value of this moment for inflexible substrates:

$$\frac{1}{\kappa} = 1 + 2(2)^{1/2}\tanh\left\{\left(\frac{3}{2}(1-\nu_s^2)\right)^{1/2}\frac{l_a}{2d}\left(\frac{F}{E_s bd}\right)^{1/2}\right\} \qquad (6.18)$$

where ν_s is the Poisson's ratio of the substrate and d is the thickness of the substrate. Thus, κ depends upon the geometry of the joint, the elastic properties of the substrates and the stress on the substrates. The value of κ is unity for undeformed substrates but for increasing load, especially with 'flexible' substrates, the value of κ will diminish towards zero as the limit. The underlying reason for this decrease in the value of κ may be seen from Fig. 6.11b and arises because as the load is increased the substrates bend, assuming joint fracture does not intervene, and so enable the overlap to rotate. This obviously moves the line of action of the applied load closer to the centre line of the substrates, so reducing the bending moment. Thus, the value of κ falls and the predicted stress concentrations in the joint decrease. It is noteworthy that it is therefore predicted that 'prebending' the substrates, to assume a configuration similar to that described above, prior to testing the joint will lead to stronger joints. This is indeed true [52] but is difficult to use in practice, where, to reduce the adverse effects of the eccentricity of loading, a more typical solution is to increase the overlap length, l_a, and/or use a double lap or a modified lap joint, as discussed below.

Goland and Reissner considered two cases:

(a) Where the adhesive layer is extremely thin and of similar elastic stiffness to the substrates, so that its deformations are of little importance, for example, adhesive bonded wooden joints. This was quantified by the condition that $h_a/d \ll E_a/E_s$.

(b) Where the adhesive layer is thin but its deformation makes a significant contribution to the stress distribution in the joint, as in bonded metal-to-metal joints. This was quantified by the condition that $h_a/d \gg E_a/E_s$.

They established the quantitative limits for these assumptions from strain–energy considerations, which later work [53] actually suggested were somewhat conservative. For the first case, Sancaktar and Lawry [52] have recently used a photoelastic technique to examine the elastic tensile stresses and reported good

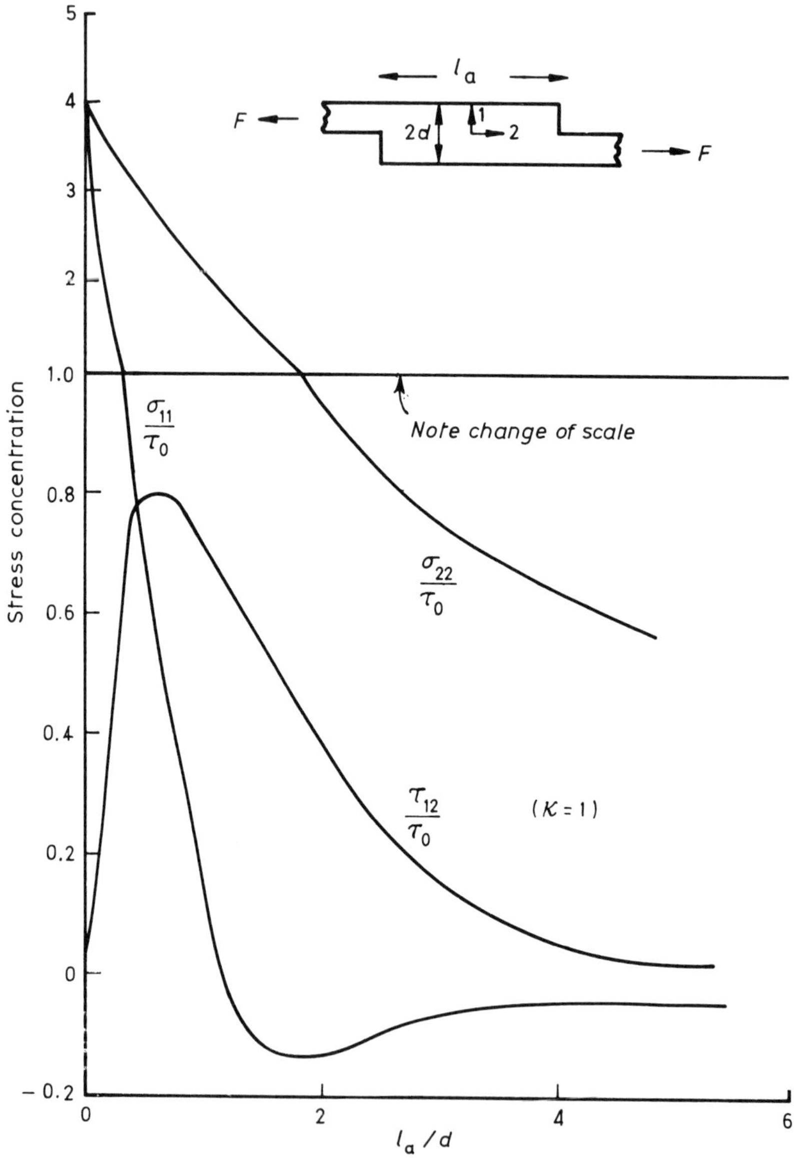

Figure 6.12 Stress distribution along the shear plane as a function of distance from the edge of the overlap. The bending moment factor, κ, is equal to one [3].

agreement with the theoretical predictions of Goland and Reissner. For the second case, Ishai *et al.* [54] have also confirmed that for joints which do exhibit essentially only elastic behaviour the analytical solutions of Goland and Reissner are in good agreement with the experimental results for both the shear and tensile stress distributions in the adhesive layer.

For Goland and Reissner's first case, the stress distributions are shown in Fig. 6.12 and the variations of the maximum stresses with the bending moment factor, κ, are shown in Fig. 6.13. As may be seen, the transverse (out-of-plane) tensile stresses, σ_{11}, which operate perpendicular to the plane of the joint are most significant and are highest at the ends of the overlap joint; hence such induced stresses are often referred to as cleavage or peel stresses, or sometimes

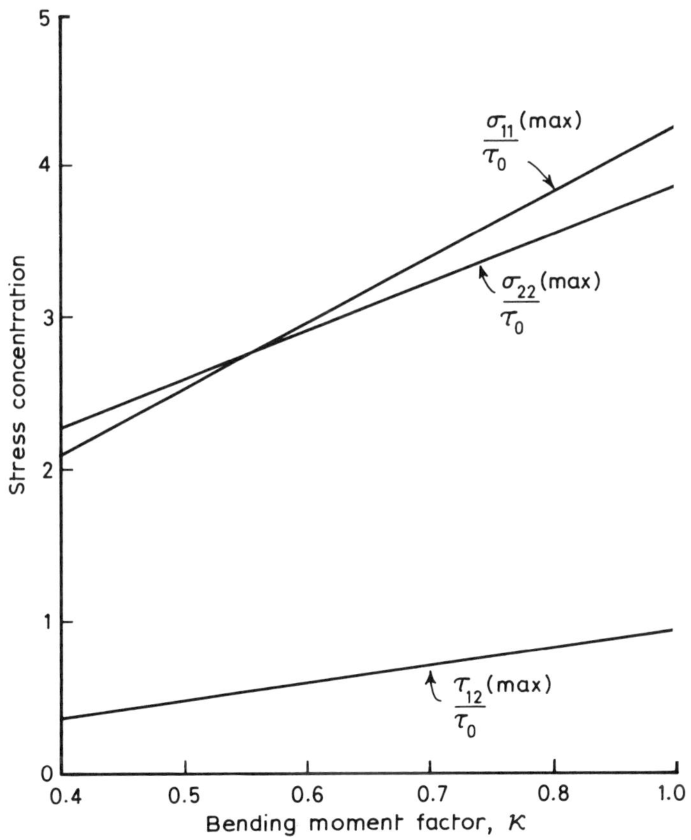

Figure 6.13 Reduction of the maximum stresses in the shear plane of a single lap joint resulting from bending of the substrates, i.e. diminishing value of κ [3].

as 'tearing' stresses since they act to tear the joint apart. Note that, as discussed above, Fig. 6.13 reveals that all the various stresses decrease as the bending moment factor, κ, decreases, i.e. as the substrates bend.

For their second case:

$$\frac{\tau_{12}(\max)}{\tau_o} = \frac{1+3\kappa}{4}(2\Delta)^{1/2}\coth(2\Delta)^{1/2} + \frac{3}{4}(1-\kappa) \qquad (6.19)$$

and tensile stresses, σ_{11} and σ_{22}, are also present in the adhesive layer. Thus, for relatively long overlaps and $\kappa = 1$, Equations 6.14 and 6.19 reveal that the shear stress concentration factor predicted by Goland and Reissner will be twice as large as that predicted from the analysis of Volkersen. This arises, of course, because the Volkersen analysis takes no account of stress concentrations caused by eccentricity of the loading path.

Figure 6.14 Schematic representation showing interlaminar failure of fibre composite substrates arising from the transverse (out-of-plane) tensile stresses, σ_{11}, which occur due to the eccentricity of the loading path, and the relatively poor transverse strength of fibre composites.

The transverse (out-of-plane) tensile, or cleavage, stresses, σ_{11}, introduced by the eccentricity of the loading path are extremely important and are invariably a major factor in the fracture of single lap joints. Their influence can be readily demonstrated in several ways. For example, firstly, they frequently lead to joint fracture by enabling cracks to initiate and propagate from the ends of the overlap. Secondly, with metal-to-metal joints the transverse tensile stresses are often sufficiently high to exceed the yield stress of the metal substrates, and so after joint fracture the substrates are observed to be plastically (i.e. permanently) deformed. Thirdly, in the case of bonding fibre composites, the presence of the transverse tensile stresses coupled with the low interlaminar tensile strength of composite laminates, frequently causes premature joint failure by delamination of the substrates, as shown schematically in Fig. 6.14. Indeed, it is considered [4,55,56] that single lap joints should not be used as a basic design for load-bearing joints with composites unless attached to a moment-resistant support or a very long overlap is used to nullify the effects of the eccentricity in the loading path.

6.4.4.4 Later work (elastic analyses)

The pioneering work of Volkersen and Goland and Reissner clearly established the basic stress distribution in a single lap joint when the materials forming the joint are considered to be elastic. However, even setting aside any non-linear behaviour of the adhesive, several features were incorrect or overlooked in these early analyses. For example, (i) the shear and transverse tensile stresses are assumed to be uniform across the thickness of the adhesive layer, (ii) the longitudinal (in-plane) tensile stresses, σ_{22}, in the adhesive layer cannot always be deduced, (iii) the shear strains in the substrate are neglected, (iv) only stresses in two dimensions are considered and (v) the effect of features such as the presence of adhesive spew fillets are ignored. Several workers [57–62], using both closed-form analytical and finite-element analyses, have subsequently developed elastic analyses which take into account the above features, although, in fact, the results of Goland and Reissner are found to be relatively accurate and have stood the test of time extremely well.

Adams and Peppiatt [60] have considered the problem of the in-plane transverse stresses and to ascertain the magnitude of such stresses they have used experimental models and analytical and finite-element analyses solutions of the Volkersen theory, but in three dimensions. They demonstrated that Poisson's ratio strains generated in the substrates cause shear stresses, τ_{13}, in the adhesive layer and tensile stresses, σ_{33}, in the substrate acting transverse to the direction of the applied load, but in the plane of the joint. For metal-to-metal joints the transverse shear stress, τ_{13}, has a maximum value of about one-third of the maximum longitudinal shear stress, $\tau_{12}(\max)$, and this occurs at the corners of the overlap. This, therefore, enhances the shear stress concentration which exists at this point due to the effects described above. Bonding substrates of dissimilar stiffness produces greater stress concentration in the adhesive than when similar substrates are employed.

Adams and Peppiatt [61] have also considered the effect of a spew fillet being present at the ends of the overlap, since this is the situation in practice. Thus, unlike previous analyses where the adhesive has been modelled with a square, 90°, edge (Fig. 6.15a), they used a finite-element analysis to model the adhesive with a triangular 45° spew fillet (Fig. 6.15b). They found that the stresses in the fillet are largely tensile and that the maximum principal tensile stresses occur near the corner of the substrate at approximately 45° to the axis of loading, as indicated in Fig. 6.15b. They concluded that, although the presence of the spew fillet reduces the stress concentrations compared to a square adhesive edge, failure in a lap joint is initiated by the still relatively high tensile stresses within the spew. With good intrinsic adhesion, a crack initiates near the corner of the substrate and runs out to the free edge and towards the interface, where it travels in the adhesive but close to the interface. It meets a crack running in the opposite direction and results in the typical failure pattern shown in Fig. 6.16.

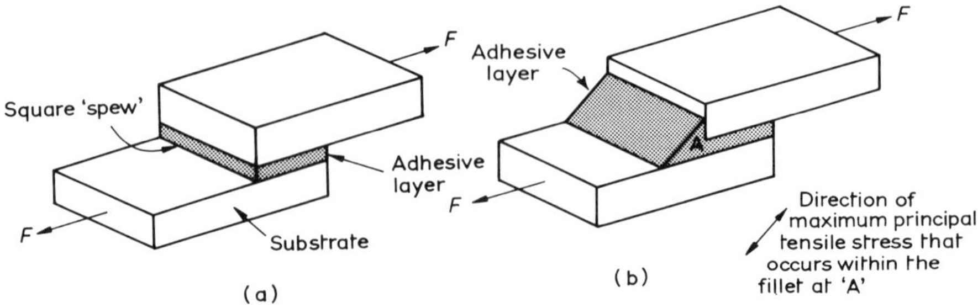

Figure 6.15 Diagram of lap joints showing different types of adhesive edge. (a) Square (90°) edge, (b) 45° spew fillet [61].

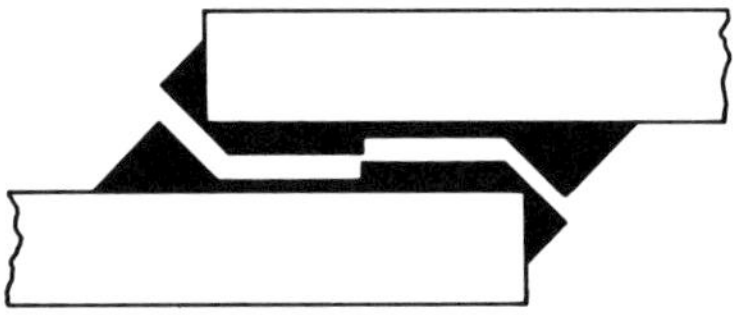

Figure 6.16 Schematic representation of typical locus of joint failure in single lap joints upon testing a good adhesive system.

If, on the other hand, the intrinsic adhesion is poor, then the tensile stresses in the spew will tend to pull the adhesive spew away from the substrate, rather than causing the spew itself to crack.

6.4.4.5 Elastic–plastic analyses

The above analyses all consider the substrates and adhesive to behave as elastic materials; however, most of the adhesives commonly employed exhibit elastic–plastic behaviour, especially under shear loads as shown in Fig. 6.3. Several groups of workers [4,6,55,56,63–65] have recently sought to include this behaviour in their calculations. The most extensive studies have been conducted by Hart-Smith [4,55,56,66] who has used closed form analytical solutions, employing iterative solutions on a digital computer for the more complicated joint configurations. These studies show that inclusion of adhesive plasticity under shear loading in the analysis may decrease the stress concentrations substantially, and thus increase the predicted joint strengths, compared with a solely elastic analysis. These analyses are similar to those available from the Engineering Sciences Data Unit [63,64,67] and both model the adhesive shear behaviour as an elastic–plastic material, as shown in Fig. 6.17a, although they use somewhat different idealizations and definitions for the shear be-

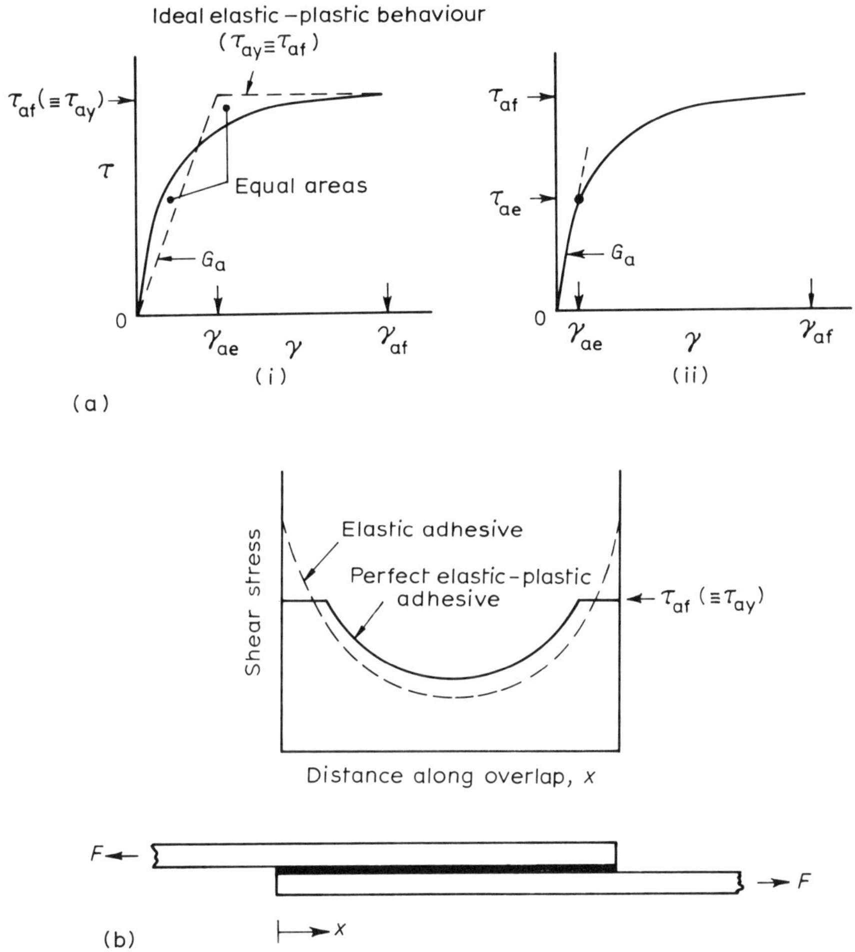

Figure 6.17 Elastic–plastic behaviour of the adhesive. (a) Models of the elastic–plastic shear stress versus strain curve of the adhesive for use in stress analysis calculations: (i) the dashed line is the representation suggested by Hart-Smith [4,55]; i.e. perfect elastic–plastic behaviour, (ii) representation suggested by ESDU [64,67]. (b) Schematic representation of the effect of elastic–plastic, as opposed to solely elastic, behaviour of the adhesive on the stress distribution in a single lap joint.

haviour. Such properties are typically measured using the napkin-ring or thick-adherend shear test, as described in Section 6.4.2. The failure criterion commonly assumed is that joint fracture results when the maximum shear *strain* at the ends of the overlap attains the value of the shear strain, γ_{af}, at fracture of the adhesive. As pointed out by Adams and Wake [68], the shear strain distribution is not simply a multiple of the low-load case, since the assumption of a limiting (plastic) shear stress in the adhesive layer causes a distortion to the elastic shear-lag theory. The effect of assuming elastic–plastic behaviour of the adhesive, compared to solely elastic, on the shear stress and strain distribution is shown schematically in Fig. 6.17b. Since such analyses do not directly take into account the out-of-plane tranverse tensile stresses, σ_{11}, they are more accurate for the double lap joint when such stresses are far lower. Hence, we will return to these analyses in Sections 6.4.5 and 6.4.6.

Considering the out-of-plane tranverse tensile stresses, σ_{11}, in the case of the single lap joint, then obviously, from the above comments, they are most significant and cannot be ignored. Therefore, to accompany his elastic–plastic analysis of the shear stresses Hart-Smith [66] derived solutions for these transverse stresses, but he assumed only elastic behaviour of the adhesive since this was considered to model the adhesive behaviour in tension sufficiently accurately (see Fig. 6.3). For a single lap joint with identical substrates, the maximum value, $\sigma_{11}(\max)$, in the adhesive at the ends of the overlap is given by:

$$\sigma_{11}(\max) = M_e \left(\frac{E_a}{2h_a D_b} \right)^{1/2} \tag{6.20}$$

where the adhesive is assumed to be isotropic, D_b is the bending stiffness per unit width of the substrates $[D_b = E_s d^3/12(1-\nu_s^2)b]$ and M_e is the bending moment per unit width in the substrates at the ends of the overlap. M_e is given by:

$$M_e = \frac{\kappa F(h_a + d)}{2b} \tag{6.21}$$

Hart-Smith's results are in good agreement with those of Goland and Reissner (Section 6.4.4.3), but they do suggest that the latter's results tend to over-estimate somewhat the transverse tensile stresses. Hart-Smith also showed that if the thickness of the substrates are not identical then these stresses increase in intensity. In the practical design of joints the philosophy of Hart-Smith was to eliminate, or at least reduce greatly, the transverse tensile stresses and this was achieved by such methods as using very long overlaps or double lap joints of various designs, as discussed below.

6.4.4.6 Mechanical properties

Geometry effects – from the above discussions, and in particular Equations 6.15, 6.17 and 6.19, it is evident that the shear stress concentrations in the adhesive layer near the ends of the overlaps increase with increasing overlap length. This is directly reflected by the observation that the average applied shear stress, τ_f, at joint fracture decreases with increasing overlap length, l_a, as

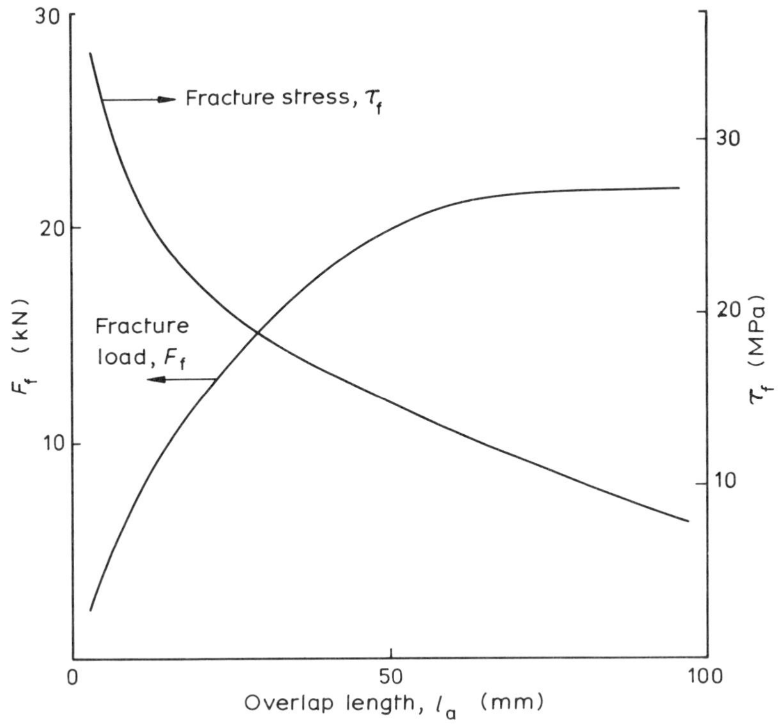

Figure 6.18 Experimental fracture load, F_f, and stress, τ_f, of single lap joints of modified phenolic/aluminium alloy as a function of overlap length, l_a[51,69].

shown in Fig. 6.18 [51,69]. However, the practically important measure of joint strength is the load a joint can carry and the breaking load, F_f, at joint fracture increases up to an approximately constant, plateau, value as l_a is increased; recall that these parameters are related by $\tau_f = F_f/b \cdot l_a$, where b is the width of the joint.

At first sight the data in Fig. 6.18 might lead to a designer choosing the minimum overlap length necessary to carry his design load safely. However, if too short an overlap length is employed high stresses will exist in the centre of the overlap upon loading and thus make the joint more susceptible to creep and environmental attack (see Sections 8.3.2.3 and 8.3.5). The overlap should therefore be sufficiently long to ensure that these stresses are elastic in nature and low in magnitude. Further, if a ductile adhesive is used then a relatively long overlap (e.g. l_a:d ratio of about 80:1 compared to a typical ASTM D 1002 lap joint with a value of about 8:1) will reduce [4] the eccentricity of the load path (see Equation 6.18). Hence, from Equations 6.19, 6.20 and 6.21, the shear stresses and the transverse tensile stresses, σ_{11}, will be reduced and the

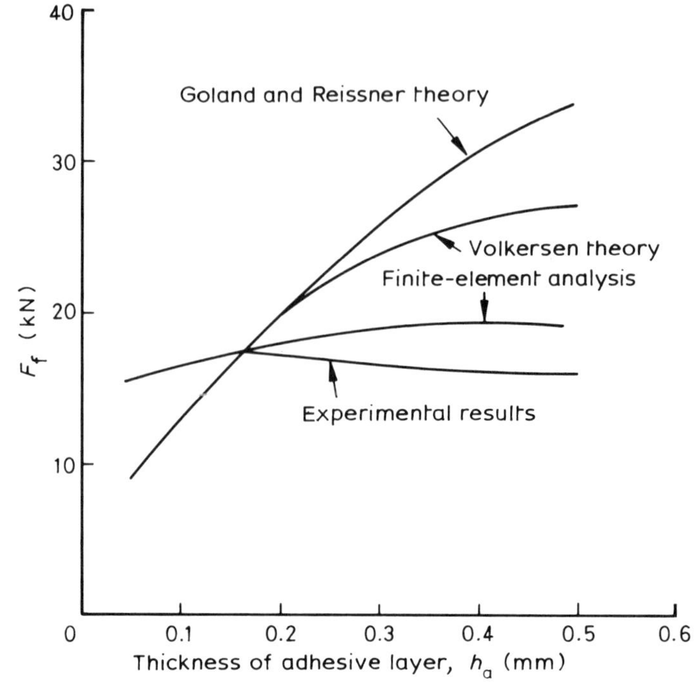

Figure 6.19 Influence of thickness, h_a, of adhesive layer upon experimental and theoretical fracture loads of epoxy/aluminium alloy single lap joints [61].

plateau breaking load, F_f(max), will be greater, and actually continue to rise somewhat with increasing l_a. Now, increasing the thickness of the substrates results in a lower shear stress concentration (see Equation 6.15) but may increase the transverse tensile stress, σ_{11} (see Equations 6.18, 6.20 and 6.21). Thus, increasing the thickness of the substrates results in higher joint strengths, up to a certain limit.

The theoretical analyses outlined above predict that the stress concentrations will decrease and hence that the fracture stress, τ_f, of lap joints will increase as the thickness, h_a, of the adhesive layer increases. Generally, if all other joint parameters are held constant the value of τ_f is predicted to be proportional to the reciprocal of the square-root of the thickness, h_a, provided that the thickness is relatively small. Figure 6.19 shows [61] the predictions for three different analyses for an epoxy/aluminium alloy single lap joint and all suggest that the breaking load will increase as the value of h_a increases. However, the experimental results for the joints, in accord with other work [38, 69–71], shows that the actual fracture load, F_f, does not increase with increasing h_a and may even fall slightly.

Table 6.3 Effect of increasing the toughness of a brittle polybismaleimide adhesive on the single lap shear strength of steel joints [72–75].

Rubber concentration (phr)	Fracture strain, ϵ_{af} (%)	Fracture energy (J/m^2)	Joint strength (MPa)
0	0.5	15	4
10	0.9	45	8
30	2.0	140	18
50	2.2	180	24
100	3.2	460	26

(a) The test temperature was 20 °C.
(b) The glass transition temperature of polybismaleimide was 260 °C.

Adhesive properties – obviously the properties of the adhesive play a major role in influencing the joint strength. Considering the structural adhesives, then a ductile adhesive with a large shear strain–energy density at fracture, invariably reflected in a high value of γ_{af} (see Fig. 6.3), will lead to high joint strengths. To withstand the tensile stress concentrations then a good resistance to crack propagation, as reflected by a high tensile fracture strain, fracture energy, G_{Ic}, and peel strength are desirable properties. The adhesive also needs to possess a reasonably high yield stress, to avoid excessive creep in the joint. Some of these aspects are illustrated in Table 6.3 where Kinloch *et al.* [72–75] have toughened a basically very brittle polybismaleimide adhesive by incorporating a rubber into the polybismaleimide to induce the formation of a second polymeric phase in the material. Note that as the fracture strain and the fracture energy of the material increases the single lap joint strength rises dramatically.

Obviously, if only very thin or inherently weak substrates are to be bonded, or only low operating loads are envisaged, then there is no need to select an adhesive with the very best properties, and the choice may rest upon other features, such as cost, ease of application, hardening temperature, etc.

Now, often all the desirable properties of the adhesive are required to be maintained over a wide range of operating temperatures. The effect of increasing the test temperature on the mechanical properties of structural adhesives is essentially to decrease the modulus, yield stress (if appropriate) and fracture stress but to increase the strain at fracture [4,7,9]. As evident from the above discussions on stress concentrations and failure criteria, these various effects lead to a complex dependence of the lap shear strength upon test temperature and some typical results for various adhesive compositions is shown in Fig. 6.20 [76,77]. It should be noted that the strengths shown are for the joint exposed at the test temperature for only a relatively short time. Long exposure at the higher temperatures might well lead to chemical degradation of

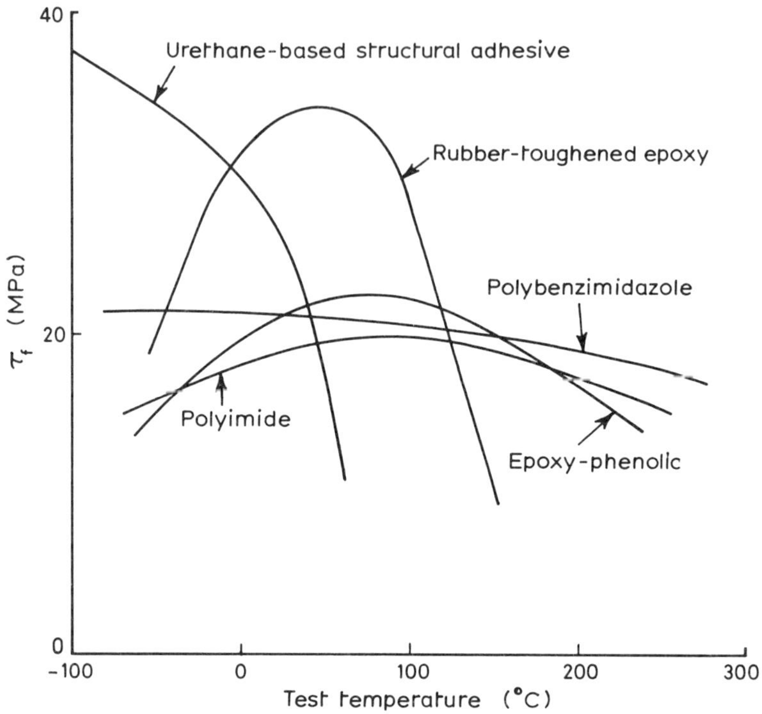

Figure 6.20 Typical strengths of single lap joints loaded in tension using various structural adhesives as a function of test temperature [76,77].

the adhesive occurring and so result in significant losses in joint strength as a function of time at the test temperature. The rubber-toughened epoxy gives a very high single lap shear strength at room temperature, reflecting its good ductility in shear and good toughness and good resistance to crack propagation under tensile loading at this temperature. But it has a relatively low glass transition temperature (about 120 °C) and therefore the joint strength is poor at elevated temperatures. The combination of these good mechanical properties with a high glass transition temperature is a current aim of many adhesives technologists. The epoxy–phenolic, polyimide and polybenzimidazole adhesives exhibit the best high temperature strengths of those shown, but at the sacrifice of some lap shear strength compared to the rubber-modified epoxy, and they are also brittle and so possess a poor resistance to cleavage and peel loads. Cotter [76] concluded that the epoxy–phenolic formulations are available for long-term use to about 175 °C and are capable of short excursions to temperatures as high as 250 °C. The particular polyimide composition shown may be used for very long times at 260 °C and maintains a useful strength at

425 °C for about 1 h. Polybenzimidazole adhesives have a higher temperature ceiling than polyimide formulations but are markedly susceptible to oxidative degradation above 260 °C. The relationships between low-temperature performance and the structure and formulation of the adhesive have also been considered [78,79], and the polyurethane adhesives are particularly good at very low test temperatures.

Finally, increasing the rate of testing of structural adhesive joints is generally equivalent to decreasing the test temperature, although at a given temperature the test rate usually needs to be changed by an order of magnitude, or more, to induce major changes in the mechanical properties of the joints. Again the importance of the joint design must be emphasized since the joint geometry and choice of substrate materials will greatly influence the measured properties of the joint [80,81]. For example, under impact test conditions it may be possible to increase greatly the energy absorbing capabilities of the joint by judicious design of the joint so as to optimize energy dissipative deformations, e.g. plastic flow, in the substrates, in addition to selecting a tough adhesive.

6.4.5 Double lap joints

The double lap joint is shown schematically in Fig. 6.21a and its symmetry results in very much reduced out-of-plane normal stresses, σ_{11}, across the adhesive layer. However, even in symmetrical double lap joints, internal bending moments cause normal stresses across the adhesive layer, as shown in Fig. 6.21b. This arises because the load is applied through the adhesive to the substrates away from their neutral axes and this effect was first considered in the case of elastic materials by Volkersen [82], and subsequently by Adams and Peppiatt [61]. The latter authors employed an elastic finite-element analysis and modelled the joints with an adhesive spew fillet. They reported that the ratio of maximum principal tensile stress in a double lap joint to that in a single lap joint (equivalent in dimensions to half the double lap) is 0.54, i.e. the use of a single lap joint results in an increase in the bending moment which increases the principal tensile stress by approximately a factor of two. This was in good agreement with experimental observations where the ratio of the half-failure load of a double lap joint to the failure load of the single lap joint was 1.92.

Several authors [4,6,8,55,56,63–65,83] have considered the effect of adhesive plasticity and have shown that, as for the single lap joint, such behaviour reduces the calculated stress concentrations. Hart-Smith has conducted an extensive examination of this effect and Fig. 6.22a illustrates some typical results. As for the single lap joint, the load the joint can withstand increases as the overlap length, l_a, increases for short overlaps but no further strength is to be gained beyond a certain point. It will be recalled that the reason behind this observation lies with the stress distribution in the joint, i.e. the maximum stress and strain concentrations are located at or near the ends of the overlap, and

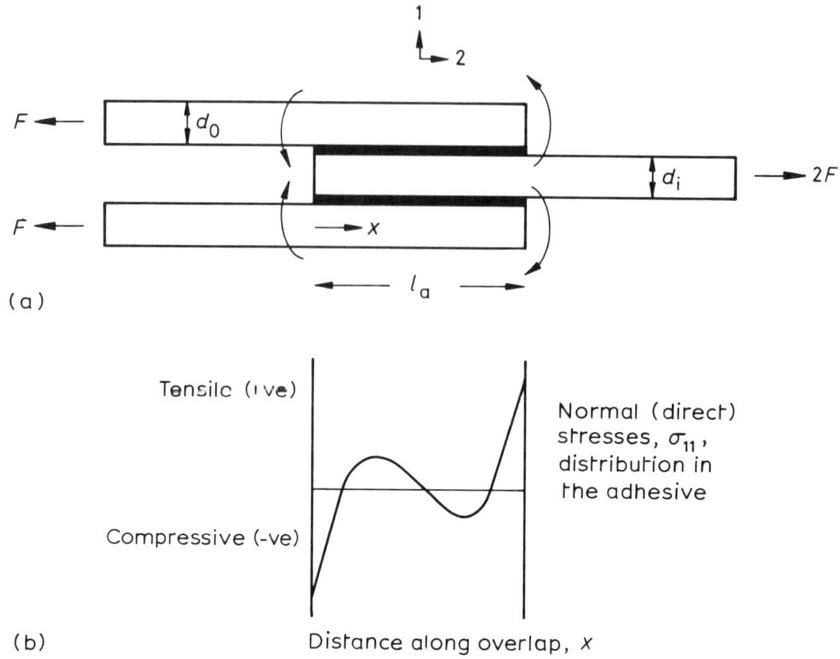

Figure 6.21 Double lap joints. (a) Schematic representation of a double lap joint showing bending moments induced in the outer substrates, (b) diagram of distribution of normal stresses, σ_{11}, along the length of the overlap which arise from the bending moments.

this is illustrated schematically in Fig. 6.22b. Joint failure is assumed to occur when the maximum shear strain, γ_{af}, of the adhesive is attained in these regions. The plateau strengths for a perfect elastic–plastic adhesive are given by:

$$\frac{\tau_0(2\Delta)^{1/2}}{2\tau_{af}} = \left(1 + 2\frac{\gamma_{af}}{\gamma_{ae}}\right)^{1/2} \tag{6.22}$$

Hart-Smith has also [4,66,83] considered the out-of-plane transverse tensile stresses, σ_{11}, which are developed across the adhesive layer and which are at a maximum at the ends of the overlap of the outer substrates. The maximum value of these stresses, $\sigma_{11}(\max)$, is given in Fig. 6.23 and the ratio $\sigma_{11}(\max)/\tau_{12}(\max)$ is given by:

$$\frac{\sigma_{11}(\max)}{\tau_{12}(\max)} = \left[\frac{3(1-\nu_s^2)\,E_a}{E_s}\right]^{1/4} \left(\frac{d_o}{h_a}\right)^{1/4} \tag{6.23}$$

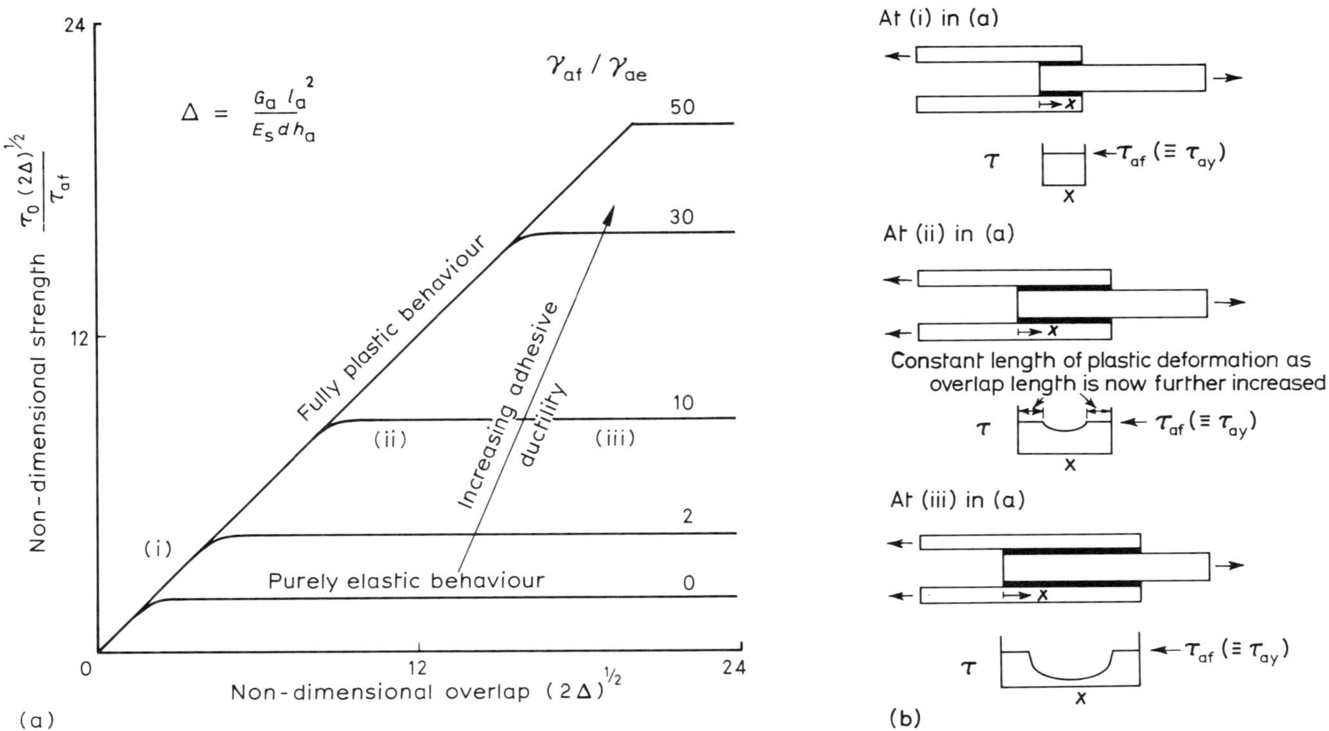

Figure 6.22 (a) Theoretical shear strength of balanced (i.e. $d_i = 2d_o$) double lap joints as a function of the degree of plasticity exhibited by a perfect elastic–plastic adhesive (i.e. as measured by γ_{af}/γ_{ae}) [4,55]. (b) Schematic representation of influence of overlap length, l_a, on the distribution of shear stress in the balanced double lap joints; relevant value of l_a shown in Fig. 6.22a for γ_{af}/γ_{ae} of 10. Note that once on the plateau in Fig. 6.22a the main effect of increasing l_a is to increase the length and depth of the elastic trough in the middle of the overlap [4,55].

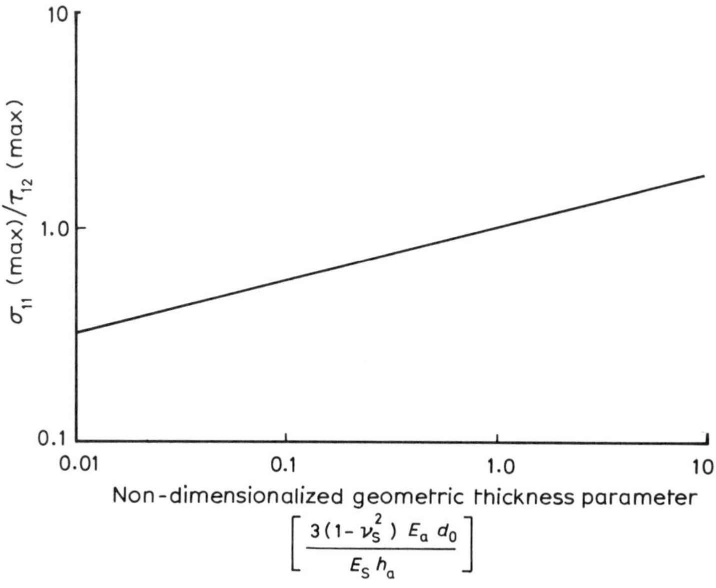

Figure 6.23 Level of transverse, σ_{11}, stresses in a double lap joint [55].

where at joint failure $\tau_{12}(\text{max}) = \tau_{\text{af}}$. It may be seen that to decrease the value of $\sigma_{11}(\text{max})/\tau_{12}(\text{max})$ the design modifications that are required are a reduction in the tip thickness of the outer substrate and a local increase in the adhesive layer thickness, as discussed below.

Thus, double lap joints behave in a similar fashion to single lap joints but have much reduced transverse stresses, σ_{11}, acting. As noted above, this leads to higher joint strengths than simply expected from a doubling of the bonded area. It also means that the analyses which take no account of such stresses do in fact model the failure loads reasonably well in many instances, especially those which allow the adhesive to behave as an elastic–plastic material. However, it is still the σ_{11} stresses which tend to be those which cause joint fracture and, as for the single lap joint, this is particularly the case when bonding fibre composite substrates due to their low transverse, interlaminar, strengths. Generally to reduce the stress concentrations in lap joints, particularly these tensile stresses, various modified types of lap joint have been described in the literature and are to be found in many practical joint designs; such joints are described in the following section.

6.4.6 Modified lap joints

Various modifications have been suggested to both the basic single and double lap-shear joints in order to decrease their stress concentrations and so raise

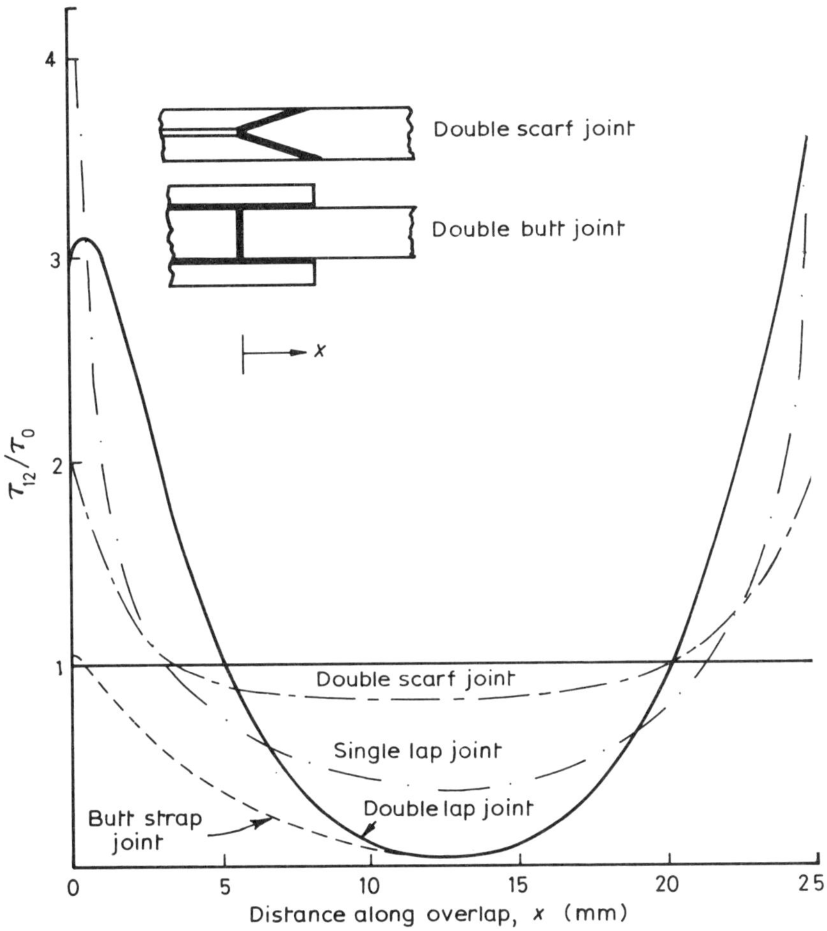

Figure 6.24 Shear stress concentrations for various lap joint designs, elastic analyses for typical epoxy adhesive/aluminium alloy joints [6].

their joint strength. Some typical examples from the work of Adams [6] are shown in Fig. 6.24.

The scarf joint [4,6,51,55,83,84], and its variation the stepped scarf joint [55,63,64,85], are good examples of joints with low stress concentrations and also with zero weight penalty. Scarfing results in the differential strains, and hence the resulting shear stress concentrations, at the ends of the joint being considerably reduced. It also lowers the tensile stresses, σ_{11}, in the adhesive. These effects are most marked in the case of the scarf single lap joint, where this design greatly reduces the eccentricity of the loading path. Another variation on this theme is to taper the substrates on the outside of the joint

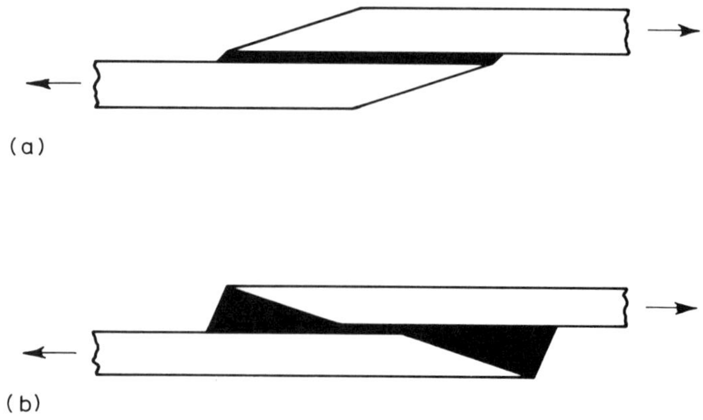

Figure 6.25 Examples of lap joints with tapered substrates.

[8,51,85–88], as shown in Fig. 6.25a. This reduces the shear stress concentrations but for the single lap joint obviously has no effect on decreasing the eccentricity of the loading path. Furthermore, Thamm [87] has reported that tapering the substrates only yields a substantial increase when the tapering is complete, i.e. the substrates are sharpened as far as possible to an edge. Partial sharpening, say to about half the original thickness, for example, is practically valueless from the strength point of view. In some instances the tapered substrates are reversed inwards, as shown in Fig. 6.25b. This also results in a thicker adhesive layer at the ends of the overlap which may also assist in decreasing both the shear and tensile stress concentrations (see Equations 6.15, 6.20 and 6.23 and, for example, Figs 6.22 and 6.23).

The double-butt strap lap joint has been studied in detail by Adams *et al.* [10], Hart-Smith [55], Wright [89] and Sage [90] and the last three were particularly concerned with employing this joint geometry to bond carbon fibre reinforced composite materials. Sage [90] considered various modifications to the double butt-strap lap joint, such as tapering and radiusing the outside of the straps, and obtained joints which failed in the substrate completely, outside the joint area, and were therefore 100% efficient for strength. (The joint efficiency being defined as the fracture load of the joint divided by the load necessary to fracture the weakest of the substrates, expressed as a percentage.) However, this design does incur a weight penalty.

Many of the points raised in this and the previous sections are illustrated by recent studies by Kinloch and co-workers [8]. They examined the strength of a high tensile strength steel substrate bonded to a carbon fibre substrate in a double lap joint design using a toughened epoxy adhesive. The strength of such joints was ascertained by both experimental and theoretical techniques. The material properties used in the theoretical studies are given in Table 6.4 and

Table 6.4 Material properties used in theoretical stress analysis and failure predictions of high-tensile steel/carbon fibre composite [8]

Property	Value
(a) Adhesive (elastic–plastic: a single-part rubber-toughened epoxy)	
Young's modulus, E_a	3.05 GPa
Tensile fracture stress, σ_{af}	82 MPa
Tensile fracture strain, ϵ_{af}	4.75%
Shear modulus, G_a	1.13 GPa
Shear yield stress, τ_{ay}	54 MPa
Shear (plastic) fracture, γ_{af}	35%
Poisson's ratio, ν_a	0.35
(b) High-tensile steel substrates (elastic)	
Young's modulus, E_s	210 GPa
Poisson's ratio, ν_s	0.29
(c) Unidirectional carbon fibre composite substrate	
Longitudinal tensile modulus, E_{cl}	140 GPa
Transverse tensile modulus, E_{ct}	7 GPa
Interlaminar shear modulus, G_c	4.5 GPa
Poisson's ratio, ν_c	0.3
Longitudinal tensile fracture stress	1400 MPa
Interlaminar, transverse fracture stress, σ_{ctf}	40 ± 6 MPa

the results from both the experimental and theoretical studies shown in Table 6.5. Several noteworthy features are evident. Firstly, the theoretical predictions from assuming that the fracture occurs in the adhesive layer by a shear failure are extremely optimistic. This is because the out-of-plane transverse stresses, σ_{11}, are sufficiently high in all cases that the locus of joint failure is by some form of cracking due to the presence of these stresses. Indeed, shear fracture in the adhesive layer was never observed, although the analysis of the shear stress distribution was of assistance in deciding the optimum overlap length. Secondly, in joint design 1 (a 90° fillet) (Fig. 6.26 and Table 6.5) the σ_{11} tensile stresses are predicted to lead to interlaminar fracture in the fibre composite substrate. This was the observed failure mode of the joints and occurred at a relatively low fracture load. Thirdly, tapering the outer steel substrates to a sharp edge, as in joint designs 2 and 3 was predicted to have little effect on the fracture load, since a 90° fillet was retained. This again was in accord with experimental observations, and the locus of joint failure remained by delamination of the fibre composite. Fourthly, however, by combining the idea of tapering the outer steel substrates and increasing the local adhesive thickness, with the incorporation of an angled spew fillet, the σ_{11} stresses are greatly reduced. The reduction in the σ_{11} stresses leads to far higher joint strengths, and a predicted change in the locus of failure from delamination in

Table 6.5 Comparison of experimental and theoretical studies on double lap joints of various designs of high-tensile steel-carbon fibre reinforced plastic [8]

Joint design (see Fig. 6.26)			Experimental fracture load per unit width (MN/m)	Theoretical predictions for various possible failure modes (MN/m)		
Number	Description	Fillet angle		Shear failure in the adhesive	Interlaminar failure in composite	Tensile failure in the adhesive
1	Basic	90°	0.93	6.4	1.05†	1.6
2	Outside taper	90°	0.89	>6.4	1.08†	—
3	Inside taper	90°	0.94	>6.4	1.10†	—
5	Inside taper and angled fillet	45°	2.72	>6.4	3.53†	4.0
5	Inside taper and angled fillet	30°	3.05	>6.4	7.44	3.3†
5	Inside taper and angled fillet	17°	2.80	>6.4	9.08	2.4†

† The lowest predicted fracture load is underlined.
(a) The fracture load for failure in the composite, away from the bonded overlap, is 4.2 MN/m.
(b) Failure criteria assumed were: (i) For shear failure, when γ_{12} (max) = γ_{af}. (ii) For interlaminar composite failure, when σ_{11} (max) = σ_{ctf}. (iii) For tensile failure in adhesive, when ϵ_{11} (max) = ϵ_{af}.

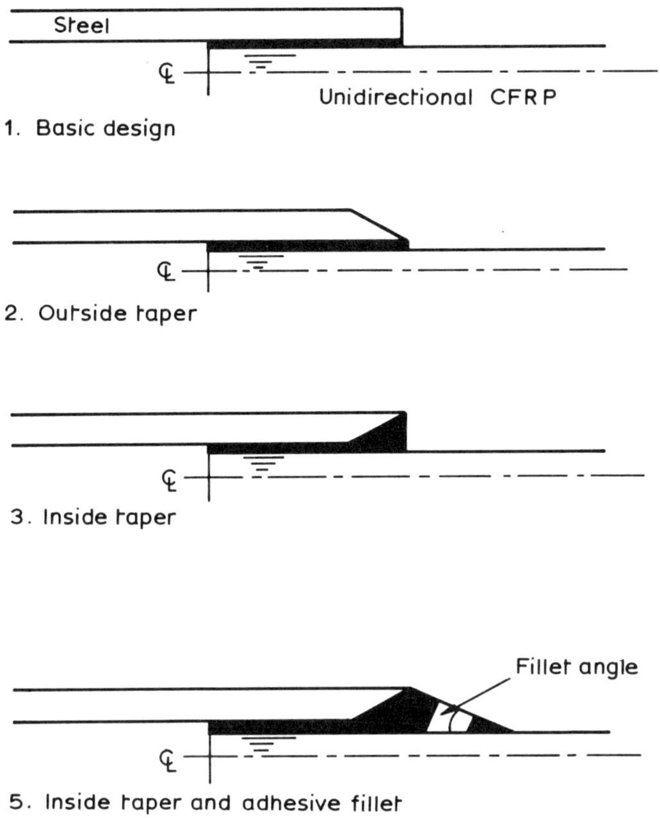

Figure 6.26 Designs of steel/carbon fibre composite double lap joints (see Tables 6.4 and 6.5 for data) [8].

the composite substrate to cracking in the spew fillet. (The 90° fillet acts as a step of course, which is as damaging as a crack from the viewpoint of causing stress concentrations! The angled spew fillet shape was formed by casting the adhesive layer, using a rubber mould at the ends of the joint.) Finally, the major practical feature to emerge, is that the modification of the double lap joint has resulted in a threefold increase in joint strength and has given a joint efficiency of approximately 73%.

6.4.7 Peel joints

The static stress distribution that results when a flexible member is peeled away from an adhesive layer supported on a rigid substrate has been considered by a number of authors [91–102]. These analyses have considered the adhesive and

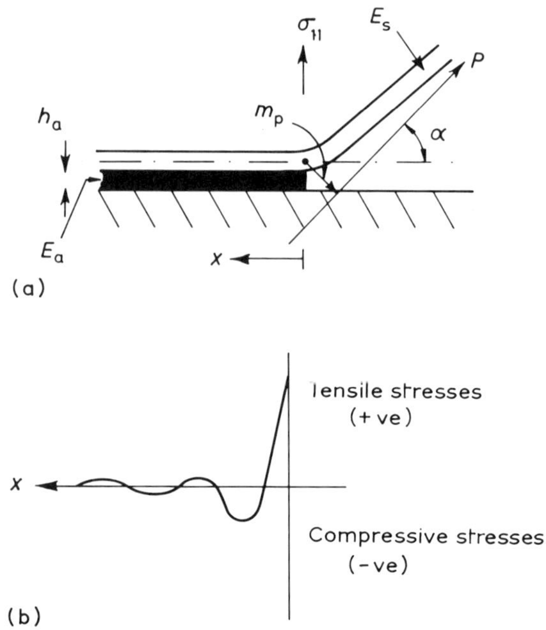

Figure 6.27 (a) Schematic diagram of a peel test. (b) Distribution of normal stresses in the adhesive layer ahead of the advancing peel front [92].

substrates as linear–elastic materials and deduced the tensile and shear stresses in the adhesive layer. Kaelble [92] has shown that the tensile, or cleavage, stress, σ_{11}, at a distance, x, in the adhesive layer (Fig. 6.27a) is given by:

$$\sigma_{11} = \sigma_p(\cos\beta_p x + \kappa_p \sin\beta_p x)\exp(\beta_p x) \tag{6.24}$$

where

$$\beta_p = \left(\frac{E_a b}{4E_s l_e h_a}\right)^{1/4} \tag{6.25}$$

$$\kappa_p = \frac{\beta_p M_p}{\beta_p M_p + \sin\alpha} \tag{6.26}$$

σ_p is the boundary cleavage stress in the adhesive at $x = 0$, i.e. at the peel front, l_e is the moment of inertia of flexible substrate cross section, M_p is the moment arm of the peel force (= peel energy × distance, Pm_p), α is the peel angle, and β_p is a constant.

Equation 6.24 assumes that the value of σ_{11} in the adhesive layer is constant through the adhesive layer thickness and across the width of the joint. The former assumption is not correct, as evidenced by the locus of failure in peel joints frequently being close to one of the substrate/adhesive interfaces, which indicates higher stresses near the interface compared to the remainder of the adhesive layer. The above analysis also does not consider the adhesive forming filaments across the peel front. Many pressure-sensitive adhesives behave in this fashion and undergo 'stringing' or 'legging' behind the advancing peel front. Several [42,102] authors have shown that this behaviour may contribute significantly to the measured peel force per unit width of joint and have offered modifications to the above analysis to take account of this feature. Returning to Equation 6.24, this expression predicts that the distribution of cleavage stress along the joint in advance of the peel front is a highly damped harmonic function involving alternating regions of tension and compression, as illustrated in Fig. 6.27b. Whether such a stress distribution really exists is open to debate but certainly a region of compression exists just in advance of the peel front [92,94]. As commented by Wake [103], this is easily demonstrated with a piece of glass plate and a pressure-sensitive adhesive tape. If the tape is adhered to the underside of the glass over a short length only and slowly removed, then the unattached portion of the tape will progressively adhere due to the action of the compressive stresses in advance of the peel front forcing the unattached portion in contact with the glass substrate.

Kaelble has also shown that peel force, P, per unit width (which is equivalent to an energy per unit area) is given by:

$$P = \frac{\sigma_p^2 h_a \kappa_p^2}{2E_a(1-\cos\alpha)} \qquad (6.27)$$

However, Gent and Hamed [104] have shown that the theory of small bending deformations used to derive Equation 6.24 is only valid to peel mechanics when $\beta_p m_p >> \sin\alpha$, i.e. when $\kappa_p \cong 1$. The results from these analyses have not been widely used in the interpretation of peel test data and the fracture mechanics approach, based upon an energy balance argument, which avoids the necessity for developing a detailed stress analysis has been far more widely applied. This approach is considered in Section 7.3.2.

More recently Crocombe and Adams [105] have re-examined the problem of defining the stress distribution in the peel test. They have employed an elastic, but large-displacement, finite-element analysis and reported that initial failure is caused by the principal tensile stresses in the adhesive driving the crack towards the interface between the adhesive and the flexible substrate. Subsequent propagation was found to occur at a critical applied bending moment for a particular adhesive and substrate, independent of peel angle. Further, the load measured by the peel test was not proportional to the actual tensile strength of the adhesive; a small increase in the adhesive strength caused a much larger increase in the applied peel load.

6.4.8 Miscellaneous joint geometries

Adhesive bonding provides a convenient and light-weight method of assembling structures consisting of thin-walled tubes joined by a bonded overlap and the stress distribution in such joints has been considered by several authors [53,106–109]. Adams and Peppiatt [108] and Nagaraja and Alwar [109] have employed finite-element analysis techniques to examine such tubular joints. The former authors examined the cases of both axial and torque loading and also considered the effects of *partial* sharpening of the substrates to form a tubular scarf lap joint. They concluded that, in agreement with Thamm [87], the reductions in stress concentrations obtained did not justify its manufacture. Nagaraja and Alwar [109] assumed that the adhesive behaved in a non-linear manner and found that the calculated stress concentrations were lower than those for the purely elastic case; maximum stress concentrations of about 1.3 were found for the dimensions and joint materials studied.

As stated in Section 6.4.2, an annular butt joint, or napkin-ring, specimen tested in shear minimizes the variation of shear stress in the adhesive and has been used by many workers to assess the shear strength and shear stress versus strain behaviour of adhesives [14,17,111–115] and is also listed for this purpose as ASTM E 229 (see Table 6.1). The independence of the measured strength upon specimen geometry has been substantiated by Bryant and Dukes [112] and Foulkes *et al.* [110] and the shear stress distribution calculated using linear elasticity theory is:

$$\tau_{z\theta} = \frac{2T_q r_o}{\pi(r_{oo}{}^4 - r_{oi}{}^4)} \tag{6.28}$$

where $\tau_{z\theta}$ is the shear stress at a radius r_o caused by an applied torque T_q, and where r_{oi} and r_{oo} are the inner and outer radii of the annulus, respectively. Adams *et al.* [10] used elastic finite-element analysis to confirm that the shear stress does indeed increase linearly from zero at the centre to a maximum at the outside of a solid butt joint loaded in torsion. However, when they repeated the analysis allowing for an adhesive spew, a greater shear stress concentration occurred at the interface at the substrate edge, although the general level of shear stress was lower than in a similar joint with no spew fillet. As remarked previously, they concluded that if this type of joint was to be used to obtain the shear stress versus strain curve of an adhesive, it should be used without a spew fillet; not always an easy task in practice. Now, at failure the maximum stress which occurs at the outer circumference of the joint is the shear fracture stress, τ_{af}, and is therefore given by:

$$\tau_{af} = \frac{2T_q r_{oo}}{\pi(r_{oo}{}^4 - r_{oi}{}^4)} \tag{6.29}$$

However, if the failure of the adhesive involves significant plastic deformation then a more uniform stress results through the width of the annulus and the fracture stress, τ_{af}, is now given by [116]:

$$\tau_{af} = \frac{3T_q}{2\pi(r_{oo}^3 - r_{oi}^3)} \qquad (6.30)$$

Another form of common practical joint, especially with anaerobic adhesives, is the pin or stud bonded into a hole in a solid block. This type of joint has been examined by Lees [1,117] for the pin either subjected to tension or torsion loads. In both cases a shear stress concentration occurs at the end of the joint where the pin first engages into the block. After a certain length of engagement, the value of this stress concentration does not decrease any further with longer engagement lengths as shown in Fig. 6.28. Hence little is to be gained by increasing the engagement length thereafter, but the stresses may be reduced, of course, by increasing the diameter of the pin. Similar observations are recorded for a pin and collar joint, again a common form of design with anaerobic adhesives especially. The exact value of the maximum stress concentration depends greatly upon the shear properties of the adhesive; a low modulus adhesive giving the lowest stress concentrations. However, as for lap joints, whether such an adhesive gives the best joint performance will depend upon whether the fracture properties of the more flexible adhesive are inferior to those of the stiffer adhesive and, if so, by how much.

6.4.9 Internal stresses

Adhesives often operate with some additional stress in the joint arising from shrinkage of the adhesive relative to the substrates. The main reason for the shrinkage comes from equilibrium contact between the adhesive and substrates being established at temperatures above the subsequent operating temperature of the joint. Thus, since the adhesive and substrates usually have different coefficients of thermal expansion, thermal strains are introduced upon cooling. Other events such as loss of solvent, polymerization reaction, crosslinking, etc. may also be accompanied by volume contractions but are usually considered to be of secondary importance.

Experimental work [118–125], especially that using photoelastic techniques, has established the presence of residual stresses in joints but the results have often not been quantitative. For the simpler case of polymeric films coated onto metallic substrates, photoelastic techniques and a method based upon the bimetallic strip principle have often been employed [126–131]. Using the latter method, Danneberg [130] showed that for a wide range of epoxy-based coatings on an aluminium substrate thermal contraction was a major cause of internal stress and that the stresses generated were of the order of 0.08

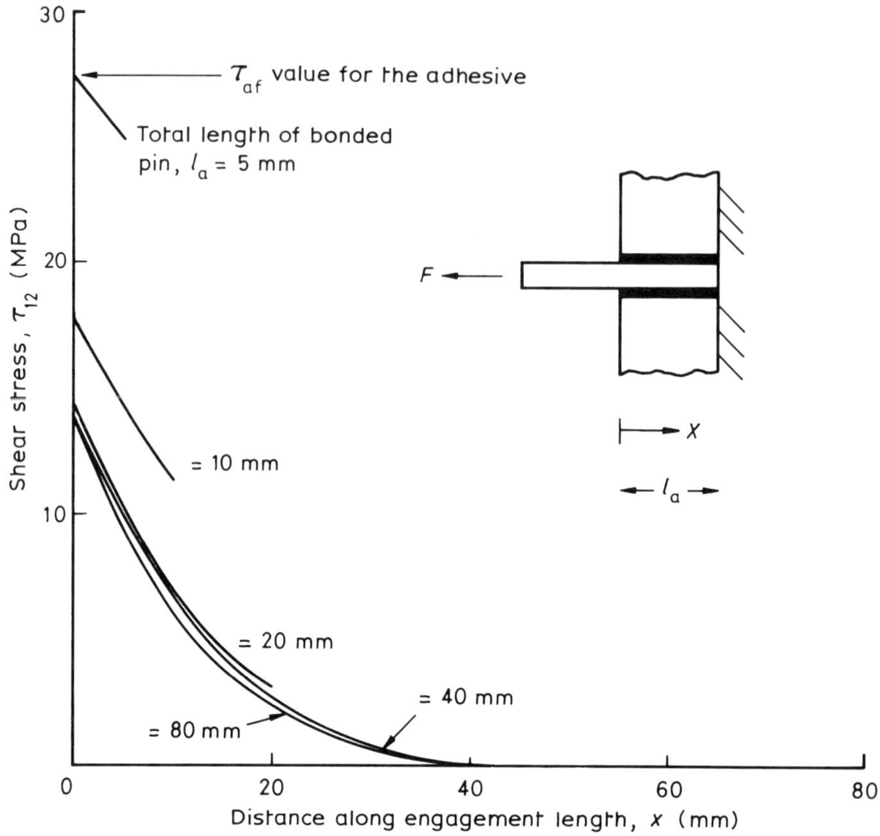

Figure 6.28 Shear stress distribution along the engagement length for a steel pin, subjected to a 4 kN axial load, bonded into a steel block with an epoxy adhesive with various values for the bonded engagement length [117].

MPa/°C. More recently Hahn [125] has studied residual stresses using a very sensitive electrical strain measuring technique and, although the method was not quantitative, he clearly showed that residual stresses were present in the interfacial regions of the adhesive layer.

The theoretical analysis of the internal stresses in an adhesive joint is beset with great difficulties, even if it assumed that the adhesive is perfectly elastic. Bikerman [132], Wake [133], Harrison and Harrison [134], Carlson and Sapetta [135] and Inoue and Kobatake [123] have made this assumption and attempted theoretical predictions. Some of the main points to emerge are that the in-plane normal stresses, σ_{22}, are tensile in nature and are greatest and uniform near the interface, and may be given under these conditions by:

$$\sigma_{22} = E_a \Delta\alpha_t \Delta T \tag{6.31}$$

where $\Delta\alpha_t$ denotes the difference in linear thermal expansion coefficient between the adhesive and substrate and ΔT is the difference between the stress-free temperature, often taken as the glass transition temperature of the adhesive, and the temperature of interest. However, as one proceeds towards the centre line (normal to the plane of the adhesive layer) they increase gradually from zero at the edges to the above maximum value at the centre line of the joint. As a result of the Poisson's strains from the above effects and the constraint imposed by rigid substrates, the normal stress, σ_{11}, is compressive near the edges of the joint but becomes small and tensile in character away from the joint edges. The maximum value of the residual shear stress, τ_{12}, in the adhesive layer occurs at, or near, an interface and near the ends of the joint, being zero at the centre, and the value of τ_{12}(max) is much higher for an adhesive possessing a Poisson's ratio of 0.33, as opposed to 0.5.

Now all these analyses assume elastic behaviour and whilst the distribution of stress concentrations from such workers as Inoue and Kobatake appear to be intuitively of the expected form and magnitude, and were confirmed by photoelasticity techniques, the quantitative results from many other calculations appear to be unduly pessimistic. For example, the results from Carlson and Sapetta [135] appear to yield extremely high values, with the calculated residual stresses alone almost equal to the strength of the adhesive. Indeed, τ_{12}(max) values which have been calculated [136] from the Carlson and Sapetta analysis for an aluminium/epoxy/carbon fibre composite joint suggested that at low temperatures this joint would readily fail solely because of the high internal stresses arising from thermal contraction. In practice, the joint possessed a relatively high strength. A more realistic, but still somewhat pessimistic, prediction of these experimental results is achieved if the analysis of Hart-Smith [55] is employed. This assumes elastic–plastic behaviour of the adhesive and increased adhesive plasticity or adhesive layer thickness results in lower internal stresses. However, still no allowance is made for stress relaxation of the adhesive.

Thus, to summarize, the magnitude and importance of internal stresses in adhesive joints is a frequently discussed topic and certainly is one which would benefit from further studies. On the practical front, most analyses greatly overestimate such stresses, largely because most of the analyses are elastic and make no allowance for stress relaxation of the adhesive. However, where such stresses do prove to be detrimental, techniques such as employing a more flexible adhesive, the use of a thicker adhesive layer, using a lower hardening temperature for the adhesive and using a longer and slower cooling cycle can usually be employed to eliminate the problem.

6.4.10 Comparison of joint test geometries

To illustrate further many of the points discussed above it is of interest to compare the results from some of the test geometries that have been considered in the separate sections.

Considering first data for a rubbery adhesive, Gent [42] has employed a fracture mechanics analysis to explain and predict joint strengths of a typically rubbery adhesive bonding relatively rigid substrates tested in tension, shear and peel. With the reasonable values of $E_a = 2$ MPa, intrinsic flaw size, $a = 0.1$ mm and fracture energy, $G_c = 10$ kJ/m^2 he predicted the dependence of the fracture stress, σ_f, of axially loaded butt joints upon the shape factor, r_o/h_a of the adhesive layer from employing Equation 6.11, assuming $\eta_c = 3$. The results are shown in Table 6.6. Also given are the calculated peel energy, P, obtained from Equation 7.41 and shear strength, τ_f, from Equation 6.32 below:

$$\tau_f = \eta_c \left(\frac{2E_a G_c}{3\pi a} \right)^{1/2} \tag{6.32}$$

As may be seen, the shear strength, τ_f, is always less than the tensile butt joint strength, σ_f, and the latter rises to much greater values for thin adhesive layers. However, the relative values of σ_f, τ_f and P depend upon the values of E_a, the intrinsic flaw size, a, and the local stress concentration effects. Ratios obtained with a particular adhesive/substrate combination will not therefore generally hold for others.

The physical properties of a modern structural adhesive examined in the form of both bulk specimens and adhesive joints are shown in Table 6.7. The results clearly illustrate many of the aspects of joint design discussed above and again demonstrate that the strength of an adhesive joint depends greatly upon the design employed. Examining some of the many aspects to be found in Table 6.7: firstly the butt joint strength is significantly lower than the bulk tensile strength due to the stress concentrations in the joint (Section 6.4.3). However, the torsional shear strength is the same as the bulk shear strength of the adhesive (see Sections 6.4.2 and 6.4.8). The single lap-shear strengths are much lower than the above shear strengths due to the severe stress concentrations that occur in such joints (Section 6.4.4). However, they may be reduced by using more rigid substrates, hence the higher single lap-shear strength when steel substrates are used as opposed to aluminium alloy substrates. The peel strength for the aluminium joints decreases as the peel angle is increased (Section 6.4.7) and the far higher 90° peel strength recorded for the aluminium joints, compared to the steel joints, reflects the extensive plastic yielding which occurs as the aluminium strip is peeled away, whilst the steel substrate remains elastic.

Table 6.6 Strength of rubbery adhesive/rigid substrate joints tested in tension, shear and peel

Joint geometry	Shape factor: r_0/h_a	Strength
Butt joint	0.1	$\sigma_f = 0.4$ MPa
Butt joint	1.0	$\sigma_f = 2.5$ MPa
Butt joint	10	$\sigma_f = 11$ MPa
Butt joint	20	$\sigma_f = 21$ MPa
Shear joint	—	$\tau_f = 2$ MPa
Peel joint ($\alpha = 90°$)	—	$P = 10$ kN/m (or kJ/m^2)

Table 6.7 Physical properties of a one part (hot-cured) rubber toughened epoxy adhesive measured in bulk and in various adhesive joints

Test	Substrates	Comments	Property
Bulk properties†			
Young's modulus			$E_a = 3.05$ GPa
Tensile fracture stress			$\sigma_{af} = 82$ MPa
Shear fracture stress			$\tau_{af} = 58$ MPa
Glass transition temperature			$T_g = 120$ °C
Adhesive joints†,‡			
Axially loaded butt	Steel	$h_a = 0.5$ mm	$\sigma_f = 59$ MPa
Torsional shear	Aluminium	$h_a = 0.5$ mm	$\tau_f = 61$ MPa
Single lap loaded in tension	(a) Aluminium	$h_a = 0.5$ mm, $d = 1.6$ mm, $l_a = 12.7$ mm	$F_f = 9$ kN, $\tau_f = 28$ MPa
	(b) Steel	As above	$F_f = 12.3$ kN, $\tau_f = 38$ MPa
Peel: 90° peel angle	(a) Aluminium		$P = 5$ kN/m
	(b) Steel		$P = 0.6$ kN/m
135° peel angle	Aluminium		$P = 4$ kN/m
Precracked, tapered double cantilever beam	Steel		$G_{Ic} = 0.9$ kJ/m^2
Precracked, compact shear	Steel		$G_{IIc} = 2.2$ kJ/m^2

†Tests conducted at 23 °C and a moderate rate.
‡All joints failed by cohesive fracture through the adhesive layer.

6.5 NON-DESTRUCTIVE TESTING

6.5.1 Introduction

The objective of conducting non-destructive tests during the manufacture and subsequent service life of adhesive joints is to ensure that the joints are acceptable for carrying the design loads under the specified operating conditions. It appears to the author that engineers always wish to undertake somewhat more non-destructive testing of adhesive joints than commonly employed in any other form of joining technology. In some ways this is not surprising. One cannot 'see' the level of interfacial adhesion forces that are present and even if the level of adhesion forces are initially sufficient to give a strong joint it is known that the disruption of such forces, via migration of water to the interface, for example, may subsequently occur and lead to weak joints. Hence, it is somewhat unfortunate that no current non-destructive test can in fact detect the level of intrinsic adhesion. Furthermore, many of the non-destructive tests that are used to assess the 'quality' of an adhesive joint basically function as sensitive void detectors; any direct relation between number and size of voids in the adhesive layer, or at the interface, and the strength of adhesive joints is extremely complex, as discussed below.

Therefore, the general philosophy that must be adopted with critical adhesive joints is, during the bonding operation, to use a combination of destructive tests on process control specimens and non-destructive tests. The latter may be used to inspect (i) the pretreated substrates prior to bonding, and (ii) the finished bonded joint. The non-destructive tests will *not* yield direct values of joint strength. Most will indicate whether the bonding operation has been performed to a similar quality as on previous occasions, whilst a few techniques will give this information and more quantitative data concerning the thickness and modulus of the adhesive layer. These various methods should form part of a good quality control system that ensures that all materials and processes are within the specified limits. During the service life of the bonded component non-destructive tests may again be employed, although often the type of test that can be conducted will be limited by the accessibility of the joint and the possible need to transport the non-destructive test equipment to the bonded structure.

Some of the non-destructive tests that are commonly used for the inspection of both the pretreated substrates and the finished joint are briefly outlined below, and more detailed reviews may be found in the literature [137–143].

6.5.2 Inspection of pretreated substrates

The aim of inspecting the surface of the substrate after any pretreatment process has been conducted, but prior to bonding, is to ensure that the treatment has been correctly carried out, and hence that good adhesion to the substrate should result. The techniques commonly used attempt to detect the presence of any surface contamination and, if a metallic substrate, to estimate the thickness of the oxide layer. Obviously, modern surface analytical methods such as X-ray photoelectron spectroscopy may readily undertake these tasks, but are not non-destructive with respect to the specimen needed and the size of a typical bonded component, and are difficult to apply on a production line!

A crude check on the cleanliness of the substrate surface may be made using the 'water-break' test. This test depends upon the observation that a technologically clean surface, such as a pretreated plastic, fibre composite, ceramic or metallic substrate, will frequently hold a continuous film of water. Thus, a break in the water film indicates a soiled area. Since the danger exists that a continuous film of water may form if a layer of the cleaning solution remains on the surface, care must be taken to ensure that the surface is thoroughly rinsed. In a pretreatment schedule this test may be assessed at each rinsing stage in the process by watching the substrate surface for discontinuities in the water film as the substrate is removed from the rinsing tank. A more sophisticated method is to measure the contact angle of a liquid on the pretreated surface, but this method has not gained wide acceptance.

Instruments are available [137,144] which give a measure of the surface cleanliness by detecting changes in electron emission energy or surface impedance. Practical measurements of the electron emission energy or Volta potential are carried out using Kelvin's dynamic condenser method. A gold reference plate, insulated from environmental influence, is placed close to and in electrical contact, via a resistor, with the substrate surface. An electrical field is created in the air gap between the reference plate and the surface. If the reference plate is moved relative to the treated surface a charge will be generated through the resistor and, if the displacement of the reference plate is sinusoidal, the signal may be shown on an oscilloscope. A compensator circuit is made in such a way that this signal can be reduced to zero by adjusting the charge on the reference plate to a sufficient magnitude. This compensating charge is called the Volta potential and is used as a quantitative parameter for checking the cleanliness of a pretreated surface. In some instances it has been found that if the pretreatment process is changed from its known optimum specification then a relationship may be established between the measured surface potential and the peel strength, as indicated in Fig. 6.29.

The thickness of the oxide layer generated on a metallic substrate during the pretreatment process is a good check on the general quality of the treatment procedure. Sensitive equipment is available to measure the surface impedance which can be correlated with the thickness of the oxide layer. However, with some pretreatments such as phosphoric acid anodizing of aluminium alloy, the oxide thickness formed is of the right order that it may be tested by its colour upon scattering polarized light.

6.5.3 Inspection of adhesive joints

The basic aim of any non-destructive test of an adhesive joint must be the direct correlation of some parameter measurable by the test with the failure property to be measured or predicted, without impairing the effectiveness of the bonded part for its intended application [145].

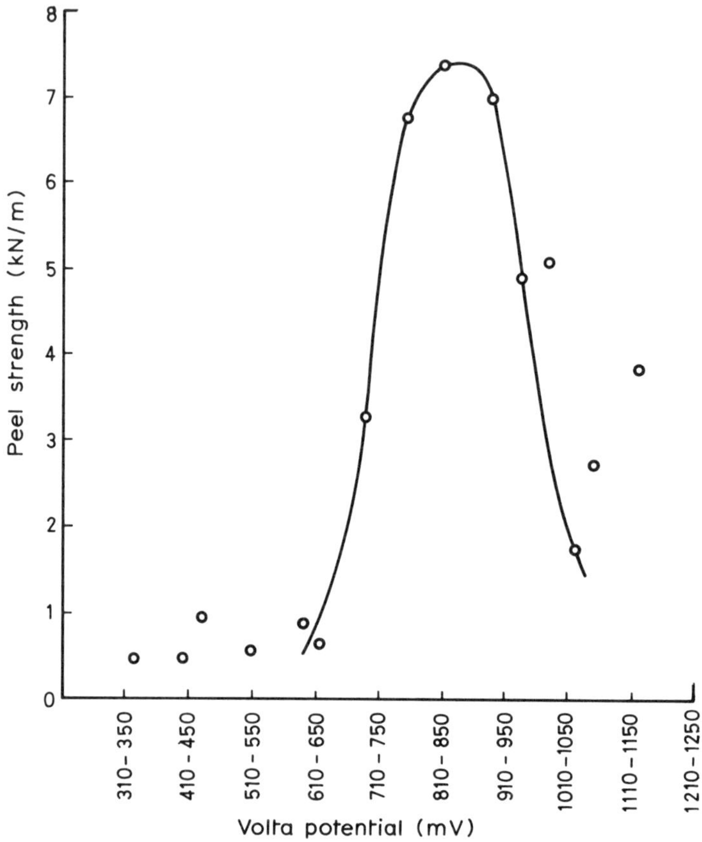

Figure 6.29 Relationship between peel strength of bonded aluminium alloy joint and the surface condition of the substrates prior to bonding as ascertained by measuring the surface potential [139].

However, it appears that none of the non-destructive tests currently employed directly correlate with any critical failure property. Most industrial test techniques such as through-transmission and pulse-echo ultrasonics, sonic vibration techniques, X-ray radiography, thermal inspection methods, holography, liquid penetrants, etc. basically attempt to find defects in the joint. Such defects may arise from several sources. Some defects arise from porosity, cracks or voids in the adhesive layer or at the interface and are typically filled with air; they will simply be referred to as voids in the present discussions. However, during the service life of the joint such voids may fill with water which makes them far more difficult to detect; since, for example, water has a much higher acoustic impedance than air. Also, 'zero-volume' voids, or debonds, may occur when the adhesive and substrate are in contact but no

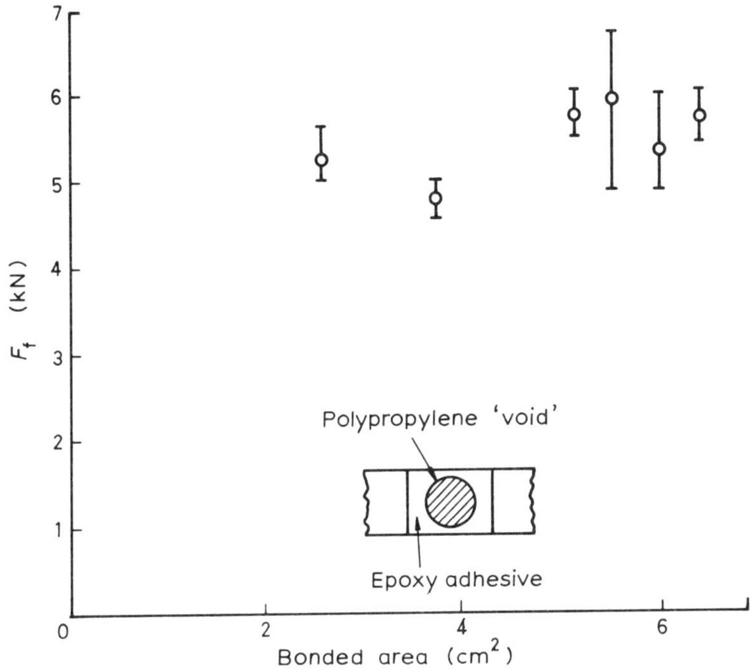

Figure 6.30 The breaking force for epoxy/aluminium alloy single lap joints containing various size 'voids' [146].

intrinsic adhesion forces are present. A few techniques may find such defects but, as mentioned above, as long as some intrinsic adhesion forces are present, albeit extremely weak ones, such defects are impossible to detect using current techniques. Other defects may arise from regions of low strength in the adhesive layer due, for example, to inadequate hardening of the adhesive having occurred. But again only a few techniques are capable of detecting such defects. Thus, most non-destructive test methods are basically sensitive void detectors but, as commented above, any correlation between number and size of voids and the strength of adhesive joints is extremely complex.

This last statement is illustrated by work of Schonhorn *et al.* [146] who have measured the strength of single lap-shear joints consisting of aluminium alloy substrates bonded with an epoxy adhesive but they inserted polypropylene discs into the central region of the overlap instead of adhesive, the polypropylene discs acting as 'voids'. As may be seen from Fig. 6.30, even relatively large 'voids' had no significant effect on the measured fracture load. This arises of course from the maximum, critical, stress concentrations occurring near the ends of the overlap and not in the central region, as discussed in Section 6.4.4. The importance of this result is that it clearly demonstrates that the presence of voids does not necessarily imply weak joints. Therefore, non-destructive techniques which solely act as void detectors usually provide no information about any critical failure property and even the absence of detectable voids does not, of course, guarantee against weak joints or premature failure. To provide more quantitative information the relationships between size, number and location of voids and joint strength must be deduced from such concepts as continuum fracture mechanics (see Chapter 7). However, what useful purpose such techniques undoubtedly do serve is to indicate that when relatively large defect areas are detected, all is not well with the quality of the materials, or bonding process. They thus act as a quality control check, as commented above.

6.5.3.1 Void detectors

Considering some of these techniques which are essentially void detectors in a little more detail, the monitoring of ultrasonic echoes forms the basis of several widely used non-destructive test methods. If a void is assumed to contain air or any other low-density medium then it will have a very low acoustic impedance relative to the adhesive or substrates and at an interface an incident pulse of ultrasound, typically in the frequency range 1 to 25 MHz, will be totally reflected leaving negligible energy to be transmitted through the void. The transmitted or reflected energy may therefore be used to indicate the presence of such a defect. In most cases piezoelectric transducers are used: these transform electrical into mechanical vibrations and vice versa. There are two distinctly different methods: through-transmission and the pulse-echo method.

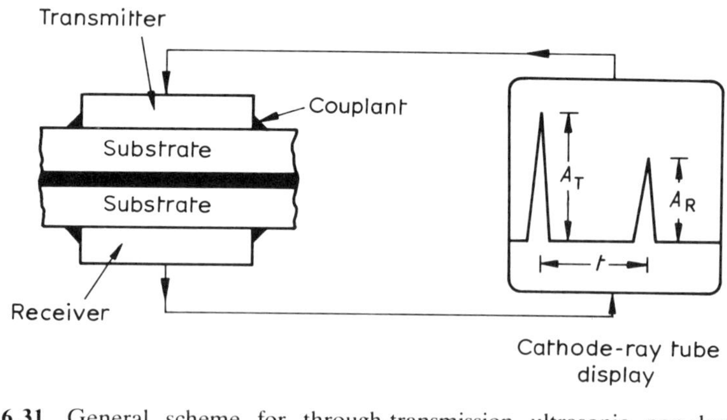

Figure 6.31 General scheme for through-transmission ultrasonic non-destructive testing. An ultrasonic signal is sent from the transmitter to the receiver through the bonded component. If the received signal, A_R, is very small compared to the transmitted signal, A_T, then it may be concluded that most of the energy has been reflected by an area of low acoustic impedance, i.e. a void may be present [138].

The through-transmission technique (Fig. 6.31) uses a separate transmitter and receiver on either side of the joint and therefore requires the two probes to be exactly aligned. This practical difficulty may be avoided with the pulse-echo technique in which the same transducer is used as transmitter and receiver (Fig. 6.32). The single transducer emits pulsed ultrasound and in the dead periods between emitted pulses it receives the reflected energy signals. Both techniques need the transducer(s) to be satisfactorily coupled to the adhesive joint and this may be achieved by (i) the transducer(s) being held onto the joint with grease, (ii) immersing the transducer(s) and joint in water, (iii) using a search-wheel,

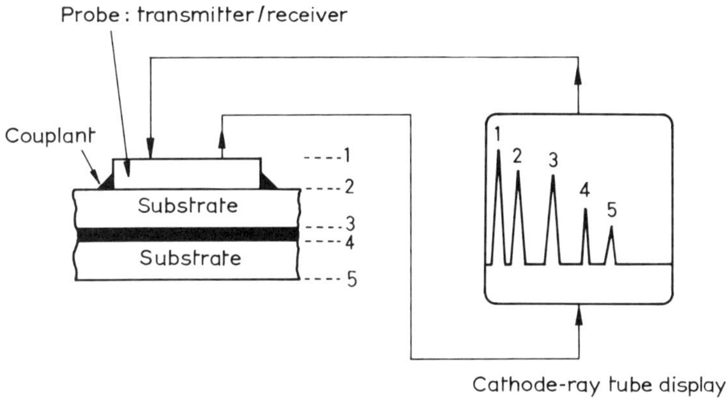

Figure 6.32 General scheme for pulse-echo ultrasonics. An ultrasonic pulse is sent into the bonded component and is received by the same transducer at different times after returning from the different interfaces. Where echoes are missing from the lower interfaces it is assumed that a void is present [138].

which is continually soaked in water, in contact with the joint or (iv) propagating the ultrasound along a water jet which is directed at the joint. Another problem that may arise, particularly for components which may be inspected from one side only, is that with honeycomb structures the air in the pores of the honeycomb hinders the ultrasound from reaching the lower interface. With both techniques the instrumentation may be arranged such that a signalling device is triggered when the received signal is at an unacceptably low level. Thus, using an *x–y* recorder, it is possible to make a complete facsimile recording, or 'C-scan' of the component which indicates all those joint areas yielding a reception signal below the preset minimum.

A range of sonic vibration techniques may be used for non-destructive testing of adhesive joints [143,147,148]. Most of these depend upon a void causing a local change in stiffness and hence a change in vibrational properties of the structure. They involve applying vibration excitation at a test point and measuring the input force and/or the vibration response. They can detect planar voids parallel to the surface being inspected and the size of defect which can be detected is again related to the wavelength employed. As the frequency increases, the wavelength decreases, and the minimum detectable size of void is reduced. Instruments using sonic vibrations operate typically at frequencies of up to 20–30 kHz, so they will not be able to find defects as small as those detectable at ultrasonic frequencies (up to about 25 MHz), but the former non-destructive testing methods are often faster than ultrasonic techniques and do not require a couplant. The sonic techniques are most sensitive to voids close to the surface of a stiff structure and are therefore well suited to the inspection of honeycomb structures. The sonic methods essentially fall into three categories: impedance methods, coin-tap and membrane resonance.

The impedance method has been used for many years in the Soviet Union and the typical equipment uses measurements of the local impedance of a structure; where the point impedance is given by F/v_s, where F is the harmonic force input to the structure and v_s is the resultant velocity of the structure. Measurements are generally taken at a single preset frequency, typically between 1 and 10 kHz, and as the probe is moved from a good to a defective area the impedance decreases. The coin-tap test is one of the oldest methods and is still widely employed, the change in sound when good and defective areas are tapped being due to a change in the frequency content of the force input. The sound produced on a defective structure, which therefore possesses a relatively low stiffness and impedance, is at a lower frequency and hence the structure sounds 'dead'. More recently techniques have been developed which employ an automatic, instrumented hammer designed to give a single, reproducible impact to the joint. The frequency spectrum of the impulse on the area under test is compared with that of an impulse, with the same hammer, on an area of the joint that is known to be good and void free. Data from the good joint is stored in the testing instrument so that it may be compared to that from the local area under test and so give an indication of the quality of this area. The membrane resonance techniques work on the principle that a planar void can be modelled as a plate restrained around the edges by the surrounding structure. This plate can resonate, the first mode being the membrane response in which the motion is similar to that of a diaphragm. As the frequency of vibration excitation approaches the membrane resonant frequency, the impedance of the system is reduced and the response amplitude obtained for a given force input increases. Therefore, at frequencies around the membrane resonance of the layer, above a void, the vibration response for a given force will be greater than that obtained in a sound region of the structure. Thus, a non-destructive test may be based upon response measurements alone, assuming the input force to be approximately constant. The eddy-sonic method is a membrane resonance method where an electromagnetic probe excites vibrations in the top substrate whilst a microphone in the probe detects the response. This method overcomes the problems of coupling transducers to the bonded component to be tested but part of the structure must usually be electrically conducting to support the eddy currents. The typical operating frequencies of instruments of the membrane resonance type is 20 to 30 kHz, so the minimum detectable void diameters are in the range 20 to 40 mm.

Radiography can generally detect density variations by means of attenuation of rays passing through the structure. However, this technique cannot be readily employed when a material with high absorption is used as either substrate. Thus, with metallic substrates, the adhesive cannot normally be seen on X-radiographs, unless the adhesive contains metallic fillers. However, the density of fibre-reinforced plastic substrates is of a similar order to the adhesive and so soft X-rays may be used. Neutron radiography is very suitable for observing density variations in adhesive layers, although the source of neutrons is somewhat limited!

There are a number of thermal inspection methods based upon variations in thermal conductivity, thermal emissivity or thermal capacity and the most important are the infrared and liquid crystal methods. Although, in theory, such techniques will detect changes in the thickness and density of the adhesive layer, in practice they detect changes in the above properties that arise from the presence of relatively large voids.

Finally, holographic interferometry has the ability to measure changes of the outer surface of a body of the order of half a wavelength which in the optical case, is less than about 0.25 μm. Thus, stresses are applied to a joint in order to obtain such displacements in an outer substrate that is covering a voided or debonded area compared to that over a well-bonded, good region. This method is particularly useful for examining thin fibre composite substrates, where only very low applied stresses are necessary, but obviously local surface deformations of the required order are more difficult to generate when thick metal (i.e. stiffer) substrates are present. The applied stresses are generated via thermal, vibration or pressure stressing and under the correct operating conditions debonds may be observed by this technique, as shown [149] in Fig. 6.33.

It is difficult to state exactly the size of minimum voids that the above non-destructive techniques will detect; it depends upon such factors as the details of the technique, type of joint, location and orientation of void and whether the void is filled with air or water. However, air-filled voids of the order of somewhat less than a few square millimetres to a few square centimetres represent the typical minimum size that are readily detectable.

6.5.3.2 Techniques yielding more basic information

As mentioned above, a few non-destructive test methods do yield more information than merely the presence or absence of voids above a certain size.

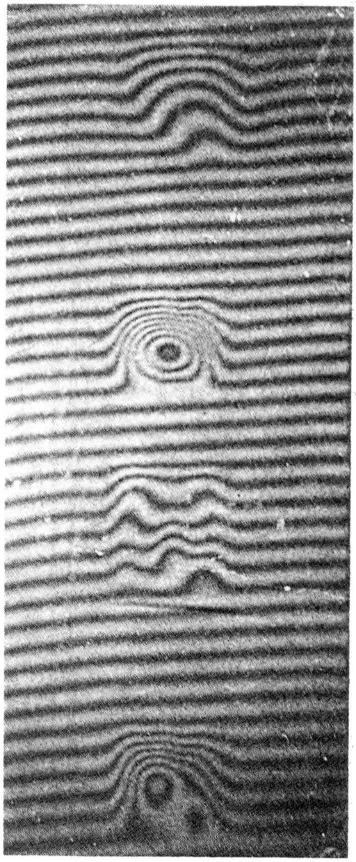

Figure 6.33 An example of holographic non-destructive testing. The test specimen was an adhesively bonded carbon fibre laminate honeycomb structure and a vacuum was applied to cause increased displacement of the face sheet when a defect was present. Debonds were present at the interface between the face sheet and honeycomb core (below the disturbance in the regular pattern that can be seen second from top and bottom) and crushed honeycomb core was also present (top and third from top) [149].

One such method is the resonance ultrasonic technique and it is the basis of a very widely used commercial non-destructive test method: the Fokker Bond Tester. It depends upon the fact than an elastic wave propagating within an adhesive bond is reflected and refracted at the substrate/adhesive interfaces and, if the correct frequency is selected, the multiple interference processes will give rise to resonance. Any changes in the thickness or elastic modulus of the substrates or adhesive will often give rise to changes in the frequency of resonance and to changes in the damping in the adhesive, which will change the amplitude of the resonance. As seen from the previous discussions in the present chapter, the strength of an adhesive joint does depend upon the values of these parameters. In many adhesive joints the thickness and modulus of the substrates may be assumed to be reliably constant; hence, any change in the frequency or amplitude of the resonance wave may relate not only to the presence of voids (which reduce the stiffness and density of the adhesive layer) but also to the modulus and thickness of a non-voided adhesive layer and hence, albeit indirectly, to the strength of the joint. One such instrument widely used in industry is the Fokker Bond Tester which works between 30 kHz and 1 MHz and measures differences in resonance frequency and the amplitude of the standing waves. Usually a series of joints differing in strength, typically achieved by varying the thickness of the adhesive layer, is prepared and examined with the Fokker Bond Tester. These non-destructive tests are correlated with subsequent destructive tests and a calibration curve obtained. A typical such relationship is shown in Fig. 6.34. Once the required strength of the joint is known these calibration curves serve to establish acceptance limits based upon readings of the instrument, which may now be used for quality control.

A second technique which has the potential of giving basic information on the properties of adhesive joints is ultrasonic spectroscopy. The Fokker Bond Tester described above does use a spectroscopic approach but the concept of conducting a detailed ultrasonic spectral analysis has been examined by Rose and Meyer [150] and Lloyd and Brown [151]. The latter authors used a pulse-echo method with a broad-band ultrasonic transducer to emit short (about 0.2 μs) pulses with a frequency of 2 to 12 MHz and using the same transducer to receive the echoes from the interfaces. The differences compared to the pulse-echo ultrasonic 'void-detectors' described previously are that the ultrasonic spectroscopy method uses broad-band, rather than the conventional narrow-band ultrasound, and that the reflected echoes are analysed in terms of obtaining a spectrum of amplitude versus frequency across a wide range of frequencies, i.e. about 2 to 14 MHz. It is suggested that the quality of the joint may be assessed since differences in the spectra so recorded may be associated with features such as thickness of adhesive layer and, more notably, with the degree of intrinsic adhesion and the presence of ingressing moisture at the adhesive/substrate interface. However, at the time of writing such techniques have not really fulfilled their early promise and have yet to be commercially introduced into a practical bonding operation.

Finally, non-destructive tests have been developed based upon the phenomenon of acoustic emission. Such an emission is a stress wave derived when some materials undergo an internal change induced by local stress. The basic idea behind using acoustic emission as a quantitative test is that some of the energy expended in rupturing an adhesive joint is converted into the energy of acoustic emissions which arise prior to failure at low applied stress levels. Thus, if there are sufficient, characteristic emissions at an acceptably low applied

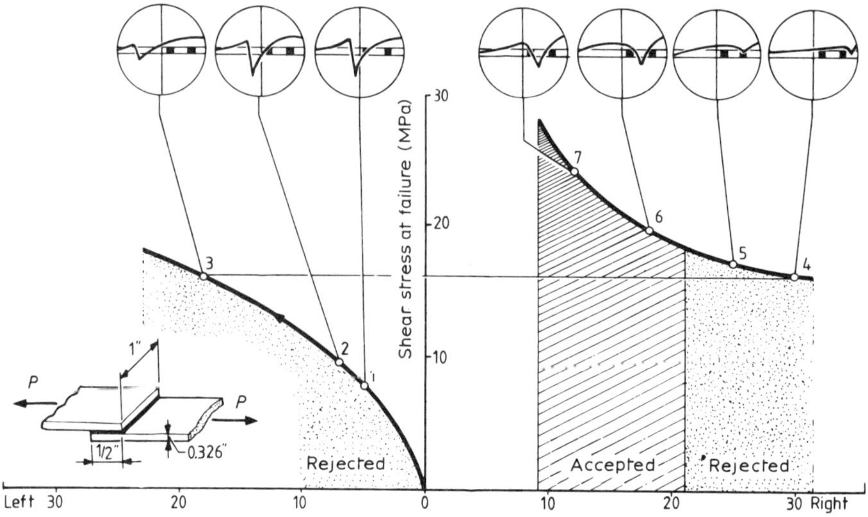

Figure 6.34 The typical relationship between single lap-shear strength and response from the Fokker Bond Tester 'A-scale'. With strengths increasing from zero up to 50% of the maximum, the resonance peak moves to the left; further increases in the quality of the joints brings the peak back to the far right-hand side of the oscilloscope, whereupon it continues to move to the left as the strength increases to a maximum [138].

stress then acoustic monitoring has the basic requirements of a quantitative non-destructive test. The work of Curtis and Speake [141,152,153] has been some of the most extensive and they have developed a broad-band (10 kHz to 2 MHz) acoustic emission spectrum and energy analysis system and used it to study the fracture of single lap joints loaded in tension. They measured the cumulative acoustic emission at fracture per unit volume of adhesive and found that for the two structural adhesives that were examined the weaker the joint, obtained by changing the thickness of the adhesive layer, then the more emissions were released. Further, for the more brittle adhesive, which gave rise to a greater intensity of emissions, the cumulative acoustic emission energy per unit volume appeared to be inversely proportional to the lap-shear strength raised to the power of about five. However, several practical problems remain to be resolved, such as how to apply a 'standard' stress level and to establish that a low applied stress level does indeed cause characteristic acoustic emissions.

6.6 CONCLUDING REMARKS

The essential nature of stress concentrations in adhesive joints and the basic criteria to design bonded structures have been established for many years.

However, the present chapter has clearly revealed that over the last decade there have been considerable advances in our ability to measure the basic engineering properties of the adhesive and to couple this data with increasingly realistic analyses on the nature and magnitude of the stresses in joints. Such studies have been used to design more efficient joints in several advanced bonded structures and two examples are the PABST (Primary Adhesively Bonded Structures Technology) programme [13] where the aerospace industry used adhesives as a major primary method of fastening and the ECV-3 (Energy Conservation Vehicle Mark 3) programme [154] where British Leyland exhibited similar faith in adhesives for constructing a low fuel consumption motor car.

However, there are still many areas where further work is required. There is a need to confirm that the basic engineering properties that are currently being measured by different methods are yielding similar values and that when tests on the 'bulk' adhesive are undertaken that they give representative properties of the adhesive as a thin layer in a joint. There is also a need to model the complete stress distributions in complex joint designs more readily and accurately, and to be able to undertake conveniently parametric design and fracture prediction studies. In connection with this last requirement, the failure criteria necessary for predicting joint fracture have yet to be established firmly.

Finally, there is an urgent need, as has existed for at least the last two decades, for improved non-destructive tests which are capable of giving direct correlations with joint fracture criteria. This need will only be met by interfaces being established between the non-destructive test specialists and the stress and fracutre analysts, since there are important problems to be resolved in both areas.

REFERENCES

1. Lees, W. A. (1984) *Adhesives in Engineering Design* Design Council, London.
2. Volkersen, O. (1938) *Luftfahrtforsch* **15**, 41.
3. Goland, M. and Reissner, E. (1944) *J.Appl. Mech.* **2**, A-17.
4. Hart-Smith, L. J. (1981) *Developments in Adhesives–2* (ed. A. J. Kinloch), Applied Science Pub., London, p. 1.
5. Zienkiewicz, O. (1982) *The Finite Element Method* McGraw-Hill Book Co., London.
6. Adams, R. D. (1981) *Developments in Adhesives–2* (ed. A. J. Kinloch), Applied Science Pub., London, p. 45.
7. Schjeldrup, H. C. and Jones, W. B. (1978) *Adhesives Age* **21**(2), 35.
8. Adams, R. D., Atkins, R. W., Harris, J. A. and Kinloch, A. J. (1986) *J. Adhesion* **20**, 29.
9. Kinloch, A. J. and Young, R. J. (1983) *Fracture Behaviour of Polymers* Applied Science Pub., London, p. 286.
10. Adams, R. D., Coppendale, J. and Peppiatt, N. A. (1978) *J. Strain Anal.* **13**, 1.
11. Kreiger, R. B. (1973) *SAMPE Q* **5**(1), 643.

12. Kreiger, R. B. (1975) *Stiffness Characteristics of Structural Adhesives for Stress Analysis in Hostile Environments*, Report from American Cyanamid, USA.
13. Jurf, R. A. and Vinson, J. R. (1985) *J. Mater. Sci.* **20**, 2979.
14. Thrall, E. W. (1979) *Bonded Joints and Preparation for Bonding*, AGARD Lecture Series No. 102, Paper 5.
15. Kreiger, R. B. (1985) *Adhesives Age* **28**(11), 26.
16. Kinloch, A. J. and Taig, C. M. (1987) *J. Adhesion* **21**, to be published.
17. Stringer, L. G. (1985) *J. Adhesion* **18**, 185.
18. Althof, W.,Klinger, G., Neumann, G. and Schlothauer, J. (1979) *Environmental Effects on the Elastic–Plastic Properties of Adhesives in Bonded Metal Joints*, Royal Aircraft Establishment Translation No. 1999, Farnborough, UK.
19. Knollman, G.C. (1985) *Int. J. Adhesion Adhesives* **5**, 137.
20. Knollman, G. C. and Hartog, J. J. (1985) *J. Adhesion* **17**, 247.
21. Post, D., Czarnek, R., Wood, J. D. and Joh, D. (1988) *ASTM, STP (International Symposium on Adhesively Bonded Joints*, Baltimore, USA, 1986) to be published.
22. Alwar, R. S. and Nagaraja, Y. R. (1976) *ASTM, STP J. Adhesion* **7**, 279.
23. Kuenzi, E. W. and Stevens, G. H. (1963) *Determination of Mechanical Properties of Adhesives for use in the Design of Bonded Joints*, FPL Report No. 011, US Forest Products Laboratory.
24. Gent, A. N. and Lindley, P. B. (1958) *Proc. Roy. Soc.* **A249**, 195.
25. Gent, A. N. and Lindley, P. B. (1959) *Proc. Inst. Mech. Eng.* **173**, 111.
26. Holownia, B. P. (1972) *J. Strain Anal.* **7**, 236.
27. Anderson, G. P., Bennett, S. J. and DeVries, K. L. (1977) *Analysis and Testing of Adhesive Bonds*, Academic Press,New York, p. 199.
28. Kinloch, A. J. and Young, R. J. (1983) *Fracture Behaviour of Polymers*, Applied Science Pub., London, p. 370.
29. Rutherford, J. L., Bossler, F. C. and Hughes, E. J. (1968) *Rev. Sci. Instruments* **39**, 666.
30. Adams, R. D. and Coppendale, J. (1976) *J. Mech. Eng. Sci.* **18**, 149.
31. Bryant, R. W. and Dukes, W. A. (1967) *Aeronautic and Space Engineering and Manufacturing Meeting*, Los Angeles, October, Paper No. 670855, Society for Aerospace Engineering, New York.
32. Meissner, H. P. and Baldauf, G. H. (1951) *Trans. ASME* **73**, 697.
33. Bredzs, N. and Schwartzbart, H. (1956) *Welding J. (Welding Res. Suppl.)* **35**, 610.
34. Meissner, H. P. and Merrill, E. W. (1948) *ASTM Bull.* **151**, (March) 80.
35. Lososki, S. W. and Kraus, G. (1955) *J.Polymer Sci.* **18**, 359.
36. Dietz, A. G. H. (1957) *ASTM, STP* **194**, 19.
37. Bikerman, J. J. and Huang, C. R. (1959) *Trans. Soc. Rheol.* **3**, 5.
38. Bikerman, J. J. (1968) *The Science of Adhesive Joints*, Academic Press, New York, p. 273.
39. Dukes, W. A. and Bryant, R. W. (1969) *J. Adhesion* **1**, 48.
40. Wake, W. C. (1959) *Trans. IRI* **35**, 145.
41. Wake, W.C. (1961) *Adhesion* (ed. D. D. Eley), Oxford University Press, Oxford p. 191.
42. Gent, A. N. (1974) *Rub. Chem.Technol.* **47**, 202.
43. Gent, A. N. (1974) *J. Polymer Sci.* **A2**, 283.
44. Hilton, P. D. and Gupta, G. D. (1973) *Design and Engineering Conference* Philadelphia, April 1973, Paper No. 73-De-21, ASME, Philadelphia.
45. Williams, M. L., Landel, R. F. and Ferry, J. D. (1955) *J. Am. Chem. Soc.* **77**, 3701.
46. Doolittle, A. K. (1952) *J. Appl. Phys.* **23**, 236.

47. Kinloch, A. J. and Young, R. J. (1983) *Fracture Behaviour of Polymers*, Applied Science Pub., London, p. 373.
48. Arin, K. and Erdogan, F. (1971) *Int. J. Eng. Sci.* **9**, 213.
49. Adams, R. D. and Coppendale, J. (1979) *J. Adhesion* **10**, 49.
50. Jennings, C. W. (1972) *J. Adhesion* **4**, 25.
51. Mylonas, C. and DeBruyne, N. A. (1951) *Adhesion and Adhesives* (eds N. A. DeBruyne and R.Houwink), Elsevier, Amsterdam, p. 91.
52. Sancaktar, E. and Lawry, P. O. (1980) *J. Adhesion* **11**, 233.
53. Lubkin, R. L. and Reissner, E. (1956) *Trans. ASME* **78**, 1213.
54. Ishai, O., Peretz, D. and Gali, S. (1977) *Expt. Mech.* **17**, 265.
55. Hart-Smith, J. L. (1974) *Analysis and Design of Advanced Composite Bonded Joints* NASA Report No. CR- 2218, Washington DC, USA.
56. Hart-Smith, J. L. (1980) *Further Developments in the Design and Analysis of Adhesively-Bonded Structural Joints*, Douglas Aircraft Report No. 6922, Long Beach, CA, USA.
57. Allman, D. J. (1977) *Q. J. Mech. Appl. Math.* **30**, 415.
58. Renton, W. J. and Vinson, J. R. (1975) *J. Adhesion* **7**, 175.
59. Renton, W. J. and Vinson, J. R. (1977) *J. Appl. Mech.* **44**, 101.
60. Adams, R. D. and Peppiatt, N. A. (1973) *J. Strain Anal.* **8**, 134.
61. Adams, R. D. and Peppiatt, N. A. (1974) *J. Strain Anal.* **9**, 185.
62. Tuttle, M. E., Barthelemy, B. M. and Brinson, H. F. (1984) *Expt. Techniques* **8**, 31.
63. ESDU (1978) Engineering Sciences Data Unit Report No. 78042, London, UK.
64. ESDU (1979) *Inelastic Shear Stresses and Strains in the Adhesives Bonding Lap Joints Loaded in Tension or in Shear*, Engineering Sciences Data Unit Report No. 79016, London, UK.
65. Grant, P. J. (1978) *Jointing in Fibre Reinforced Plastics* IPC Press, Guildford, UK, p. 41.
66. Hart-Smith, J. L. (1985) *ASTM, STP* **876**, 238.
67. Engineering Sciences Data Unit, 251–259 Regent Street, London, W1R 7AD, UK.
68. Adams, R. D. and Wake, W. C. (1984) *Structural Adhesive Joints in Engineering*, Applied Science Pub., London, p. 54.
69. DeBruyne, N. A. (1944) *Aircraft Eng.* **16**, 115.
70. DeBruyne, N. A. (1944) *Aircraft Eng.* **16**, 140.
71. Wellinger, K. and Rembold, U. (1958) *VDI Zeitschrift* **100**, 41.
72. Kinloch, A. J. and Shaw, S. J. (1982) UK Patent Application 8,236,849.
73. Kinloch, A. J. and Shaw, S. J. (1984) UK Patent Application 8,416,024.
74. Kinloch, A. J., Shaw, S. J. and Tod, D. A. (1984) *Adv. Chem. Series* **208**, 101.
75. Shaw, S. J. and Kinloch, A. J. (1985) *Int. J. Adhesion Adhesives* **5**, 123.
76. Cotter, J. L. (1973) *Rev. High Temp. Mater.* **3**, 277.
77. Bolger, J. C. (1973) *Treatise on Adhesion and Adhesives*, Vol. 3 (ed. R. I. Patrick), Marcel Dekker, New York, p. 1.
78. Yaeger, R. E. (1969) *Appl. Polymer Symp.* **3**, 369.
79. Sandler, S. R. and Berg, F. R. (1965) *J. Appl. Polymer Sci.* **9**, 3909.
80. Beevers, A. and Ellis, M. D. (1984) *Int. J. Adhesion Adhesives* **4**, 13.
81. Harris, J. A. and Adams, R. D. (1985) *Proc. Inst. Mech. Eng.* **199**, 121.
82. Volkersen, O. (1965) *Const. Met.* **4**, 3.
83. Hart-Smith, L. J. (1972) *Design and Analysis of Adhesive-Bonded Joints*, Douglas Paper No. 6059A, Douglas Aircraft Co., Long Beach, CA, USA.
84. Adkins, D. W. and Pipes, R. B. (1985) *Comp. Sci. Technol.* **22**, 209.
85. Erdogan, F. and Ratwani, M. (1971) *J. Adhesion* **5**, 378.

86. Adams, R. D., Chambers, S. H., Del-Strother, P. J. A. and Peppiatt, N. A. (1973) *J. Strain Anal.* **8**, 52.
87. Thamm, F. (1976) *J. Adhesion* **7**, 301.
88. Cherry, B. W. and Harrison, N. L. (1970) *J. Adhesion* **2**, 125.
89. Wright, M. D. (1978) *Composites* **9**, 259.
90. Sage, G. N. (1979) *Adhesion–3* (ed. K. W. Allen), Applied Science Pub., London, p. 123.
91. Kaelble, D. H. (1959) *Trans. Soc. Rheol.* **3**, 161.
92. Kaelble, D. H. (1960) *Trans. Soc. Rheol.* **4**, 45.
93. Kaelble, D. H. (1965) *Trans. Soc. Rheol.* **9**, 135.
94. Kaelble, D. H. and Ho, C. L. (1974) *Trans. Soc. Rheol.* **18**, 219.
95. Bikerman, J. J. (1957) *J. Appl. Phys.* **28**, 1484.
96. Inoue, Y. and Kobatake, Y. (1959) *Appl. Sci. Res.* **A8**, 321.
97. Gardon, J. L. (1963) *J. Appl. Polymer Sci.* **7**, 625.
98. Gardon, J. L. (1963) *J. Appl. Polymer Sci.* **7**, 643.
99. Spies, G. J. (1953) *Aircraft Eng.* **30**, (March), 2.
100. Jouwersma, J. (1960) *J. Polymer Sci.* **45**, 253.
101. Mylonas, C. (1965) *Proceedings of the 4th International Congress on Rheology*, Providence, USA, Part 2, Wiley Interscience, New York, p. 423.
102. Niesiolowski, F. and Aubrey, D.W. (1981) *J. Adhesion* **13**, 87.
103. Wake, W. C. (1982) *Adhesion and the Formulation of Adhesives*, Applied Science Pub., London, p. 143.
104. Gent, A. N. and Hamed, G. R. (1975) *J. Adhesion* **7**, 91.
105. Crocombe, A. and Adams, R. D. (1981) *J. Adhesion* **12**, 127.
106. Kukovyakin, N. M. and Skoryi, I. A. (1972) *Russ. Eng. J.* **52**(4), 40.
107. Terekhova, L. P. and Skoryi, I. A. (1973) *Strength Mater.* **4**, 1271.
108. Adams, R. D. and Peppiatt, N. A., (1977) *J. Adhesion* **9**, 1.
109. Nagaraja, Y. R. and Alwar, R. S. (1979) *J. Adhesion* **10**, 97.
110. Foulkes, H., Shields, J. and Wake, W. C., (1970) *J. Adhesion* **2**, 254.
111. Bryant, R. W. and Dukes, W. A. (1965) *Br. J. Appl. Phys.* **16**, 101.
112. Bryant, R. W. and Dukes, W. A. (1966) *Appl. Polymer Symp.* **3**, 81.
113. Humpidge, R. T. and Taylor, B. J. (1967) *J. Sci. Instrum.* **44**, 457.
114. McCarvill, W. T. and Bell, J. P. (1974) *J. Adhesion* **6**, 185.
115. Bossler, F. C., Franzblau, N. C. and Rutherford, J. L. (1968) *J. Phys. E* **1**, 829.
116. Nadai, A. (1931) *Plasticity: A Mechanics of the Plastic State of Matter*, McGraw-Hill, New York, p. 128.
117. Lees, W. A. (1985) *Mater. Design* **6**, 117.
118. Mylonas, C. (1948) *Proceedings of the VIIth International Congress on Applied Mechanics*, Vol. 4, London, p. 137.
119. DeLollis, N. J. and Montoya, O. (1972) *J. Appl. Symp.* **19**, 417.
120. Swanson, F. (1977) *Polymer Eng. Sci.* **17**, 122.
121. Cherry, B. W. and Thompson, K. W. (1977) *Adhesion–1* (ed. K. W. Allen), Applied Science Pub., London, p. 251.
122. Garf, S. E., Myshko, V. I. and Iuchenko, R. A. (1974) *Sintez-i-Fizikv-Khinya Polimerov* **13**, 162.
123. Inoue, Y. and Kobatake, Y. (1958) *Appl. Sci. Res.* **A7**, 53.
124. Sargent, J. P. and Ashbee, K. H. G. (1982) *Ind. Eng. Chem. Prod. Res. Dev.* **21**, 302.
125. Hahn, H. T. (1984) *J. Adhesion* **17**, 21.
126. Elm, A. C. (1956) *Official Digest* **28**, 752.
127. DeBruyne, N. A. (1956) *J. Appl. Chem.* **6**, 303.

128. Inoue, Y. and Kobatake, Y. (1958) *Appl. Sci. Res.* **A7**, 314.
129. Inoue, Y. and Kobatake, Y. (1958) *Kolloid Z.* **18**, 18.
130. Danneberg, H. (1965) *SPE J.* **21**, 669.
131. Gusman, S. (1962) *Official Digest* **34**, 884.
132. Bikerman, J. J. (1968) *The Science of Adhesive Joints*, Academic Press, New York, p. 192.
133. Wake, W. C. (1959) *Trans IRI* **35**, 145.
134. Harrison, N. L. and Harrison, W.J. (1972) *J. Adhesion* **3**, 195.
135. Carlson, J. A. and Sapetta, L. P. (1967) *Adhesives Age* **19**(12), 26.
136. Stone, M. H. (1980) *Adhesion and Adhesives*, Plastics and Rubber Institute, London, p. 10.1.
137. Schliekelmann, R. J. (1972) *Non-Destructive Test.* **5**, 79.
138. Schliekelmann, R. J. (1972) *Non-Destructive Test.* **5**, 144.
139. Schliekelmann, R. J. (1979) *Bonded Joints and Preparation for Bonding*, AGARD Lecture Series No. 102, AGARD, France, p. 8.1.
140. Segal, E. and Rose, J. L. (1980) *Research Techniques in Nondestructive Testing IV* (ed. R. S. Sharpe), Academic Press, London, p. 275.
141. Curtis, G. J. (1982) *Ultrasonic Testing* (ed. J. Szilard), John Wiley, New York, p. 495.
142. Hagemier, D. J. (1985) *Adhesive Bonding of Aluminium Alloys* (eds E. W. Thrall and R. W. Shannon), Marcel Dekker Inc., New York, p. 337.
143. Guyott, C. C. H., Cawley, P. and Adams, R. D. (1986) *J. Adhesion* **20**, 129.
144. Kim, D. H. and Sutliff, E. F. (1978) *SAMPE Q.* July, 59.
145. Dukes, W. A. and Kinloch, A. J. (1974) *Non-Destructive Test.* **7**, 324.
146. Schonhorn, H., Wang, T. T. and Ryan, F. W. (1972) *J. Appl. Polymer Sci.* **16**, 1901.
147. Adams, R. D. and Cawley, P. (1985) *Research Techniques in Nondestructive Testing* (ed. R. S. Sharpe), Academic Press, London, p. 303.
148. Cawley, P. (1984) *NDT Int.* **17**, 59.
149. Kersch, L. A. (1974) *Holographic Non-Destructive Testing* (ed. R. K. Erf), Academic Press, New York, p. 303.
150. Rose, J. L. and Meyer, P. A. (1973) *Mater. Eval.* **31**, 109.
151. Lloyd, E. A. and Brown, A. F. (1978) *Adhesion–2* (ed. K. W. Allen), Applied Science Pub., London, p. 133.
152. Speake, J. H. and Curtis, G. (1973) *Plastics Rubber: Mater. Applic.* 193.
153. Curtis, G. (1975) *Non-destructive Test.* **8**, 249.
154. King, C. S. A. (1984) *Proc. Instn. Mech. Eng.* Part D, No. 3, 21.

CHAPTER SEVEN

Fracture mechanics of adhesive joints

7.1 INTRODUCTION

Adhesive joints usually fail by the initiation and propagation of flaws and, since the basic tenet of continuum fracture mechanics is that the strength of most real solids is governed by the presence of flaws, the application of such theories to adhesive joint failure has received considerable attention. The main aims of the various theories are to analyse mathematically the loads at which the flaws propagate and describe the manner in which they grow. The source of naturally occurring flaws, sometimes referred to as 'intrinsic flaws', may be voids, cracks, dirt particles, additive particles, inhomogeneities in the adhesive, etc. which may be initially present at a critical size or develop during the fracture test.

In many ways the use of fracture mechanics to study the failure of joints is a complementary approach to that of mapping the nature and magnitude of the stresses, which was discussed in the previous chapter. Both approaches have their advantages and disadvantages. Fracture mechanics has proved to be particularly useful for such aspects as characterizing the toughness of adhesives, identifying mechanisms of failure and estimating the service life of 'damaged' structures – the 'damage' being in the form of cracks, air-filled voids, debonds, etc. and having arisen, for example, from environmental attack, fatigue loading, subcritical impact loads. On the other hand, a detailed knowledge of the stress distribution in a bonded structure has proved to be of particular value in initial design studies and for interpreting the effects of geometric parameters, such as the effect of overlap length in lap joints.

Essentially, continuum fracture mechanics is the study of the strength of a structure which contains a flaw, usually considered as an elliptical crack. The theories were originally developed for the cohesive fracture of materials, and several reviews [1–4] have been published on this subject, but a large body of work has also been reported on extending the principles of fracture mechanics to the failure of adhesive joints and such studies are discussed in the present chapter.

The theoretical aspects will first be considered and, as well as the case of a crack in the adhesive layer, the important but complex problem of cracks at, or very near, interfaces will be discussed in depth. This is followed by a review of the experimental techniques used to measure parameters such as the fracture energy, G_c and the fracture toughness, K_c. Various aspects such as the effects of the microstructure of the adhesive, the geometry of the joint and the test rate and temperature will then be discussed with the emphasis on attempting to correlate and understand the extensive experimental data that has been accumulated over the last decade, or so. However, the use of fracture mechanics in studying creep and fatigue fracture and in identifying the mechanisms of environmental attack and predicting the service life will be discussed later in Chapter 8.

On the use of nomenclature, it should be noted that there are obviously many failure paths available to a crack in an adhesive joint, and frequently the adhesive is also tested as a bulk, homogeneous material. To avoid the use of too many subscripts, etc. the symbols G and K will be used freely to represent any of the possible options mentioned above and, hopefully, the text will make clear the type of test and locus of failure. Where confusion might arise then a suitable subscript or bracketed description will be used in that part of the text.

7.2 THEORETICAL CONSIDERATIONS

7.2.1 Introduction

Two main, interrelatable, conditions for fracture are proposed. Firstly, the energy criterion arising from Griffith's [5], and later Orowan's [6], work which supposes that fracture occurs when sufficient energy is released (from the stress field) by growth of the crack to supply the energy requirements of the new fracture surfaces. The energy released comes from stored elastic or potential energy of the loading system and can, in principle, be calculated for any type of test piece. This approach, therefore, provides a measure of the energy required to extend a crack over unit area and this is termed the fracture energy or critical strain energy release rate and is denoted as G_c. Secondly, Irwin [7] found that the stress field around a sharp crack in a linear-elastic material could be uniquely defined by a parameter named the stress intensity factor, K, and stated that fracture occurs when the value of K exceeds some critical value, K_c. Thus, K is a stress field parameter independent of the material whereas K_c, often referred to as the fracture toughness, is a measure of a material property.

A basic aim of fracture mechanics is to identify fracture criteria such as G_c and K_c and others discussed later, which are independent of the geometry of the cracked body and, ideally, even of the test conditions, e.g. test temperature.

Thus, fracture mechanics should greatly assist in developing a more fundamental understanding of the fracture process and should be of considerable use in the practical areas of data specification and engineering design, particularly in connection with service life prediction studies.

7.2.2 Energy balance approach

7.2.2.1 Basic principles

The energy criterion for fracture is simply an extension of Griffith's hypothesis which describes quasi-static crack propagation as the conversion of the work done, W_d, by the external force and the available elastic energy stored in the bulk of the specimen, U, into surface free energy, γ_m. It may be written:

$$\frac{\partial(W_d - U)}{\partial a} \geq \gamma_m \frac{\partial A}{\partial a} \tag{7.1}$$

where ∂A is the increase in surface area associated with an increment of crack growth ∂a. For a crack propagating in a lamina of thickness, b, the criterion becomes:

$$\frac{1}{b} \cdot \frac{\partial(W_d - U)}{\partial a} \geq 2\gamma_m \tag{7.2}$$

Now initial work by Orowan [6], Rivlin and Thomas [8] and Berry [9] examined this criterion with respect to metals, crosslinked rubbers and poly(methyl methacrylate) and all found that the energy required to cause crack growth was far greater than twice the surface free energy. There are two main reasons for this discrepancy. Firstly, as discussed in Section 2.2, the value of $2\gamma_m$ (equivalent to W_A in the case of crack growth along an interface) only reflects the energy required to rupture secondary bonds such as van der Waals forces; crack growth in materials, and indeed along interfaces, often necessitates the rupture of stronger forces such as primary bonds. The energy required solely for rupturing intrinsic bonding forces, such as secondary and primary bonds, is termed the intrinsic fracture energy – this will be denoted by the symbol G_o and if the material is perfectly elastic then G_o is the appropriate energy requirement. However, secondly, fracture in even the most brittle of adhesives invariably involves localized viscoelastic and/or plastic energy dissipative processes occurring where high strains are experienced. Indeed, even in the most brittle of adhesives, e.g. phenolic resins, this is the situation in the regions around the crack tip. Such micromechanisms usually represent the main source of energy absorption in the material and, for example, the microstructure of the rubber-modified structural adhesives is designed to maximize such processes, and thereby impart high toughness, as discussed later in Section 7.6.3.2.

However, if it is assumed that energy dissipation around the crack tip occurs in a manner independent of both the test geometry and the way in which the forces are applied to the specimen then $2\gamma_m$ may be replaced in Equation 7.2 by the symbol G_c. The value G_c encompasses all the energy losses incurred around the crack tip and is, therefore, the energy required to increase the crack by unit length in a specimen of unit width. Hence, the fracture criterion becomes:

$$\frac{1}{b}\cdot\frac{\partial(W_d - U)}{\partial a} \geq G_c \qquad (7.3)$$

For bonded structures exhibiting bulk linear-elastic behaviour, i.e. away from the crack tip regions, they obey Hooke's Law. Equation 7.3 may be expressed [1–4] as:

$$G_c = \frac{F_c^2}{2b}\frac{\partial C}{\partial a} \qquad (7.4)$$

where F_c is the load at the onset of crack propagation and C is the compliance of the structure and is given by displacement/load, i.e. v/F. Further, and most importantly, for an infinitesimally small amount of crack growth this equation is equally valid for a cracked body under fixed-extension or constant-load conditions. This equation is the foundation of many calculations of G_c, since if C is determined as a function of a, either experimentally or theoretically by an analytical or numerical technique, then $\partial C/\partial a$ may be found. An example of this is shown in Fig. 7.1 where the compliance, C, of two edge-bonded plates containing an interfacial crack has been determined [10] as a function of the non-dimensionalized crack length, a/D, where D is the depth of the specimen (Fig. 7.1a). Both an experimental technique employing very sensitive omega strain gauges and a theoretical method employing a numerical boundary integral equation analysis has been used. (The boundary integral equation method is similar to the finite element method but only the boundary of the solution domain need be discretized; so allowing, for example, greater resolution of stresses and faster program execution times.) The resulting relation between $\partial C/\partial a$ and a/D is shown in Fig. 7.1b and there is excellent agreement between experiment and theory. Thus, as shown later, if the load, F_c, at the onset of crack growth is measured for a known crack length the value of G_c may be deduced.

However, the evaluation of Equation 7.3 is not necessarily restricted to a linear-elastic fracture mechanics (LEFM) approach, i.e. it is not restricted to those materials or structures which exhibit linear-elastic behaviour. For example, Rivlin and Thomas [8] have employed this approach to characterize

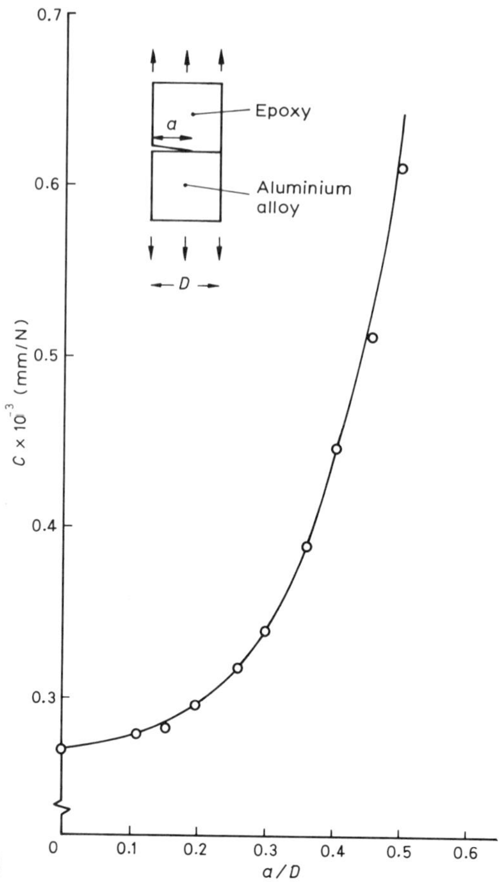

(a)

Figure 7.1 (a) Compliance, C, versus the non-dimensional crack length, a/D, for a bimaterial single-edged notched specimen with an interfacial crack. The points are experimental and the solid curve is theoretical using boundary integral equation analysis [10]. (b) Change of compliance with crack length, $\partial C/\partial a$, versus the non-dimensional crack length, a/D, for a bimaterial single-edged notched specimen with an interfacial crack. The points are experimental, and the solid curve is theoretical using boundary integral equation analysis [10].

the tearing of crosslinked rubbers which possess bulk non-linear, but reversible elastic, properties and Gent and Kinloch [11] used Equation 7.3 to determine the adhesive fracture energy of a non-linear elastic rubbery adhesive adhering to a poly(ethylene terephthalate) substrate. This work, and later studies, are discussed below.

More recently, Ahagon *et al.* [12], Thomas and Kadir [13] and Kinloch and Tod [14] have suggested that, even for rubbers which show significant internal

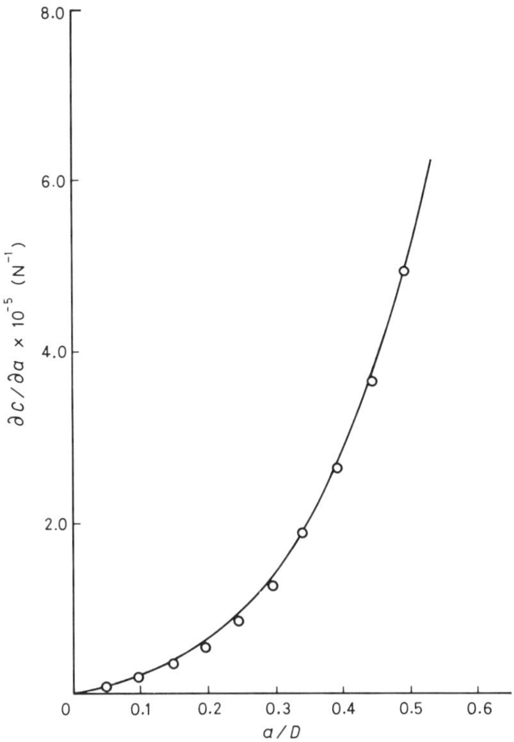

(b)

Figure 7.1—continued

energy dissipation outside the immediate crack-tip regions, Equation 7.3 may still be used. However, the stored strain energy available for crack propagation should not now be taken as the input energy but rather the recoverable strain energy density deduced from the stress–strain relation upon retraction from the deformed state, i.e. the input energy minus the hysteresis (loss) energy. Other approaches to this problem of analysing non-linear and inelastic behaviour of the material well away from the crack-tip regions include the J-contour integral method [3,4] and that of Andrews [15] who has developed the generalized theory of fracture mechanics.

Advantages of the energy balance approach are firstly that from Equations 7.3 and 7.4 an unambiguous value of G_c may be deduced. This is true whether a thin or thick adhesive layer is employed and whether the crack propagates deep in the adhesive layer or at, or near, an interface. As will be seen later, determination of the fracture toughness, K_c, of a joint is not always so straightforward. Secondly, the value of G_c may be related to the intrinsic bonding forces in the material or acting across the interface, as discussed below.

7.2.2.2 *Relationships between G_c, G_o and W_A*

The relationships between G_c, G_o and W_A were briefly outlined in Chapter 3 (Section 3.5.2.2) but will be developed and discussed more fully here. From the first law of thermodynamics (ignoring dynamic effects such as energy radiation by stress waves) then:

$$G_c = G_o + \psi \tag{7.5}$$

where ψ is the energy dissipated in viscoelastic and plastic deformations at the crack tip. As mentioned above, the value of ψ is usually the major contribution to the value of G_c and it is this parameter which frequently results in the measured value of G_c being highly dependent upon the rate and temperature of testing. Now, from dimensional and intuitive arguments Gent and Kinloch [10], Andrews and Kinloch [16–18] and Gent and Schultz [19] proposed the simple relationship:

$$\psi = G_o f(\dot{a}, T, \epsilon) \tag{7.6}$$

where f is a function, the value of which depends upon the crack growth rate, $\dot{a}$, the test temperature, T, and the strain level, ϵ. Now let:

$$\Phi_v(\dot{a}, T, \epsilon) = 1 + f(\dot{a}, T, \epsilon) \tag{7.7}$$

then, combining Equations 7.5 and 7.7 yields:

$$G_c = G_o \Phi_v(\dot{a}, T, \epsilon) \tag{7.8}$$

where Φ_v is a loss function. The dependence of the energy loss term, ψ, upon the intrinsic fracture energy, G_o, as shown in Equation 7.6, is particularly intriguing. This dependence arises because material around the crack tip can only be subjected to stress, and hence energy losses, whilst the chemical bonds, physical bonds and other intrinsic forces acting ahead of the crack tip remain unbroken. An important aspect of Equation 7.8 is that when viscoelastic and plastic energy losses are negligible, i.e. $\Phi_v\ (\dot{a}, T, \epsilon) \rightarrow 1$ and $f(\dot{a}, T, \epsilon) \rightarrow 0$, then the measured value of the fracture energy is equal to G_o.

Now, in an adhesive joint the failure plane may be situated either in the materials forming the joint or along the interface, or wander between these possible paths. Thus, the value of the intrinsic adhesive fracture energy, G_o, may be expressed as a weighted mean of the various possible failure paths:

$$G_o = iG_o\,(\text{interfacial}) + b'\,\mathrm{G_o}\,(\text{adhesive}) + sG_o\,(\text{substrate}) \tag{7.9}$$

where G_0 (interfacial), G_0 (adhesive) and G_0 (substrate) are the intrinsic fracture energies for interfacial, cohesive-in-adhesive and cohesive-in-substrate failure, respectively, and i, b' and s are the respective area fractions of interfacial failure, failure in adhesive and failure in substrate.

For joints which exhibit solely interfacial failure then obviously $i = 1$ and, from Equation 7.9, G_0 takes the same value as G_0 (interfacial). Hence, if only secondary bonds are operating across the interface then the value of G_0 should be equal to the value of the thermodynamic work of adhesion, W_A, as may be seen from the data in Table 3.5. Alternatively, if stronger interfacial forces are present, such as primary bonds, then the value of G_0 is much greater than W_A, as may be seen from the results in Fig. 4.3.

Finally, it should be emphasized that it may indeed be true to state that the value of the energy loss term, ψ, depends upon the intrinsic fracture energy, G_0; since material around the crack tip can only be subjected to stress, and hence energy losses, whilst the intrinsic forces acting ahead of the crack tip remain unbroken. However, whether the quantitative relationships suggested in Eqns 7.6–7.8 are always of this form has yet to be firmly established. In the simple case of interfacial failure involving the rupture of secondary bonds and a model viscoelastic adhesive then the relationships, as stated in these equations, do appear to be valid. Also, they appear to hold for cases of joint failure again involving model viscoelastic adhesives but with the locus of failure near the interface, or in the bulk adhesive, and with the fracture event requiring the rupture of primary bonds. However, in the case of, say, joint fracture involving an epoxy adhesive, where *plastic* deformation occurs at the crack tip, it is not clear whether the simple relations outlined above are still appropriate. For example, in such a case there may well be a minimum value for the strength of the intrinsic adhesion forces that are necessary in order for the adhesive's plastic yield stress to be attained. Hence, the above relationships would need to be suitably modified. Establishing such relationships is a current challenge for those working in this area.

7.2.3 Stress intensity factor approach

7.2.3.1 Basic principles

The problem of a crack situated well away from any interface will be considered first, followed by the far more complex problem of a crack at, or very close to, an interface.

'Bulk' material – Figure 7.2 shows a sharp crack in a uniformly stressed, infinite, homogeneous lamina and, assuming Hookean behaviour and infinitesimal strains, i.e. LEFM, Westergaard [20] has developed certain stress-functions solutions which relate the local stress concentration of stresses at the

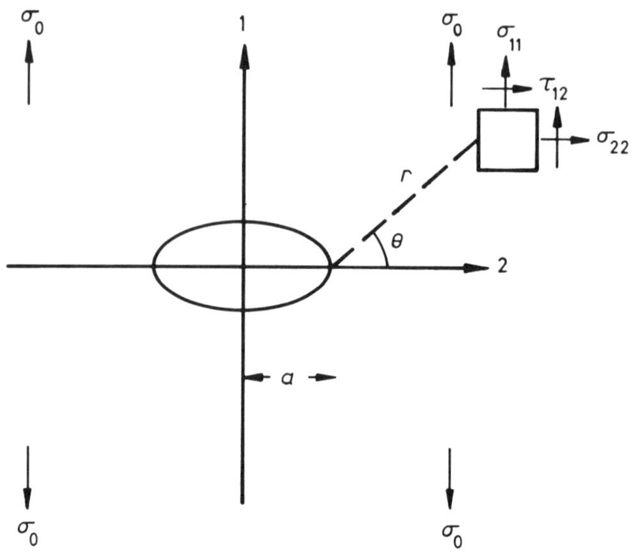

Figure 7.2 Sharp crack in a uniformly stressed, infinite lamina.

crack tip to the applied stress, σ_o. For regions close to the crack tip the solutions take the form:

$$\sigma_{ij} = \sigma_o \left(\frac{a}{2r}\right)^{1/2} f_{ij}(\theta) \tag{7.10}$$

where σ_{ij} are the components of the stress tensor at a point, r and θ are the polar coordinates of the point, taking the crack tip as the origin, and $2a$ is the length of the crack.

Irwin [7] modified this solution to give:

$$\sigma_{ij} = \frac{K}{(2\pi r)^{1/2}} f_{ij}(\theta) \tag{7.11}$$

The parameter K is the stress intensity factor and relates the magnitude of the stress intensity local to the crack in terms of the applied loadings and geometry of the structure in which the crack is located.

A crack may be stressed in three different modes, denoted I, II and III, as depicted in Fig. 7.3; superposition of the three modes constitutes the general case of crack loading. The cleavage or tensile-opening mode, mode I, is technically the most important since it is the most commonly encountered and usually the one which most often results in failure. However, crack propagation in joints is often constrained to the adhesive layer, regardless of the orientation of the adhesive layer to the applied loads. Thus, in the case of joint fracture attention must sometimes be given to modes II and III, as well as mode I.

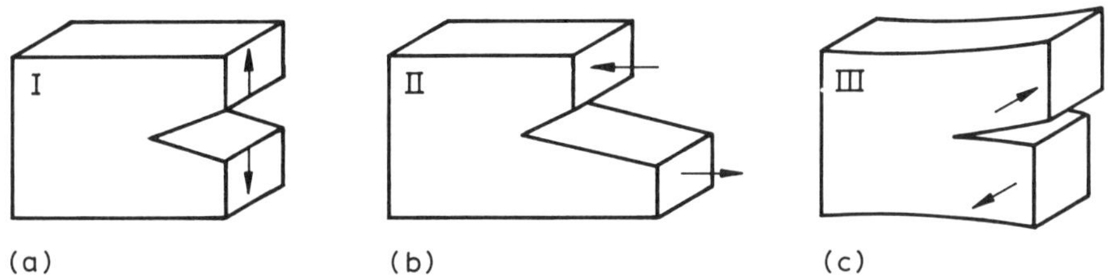

Figure 7.3 Modes of loading. (a) Cleavage or tensile-opening mode: mode I. (b) In-plane shear mode: mode II. (c) Antiplane shear mode: mode III.

For the mode I case the crack tip stresses may be developed from Equation 7.11 to yield the expressions below, and it should be noted that analogous expressions may be developed for the other two modes.

$$\begin{Bmatrix} \sigma_{11} \\ \tau_{12} \\ \sigma_{22} \end{Bmatrix} = \frac{K_\mathrm{I}}{(2\pi r)^{1/2}} \cos(\theta/2) \begin{Bmatrix} 1 + \sin(\theta/2)\sin(3\theta/2) \\ \sin(\theta/2)\cos(3\theta/2) \\ 1 - \sin(\theta/2)\sin(3\theta/2) \end{Bmatrix} \qquad (7.12)$$

and

$$\sigma_{33} = 0 \qquad \text{(plane stress), or} \qquad (7.13)$$

$$\sigma_{22} = \nu(\sigma_{11} + \sigma_{22}) \qquad \text{(plane strain), and} \qquad (7.14)$$

$$\tau_{23} = \tau_{13} = 0 \qquad (7.15)$$

Note that in the plane $\theta = 0°$ the shear stress is zero hence for $\theta = 0°$ the stresses $\sigma_{11}, \sigma_{22}, \sigma_{33}$ are the principal stresses σ_1, σ_2 and σ_3, respectively.

From Equation 7.12 it is evident that as $r \rightarrow 0$ then the stress $\sigma_{ij} \rightarrow \infty$ and hence stress alone does not make a reasonable local fracture criterion. Therefore, since the level of K_I uniquely defines the stress field around the crack, Irwin [7] postulated that the condition:

$$K_\mathrm{I} \geq K_\mathrm{Ic} \qquad (7.16)$$

represented a fracture criterion, where K_Ic is a critical value for crack growth in the material and, as such, is a material property and often termed the fracture toughness. The units of K are $\mathrm{MN/m^{3/2}}$ and it should be emphasized that it is a factor which characterizes the *intensity* of the stress field ahead of a crack. It is not a stress concentration, η_c, which may be defined by σ_{ij}/σ_o. The power of the stress intensity factor approach is that for any mode I problem K_I may always be expressed in the form:

$$K_\mathrm{I} = Q\sigma_o a^{1/2} \qquad (7.17)$$

and, hence, K_{Ic} by:

$$K_{Ic} = Q\sigma_c a^{1/2} \tag{7.18}$$

where σ_c is the applied stress at the onset of crack growth and Q is the geometry constant. To deduce values of K_I and K_{Ic}, and so examine the criterion for fracture embodied in Equation 7.16, obviously requires a knowledge of the geometry factor, Q. Values of Q may be experimentally or theoretically ascertained and such methods are considered in the following section.

Cracks at, or near, interfaces – The above has considered the aspect of cracks located in 'bulk' material, but a second important case is that of cracks at, or very close to, a bimaterial interface. However, an immediate problem arises: namely, that when the joint is subjected to solely tensile loads applied normal to the crack, which is located along or parallel to the interface, then these will induce both tensile *and* shear stresses around the crack tip. Therefore, both K_{Ii} and K_{IIi} terms are needed to describe the stress field; the subscript i indicating a crack at, or near, the interface. Similarly, an applied pure shear stress will also induce both such terms. However, these K_{Ii} and K_{IIi} terms no longer have the clearly defined physical significance, as for the bulk material case and illustrated in Fig. 7.3. Mathematical modelling has shown [21–27] that, for linear-elastic materials, the local stresses ahead of the crack tip at a bimaterial interface are proportional to:

$$\frac{f(K_{Ii}, K_{IIi})}{(2\pi r)^{1/2}} \begin{Bmatrix} \sin \\ \cos \end{Bmatrix} (\zeta \ln r) \tag{7.19}$$

where ζ is a 'bimaterial constant' and is a function of the moduli and Poisson's ratios of the two materials forming the interface and is given by:

$$\zeta = \frac{1}{2\pi} \ln \left[\left(\frac{\beta_a}{G_a} + \frac{1}{G_s} \right) \bigg/ \left(\frac{\beta_s}{G_s} + \frac{1}{G_a} \right) \right] \tag{7.20}$$

where G_a and G_s are the shear moduli of the adhesive and substrate, respectively, β_a and β_s are functions of the Poisson's ratios of the adhesive, ν_a, and substrate, ν_s, such that:

$$\begin{aligned} \beta_j &= 3 - 4\nu_j \text{ for plane strain and} \\ \beta_j &= (3 - \nu_j)/(1 + \nu_j) \text{ for plane stress} \\ j &= \text{a or s as required.} \end{aligned} \tag{7.21}$$

Thus, it may be seen from Equation 7.19 that the singular behaviour of the stresses is again proportional to the inverse square root of the distance, r. But, unlike the bulk material case, a major consequence of the above relationship is that very close to the crack tip the stresses are oscillatory and have the highly improbable property of changing signs with increasing frequency as $r \to 0$. Even more unlikely, the crack face displacements also oscillate and near the crack tip interfere, giving a physically impossible solution. However, it may be argued that, since such oscillatory characteristics are an artefact of the analysis and are in fact confined [23,26] to a distance very, very close to the crack tip, the solutions to this relationship do model the near- and far-field stresses reasonably adequately. Nevertheless, a further complication is that the analysis results in a logarithmic term of a dimensional parameter, r. Thus the crack-tip stresses, and values of K, become a function of the measuring units of r. More recently, various modifications have been proposed [28–33] to this analysis to remove this arbitrary length parameter and thereby resolve these problems of unrealistic oscillating stresses and displacements and the values of K_{Ii} and K_{IIi} being a function of the dimensions used for r. Typically this has been achieved by scaling to another distance parameter, but such modifications are either forced to break one or other of the boundary conditions or have to assume some unrealistic condition. For example, some of the modifications which have been suggested are based upon assumptions such as closed crack tips in frictionless contact, an interfacial crack where no slip occurs along the interface ahead of the crack and non-stress-free crack faces.

From Equation 7.19 the values of K_{Ii} and K_{IIi} for an interfacial central crack of length $2a$ in an infinite sheet under only tensile applied stresses, σ_{o}, are given by [24]:

$$K_{\text{Ii}} = \sigma_{\text{o}} \frac{\{(2\pi)^{1/2}[\cos(\zeta \ln 2a) + 2\zeta \sin(\zeta \ln 2a)]\}}{\cosh(\pi\zeta)} (a)^{1/2} \tag{7.22}$$

or,

$$K_{\text{Ii}} = \sigma_{\text{o}} Q_{\text{Ii}} a^{1/2} \tag{7.22a}$$

and

$$K_{\text{IIi}} = \sigma_{\text{o}} \frac{\{-(2\pi)^{1/2}[\sin(\zeta \ln 2a) - 2\zeta \cos(\zeta \ln 2a)]\}}{\cosh(\pi\zeta)} (a)^{1/2} \tag{7.23}$$

or,

$$K_{\text{IIi}} = \sigma_{\text{o}} Q_{\text{IIi}} a^{1/2} \tag{7.23a}$$

Obviously, as indicated, these equations may be simply recast into an equation of the general form $K = Q\sigma_{\text{o}} a^{1/2}$, as in the previous case of a crack in a bulk material, but now of course there are both K_{Iic} and K_{IIic} terms. However, there will now be two geometry factors: Q_{Ii} and Q_{IIi}, respectively, and these geometry factors may also be a function of the bimaterial constant. Further, and of greater concern, there is still a logarithmic term of a dimensional parameter, a, present in Equations 7.22 to 7.23a. This leads to the values of Q, and hence K, being dependent upon the units of a and makes it difficult to evaluate the mode I and II contributions independently of each other. One method is to use one of the modifications which have been suggested to the basic stress analysis but, as commented previously, these all either violate a boundary condition or introduce an unrealistic condition. Also, even if the mode I and II contributions are evaluated independently the question that immediately arises is how to combine them so as to propose a fracture criterion that takes both into account.

These last two aspects have led several workers [10,34,35] to suggest that for cracks at, or near, an interface a combined interfacial stress intensity factor, K_{ic}, for crack growth may be defined such that:

$$K_{ic} = (K_{Iic}^2 + K_{IIic}^2)^{1/2} \tag{7.24}$$

The main advantage of this approach, apart from combining the mode I and II contributions into a usable fracture criterion, is that by taking the root of the sum of the squares the length term is eliminated. Hence, the values of K_{ic} may be ascertained without the ambiguity of the stress intensity factor being a function of the units of the length parameter and without needing a modification which breaks one of the boundary conditions and merely leads to a somewhat different physically unrealistic model. For the case of solely an applied tensile stress, then:

$$K_{ic} = Q_i \sigma_c a^{1/2} \tag{7.25}$$

where Q_i is now the relevant geometry factor which may be determined by several techniques as discussed below, although it may still be a function not only of geometric parameters but also of the moduli and Poisson's ratios of the materials forming the interface, possibly via the bimaterial constant, ζ. However, there is only a limited amount of evidence that the above proposals are experimentally correct or that the suggested fracture criterion does adequately predict the fracture behaviour of adhesive joints with cracks at, or near, an interface.

Finally, it should be noted that the presence of an interface may influence the stress field ahead of a crack even when the crack is located in the centre of the adhesive layer, i.e. at first sight not near an interface. Wang *et al.* [36] have used a finite element analysis and shown that for a crack in the centre of an adhesive layer the stress field, except very close to the crack tip, is markedly different to that predicted from Equation 7.12 or 7.19 (page 273/4). This is shown in Fig. 7.14 (page 273) and has important consequences, as discussed later.

It is evident from even this brief discussion of applying the stress intensity factor approach to cracks in adhesive joints that many problems arise and have yet to be fully resolved. Therefore, it is not surprising that many workers have chosen to adopt the energy balance approach when studying crack growth in joints, and for linear-elastic behaviour deduced a value of G_c(joint) from Equation 7.4 via a theoretical or experimental determination. Although, for engineering design and life-prediction the stress intensity factor approach could possibly be of greater benefit if some of the above aspects were clarified and validated.

7.2.3.2 The geometry factor, Q

To evaluate the stress intensity factors from the above relationships requires the geometry factor(s) to be ascertained.

For a crack in a homogeneous material, the geometry factor, Q may be expressed as a non-dimensional function of crack length (i.e. a/D) and structural geometry and is often in the form of a finite series. There are many techniques for obtaining values of Q. For example, (i) by direct methods where equations relating the crack-tip stresses or displacements to the stress intensity factor are solved by closed-form analytical methods, collocation procedures or numerically for the geometry of interest and compared to the infinite plate case (Equation 7.12) or (ii) by indirect methods such as from a knowledge of the value of $\partial C/\partial a$ (either ascertained experimentally or theoretically) and then using equations such as Equations 7.4, 7.18 and 7.33 to relate Q to $\partial C/\partial a$. For a crack in a homogeneous infinite plate the value of Q is $\pi^{1/2}$ and thus Q may be considered as a suitable function to make the appropriate surface forces zero. For cracks in a wide range of geometries there are several handbooks [37–39] which provide expressions for Q.

For cracks in an adhesive layer in a joint, the value of Q is generally a function of the ratio of the moduli, E_a/E_s, and the thickness of the adhesive layer, h_a, in addition to being dependent upon the overall geometry of the joint. In many geometries where the elastic energy available for crack growth is largely stored in the substrates, such as the double cantilever beam (DCB), double torsion and compact tension, for example, the value of Q(joint) is given by the value of Q for the homogeneous specimen multiplied by $(E_a/E_s)^{1/2}$. An alternative way of visualizing this dependence is to consider the adhesive layer as being relatively thin compared to the thickness of the arms of the substrates, so that the contribution to the compliance from the adhesive is negligible. For such geometries with the typical adhesive layer thicknesses of about 0.1 to 0.5 mm, used by industry, this is in fact a reasonable assumption, but the range of thickness values over which this simple assumption and modification is valid does depend upon the exact geometry and modulus values. For example, in the case of the simple or tapered double cantilever beam geometry (Fig. 7.8, page 286) with structural adhesives bonding metallic substrates this assumption is valid up to values of h_a of several millimetres. However, when the adhesive layer does contribute to the compliance, such as in a butt joint with even a moderately thick adhesive layer, the value of Q(joint) will approach the value for a homogeneous material. If doubts exist about the exact value of Q for a given joint then the value must be ascertained using a numerical analysis or experimentally, via a compliance calibration [10,36,40–42].

It should be noted that in the literature the above details for determining the value of the geometry factor for a joint are often overlooked. The homogeneous value for a given joint geometry is simply used without modification and, for many joint geometries with the typical range of thicknesses of adhesive layer used in industry, this leads to incorrectly high values of K_c (joint), although this incorrect value of K_c is often used, directly or indirectly, to derive an apparently reasonable value of G_c (joint), since the incorrect term E_s, instead of the appropriate value E_a, is employed in Equation 7.35 (page 283), and the two errors cancel each other.

For cracks at, or near interfaces, the picture is far more complex. For example, some workers [43] have used Equations 7.22 and 7.23 to evaluate values of Q_{Ii} and Q_{IIi}, recognizing that the values are functions of the units of the length parameter but arguing that the trends will still be correct. Others [44] appear to have neglected the logarithmic terms in these equations altogether. Notwithstanding these approximations, Erdogan and co-workers [45–49], Lin and Mar [50] and Wang and Yau [51] have developed extremely complex mathematical solutions to obtain values of Q_{Ii} and Q_{IIi} for specific geometries and materials, but again usually making modifications to the basic stress analysis (Equation 7.19) which violate a boundary condition. For example, in the case of an infinite butt joint consisting of aluminium substrates bonded with an epoxy adhesive with a crack in the centre of a very thin adhesive layer the values of Q_{Ii} and Q_{IIi} are shown in Fig. 7.4. They clearly reveal that as the thickness, h_a, of the adhesive layer is decreased the values of Q_{Ii} and Q_{IIi} fall. As argued in Section 6.4.3.2, this means that from Equations 7.22 and 7.23 the values of the *applied* stress intensity factors K_{Ii} and K_{IIi} will decrease for a given value of the applied stress, σ_o. Therefore, assuming failure occurs when the applied stress intensity factor values attain constant critical values, the fracture stress, σ_f, of the joint will rise as the adhesive layer becomes thinner (see Fig. 6.8). As indicated above, many of these determinations, whilst taking account of the bimaterial interface, are solutions for infinite plates. If this is the case, then the geometry factor for a finite joint geometry, i.e. taking edge (a/D) effects into account as well as the bimaterial interface, may be obtained by multiplying the joint infinite plate factor by the geometry factor for a homogeneous finite plate. The value of this parameter may simply be taken from the handbooks.

As discussed above, many of the most severe problems associated with a stress intensity factor approach to cracking along, or close to, interfaces may be circumvented if a combined interfacial stress intensity factor, K_i, is defined and therefore only a single geometry factor, Q_i, has to be determined. This approach has been adopted by several workers [10,34,35,52] and values of Q_i

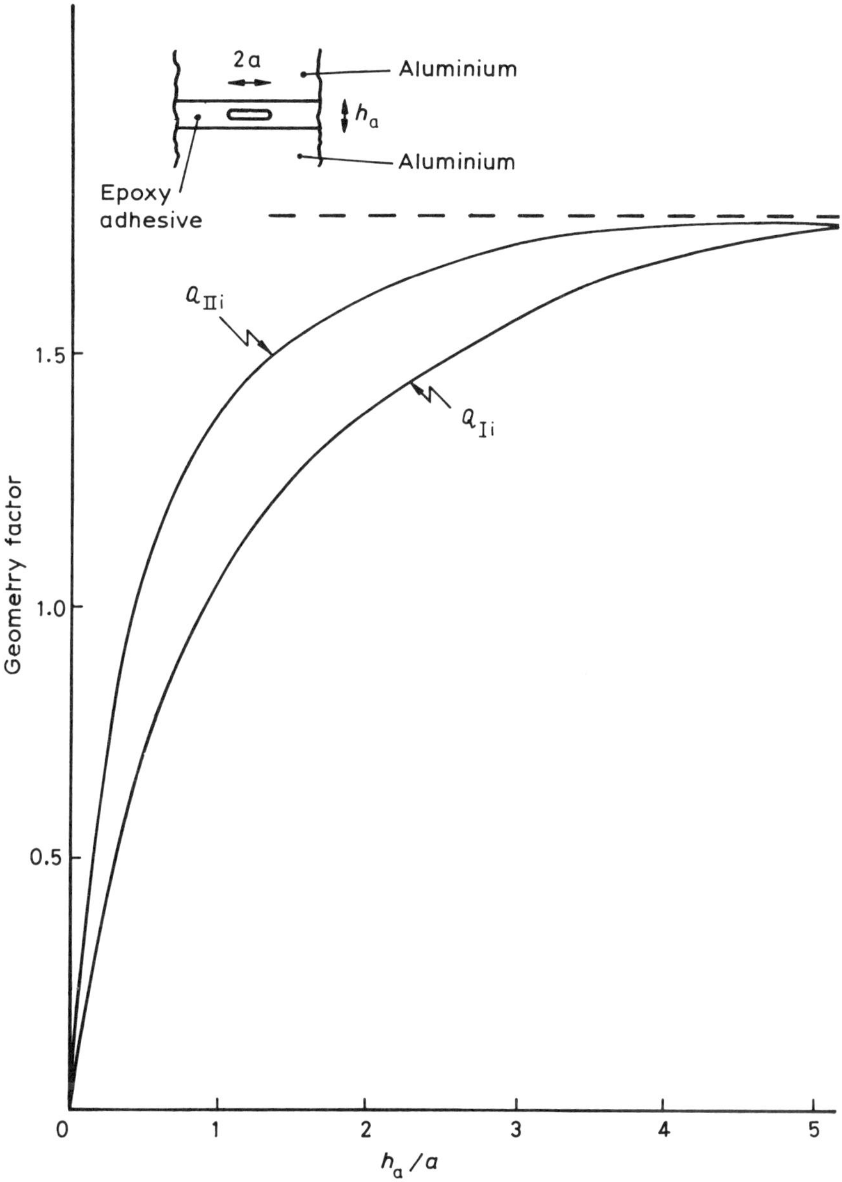

Figure 7.4 Geometry factors, Q_{Ii} and Q_{IIi}, as a function of h_a/a for a crack of length, a, in the centre of an adhesive layer of thickness, h_a, in an infinite butt joint [48].

have been determined as a function of the ratio crack length:specimen depth, a/D, for bimaterial sheets containing either edge or central interfacial cracks. Finally, from Equations 7.4, 7.25 and 7.36, values of the geometry factor, Q_i, may be ascertained from values of $\partial C/\partial a$, such as that shown in Fig. 7.1b. However, this does require a knowledge of the relationship between K and G (see Section 7.2.5) and, whilst this is well established for cracks in bulk material, that for interfacial cracks has yet to be conclusively resolved.

Many of the above points are further illustrated in Section 7.3.3 where experimental aspects are considered.

7.2.3.3 Small-scale yielding

All the above relations predict that the stresses at the crack tip are singular. Clearly, therefore, in reality the stress necessary for the adhesive to undergo plastic yielding in some zone ahead of the crack tip will be exceeded. However, if this zone is assumed to be small then it will not greatly disturb the elastic stress field and the extent of the plastic zone may be defined by the elastic stresses. Thus again we return to the assumption of LEFM.

Irwin [7] suggested that if a material is considered which is elastic up to the uniaxial yield stress, σ_{ay}, and then becomes plastic, then the extent of crack tip plasticity for a crack in bulk material is, to a first approximation, as shown in Fig. 7.5. Here δ_t is the crack tip opening displacement and r_y is the radius of the circular plastic zone at the tip of a 'notional' crack. The tip of this notional crack is therefore at the centre of the plastic zone so that its length is $(a + r_y)$. The elastic stress field ahead of this notional crack may therefore be regarded

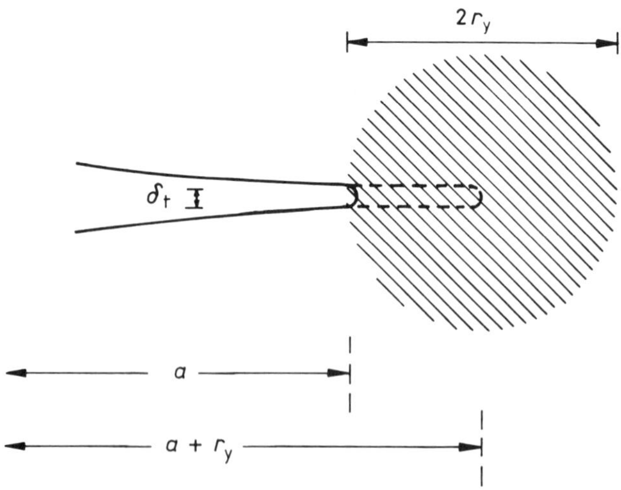

Figure 7.5 Irwin model of the plastic zone at a crack tip.

as identical to the stress distribution of a real crack of length a with the extent of plastic zone $2r_y$. For $r_y \ll a$ the above equations have been shown to be still valid, although a small correction to the crack length may be occasionally used to improve the accuracy of the calculation. This correction factor simply consists of using $(a + r_y)$ instead of a in these calculations. The size of the plastic zone radius, r_y, is given by [3]:

$$r_y'' = \frac{1}{2\pi}\left(\frac{K_I}{\sigma_{ay}}\right)^2 \qquad \text{for plane stress} \tag{7.26}$$

and the extra constraint in the case of plane strain leads to:

$$r_y' = \frac{1}{6\pi}\left(\frac{K_I}{\sigma_{ay}}\right)^2 \qquad \text{for plane strain} \tag{7.27}$$

The corresponding crack-opening displacements at the crack tip, δ_t, are:

$$\delta_t = \frac{K_I^2}{E_a \sigma_{ay}} \qquad \text{for plane stress} \tag{7.28}$$

and

$$\delta_t = \frac{K_I^2}{E_a \sigma_{ay}}(1 - \nu_a^2) \qquad \text{for plane strain} \tag{7.29}$$

7.2.4 Width effects

It is often found that, in practice, the measured value of G_{Ic} or K_{Ic} varies with the width of the specimen over a certain range of widths, and this usually arises because the state of stress near the crack tip varies from plane stress in a very thin specimen to plane strain near the centre of a wide plate. The general form of the relationship between K_{Ic} (or G_{Ic}) and specimen width, b, is shown schematically in Fig. 7.6. The value for fracture in plane strain conditions is usually less than under plane stress. This arises because the tensile stress at which a material yields is greater in a triaxial stress field (plane strain) than in a biaxial one (plane stress) and thus, in the former, a more limited degree of plasticity develops at the crack tip; cf. Equations 7.26 and 7.27. The lower, conservative, plane strain value is usually the one required for engineering design and life prediction studies. The width, b, necessary to achieve this condition is usually taken to be:

$$b \geq 2.5\left(\frac{K_{Ic}}{\sigma_{ay}}\right)^2 \tag{7.30}$$

where the value of K_{Ic} is the plane strain value.

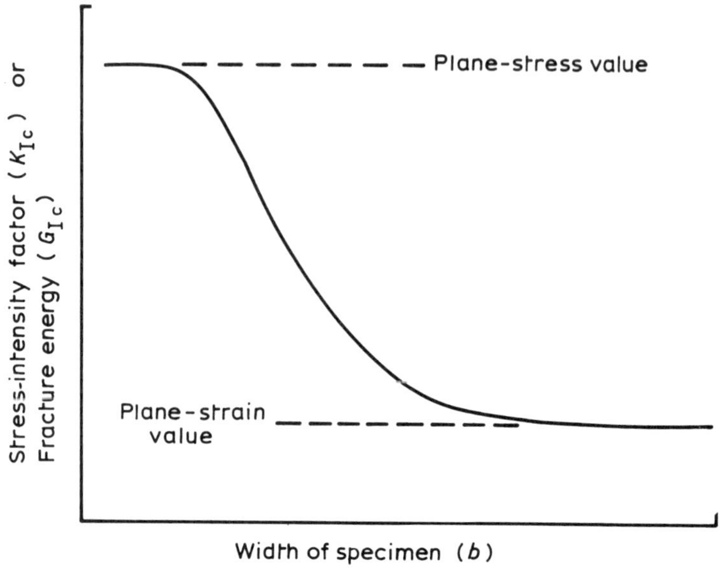

Figure 7.6 Schematic representation of the variation of the value of K_{Ic} (or G_{Ic}) with the width of a bulk specimen.

7.2.5 Relationships between G and K

In the case of LEFM, values of the fracture energy, G_c, and the stress intensity factor, K_c, may be related.

For example, for a crack in a homogeneous body a simple relationship exists between G and K, under plane strain conditions. This relationship is given by [7,53]:

$$G_c = \frac{(1-\nu^2)}{E} K_{Ic}^2 + \frac{(1-\nu^2)}{E} K_{IIc}^2 + \frac{(1+\nu)}{E} K_{IIIc}^2 \tag{7.31}$$

or equivalently:

$$G_c = G_{Ic} + G_{IIc} + G_{IIIc} \tag{7.32}$$

where, for mode I:

$$G_{Ic} = \frac{K_{Ic}^2}{E} \qquad \text{for plane stress} \tag{7.33}$$

and

$$G_{Ic} = \frac{K_{Ic}^2}{E}(1-\nu^2) \qquad \text{for plane strain} \tag{7.34}$$

For a crack in the centre of an adhesive layer, relatively distant from any interface, it has been shown [36,40,53] that the above expressions are still valid and the appropriate value of the tensile modulus, E_a, of the adhesive may be employed to correlate G (joint) and K (joint). Thus, for plane strain:

$$G_{Ic}(\text{joint}) = \frac{K_{Ic}^2(\text{joint})}{E_a}(1-\nu_a^2) \tag{7.35}$$

For the case of a crack at an interface there is no clearly established relationship but several workers [10,25,33,46,48,54] have advanced the reasonable suggestion that the appropriate value of the 'modulus' should be some weighted average of the moduli of the materials forming the interface. For example, the values of G_{ic} and K_{ic} may be related by:

$$G_{ic} = K_{ic}^2 \cdot \frac{1}{2}\left(\frac{1}{E_a} + \frac{1}{E_s}\right)\left(\frac{2\alpha_m - 1}{\alpha_m^2}\right) \tag{7.36}$$

where

$$\alpha_m = \frac{\xi + 1}{\chi + 1} \tag{7.37}$$

and

$$\xi = E_s/E_a \tag{7.38}$$

$$\chi = 1 + \frac{E_s}{2}\left(\frac{1+\nu_a}{E_a} - \frac{1+\nu_s}{E_s}\right) \tag{7.39}$$

The exact form of these relationships is important. Apart from needing to convert from one fracture criterion to the other, the equations used to determine G or K for a particular geometry are often derived using a 'modulus' term, although sometimes it may not appear in the final equation as it may cancel out. Obviously, the 'correct' modulus term must be selected, as discussed in Section 7.2.3.2.

For the case of a crack not actually at the interface, but very close to the interface, then no relations are readily available. Intuitively one might expect the appropriate modulus value to lie betweeen that given by Equation 7.35 and 7.36.

Finally, it should be noted that for non-LEFM the concept of G is still valid, as discussed earlier, and of practical benefit but the interpretation of stress intensity factors is not always clear and the above relations are not generally applicable.

7.3 EXPERIMENTAL CONSIDERATIONS

7.3.1 Introduction

A basic aim of fracture mechanics is to provide a parameter for characterizing crack growth which is independent of test geometry. In order to validate this requirement experimentally a wide range of different geometries has been developed. Further, different geometries have been developed in order to ascertain values of the various fracture parameters under modes II and III loading, as well as the more usual mode I case. It is convenient to discuss separately the methods which have been used to study relatively rigid joints, often using structural adhesives, and those which possess a fair degree of flexibility such as when a rubbery adhesive is employed; although this division is sometimes rather artificial. Sketches of some of the fracture mechanics specimens that have been used with flexible and rigid adhesive joints are shown in Figs 7.7 and 7.8, respectively.

7.3.2 Flexible joints

The energy balance approach is generally the most applicable to flexible joints, since away from the crack tip the adhesive or substrates may not exhibit linear-elastic behaviour and so the stress intensity factor is invalid. Of course, in those cases where linear-elastic behaviour is observed then either approach may in principle be adopted. Essentially, to obtain an expression for the value of the adhesive fracture energy, G_c, for a particular geometry requires a solution to the left-hand side of Equation 7.3 to be deduced.

The most common test method involves peeling a rubbery adhesive, or substrate, away from a rigid support, as shown schematically in Fig. 7.7a. A detailed analysis has been published by Lindley [55] who assumed that when a force, F_c, was applied to the rubber tab of a peel test specimen at an angle α to the rigid support substrate, the state of stress around the peel front was independent of the amount of peeling which had taken place. He considered the energy balance for a bulk elastic, but not necessarily linear-elastic, rubber before and after peeling of an amount a, i.e. when a strip of the rubber of unstrained length a is transferred from the unstrained state to the strained state. He argued that in this process (see (i) to (iv) in Fig. 7.7a):

(i) The point of application of the force, F_c, will be moved a distance $\lambda_c a$, where λ_c is the extension ratio of the rubber under the force, F_c. The work input is therefore, $F_c\lambda_c a$.

(ii) The rubber in the tab will have gained strain energy $ah_a bU_{dc}$, where U_{dc} is the strain energy density in uniaxial tension at an extension ratio λ_c, h_a is the thickness of the tab and b its width.

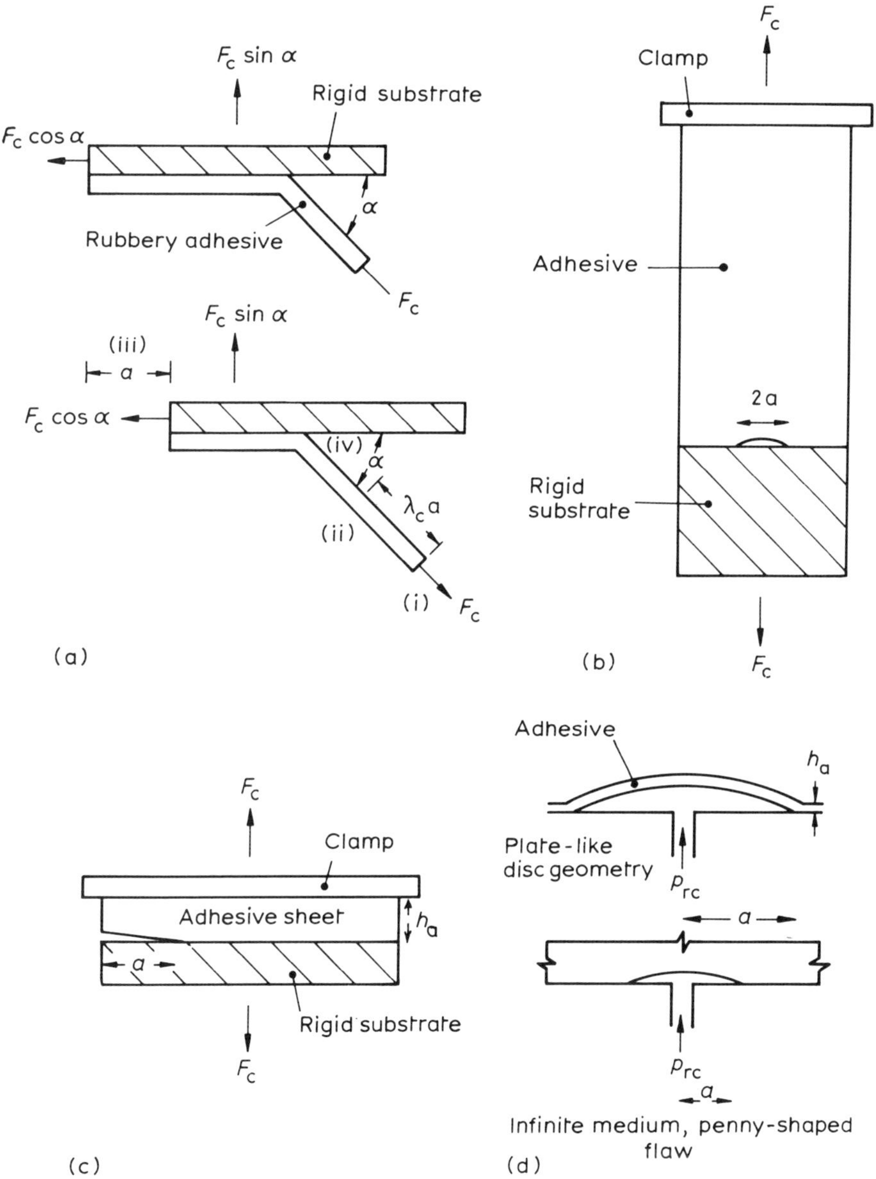

Figure 7.7 Adhesive joint geometries used for evaluating G_c for flexible joints. (a) Peel test specimen before and after peeling by an amount, *a*. (b) Simple-extension test specimen. (c) Pure shear test specimen. (d) Blister test specimen.

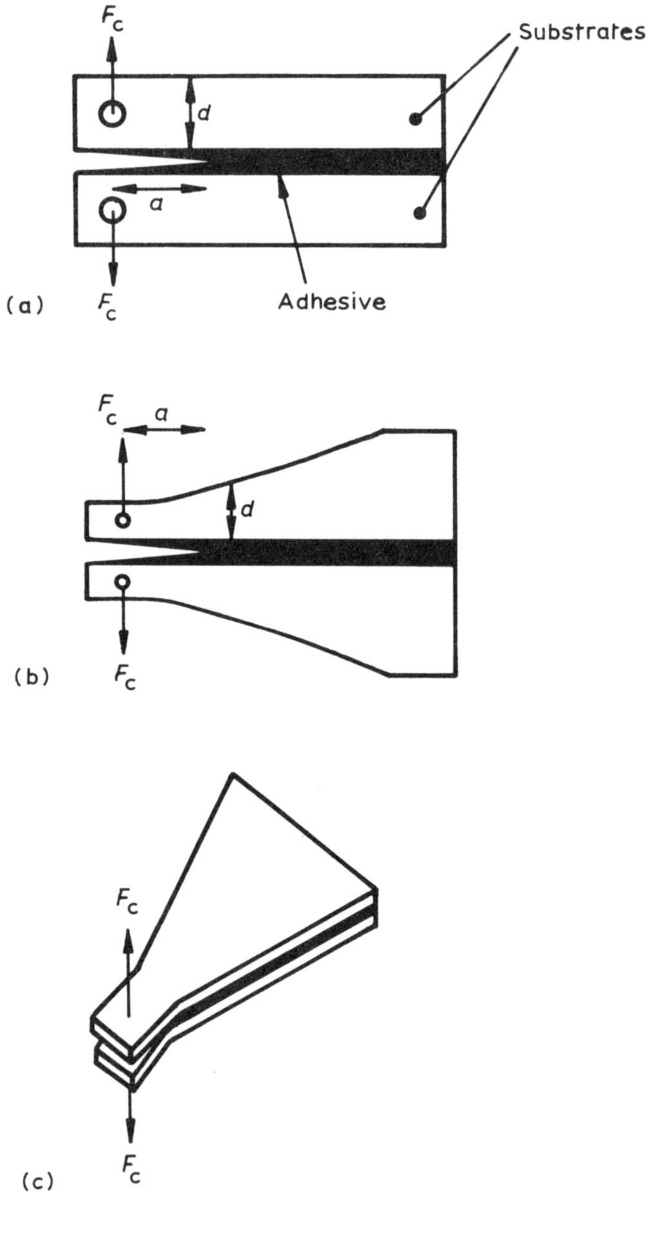

Figure 7.8 Adhesive joint geometries used for evaluating G_c and K_c for rigid joints (for details see Table 7.1).

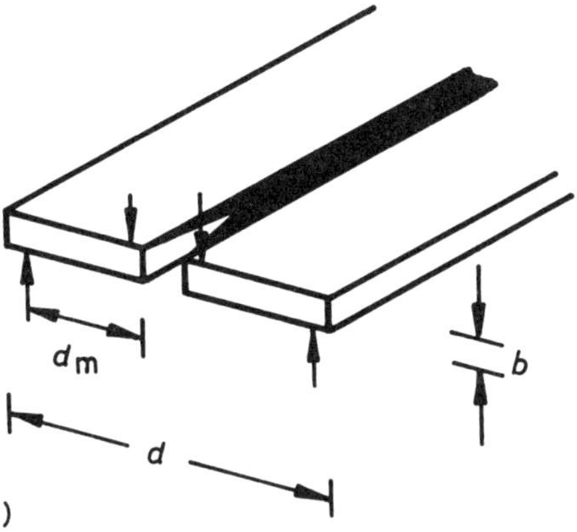

(d)

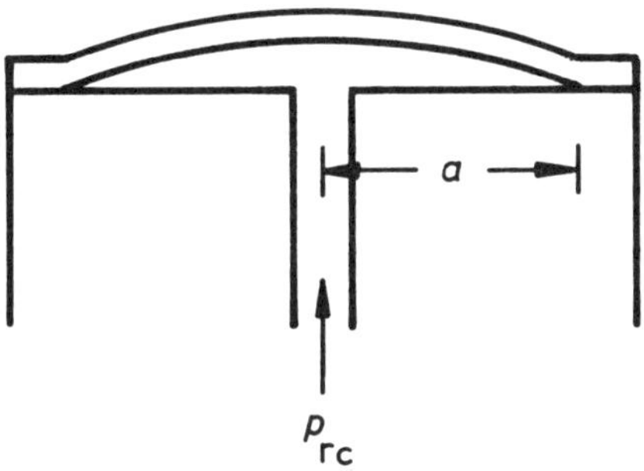

(e)

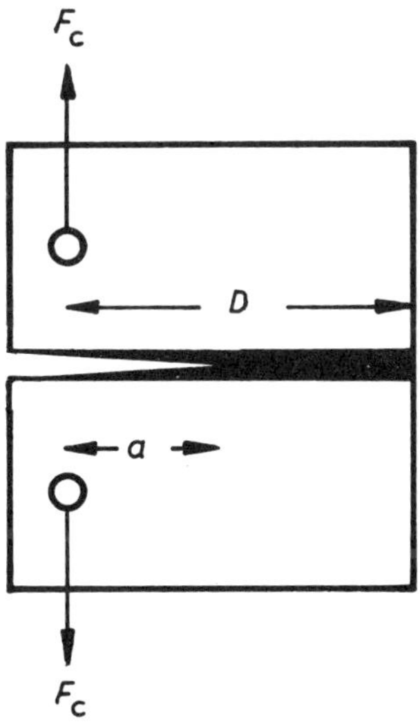

(f)

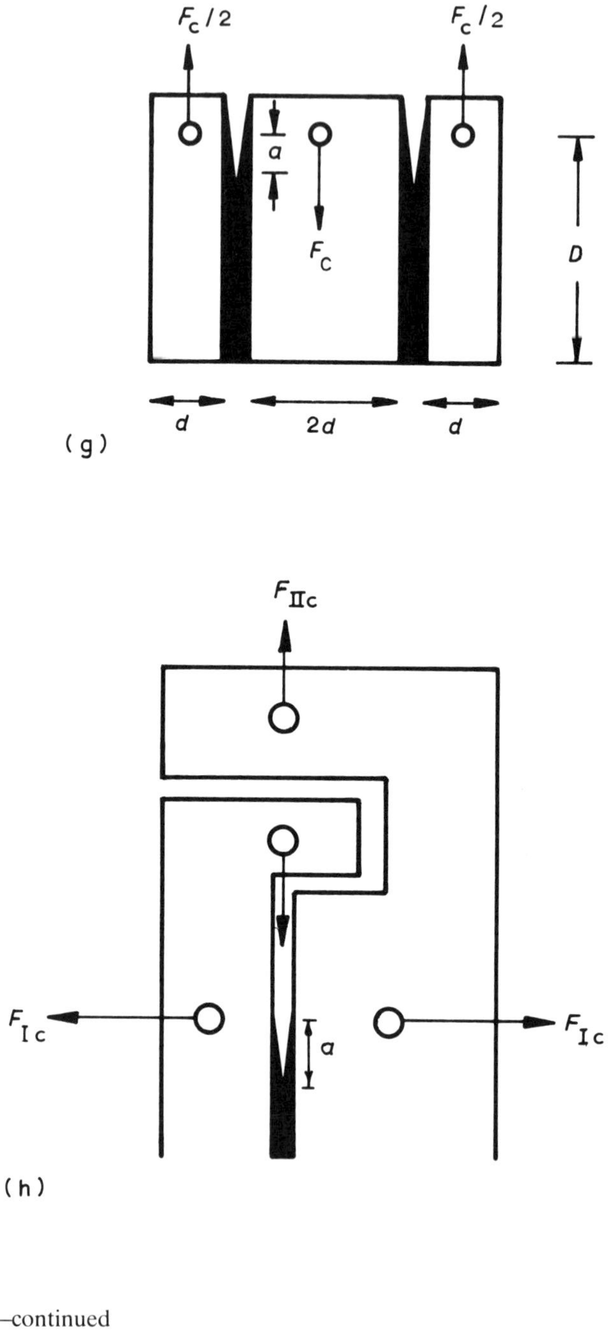

Figure 7.8—continued

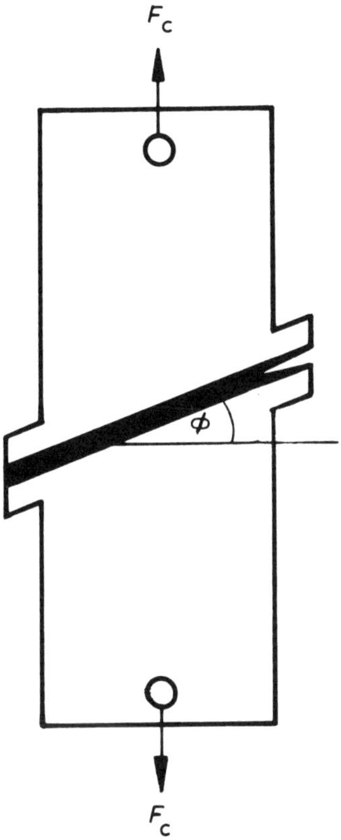

(i)

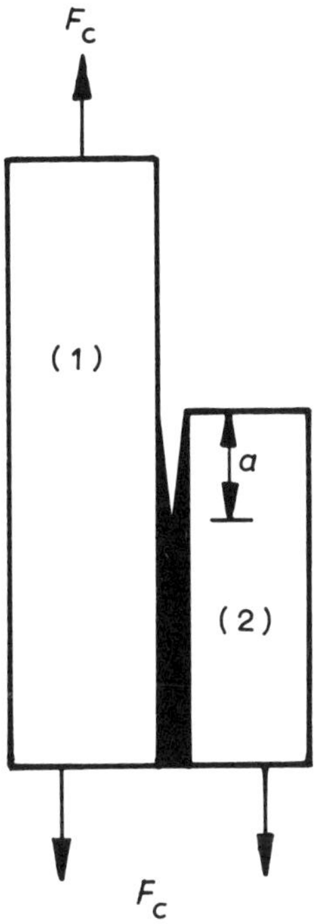

(j)

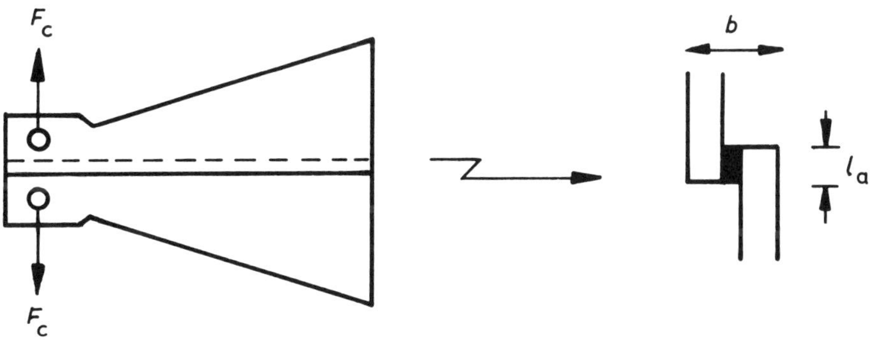

(k)

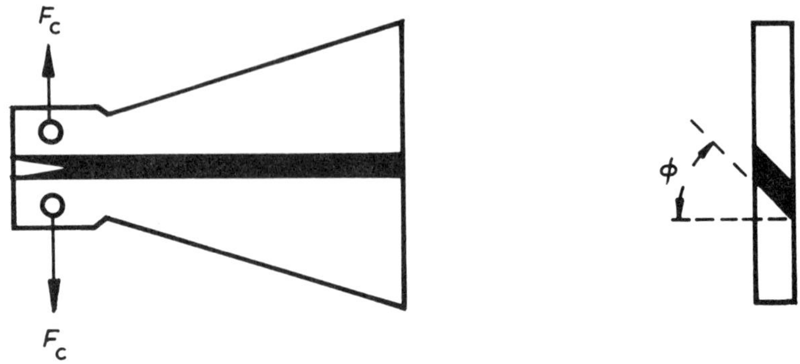

(l)

Figure 7.8—continued

(iii) A horizontal force $F_c \cos\alpha$ is necessary to maintain equilibrium. The point of application of this force will move a distance a, requiring an energy $F_c a \cos\alpha$.

(iv) An area of bond ab will have peeled requiring a total fracture energy of $G_c ab$.

If no other external work is done the input energy (i) must be balanced by those of (ii), (iii) and (iv) giving:

$$F_c \lambda_c a = a h_a b U_{dc} + F_c a \cos\alpha + G_c ab \tag{7.40}$$

Thus:

$$G_c = \frac{F_c \lambda_c}{b} - \frac{F_c \cos\alpha}{b} - h_a U_{dc} \tag{7.41}$$

If the strain in the tab can be neglected, for example if (i) the peel forces are very low or (ii) a fabric- or plastic-backed rubbery adhesive or a relatively thick plastic or metallic substrate is the peeling member, then:

$$G_c = \frac{F_c}{b}(1 - \cos\alpha) \tag{7.42}$$

or:

$$G_c = P(1 - \cos\alpha) \tag{7.43}$$

where P is the peel force per unit width, commonly termed the peel strength or peel energy. As stated above, this analysis applies to linear and non-linear elastic materials. Hata *et al.* [56] have also derived Equation 7.41, in a somewhat different form, and Deraguin and Krotova [57] and Kaelble [58–60] (see Section 6.4.7) have obtained the simplified version embodied in Equations 7.42 or 7.43.

In some instances the rubbery adhesive is supported by a plastic backing-strip or the test geometry is changed so that a flexible substrate is peeled away from a flat, supported layer of adhesive. Under such circumstances it must be recognized that the stripping member may undergo plastic yielding if the bending stresses imposed by the peel force are sufficiently large. Plastic yielding provides an energy dissipation mechanism and thus a higher peel force is required than if yielding does not occur. This mechanism must therefore be considered in any energy balance analysis and the magnitude of this additional energy dissipation has been discussed by Gent and Hamed [61,62], as discussed in detail in Section 7.5.2.

As discussed above, a major feature of the fracture mechanics argument is that the fracture energy, G_c, for a given joint, tested at a stated rate and temperature, is independent of the test geometry employed. Gent and Kinloch [11] and Andrews and Kinloch [15–18] have employed the test geometries shown in Figs 7.7a to c to verify that this was indeed true for a crosslinked styrene–butadiene rubber adhering to rigid substrates. The relations for determining G_c are [16]:

Simple-extension test specimen

$$G_c = k_6 a U_{dc} \tag{7.44}$$

Pure-shear test specimen

$$G_c = h_a U_{dc} \tag{7.45}$$

where k_6 is a numerical factor given by $\pi/\lambda_c^{1/2}$, where λ_c is the extension ratio of the rubber sheet for crack propagation. Obviously for low strains k_6 may be assumed to be a constant whose value is π. The simple-extension test geometry has also been used [63] to investigate the adhesion of epoxy resins, above their glass transition temperature, to metals.

Another configuration frequently employed for both flexible, and rigid, joints is the 'pressurized blister-test', shown in Fig. 7.7d. The specimen configuration was originally suggested by Danneberg [64], amplified in certain analytical respects by Malyshev and Salganik [25] and subsequently considerably developed by Williams and co-workers [65–69] and Anderson *et al.* [70]. In its simplest for it consists of a disc or plate which has been cast, and hence adhered, against a rigid substrate, except for a central portion of radius, a. When the unbonded region is pressurized, as by the injection of compressed air or fluid, the plate lifts off the substrate and forms a blister whose radius stays fixed until a critical pressure, p_{rc}, is attained. At this value the radius of the blister increases in size, i.e. crack growth occurs. In principle, any thickness or diameter of disc and any debond radius smaller than the disc may be used. However, analytical closed-form solutions for the displacement are currently only available for the near-limit case of a disc using plate theory and for the infinite isotropic medium adhering to a rigid substrate with a small penny-shaped debond between the two materials. Assuming linear-elastic behaviour:

For plate-like disc geometry

$$p_{\mathrm{rc}} = \left[\frac{32}{3(1-\nu_{\mathrm{a}}^2)}\left(\frac{h_{\mathrm{a}}}{a}\right)^3\right]^{1/2}\left(\frac{E_{\mathrm{a}}G_{\mathrm{c}}}{a}\right)^{1/2} \tag{7.46}$$

For an infinite medium, penny-shaped flaw ($\nu_{\mathrm{a}} = 0.5$):

$$p_{\mathrm{rc}} = \left(\frac{2\pi}{3}\right)^{1/2}\left(\frac{E_{\mathrm{a}}G_{\mathrm{c}}}{a}\right)^{1/2} \tag{7.47}$$

More recently Andrews and Stevenson [71] have expressed these equations as:

$$G_{\mathrm{c}} = \left(\frac{1}{f(h_{\mathrm{a}}/a)}\right)\left(\frac{p_{\mathrm{rc}}^2 a}{E_{\mathrm{a}}}\right) \tag{7.48}$$

and evaluated the function $f(h_{\mathrm{a}}/a)$ over a wide range of h_{a}/a values for cracks propagating both along the interface and cohesively through the adhering layer. Williams and co-workers [66,67,72,73] have also considered the case for an elastic adhesive interlayer between the plate and rigid substrate. Further, a first-order correction for the effect of viscoelasticity has been suggested [72,73] by replacing the constant elastic modulus, E_{a}, by the stress-relaxation modulus, $E_{\mathrm{a}}(t)$ evaluated at the appropriate time scale, t. This method for allowing for the time-dependent behaviour of the modulus has been used in several analyses of the fracture of polymeric materials [3]. The pressurized-blister test has been used for many diverse applications in conjunction with flexible joints, such as for assessing and examining the adhesion of paints [74], solid-propellant rocket motors [72,75] and ice [76]. However, Anderson *et al.* [69] have commented that the value of the adhesive fracture energy, G_{c}, may not be completely independent of the geometry of the blister test, since the loading mode changes from largely mode I to a combination of modes I and II if the specimen thickness to initial debond radius ratio is decreased. However, a constant mode I value of G_{c} is obtained when h_{a}/a is greater than about four.

Other geometries have also been developed to determine G_{c} for rubbery adhesives and include a cone-test specimen [69], a torsion test which results in mode III failure [66] and the adherence of spheres to rubbers [77–80].

7.3.3 Rigid joints

In recent years a large number of test specimens have been designed to determine the crack growth behaviour of structural adhesive joints and some of the designs which have been employed are shown in Table 7.1 and Fig. 7.8. The analyses derived and given in Table 7.1 all assume that the deformation in the bonded structure, away from the crack tip region, is linear-elastic, although techniques such as the J-integral method [3,4] could be employed if this condition was not fulfilled.

Several points should be noted concerning the data shown in Table 7.1. Firstly, the mode of loading which is given refers to the case of a crack in the adhesive layer. As discussed above, if the crack is at, or very close to, an interface then both modes I and II will always be present, but they no longer have the physical interpretation as indicated in Fig. 7.3. Secondly, the value of $\partial C/\partial a$ may of course always be evaluated experimentally and used in Equation 7.4 to deduce a value of G_c. This is often the best approach since it gives an unambiguous value of G_c. However, for many of the test specimens expressions have been determined for $\partial C/\partial a$ that obviate the need to conduct an experimental calibration. Examples of these expressions are given for some geometries. Such expressions typically have been derived assuming that the adhesive layer is relatively thin, i.e. the compliance of the adhesive layer may be considered to be negligible, but some work has been conducted on the effect of thicker layers of adhesive when this assumption is invalid. Thirdly, expressions for the relevant geometry factor, Q, are given in some instances. As discussed in Section 7.2.3.2 they may be evaluated either directly by analytical or numerical techniques, which often are essentially based upon solving the equations relating the stresses or displacements around the crack tip, or indirectly via a knowledge of $\partial C/\partial a$. Fourthly, values of K may be ascertained from values of G, and *vice versa*, by using the appropriate modulus value as outlined in Section 7.2.5. Finally, not all possible combinations of specimen geometry, failure mode or equations for G and K are quoted, but it is hoped that a sufficient amount of data has been given to illustrate the various approaches that have been adopted.

Considering other experimental aspects, then another important requirement in the determination of the fracture energy or fracture toughness values of rigid adhesives is to obtain a 'naturally' sharp initial or 'starter' crack. For rubbery adhesives this is not so important as even a relatively blunt initial crack will usually blunt further with the large deformations involved. However, for rigid adhesives a blunt initial crack will lead to optimistically high values of G_c or K_c being recorded. Suitable techniques for inserting naturally sharp cracks include slow, controlled pressure applied to a razor blade, possibly having precooled the specimen in liquid nitrogen, and controlled propagation of a sharp crack from an initially inserted blunt crack. The more brittle the

Table 7.1 Adhesive joint designs for determining values of G and K_c for rigid joints

Geometry	Equation	Comments	References
Double cantilever beam (DCB) (Fig. 7.8a)	$G_{Ic} = \frac{F_c^2}{2b}\frac{\partial C}{\partial a}$ (7.4) For thin adhesive layers: $\frac{\partial C}{\partial a} = \frac{8}{E_s b}\left(\frac{3a^2}{d^3} + \frac{1}{d}\right)$ (7.49) where d is the height of the substrate beam	Mode I. Substrates are simple beams and cheap. $\partial C/\partial a$ is not constant and under a constant load G_I increases as a increases. Used in ASTM D 3433	81–83
Tapered double cantilever beam (TDCB) (Fig. 7.8b)	Given by the above but may be written: $G_{Ic} = \frac{4F_c^2 m_b}{E_s b^2}$ (7.50) where: $m_b = \left(\frac{3a^2}{d^3} + \frac{1}{d}\right)$ (7.51)	Mode I. Substrates require careful machining if geometry factor, m_b, is to be constant. If this is achieved then $\partial C/\partial a$ is constant. Thus G_I is independent of crack length and this design is well suited to tests where the value of a is difficult to measure, e.g. environmental tests. Used in ASTM D 3433 and is one of the most popular test geometries.	81–83
Width-tapered beam (Fig. 7.8c)	As above	Mode I. $\partial C/\partial a$ is now constant due to a variation in width with crack length rather than beam height. Useful for composite substrates	84,85
Double torsion (Fig. 7.8d)	Could use Equation 7.4 and, for thin adhesive layers, take: $\frac{\partial C}{\partial a} = \frac{(1+\nu_s)}{E_s}\frac{2d_m^2}{k_7 d b^3}$ or express this as: $K_{Ic} = F_c d_m Q$ where d_m = length of moment arm k_7 = constant [89]	Mode I. K_I or G_I is independent of a, i.e. $\partial C/\partial a$ is constant	86–89

Table 7.1—continued

Geometry	Equation	Comments	References
	(when $d/2b = 4$, $k_7 = 0.28$) b_n = width of specimen in plane of crack, i.e. for grooved specimen $Q = Q_h(E_a/E_s)^{1/2}$ $Q_h = \left[\frac{1}{(1-\nu_s)k_7 db^3 b_n}\right]$ i.e. homogeneous value		
Blister specimen (Fig. 7.8e)	$G_c = \left(\frac{1}{f(h_a/a)}\right)\left(\frac{p_{rc}^2 a}{E_a}\right)$ (See Equations 7.46 to 7.48; and see ref. [71] for values of $f(h_a/a)$)	Largely mode I, but more mode II as a increases. Major advantage over other mode I tests is that the value of G_{Ic} is truly plane strain. Major disadvantage is that a sharp precrack cannot be inserted and a blunt precrack is employed which can lead to artificially high values of G_{Ic}. Also, it is difficult to measure the load at crack velocity unless the adhesive is transparent	25,70,71 90–92
Compact tension (Fig. 7.8f)	$K_{Ic} = \frac{F_c Q}{bD^{1/2}}$ Ref. [41] deduces the value of Q from the homogeneous case [3], i.e. $f(a/D)$, and then multiplies this by $(E_a/E_s)^{1/2}$. But ref. [42] shows that, except for $h_a \rightarrow 0$, Q is also $f(h_a, \nu_a)$	Mode I	40,41
Compact shear (Fig. 7.8g)	$K_{IIc} = \sigma_c Q a^{1/2}$ where: $\sigma_c = F_c/2bd$ Ref. [41] deduces the value of Q from the homogeneous case [93], i.e. $Q_h = f(a/D)$, and then multiplies this by $(E_a/E_s)^{1/2}$. But ref. [42] shows that, except for $h_a \rightarrow 0$, Q is also $f(h_a, \nu_a)$	Claimed that loading mode is basically mode II (mode I/II $\cong$ 0.01)	41,42

Table 7.1—continued

Geometry	Equation	Comments	References
Modified compact shear	$K_{IIc} = \sigma_c Q a^{1/2}$ For thin adhesive layers: $Q = Q_h(E_a/E_s)^{1/2}$ where: $Q_h = (d/D)[0.0325 + 14.63\,(a/D) - 46.07\,(a/D)^2 + 61.03\,(a/D)^3 - 28.86\,(a/D)^4]$	Claimed that by loading Fig. 7.8g at an angle of 8.5° away from the crack direction that a pure mode II is obtained	94
Independently loaded mixed-mode (Fig. 7.8h)	$G_{Ic} = \dfrac{4F_{Ic}^2}{b^2 d^3 E_s} \times [3(a+0.6d)^2 + d^2]$ $G_{IIc} = \dfrac{F_{IIc}^2}{b^2 d E_a}$ (Above are for thin adhesive layers)	Cleavage and shear loads can be applied independently to produce either mode I or mode II conditions, or a combination of both	95,96
Scarf joint (Fig. 7.8i)	$K_{I,IIc\phi} = \dfrac{F_c Q}{b D^{1/2}}$ Q is $f(a/D, \phi, E_a/E_s, h_a)$ and ϕ is the scarf angle	Variation in angle ϕ alters mode I/II	40,95 97–99
Cracked lap-shear (Fig. 7.8j)	$G_c = \dfrac{F_c^2}{2b^2}\left(\dfrac{1}{(E_s d)_1} - \dfrac{1}{(E_s d)_2}\right)$ where 1 and 2 refer to the substrates and $(E_s d)_2 > (E_s d)_1$	Mixed mode I/II claimed to be typical of many practical joints ($G_I/G_{II} \cong 0.2$ to 0.3), but a wide range of G_I/G_{II} values may be obtained [101]	85,100 101
Modified zero-K gradient (Fig. 7.8k)	$G_{IIIc} = \dfrac{F_{IIIc}^2}{2l_a}\,\dfrac{2k_8}{3E_s b}$ k_8 is a constant	Designed for mode III but almost certainly has an indeterminate amount of mode I and possibly also some mode II	86
Scarfed TDCB (Fig. 7.8l)	$G_c = \cos\phi \dfrac{F_c^2}{2b}\,\dfrac{\partial C}{\partial a}$ $\partial C/\partial a$ determined experimentally as $f(\phi)$ $G_{Ic} = G_c \cos^2\phi$ $G_{IIIc} = G_c \sin^2\phi$	Variation in the angle ϕ alters mode I/III ratio: as ϕ increases ratio decreases	102

adhesive, i.e. those possessing a low value of G_c or K_c, the more important it is to conduct the fracture test with the specimen containing a sharp crack.

7.4 TYPICAL VALUES OF G_c AND K_c

From the various methods discussed above values of the fracture energy, G_c, and, for LEFM, the stress intensity factor, K_c, may be obtained.

For example, Fig. 7.9 shows the results from studying crack growth in a simple-extension joint (Fig. 7.7) consisting of a crosslinked styrene–butadiene rubbery adhesive (a non-linear-elastic material) adhering to a rigidly supported poly(ethylene terephthalate) substrate [103]. From Equation 7.44, with $k_6 \cong \pi$, the data is plotted as U_{dc} versus a^{-1} and the linear relationship which is observed clearly establishes the independence of the fracture energy, G_c, upon the value of crack length employed and enables the value of G_c to be calculated.

Another example is shown in Fig. 7.10 which is for a crack at the interface between an epoxy adhesive and an aluminium alloy substrate and for this system LEFM is applicable. From Equation 7.4 and the data shown in Fig. 7.1, the value of F_c may be determined as a function of crack length and then $F_c^2/2b$ may be plotted against $(\partial C/\partial a)^{-1}$. The linear relation observed again establishes the independence of G_c upon initial crack length, a, and allows the value of G_c to be deduced [10]. It is of interest to note that in this case the crack propagated slowly along the interface before entering the sheet of epoxy. In other cases the crack propagates immediately into the epoxy [10,43,44]. These observations arise because the maximum principal tensile stress at the crack tip is not parallel to the applied tensile stress, and hence is not exactly perpendicular to the bimaterial interface. The crack, which tends to grow normal to the maximum circumferential tensile stress, has therefore a propensity to veer away from the interface. The direction the crack takes is a balance between the relative fracture energies of the interface and bulk materials and the direction of the principal tensile stress with respect to the interface.

A final example is illustrated in Fig. 7.11 and shows the experimentally determined values of compliance, C, versus crack length, a, for a tapered double cantilever beam joint (Fig. 7.8) consisting of an epoxy adhesive bonding aluminium alloy substrates. As discussed in Table 7.1, this type of joint is designed to produce a constant value of $\partial C/\partial a$ and this may be readily seen. Further, the joint obeys LEFM and for the adhesive thickness, h_a, used of 0.5 mm the adhesive layer is sufficiently thin so that the experimental value of $\partial C/\partial a$ agrees well with the theoretical value deduced from Equation 7.49 given in Table 7.1. Obviously, either from the experimental or theoretical values of $\partial C/\partial a$, the value of G_{Ic} may be calculated from Equation 7.4. For those joints which obey LEFM, in the case of cracks in the centre of the adhesive layer, values of K_c may be determined via Equation 7.35 whilst for interfacial cracks Equations 7.36 to 7.39 are more appropriate.

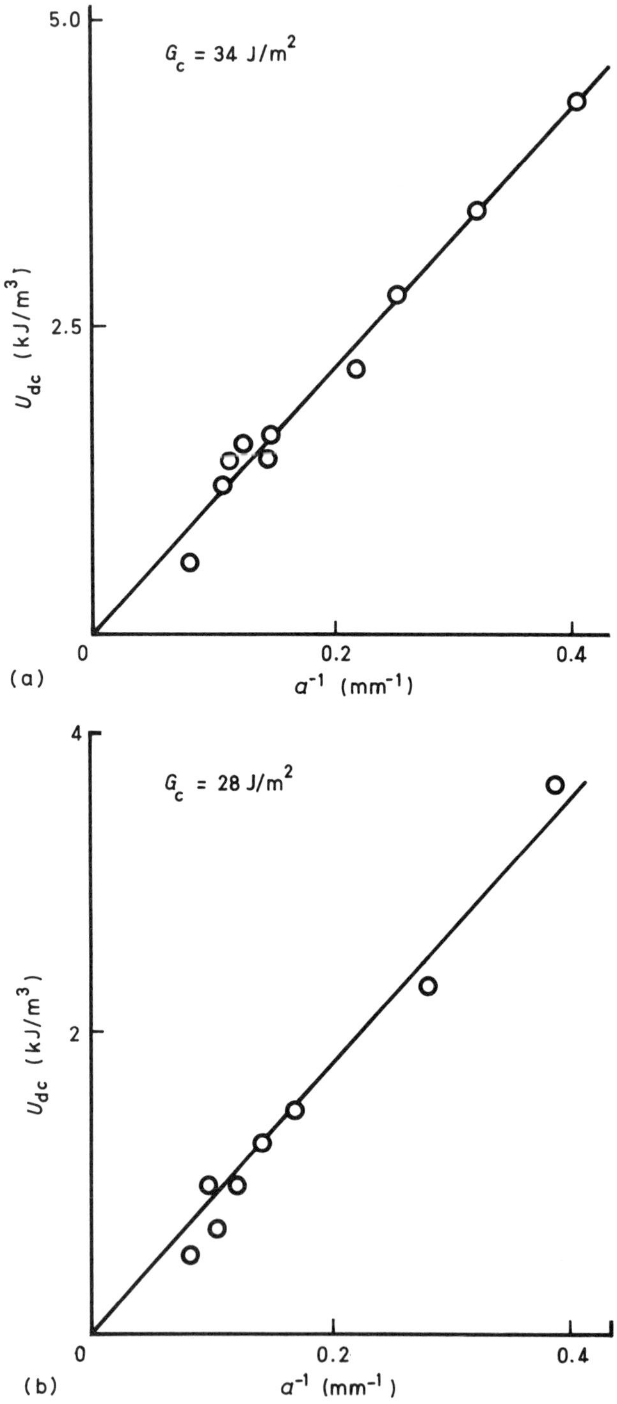

Figure 7.9 Applied strain energy density, U_{dc}, at onset of crack growth versus reciprocal of the crack length, a^{-1}, from simple-extension test specimens [103]. (a) Crosslinked SBR/fluorocarbon copolymer tested at 100 °C. (b) Crosslinked SBR/poly(ethylene terephthalate) tested at 50 °C.

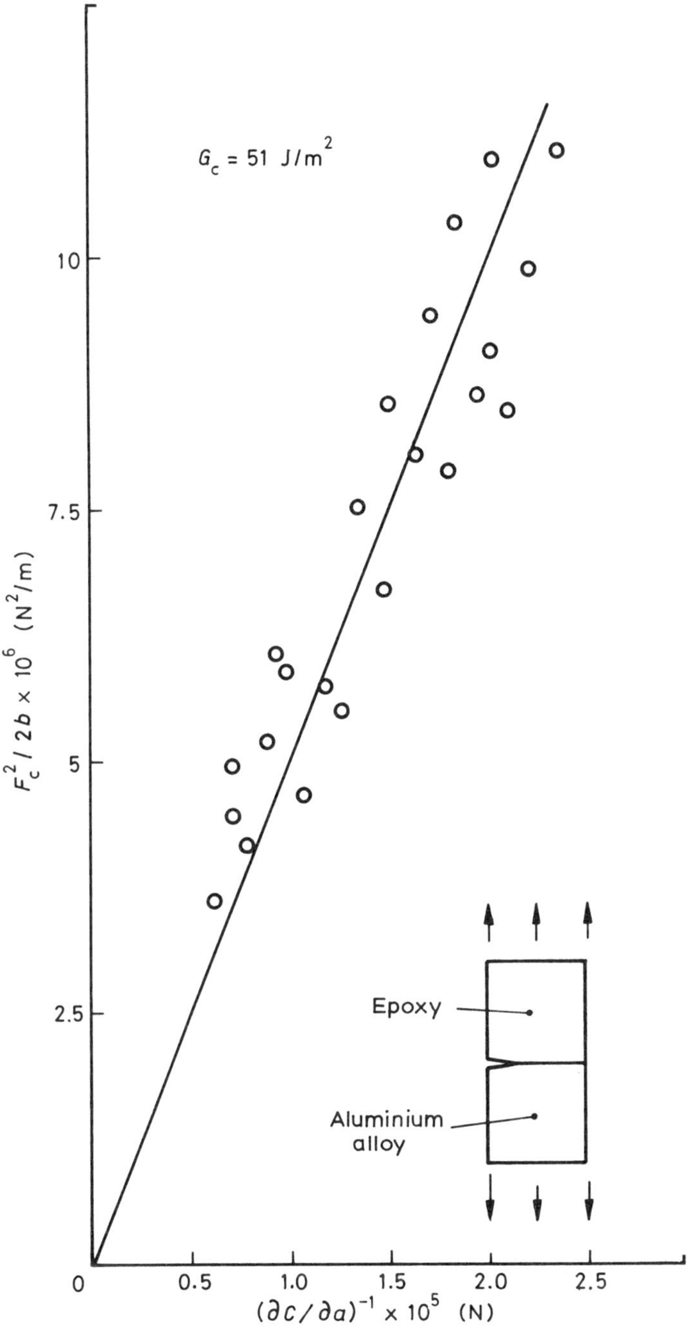

Figure 7.10 $F_c^2/2b$ versus $(\partial C/\partial a)^{-1}$ for a bimaterial single-edged notched specimen with an interfacial crack [10].

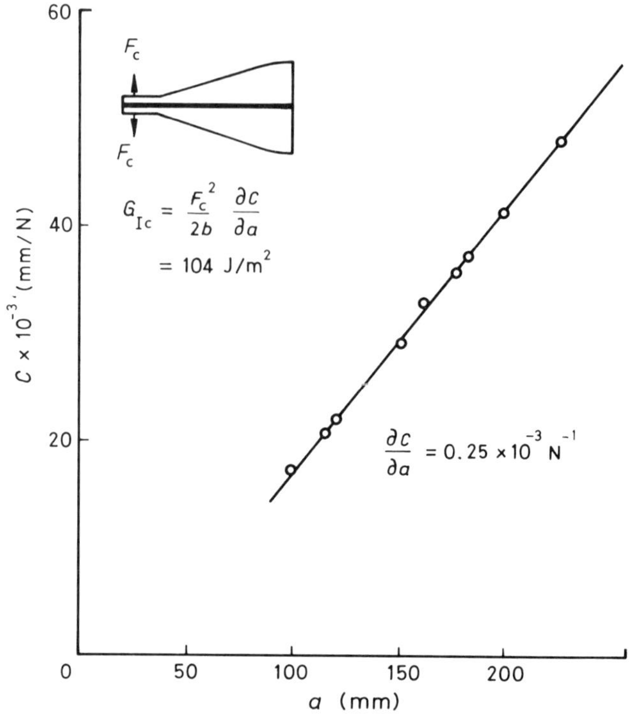

Figure 7.11 Compliance, C, versus crack length, a, for a tapered double cantilever beam (TDCB) specimen. The points are experimental, and the solid line is theoretical based on Equation 7.49.

From the previous discussions it is obviously difficult to state exact values of G_c and K_c without defining the mode of loading, test rate and temperature, etc. Nevertheless, representative values for a range of adhesive joints tested at room temperature and a moderate rate of test are given in Table 7.2, with the aim of imparting to the reader the typical values obtained and their positions relative to other materials.

7.5 EFFECT OF JOINT GEOMETRY

7.5.1 Introduction

A major feature of the continuum fracture mechanics approach is that the plane strain values of the fracture energy or the stress intensity factor for a given mode of loading should be independent of the geometry of the specimen employed for their evaluation. This is obviously required if the values of G_c and

Table 7.2 Typical values of G_c and K_c for various joints and materials

Joint	Mode	E_a (GPa)	G_c (kJ/m^2)	K_c (MN/m$^{-3/2}$)	References
SBR/PET	i	≅0.001	0.02	—	11
SBR/etched PTFE	i	≅0.001	0.06	—	16
Epoxy-TEPA/aluminium alloy	I	4.3	0.055	0.52	104
Epoxy-TEPA/aluminium alloy	II	4.3	1.45	2.7	96,104
Epoxy-TEPA/aluminium alloy (in water)	i	—	0.015	—	105
Epoxy-TETA/aluminium alloy	I	2.8	0.075	0.52	104
Epoxy-3° amine/aluminium alloy	I	2.8	0.40	1.2	104
Rubber-modified epoxy/steel	I	2.5	3.5	3.4	106
Phenolic-resin/hard maple wood	I	—	0.05	—	107
Other materials for comparison					
Material					
Rubber	I	≅0.001	0.2–10	—	3,16
Polystyrene	I	3	0.4	1.1	3
High-impact polystyrene	I	2.1	15.8 (J_{Ic})	—	3
Glass	I	70	0.007	0.7	3
Wood	I	2.1	0.12	0.5	3
Aluminium alloy	I	69	20	37	3
Steel, mild	I	210	12	50	3
Steel alloy	I	210	107	150	3

i = interfacial failure, otherwise cohesive in the adhesive layer.
SBR = styrene–butadiene rubbery adhesive.
PET = poly(ethylene terephthalate) substrate.
PTFE = etched polytetrafluoroethylene substrate.
Epoxy = diglycidylether of bisphenol A (DGEBA).
TEPA = tetraethylenepentamine hardener.
TETA = triethylenetetramine hardener.
3° amine = tertiary amine hardener.
Rubber-modified = toughened with dispersed phase of rubbery inclusions.

K_c are to be evaluated from laboratory specimens, such as those shown in Figs 7.7 and 7.8, and then used to design and predict the service life of bonded structures on the principle that the values are 'material' properties. (Allowing of course for the values to be dependent upon test rate, temperature, environment, etc.) Also, even for possibly the less rigorous application of adhesives selection and development it is a distinct advantage if the values of G_c and K_c for the joint are 'material' properties.

In the case of a rubbery adhesive the work of Gent and Kinloch [11] has shown that the adhesive fracture energy, G_c, is independent of the type of specimen. They used several of the joint geometries illustrated in Fig. 7.7 and confirmed that the value of G_c, when measured for a given value of temperature and crack velocity, is independent of the details of the test specimen.

Similarly, results have been reported for structural adhesives by several groups of workers including Trantina [40] and Kinloch and co-workers [10,87,108]. They have shown that the adhesive fracture energy, G_{Ic}, and fracture toughness, K_{Ic}, for joints, where the adhesive was a brittle, simple epoxy and the locus of failure was cohesive in the adhesive layer, are independent of test geometry and, furthermore, equivalent [108] to that of the fracture energy or fracture toughness of the bulk adhesive. Other workers [109–111] have also shown that for similar epoxy adhesives the value of G_{Ic} for the adhesive joints is not dependent upon the thickness of the adhesive layer over the common range of thicknesses employed, say from about 0.08 to 2 mm. (Again ensuring that comparisons are made when all the tests are conducted at the same rate and temperature.)

However, apart from the values of the adhesive fracture energy or fracture toughness being different when ascertained under mode I, II or III conditions as indeed would be expected, there are circumstances when the values are dependent upon test geometry. This is obviously not in accord with the principles described above and typically arises from the fact that the adhesive in the joint is present as a thin layer and the nearby presence of high modulus and high yield strength substrates interferes with the crack tip micromechanisms, compared to when testing the adhesive as a bulk material. Such geometry effects will first be considered before discussing the effects of the mode of loading.

7.5.2 Flexible peel joints

Some of the most elegant work on the effects of joint geometry upon the measured peel energy or strength, P, (i.e. $P = F_c/b$,) and the calculated adhesive fracture energy, G_c, has been reported by Gent and Hamed [61,62,112–114]. They have investigated the effects of the thickness of the adhesive layer, thickness of the flexible substrate and peel angle.

In the case of adhesive thickness, h_a, they argued that the boundary cleavage stress, σ_p, (Equations 6.24 to 6.27) actually represents the mean tensile stress throughout the adhesive in the region of the debonding peel front and not the much larger stress which acts at the actual line of separation. Thus, the degree of stress concentration at the line of separation will, in general, depend upon the thickness of the adhesive layer so that a larger mean stress will be necessary to bring about the same detachment stress in a thin layer as in a thick one. Hence, the assumption in deriving Equation 6.24 of a constant value of σ_p, in the vicinity of the line of detachment is incorrect. They suggested that σ_p is inversely proportional to $h_a^{1/2}$. Therefore, both the stress analysis approach, Equation 6.27, and the fracture mechanics approach, Equations 7.40 to 7.43, predict that the peel force, P, should be independent of the thickness of the adhesive layer.

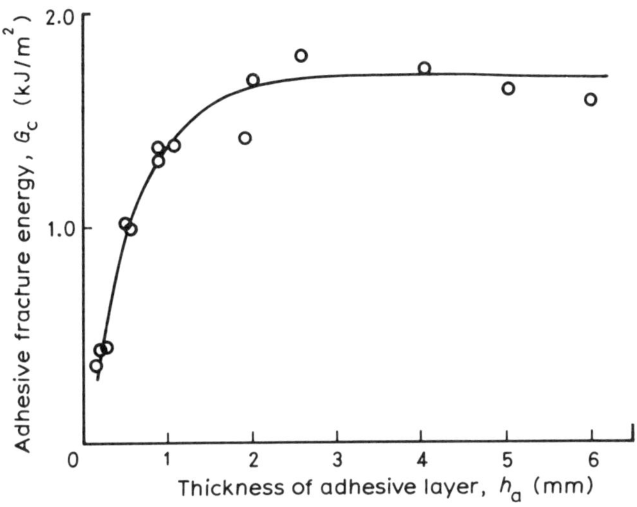

Figure 7.12 Relation between adhesive fracture energy G_c, and thickness, h_a, of adhesive layer for a poly(ethylene terephthalate) substrate detaching from a thermoplastic rubber adhesive layer at a peel angle of 180° [112].

However, it is commonly observed [112,113,115–119] that the peel strength, P, and the calculated value of G_c, increase as the thickness, h_a, of the adhesive layer increases. This is shown in Fig. 7.12 where it may be seen that this behaviour is especially pronounced when the adhesive thickness, h_a, is low but that the value of P or G_c eventually reaches a plateau value for relatively thick adhesive layers. These observations may be attributed to additional energy dissipation within the adhesive layer thereby increasing the value of ψ in Equation 7.5. Essentially as the thickness of the adhesive layer is increased in the peel test a larger volume of adhesive is subjected to deformation per unit area of detachment so that the total energy expended in peeling increases. However, at large thicknesses the energy dissipated during peeling then becomes independent of the overall thickness of the adhesive, since the dissipation process no longer involves the entire layer of adhesive. Gent and Hamed [113] suggested that the energy, ψ, dissipated in viscoelastic and plastic deformations at the crack tip may be given by:

$$\psi = \int_0^{h_a} U_{dhc}(y)\,\partial y \tag{7.52}$$

where U_{dhc} is the hysteresis or loss strain energy density, i.e. the difference between the input and recoverable strain energy densities and may be evaluated, for a given stress level, by ascertaining the area between the extension and retraction stress versus strain curves for the bulk adhesive. The value of U_{dhc} will be a function of the rate and temperature of test and the stress (or strain) levels attained through the adhesive layer. Obviously, Equation 7.41 may be modified by using Equations 7.5 and 7.52 to introduce the effect of the thickness of the adhesive layer, leading to the intrinsic fracture energy, G_o, rather than G_c, being the 'material' property. However, the practical problem with evaluating the right-hand side of Equation 7.52 is that the value of $U_{dhc}(y)$ will be a function of the stress (or strain level) and so will be a function of the distance, y, through the adhesive layer, as indicated. But the stress distribution through the adhesive layer is very difficult to ascertain, especially close to the peel front when the stress gradient will be very high. Nevertheless, it is evident that adhesives capable of yielding and filamenting to large strains will give large peel energies, provided the intrinsic adhesion or cohesion forces are capable of withstanding sufficiently high stresses so that all the adhesive layer is brought into the energy dissipating region before fracture occurs.

Considering the effect of the thickness of the flexible, peeling substrate, Gent and Hamed [61,62,112] have suggested that for a sufficiently long peeling strip the bending energy stays constant as peeling proceeds. Thus, provided the strip is perfectly elastic, no energy is dissipated within the substrate and Equations 7.40 to 7.43 need no modification. (Although, it should be recalled that if Equations 7.42 or 7.43 are used to deduce a value of G_c then the elastic strain in the tab of the detached, peeled strip has been assumed to be negligible; if this is not the case then the full equation, 7.41, must be employed.) However, if the level of adhesion is sufficiently high, or if the peeling strip is sufficiently thin, bending stresses may cause irreversible deformations of the substrate in the course of peeling and the peel energy, P, will be augmented by the work expended in plastic deformation of the substrate. Experimentally, it has been found that the relation between the measured peel strength, P, and the thickness of the peeling strip shows a maximum. This arises because, whilst a thin strip may undergo plastic yielding the total energy dissipated in this manner will be small. Hence, as the thickness of the peeling strip increases more energy is dissipated in yielding and the measured value of P rises. However, at large thicknesses the backing will not experience sufficiently high bending stresses to cause yielding so that the peel energy, P, will decrease again back to the value obtained at 'zero' thickness of backing-strip. Thus, the peel strength at both 'zero' and large thicknesses of the substrate should be equal to the value of P in the absence of plastic yielding of the flexible substrate. In terms of calculating the true value of the adhesive fracture energy, G_c, when plastic deformation of the flexible substrate is occurring this extra source of 'energy loss' must be accounted for in Equation 7.41, and subsequent equations. This may be readily undertaken for various boundary conditions by including a term on the right-hand side of Equation 7.41 and the form of such a term has been discussed by Gent and Hamed [62].

It is of interest to note that this energy term for plastic yielding of the substrates will also be a function of the degree of intrinsic adhesion or cohesion forces which are present, depending upon the locus of joint failure. For example, moderately thick substrates which may undergo plastic yielding when the intrinsic strength of the interface is high may not do so if the adhesion is weak since the peel strength will be insufficient to generate the bending forces necessary to attain the yield stress in the substrate. As an example of the nature of the energy term, for a peel angle of 90° the energy dissipation, ψ_p, from plastic yielding of the substrate is given by:

$$\psi_p \cong \frac{d^2 \sigma_{sy}}{4 r_c} \tag{7.53}$$

where σ_{sy} is the yield stress of the substrate of thickness, d, and r_c is the radius of curvature of the peeling substrate. When the radius takes its minimum feasible value, $d/2$, the corresponding maximum value of ψ_p becomes:

$$\psi_p(\max) = d\sigma_{sy}/2 \tag{7.54}$$

The thickness of the substrate when no contribution from plastic yielding results, i.e. $\psi_p \rightarrow 0$, occurs when the thickness, d, of the substrate exceeds a critical thickness, d_c; where d_c is given by:

$$d_c = \frac{6 E_s P^\circ}{\sigma_{sy}^2} \tag{7.55}$$

where P° is the peel strength in the absence of plastic yielding of the substrate.

Both fracture mechanics (Equations 7.40 to 7.43) and stress analysis (Equation 6.27) considerations lead to the conclusion that the directly measured peel energy, P, should be inversely proportional to $(1 - \cos\alpha)$, where α is the peel angle. Thus, the value of P should be a maximum when $\alpha \rightarrow 0°$ and at a minimum at $\alpha = 180°$, but the adhesive fracture energy should, of course, be constant. Many investigators [55–60,120–124] have examined this dependence and generally found that this relation is indeed followed. However, two principal discrepancies do sometimes occur: at low peel angles, below about 30°, and at high peel angles above about 150°. At low peel angles the measured peel energy, P, may be either less than [59,124] or greater than [121] expected, depending upon the adhesive/substrate combination. At high peel angles the value of P is usually greater than expected; for example, it has been observed that $2P_{180°} > P_{90°}$. Various reasons [61,62,121] have been advanced to explain these discrepancies and so account for the observation that the calculated value of G_c may be dependent upon the peel angle. All invoke the argument that at these extreme peel angles the energy dissipative mechanisms that occur in the peel test are accounted for incorrectly, or not accounted for at all, in Equations 7.40 to 7.43. For example, at low peel angles the value of G_c may not be a constant but be dependent upon the test geometry, and be somewhat higher, due to an increasing mode II (shear) contribution (see Section 7.5.4). At high peel angles the contribution from plastic yielding of the flexible substrate might be relatively high as it is forced to bend through a large angle. As stated above, this type of deformation is not taken into account in Equations 7.40 to 7.43, as they are written, so unless an appropriate term is included the value of G_c deduced from Equation 7.41, or the simplified version, Equation 7.43, will be dependent upon the peel angle.

Finally, it should be noted that similar arguments will be applicable to the case of rigid, structural adhesives bonding metallic substrates being assessed via a peel test, as discussed in Section 6.4.10.

7.5.3 Structural adhesives

7.5.3.1 Thickness of adhesive layer

As mentioned above the thickness, h_a, of the adhesive layer does not usually affect the measured value of the adhesive fracture energy for relatively brittle adhesives but for tougher adhesives this parameter may significantly affect the measured value of G_c of the joint.

Several groups of workers [106,110,111] have shown that for a toughened adhesive, typically toughened by the inclusion of a second rubbery phase, the relation between G_{Ic} for the joint and h_a is complex with G_{Ic} passing through a maximum value, G_{Icm}, at a certain thickness, h_{am}. This is illustrated in Fig. 7.13 where it may also be seen that the values of both G_{Icm} and h_{am} are a function of the rate of test, and it should be emphasized that throughout these tests the locus of joint failure was cohesive in the adhesive layer [106]. Similar observations have been recorded as the test temperature, instead of the rate, has been varied [111].

Essentially, the more complex behaviour with such tough adhesives arises from the extensive plasticity that occurs in the vicinity of the crack tip. Firstly, with relatively thin adhesive layers the presence of high modulus and high yield strength substrates, such as steel or aluminium alloy, may restrict the full volume of the plastic zone developing in the adhesive layer ahead of the crack tip. Since the toughness is largely derived from the energy dissipated, ψ, in forming the plastic zone, the adhesive fracture energy, G_{Ic}, will steadily decrease as the adhesive layer thickness, h_a, is reduced past a certain value. Secondly, however, it has been found [36] that the constraint imposed upon the adhesive layer by the rigid substrates may increase the level of the σ_{11} local tensile stresses ahead of the crack tip, compared to the homogeneous case, as illustrated [106] in Fig. 7.14. This effect will increase the length of the plastic

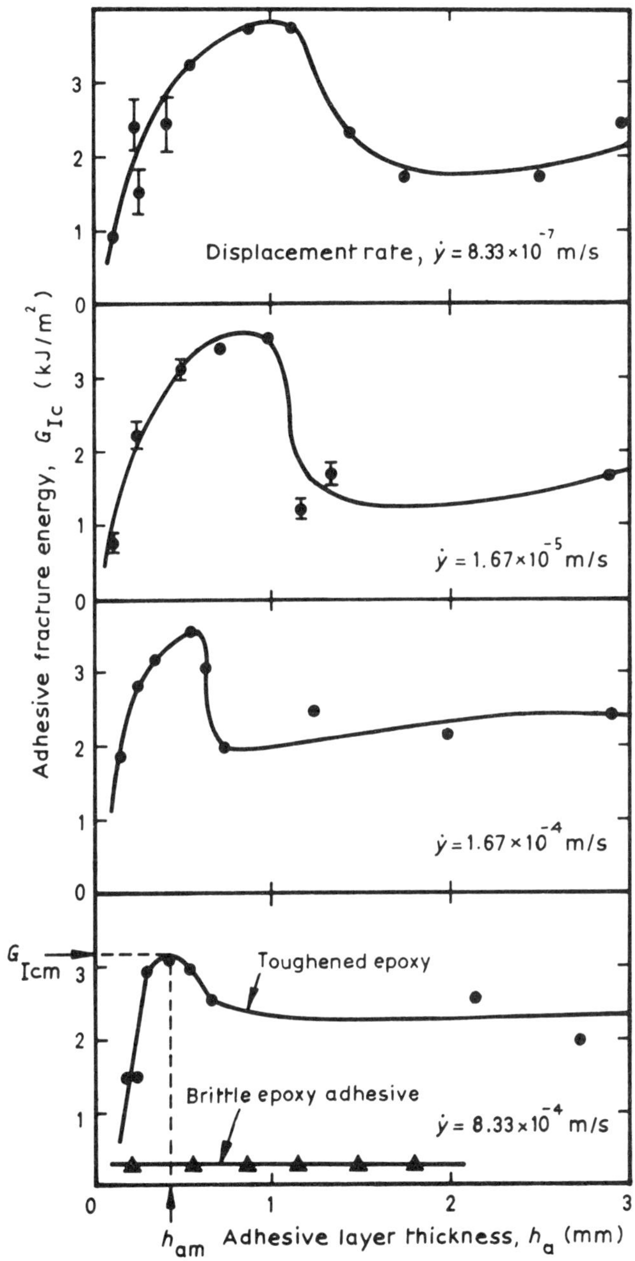

Figure 7.13 Adhesive fracture energy, G_{Ic}, as a function of the thickness, h_a, of the adhesive layer for joints consisting of steel bonded with a rubber-toughened epoxy using a TDCB specimen. Various rates of test have been employed. Note that in one case a typical result for a simple, untoughened epoxy is shown – the plastic zone at the crack tip is far smaller, compared to the tough adhesive, and no significant effect of G_{Ic} upon h_a is found [106].

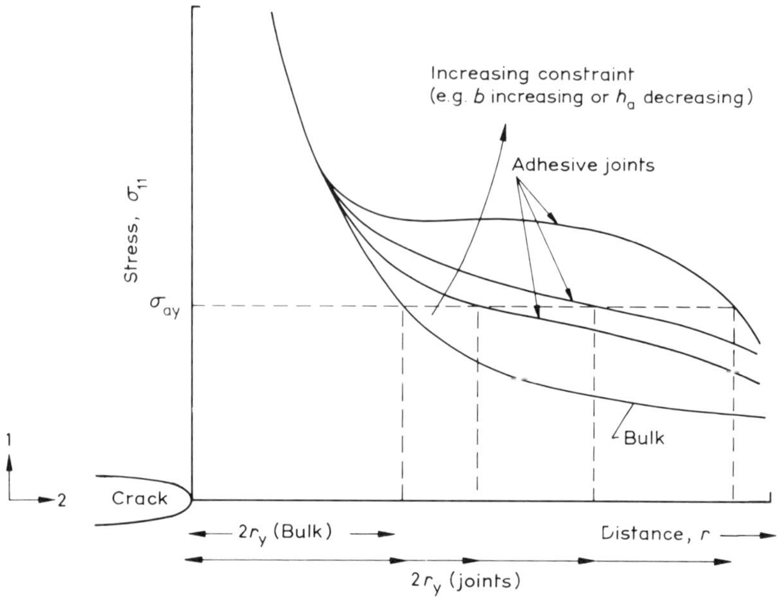

Figure 7.14 Effect of the thickness, h_a, of the adhesive layer and the width, b, of the joint on the σ_{11} stress level ahead of the crack tip. Note the increase in the length of the plastic zone in the adhesive layer for the joints with increasing degree of constraint.

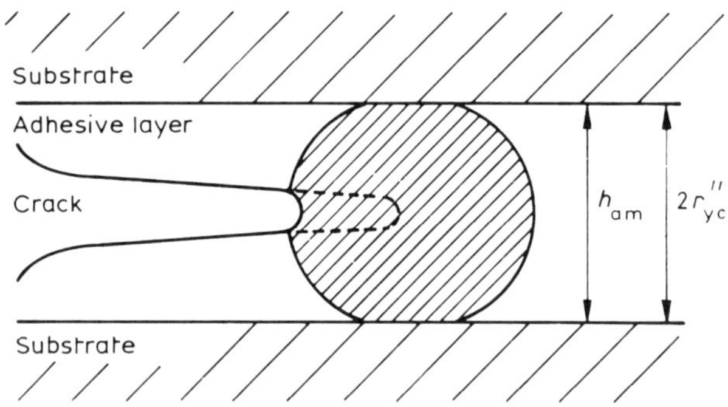

Figure 7.15 Simple elastic–plastic model for plastic deformation zone at a crack tip in the adhesive layer with high yield stress (elastic) substrates. As drawn $h_a = h_{am}$.

zone ahead of the crack tip, as has been confirmed experimentally. The degree of constraint will be greatest with thinner adhesive layers. Now these two factors combine to give the maximum in the adhesive fracture energy versus adhesive layer thickness relation. The maximum, G_{Icm}, results when the adhesive layer thickness and the plastic-zone diameter, $2r''_{yc}$ are approximately equal, as shown schematically in Fig. 7.15. Thus, from Equations 7.26 and 7.33 the adhesive thickness, h_{am}, at G_{Icm} should be expressed by:

$$h_{am} = 2r''_{yc} = \frac{1}{\pi}\frac{E_a G_{Ic}(\text{bulk})}{\sigma^2_{ay}} \tag{7.56}$$

Values of $2r''_{yc}$ calculated from Equation 7.56 are shown in Table 7.3 together with those of h_{am}, which were experimentally ascertained, and the agreement is very good.

It is interesting to note that, because of the constraint effect, the value of G_{Ic} for adhesive joints prepared using these tough adhesives may under many conditions be greater than the fracture energy of the bulk adhesive determined from tests upon cast sheets of the material, and may also be less temperature dependent, again as indicated in Table 7.3.

Finally, although the data in Figure 7.13 shows the value of G_{Ic} decreasing steadily at the thinnest adhesive layers examined, it has recently been suggested [162] that a plateau value is attained for very thin adhesive layers. Such plateau values appear to be subject to considerable scatter and, due to the extremely thin adhesive layers, would be expected to reflect the nature of the substrate's topography; since the plasticity at the crack tip is influenced by this factor (see Section 3.2 and [98]).

Table 7.3 Comparison of measured adhesive layer thickness, h_{am}, at maximum adhesive fracture energy, G_{Icm}, and calculated plastic zone diameter, $2r''_{yc}$ [106,111]

Temperature (°C)	Log (rate of test) (m/s)	G_{Ic} (bulk) (kJ/m^2)	G_{Icm} (joint) (kJ/m^2)	h_{am} (mm)	$2r''_{yc}$ (mm)
20	−6.08	2.10	3.90	1.0	0.85
20	−4.78	1.85	3.65	0.8	0.70
20	−3.78	1.55	3.55	0.55	0.49
20	−3.08	1.50	3.15	0.4	0.43
50	−4.66	4.70	2.95	1.1	1.6
37	−4.66	3.75	2.85	0.9	1.16
25	−4.66	2.70	3.85	0.6	0.57
0	−4.66	1.65	3.00	0.5	0.39
−20	−4.66	1.00	3.15	0.25	0.15
−40	−4.66	0.75	2.50	0.1	0.05

7.5.3.2 Width of joint

Kinloch and Shaw [106] have investigated the effect of joint width by employing a TDCB specimen consisting of mild-steel substrates bonded with a simple brittle epoxy or a rubber-toughened epoxy adhesive. For the former, low toughness adhesive no effect was observed. However, for the tough adhesive the relationships between G_{Ic} and adhesive layer thickness, h_a,

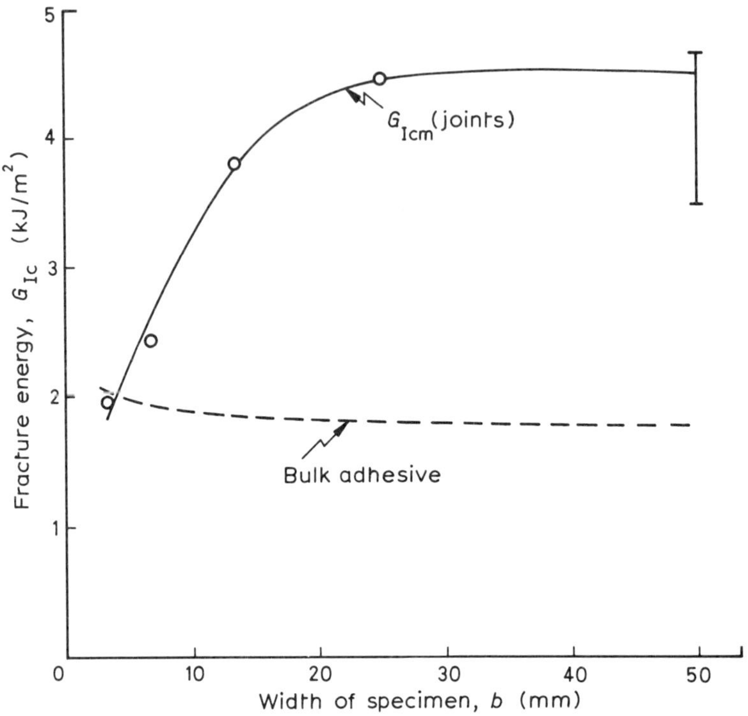

Figure 7.16 The effect of joint width upon the maximum adhesive fracture energy, G_{Icm}, for a rubber-toughened epoxy adhesive bonding steel substrates using a TDCB geometry [106].

followed the form shown in Fig. 7.13 but the values of G_{Icm} and h_{am} were a function of joint width, b. Figure 7.16 illustrates the relation between G_{Icm} and joint width, together with the bulk epoxy fracture values. It is interesting that neither the joint nor the bulk adhesive values of fracture energy follow the classical plane stress/plane strain relation with width (see Fig. 7.6). Also, G_{Icm} (joint) is usually greater than G_{Ic} (bulk) and this enhancement of the adhesive fracture energy becomes more significant as the degree of constraint on the adhesive layer is increased, i.e. as the value of b increases (see Fig. 7.14). Hence, the effect of width of joint on the adhesive fracture energy of joints with tough adhesives again appears to be due to the effect of the substrates changing the stress field around the crack tip and enhancing the extent of plasticity and energy dissipation, ψ, ahead of the crack.

Finally, the above experiments on the effects of adhesive thickness and joint width were all conducted using 'through-thickness' cracks. From some initial observations on cracks which were embedded in an adhesive layer the severe constraint imposed by the surrounding elastic material appears to restrict greatly the amount of crack-tip plasticity which develops. Thus, in such instances, the appropriate value of G_c might be significantly lower. It would, therefore, be of interest to measure the value of G_c for such tough adhesives when the crack is not a 'through-thickness' crack. If the problems of inserting sharp cracks, ensuring mode I and measuring the crack velocity could all be overcome for the blister-test geometry (see Table 7.1), then this specimen would be a useful one in this context.

7.5.4 Modes of loading

For isotropic materials mode I is the lowest energy fracture mode and, thus, a crack always propagates along a path normal to the direction of maximum principal tensile stress. Hence a crack in an isotropic plane will propagate in mode I fracture regardless of the orientation of the initial flaw with respect to the applied stress. However, this is not necessarily the case in joint fracture since crack propagation is constrained to the adhesive layer, and or interface, regardless of the orientation of the adhesive layer, except of course when the substrate has a lower toughness than the adhesive. Thus, for the fracture of joints attention must be given to joint fracture under additional loading modes.

In the case of rubbery adhesives values of the adhesive fracture energy measured under different modes of loading are shown [69,70] in Table 7.4. As expected the order of values for G_c is mode I < mode II < mode III.

Table 7.4 Values of adhesive fracture energy, G_c, of a polyurethane/poly(methyl methacrylate) interface [69,70]

Test geometry	Mode	G_c (kJ/m^2)
90° peel	I	0.032
Blister–$h_a/a > 1$	I	0.030
Blister–$h_a/a \ll 1$	II	0.072
Torsion	III	0.10

In the case of structural adhesives then joint fracture under mode II, combined mode I/II and mode III has been conducted by several groups [41,42,95–102,125–127] and some typical values are shown in Table 7.5.

For the brittle DGEBA–TEPA adhesive the mode I value is the lowest with mode II being the highest; and the mode II value may even be an underestimate since failure of the shear specimen employed did not occur by pure mode II. Indeed, it is usually extremely difficult to obtain crack growth under pure mode II conditions and, in practice, the fracture is locally in mode I.

Table 7.5 Effect of mode of loading on the adhesive fracture energy† of structural joints

Adhesive	G_{Ic}	$G_{I/IIc\phi}$ ‡ 30°	45°	60°	G_{IIc}	References
DGEBA–TEPA¶	0.07	0.15	0.13	0.15	1.45§	95,96
Rubber-toughened epoxy	3.40	0.31	0.50	0.31	3.55	99,127
Commercial rubber-toughened epoxy-film adhesive	2.30	0.18	0.90	0.20	—	99,127

† Fracture energies in kJ/m^2.
‡ Scarf joint, see Table 7.1. As the angle, ϕ, increases the amount of mode II loading increases.
§ Indcpendently loaded mixed-mode joint, see Table 7.1.
¶ DGEBA = diglyicdyl ether of bisphenol A epoxy resin.
TEPE = tetraethylenepentamine hardener.

However, this still results [128] in the apparent G_{IIc} value being considerably greater than the true value of G_{Ic}.

For the toughened structural adhesives the situation is far more complex. The mixed-mode $G_{I/IIc\phi}$ values are lower than the G_{Ic} or the G_{IIc} values, with these latter two fracture energies being approximately equal in magnitude. Why the mixed-mode value should be so much lower than the values of G_{Ic} and G_{IIc} is not immediately apparent. Bascom *et al.* [127] believe that the correct values of $G_{I/IIc\phi}$ have been deduced from the measured fracture loads, but acknowledged that the compliance calibration was only derived theoretically and assumes a very thin adhesive layer, i.e. $h_a \to 0$. This aspect is highlighted by some recent work [129] which finds that for a commercial rubber-toughened adhesive the values of G_{Ic} and G_{IIc} are approximately equal but also reports that the value of $G_{I/IIc}$ is approximately the same; although interestingly the value of $G_{I/IIc}$ was ascertained from a different test specimen, namely, the cracked lap-shear specimen (see Table 7.1).

The work of Bascom *et al.* [98,99,127] also reveals that, apart from the mixed-mode $G_{I/IIc\phi}$ values being lower than the G_{Ic} or the G_{IIc} values, with $G_{Ic} \cong G_{IIc}$, there is also a maximum in $G_{I/IIc\phi}$ at $\phi = 45°$. Assuming that these data are correct, then these features may arise for several reasons. Firstly, in the scarf-joint geometry the locus of failure is 'focused' into the interfacial region by the combined-stress loading which restrains the development of a large plastic zone at the crack tip and thus may result in $G_{I/IIc\phi} < G_{Ic}$ or G_{IIc}. This explanation is essentially the same as that invoked to account for the effects of adhesive layer thickness on the value of G_{Ic}, as discussed above. This explanation is also in accord with other results in Table 7.5, since this effect would be expected to be most severe in the case of the toughened adhesives which rely upon the development of a large plastic zone ahead of the crack tip in order to achieve their high fracture energies. (Note that this mechanical focusing [98,130] of the crack towards the interface leads to fracture occurring in the adhesive layer but very close to the interface. It provides an explanation for this type of failure without the need to invoke any 'weak boundary layer' hypothesis.) Secondly, since in the scarf joint the crack is close to the interface it is influenced by the surface roughness of the substrate, and Bascom *et al.* [98,99] have found that crack initiation is impeded by increased surface roughness and that this is most evident at $\phi = 45°$; for a smooth, polished surface no maximum in $G_{I/II\phi}(\phi = 45°)$ is observed.

Finally, it should be emphasized that the results shown in Table 7.5 for the tough adhesives do not imply that the joint strength is necessarily lower in shear than in cleavage. If the resultant applied load required for crack propagation is plotted against increasing mode II contribution then an increase in joint strength with shear loading is evident. The significance of the fracture energy anomaly is that the joint strength under increasing shear loading would be considerably higher if $G_{I/II\phi}$ was not appreciably lower than G_{Ic}.

One aspect that arises with mixed-mode loading is that of deciding a fracture criterion. A possible first choice is:

$$G_c = G_{Icj} + G_{IIcj} + G_{IIIcj} = \text{constant} \quad (7.57)$$

where the values of G_{Icj}, G_{IIcj} and G_{IIIcj} are the respective values at fracture in the joint of interest, e.g. if pure mode I then G_{Icj} is equivalent to G_{Ic}. One aspect of this equation is that it implies the equality of the pure mode I, II and III values. However, Trantina [97] has found this to be unsatisfactory since he reported that, for a brittle epoxy adhesive, the total adhesive fracture energy, G_c, increases as the amount of mode II loading in the joint is increased. The fact that Equation 7.57 is inadequate is also readily apparent from the data in Table 7.5. It has been suggested [131], therefore, that a more suitable criterion for mixed-mode fracture is:

$$\left(\frac{G_{Icj}}{G_{Ic}}\right)^m + \left(\frac{G_{IIcj}}{G_{IIc}}\right)^n = 1 \quad (7.58)$$

where the values of G_{Ic} and G_{IIc} represent the pure mode I and II values, respectively. This empirical expression allows the value of G_c calculated from Equation 7.32 to increase as the amount of mode II loading increases, in agreement with experimental observations. Work on adhesives [129] and fibre composites [132,133] has demonstrated that this expression gives a reasonably good fit to the experimental data and values of $m = 0.5$ and $n = 1$ and $m = n = 1$ have been reported.

7.6 EFFECT OF RATE AND TEMPERATURE

7.6.1 Introduction

From the previous discussions and Equation 7.5 it is evident that the measured value of adhesive fracture energy or fracture toughness typically reflects the extent of plastic and viscoelastic deformations in the region of the crack tip. In polymeric materials the extent of such deformations is highly dependent upon the test rate and temperature and, therefore, the measured values of G_c and K_c are often dependent upon these test parameters.

7.6.2 Rubbery adhesives

Andrews and Kinloch [16] have measured the adhesive fracture energy, G_c, of joints consisting of a crosslinked styrene–butadiene rubber adhering to various plastic substrates over a wide range of test rates and temperatures. In all cases the crack propagated in a stable, continuous manner along, or close to, the interface. The value of G_c was found to increase steadily with either an increase in rate of crack growth or a decrease in temperature. This reflects the dominating influence of the hysteresis occurring around the crack tip; each volume element in the path of the propagating crack is subjected to a stress–strain cycle as the crack approaches and recedes from its vicinity and energy is dissipated during this strain cycle. Now the viscoelastic hysteresis in the rubbery adhesive largely results from an internal friction, or viscosity, mechanism. Hence, at temperatures $T > T_g$, the Williams–Landel–Ferry (WLF) equation may be used to produce a single relation between G_c and rate and temperature. Such a master curve is constructed by multiplying the rate by the time–temperature shift factor, a_T. The value of a_T may be calculated from Equation 6.8, as described in Section 6.4.3.2, and the master curves produced are shown in Fig. 7.17. Note that, in accord with usual practice, the ordinate has been multiplied by the ratio T_g/T to account for the increase in retractive force with absolute temperature. Also, the master curve for cohesive fracture of the bulk rubbery adhesive is shown for comparison. There are several points of interest concerning the master curves shown in Fig. 7.17.

Firstly, the values of a_T calculated from the WLF equation yield a good master curve for data measured over a wide range of temperatures and rates. Secondly, at lower temperatures and higher rates than those shown in Fig. 7.17 the adhesive becomes glassy and the hysteresis and the fracture energy fall – the value of G_c typically being 100 to 1000 J/m^2. Thirdly, the rate–temperature dependence of the fracture energy may be modelled by Equation 7.8, namely:

$$G_c = G_o \Phi_v(\dot{a}, T, \epsilon) \tag{7.8}$$

which predicts that if the value of the intrinsic fracture energy, G_o, is changed,

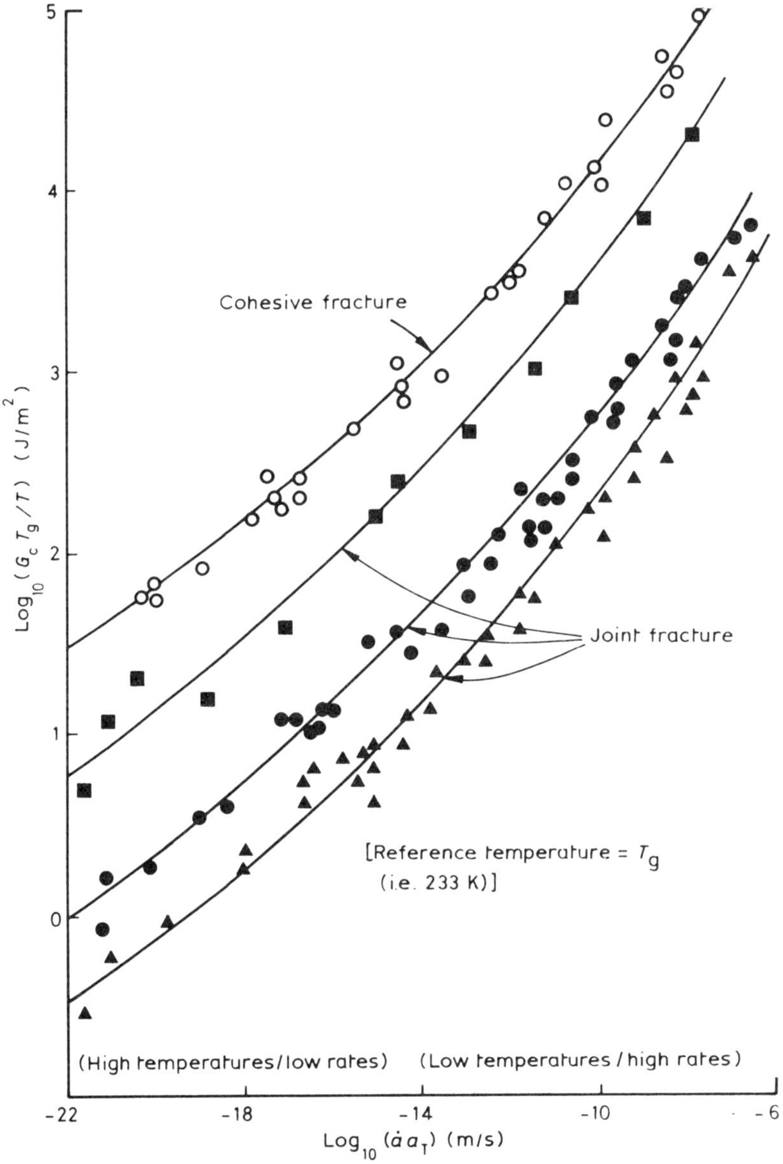

Figure 7.17 Master relationship for adhesive fracture energy, G_c, versus effective rate, $\dot{a}a_T$, of crack propagation. The points for each curve represent data from various test rates and temperatures [16]: ○, G_c for cohesive fracture of crosslinked SBR; ■, G_c for fracture of crosslinked SBR/etched fluorinated ethylene–propylene copolymer joint; ●, G_c for fracture of crosslinked SBR/treated fluorinated ethylene–propylene copolymer joint; ▼, G_c for fracture of crosslinked SBR/poly(ethylene terephthalate) joint.

by changing the nature of the interface, then the curves will be approximately parallel when log G_c is plotted against log $(\dot{a}a_T)$, as may be seen from Fig. 7.17. It should be noted that the above expression is applicable to the fracture energy, G_c. If other geometry-dependent energy-dissipative mechanisms, such as any plastic deformation of the substrate in a peel-test specimen [113,114], arise then they should be accounted for in the energy balance when determining the value of G_c. Also, the curves would be predicted to be only approximately parallel since the value of Φ_v is a function of the strain level, ϵ. This may be readily understood since the value of ψ is related to the hysteresis ratio, U_{dh}/U_d, which is well known to be dependent [135] upon the strain, ϵ. This is especially pronounced at high applied loads or at temperatures close to the T_g of the adhesive.

Fourthly, at low effective rates of test the value of $\psi \rightarrow 0$, and that of $\Phi_v \rightarrow 1$, so that the measured value of G_c is equal to G_o, the intrinsic fracture energy. In practice it is very difficult to access this region experimentally since the rates of test are very low and oxidative attack may occur during the time-scale of the test. Nevertheless, since Φ_v is also a property of the bulk adhesive material it controls the bulk fracture energy and, when the value of Φ_v is not strongly dependent upon the strain, ϵ, then the value of G_o for the joint may be obtained from:

$$G_o(\text{joint}) = G_o(\text{bulk}) \left[\frac{G_c(\text{joint})}{G_c(\text{bulk})} \right]_{(\dot{a}a_T)} \tag{7.59}$$

Thus, from measuring the values of the respective fracture energies from the curves shown in Fig. 7.17 and ascertaining the value of the intrinsic fracture energy of the bulk adhesive theoretically or from dynamic fatigue experiments [16,103], then values of the intrinsic adhesive fracture energy, G_o, may be deduced. The values so obtained for the joints studied in Fig. 7.17 are given in Table 3.5 and Fig. 4.3. Note, that if the value of Φ_v is more strongly dependent upon the strain, ϵ, level then Equation 7.58 cannot so readily be employed, since the curves shown in Fig. 7.16 are not so strictly parallel. Indeed, some degree of non-parallel behaviour may be detected in the data in Fig. 7.17 and may be seen more strongly in other adhesive systems, as discussed below. Another point of interest is that at the present time it is unclear whether the value of G_o for a joint is always directly equivalent to the number and type of secondary and chemical bonds acting across the fracture plane. It has been argued [113,134] that an allowance needs to be made for the stressing of many molecular bonds between crosslinks in the rubbery adhesive in order to rupture the bonds crossing the fracture plane. A problem in deciding this issue is that, as mentioned above, it is very difficult to attain experimentally the conditions when $\psi \rightarrow 0$, i.e. $\Phi_v(\dot{a}, T, \epsilon) \rightarrow 1$, and hence determine the value of G_o directly.

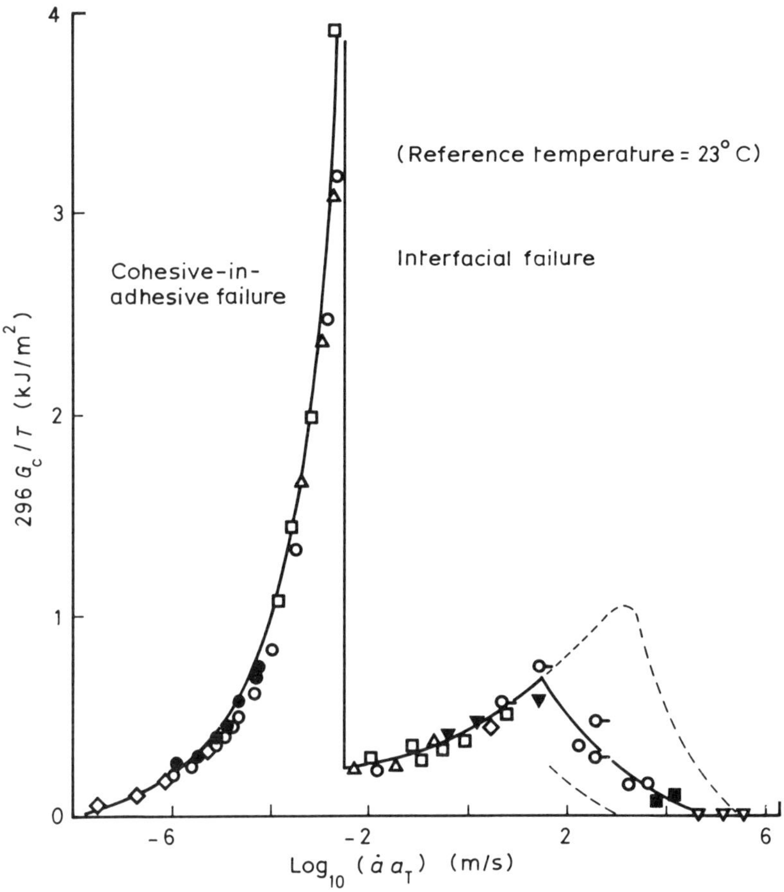

Figure 7.18 Master relationship for the adhesive fracture energy, G_c, against effective rate of crack growth, $\dot{a}a_T$, for an uncrosslinked SBR/poly(ethylene terephthalate) joint. The different symbols represent different test temperatures from −35 to +60 °C. Broken curves denote the extreme values when unstable, stick–slip peeling occurred [138].

If the rubbery adhesive is not substantially crosslinked then a far more complex relationship has been found between the adhesive fracture energy and the rate and temperature of test [136–140]. Gent and Petrich [138] have measured the adhesive fracture energy via peel tests for a cloth-backed layer of uncrosslinked butadiene–styrene rubber adhering to a poly(ethylene terephthalate) substrate. The results over a wide range of test temperatures and rates were again reduced to a single master curve and the data are illustrated in Fig. 7.18. The curve obtained is complex but the good superposition of results using the WLF relation indicates that the effects of rate and temperature are again all due to viscoelastic processes occurring within the adhesive. As the effective

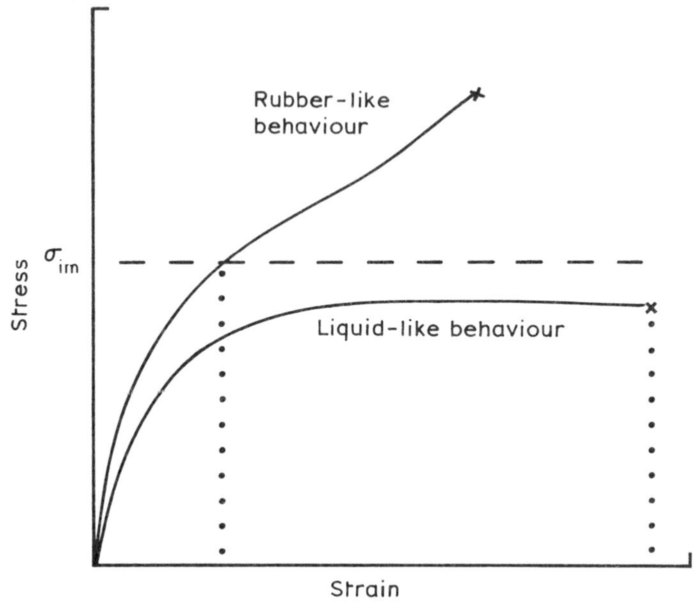

Figure 7.19 Schematic stress versus strain relationships for rubber-like (high effective rates of test, i.e. low test temperatures) and liquid-like (low effective rates of deformation, i.e. high test temperatures) behaviour of the uncrosslinked SBR adhesive. The horizontal dashed line denotes a possible level of maximum interfacial stress, σ_{im}. The vertical dotted lines denote the limiting strains attainable in each case [138].

rate of test is increased (i.e. the actual rate increased or the temperature decreased) then the value of G_c increases steadily up to a critical rate and then an abrupt transition occurs to a much lower value of G_c. Simultaneously, the locus of failure changed abruptly from cohesive failure in the rubbery adhesive to interfacial failure along the rubber/substrate interface. At higher effective rates there is a second maximum in the value of G_c which is smaller in magnitude, only involves interfacial failure and is associated with a 'stick-slip' type of fracture. 'Stick–slip', or unstable, fracture is the term used to describe the manner in which crack growth occurs when the load versus time trace has a saw-tooth appearance and the maximum value of load is associated with the crack initiating and growing rapidly and unstably down the specimen until it arrests at the point of minimum load. The values of G_c for crack initiation and arrest are also shown in Fig. 7.18. (This type of crack growth is also observed in rigid adhesives and a typical form of the load versus time, or displacement, trace is shown in Fig. 7.23, page 326.)

Gent and Petrich have explained the transition which is observed at lower rates by considering the viscoelastic behaviour of the rubbery adhesive. The stress versus strain curve for the adhesive at two different rates is shown schematically in Fig. 7.19 and the horizontal dashed line represents a possible level for the maximum stress that the interface can withstand. They assumed that the value of G_c is reflected by the energy under the stress–strain curve, up to the stress level where either cohesive or interfacial fracture occurs, i.e. they recast Equation 7.52 into the form:

$$\psi = h_a \int_0^{\sigma_m} \sigma \, \partial\epsilon \qquad (7.60)$$

where the value of σ_m is either the maximum stress, σ_{im}, the interface can bear or the tensile fracture stress, σ_{af}, of the adhesive. Thus, from Fig. 7.19 it is evident that at slow rates of test (or elevated temperatures) the adhesive exhibits liquid-like viscous deformation and the resulting adhesive fracture energy, G_c, would be high and the locus of failure would be cohesive failure in the adhesive. At somewhat faster rates (or lower temperatures) the adhesive exhibits rubber-like behaviour and the resulting value of G_c would be predicted to be low and interfacial failure would be observed. The model outlined by Equation 7.60 and Fig. 7.19 has several important consequences. Firstly, it should be noted that this transition from high to low fracture energies as the effective rate of peel is increased is not necessarily associated with a change in the locus of joint failure: whether such a change is observed depends upon the exact shape of the stress versus strain relation and the relative values of σ_{im} and σ_{af}. Secondly, the disappearance of this transition that typically results when the adhesive layer is crosslinked may be readily understood: crosslinking prevents the liquid-like flow of the adhesive and the energy under the stress–strain curve up to the failure point at low effective rates is greatly reduced. Thirdly, a practical point for the adhesives technologist is that high adhesive joint strength is obtained from adhesive materials which have large deformation energies within the stress limit set by the interfacial strength, σ_{im}.

The smaller second maximum that occurs at higher effective rates in Fig. 7.18 arises from the increasing stiffness of the adhesive layer as a result of a transition from a rubber-like to glass-like response of the adhesive, and it is also dependent upon the exact geometry employed. For example, if the adhesive layer is not forced to bend during the peel test a decline in the measured peel energy at the highest rates is recorded, as shown in Fig. 7.18; since at the effective rate associated with this maximum the adhesive layer will no longer readily bend and the contribution from the energy for plastic bending of the adhesive layer rapidly decreases. Alternatively, if the test geometry is redesigned to force the adhesive layer to bend, say through 90°, for example, then the energy necessary for this greatly increases the measured peel energy.

Similar observations to the above for the measured peel energy as a function of test rate/temperature have been recorded [136–140] for other types of adhesives, with the exact form of the relation being dependent upon the rheological behaviour of the adhesive and the magnitude of intrinsic adhesion forces at the adhesive/substrate interface.

Finally, Andrews *et al.* [141] have modified the standard peel test by inserting a spring between the crosshead of the test machine and the peeling tab. This acts as a 'soft-machine' test and the peeling tab is no longer compelled to follow the crosshead but may peel at a rate, including zero rate, determined by the adhesive joint properties and the instantaneous load. They used this test method to examine the peel behaviour of uncrosslinked pressure-sensitive surgical-tape adhesives. It was reported that the presence of the spring eliminated the stick–slip type crack growth and, further, revealed a lower bound to the value of G_c. This is not usually detected from the common, 'hard-machine', experiments and they suggested that this arises because the information on the lower bound data is contained in the initial rise of the peel force versus time curve up to the steady-state value. This part of the curve is ignored when the results are analysed: only the steady-state value is usually quoted. As other workers adopt this type of soft-machine test it will be of interest to see (i) whether such a lower bound value is indeed a true 'material' characteristic of the joint and (ii) how general is the existence of this previously unreported lower bound relation for the adhesive fracture energy.

7.6.3 Rigid adhesives

7.6.3.1 Introduction

The dependence of the type of crack growth and the measured values of the fracture energy, G_c, and fracture toughness, K_c, upon the test rate and temperature for rigid, structural adhesives is extremely complex and has only recently been elucidated. To assist in interpreting such data it is of interest to consider first the possible micromechanisms of crack growth in such adhesives.

7.6.3.2 Micromechanisms of failure

Just as in the case of rubbery adhesives, crack initiation and extension in rigid adhesives, whether along the interface or in the adhesive layer, obviously involves the rupture of intrinsic bonds. However, this mechanism is not the only energy-dissipative mechanism which is operative, since fracture in rigid polymers also invariably involves localized viscoelastic and plastic energy dissipative processes occurring in the vicinity of the crack tip. Indeed, such micromechanisms usually represent the main source of energy absorption in the material and their extent governs the value of the observed fracture energy or fracture toughness. Several different types of micromechanisms may occur at crack tips [142]:

(a) Shear yielding This micromechanism involves inhomogeneous plastic deformation of the material in the vicinity of the crack tip and occurs essentially at a contant volume. The shear yielding may be highly localized about the crack tip, involving little energy dissipation and hence result in brittle fracture. For example, this micromechanism is typically observed in the highly crosslinked brittle epoxy and acrylic resins. However, this type of crack propagation will become increasingly ductile in nature as the extent of shear deformation around the crack tip increases, effectively blunting the crack tip, and the toughness of the material generally rises. This type of behaviour is shown by many of the thermoplastic semicrystalline hot-melt adhesives such as polyamide- and polyethylene-based hot-melts. Obviously, the limiting factor is when bulk yielding occurs and fully ductile failure occurs.

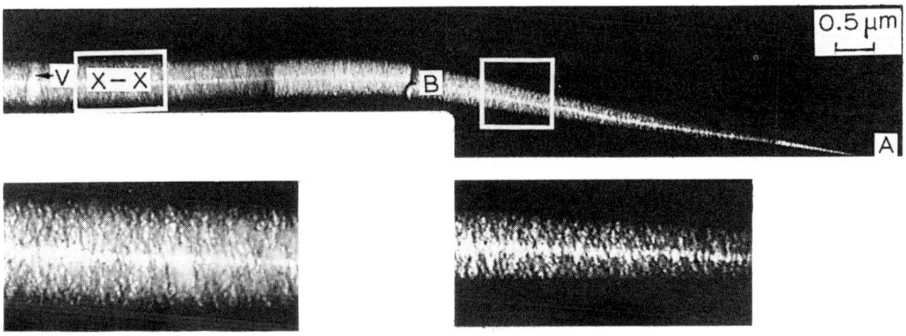

Figure 7.20 Composite electron micrograph of a typical craze formed in polystyrene. Note that it grows normal to the maximum principal tensile stress [143].

(b) Crazing Crazing also involves localized inhomogeneous plastic deformation but, unlike shear yielding, also involves cavitation of the material and therefore occurs with an increase in volume. A craze is formed when an applied tensile stress causes microvoids to nucleate ahead of the crack, or at some other point of stress concentration. The microvoids develop in a plane perpendicular to the applied stress but do not coalesce to form a true crack since they become stabilized by fibrils of plastically deformed, orientated polymeric material spanning the craze. The resulting localized yielded region consists of an interpenetrating network of voids and polymer fibrils and is known as a craze. A typical craze is illustrated in Fig. 7.20. The importance of localized crazing is that it is frequently a precursor to brittle fracture and typical adhesives which show this type of behaviour are the thermoplastic glassy adhesives such as poly(methyl methacrylate) and other thermoplastic acrylics.

(c) Multiple deformations Now although both shear yielding and crazing are energy absorbing processes they frequently lead to brittle fracture at a low G_c or K_c value because they are confined to a very small volume compared to the

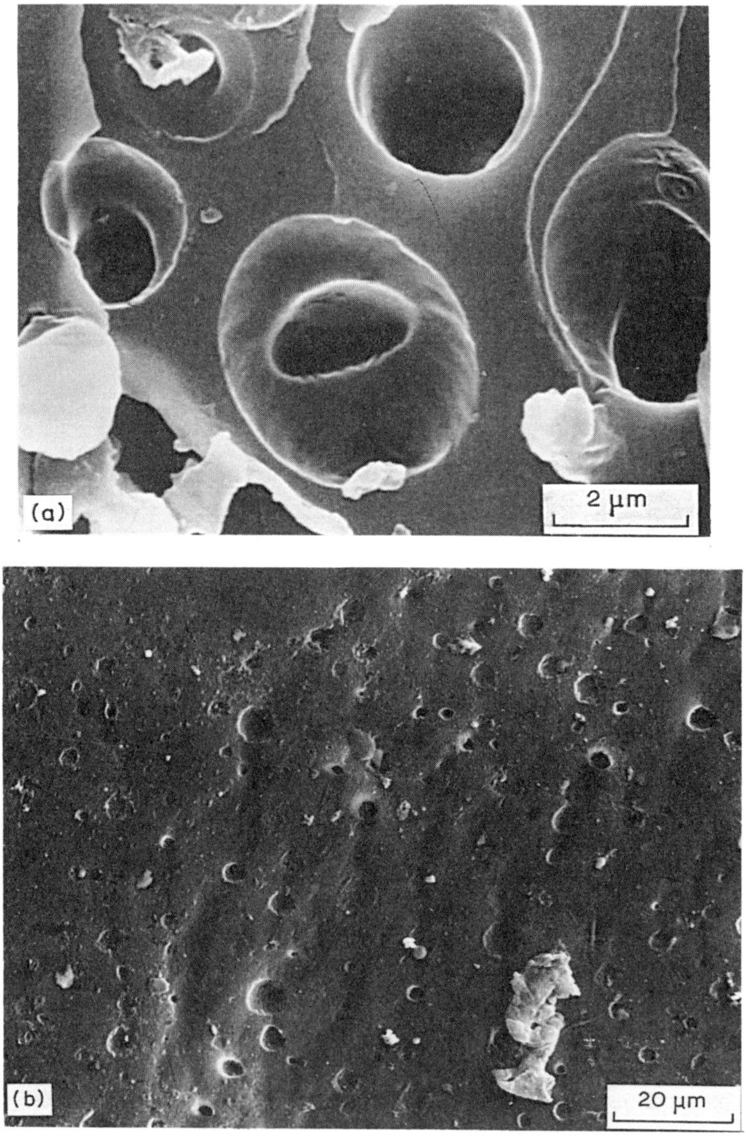

Figure 7.21 (a) Scanning electron micrograph of fracture surface of rubber-toughened epoxy adhesive showing cavitated rubber particles. (b) Scanning electron micrograph of the surface, perpendicular to the crack plane, of a rubber-toughened epoxy adhesive showing furrows caused by plastic shear bands which are initiated by the rubber particles [145,146].

size of the specimen; so that the total amount of plastic energy absorbed is low. To increase the toughness it is therefore necessary to ensure that the volume in which these, or any other energy dissipating mechanisms, occurs is sufficiently large whilst at the same time limiting the growth and breakdown of voids and crazes to prevent premature crack initiation. This aim may be attained by initiating *localized* mechanisms, such as crazing and shear yielding, but from *many* sites. This multiple-deformation mechanism has been most successfully achieved by developing multiphase adhesives, where small rubbery particles are dispersed in, but well bonded to, the matrix of the rigid resin. Important classes of such materials are the rubber-toughened epoxy and acrylic adhesives (Section 5.2). The deformation processes that occur in such adhesives have recently been discussed in detail [144,145] and may involve (i) cavitation in the rubber (Fig. 7.21a), or at the particle/matrix interface (note, *not* in the matrix as for crazing), and (ii) multiple, but localized shear yielding in the matrix initiated by the stress concentrations associated with the rubbery particles (Fig. 7.21b). The latter process is the main source of energy dissipation and the shear-yield bands may clearly be seen in Fig. 7.21b running at approximately 45° to the principal tensile stress, i.e. in the direction of the maximum shear stress. Since the properties of the matrix resin are relatively unaffected by the presence of typically about 10% w/w of small dispersed rubbery particles, this multiple-deformation toughening mechanism enables the toughness of the adhesive to be greatly increased without a significant loss in other important properties such as modulus and glass transition temperature.

(d) Crack pinning Sometimes adhesives are multiphase materials but the second phase is not rubbery in nature but is rigid, for example, the second phase might consist of glass beads, alumina or silica particles. The additional micromechanism that can now operate is crack pinning [148–151] and basically it involves the cracks being impeded by the rigid, impenetrable, well-bonded particles. This arises since when a crack meets an array of such obstacles it becomes pinned and tends to bow out between the particles forming secondary cracks. This is illustrated schematically in Fig. 7.22a and by direct observation using optical microscopy in Fig. 7.22b. Thus, new fracture surface is formed and the length of the crack front is increased due to its change of shape between the pinning positions. Now energy is not only required to create the new fracture surface but, by analogy with the theory of dislocations, energy must also be supplied to the newly formed non-linear crack front, which is assumed to possess a line energy. This latter factor particularly is suggested to lead to the enhanced crack resistance often observed when impenetrable particles are well-bonded into a brittle matrix. Since low adhesion is considered to impair the efficiency of the crack-pinning mechanism, improving the particle/matrix adhesion may result in improved toughness [147].

It should be emphasized that the above micromechanisms are frequently seen to operate together in a particular adhesive formulation. Indeed, one skill

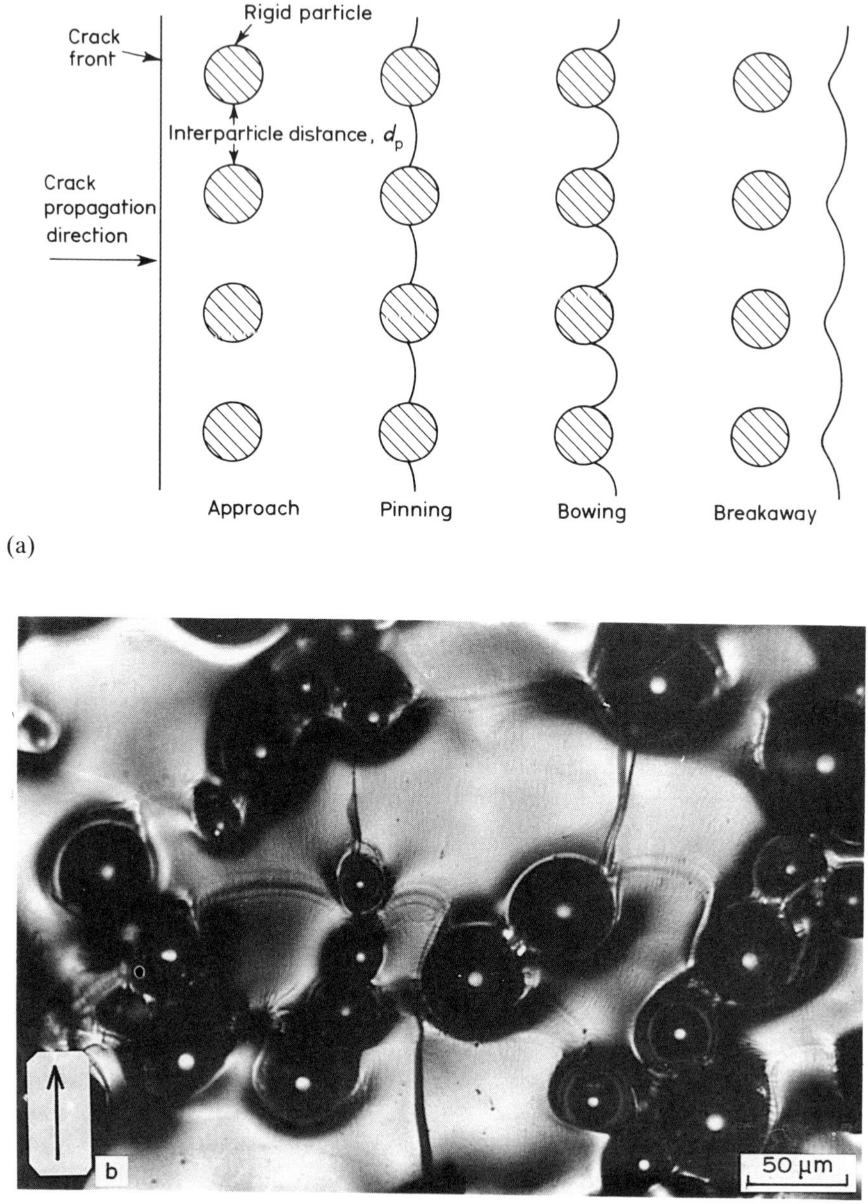

Figure 7.22 (a) Schematic representation of the crack pinning mechanism for increasing the toughness when rigid particles are present in a brittle adhesive. (b) Optical micrograph of the fracture surface of a glass-particulate filled epoxy showing the crack front pinned between glass particles. Arrow indicates direction of crack growth [146,147].

of the adhesives technologist is to be able to formulate the adhesive so that several of the above micromechanisms operate simultaneously, and encompass a large volume of material ahead of the crack tip, and also to ensure that other properties of the adhesive are not adversely effected to any significant degree.

7.6.3.3 Types of crack growth

It has been shown [108,152] that three distinct types of crack propagation may be observed in rigid adhesives. These are illustrated schematically as a function of test temperature in Fig. 7.23 and are:

(i) *Stable, brittle propagation (termed type C)* – here the crack grows in a steady controlled manner with the rate of crack propagation being dependent upon the crosshead speed of the testing machine. The fracture surfaces are relatively featureless and, since the cracks grow at low levels of G_c or K_c, stable brittle propagation may be thought of as a classic example of brittle fracture. Indeed, it is associated with the adhesive possessing a high yield stress, and hence a minimum of crack-tip plasticity occurs which results in a relatively sharp crack. The exact shape of the load versus displacement curve is somewhat dependent upon the detailed specimen geometry but the example shown for a TDCB specimen (where $\partial C/\partial a$ versus a is constant) illustrates the main features of steady crack growth.

(ii) *Unstable brittle propagation (termed type B)* – this type of crack growth is still essentially brittle in nature but the crack propagates intermittently in a stick–slip manner. This type of crack growth has a significant effect on the associated load versus displacement curve which now has a characteristic saw-tooth shape. The values of F_{ci} and F_{ca} correspond to the loads at crack initiation and arrest, respectively. The crack initiation and arrest values of G_c and K_c may be determined from the values of F_{ci} and F_{ca} and, on the fracture surface, lines corresponding to a jump/arrest event are clearly visible. This type of crack growth is associated with an increase in crack-tip plasticity compared with type C and is, therefore, typically observed at a somewhat higher test temperature (or lower test rate) when the yield stress of the adhesive is lower in value. These conditions lead to a greater degree of crack-tip blunting and hence a higher value of G_c or K_c. However, when the crack eventually propagates the rate of release of energy will be greater than that required for a stable crack, so that the crack will rapidly accelerate and unstable crack growth results. An analogous explanation is applicable to the stick–slip crack growth observed in rubbery adhesives (Section 7.6.2).

(iii) *Stable ductile propagation (termed type A)* – again the crack grows steadily in a controlled manner and for the TDCB specimen the load for crack

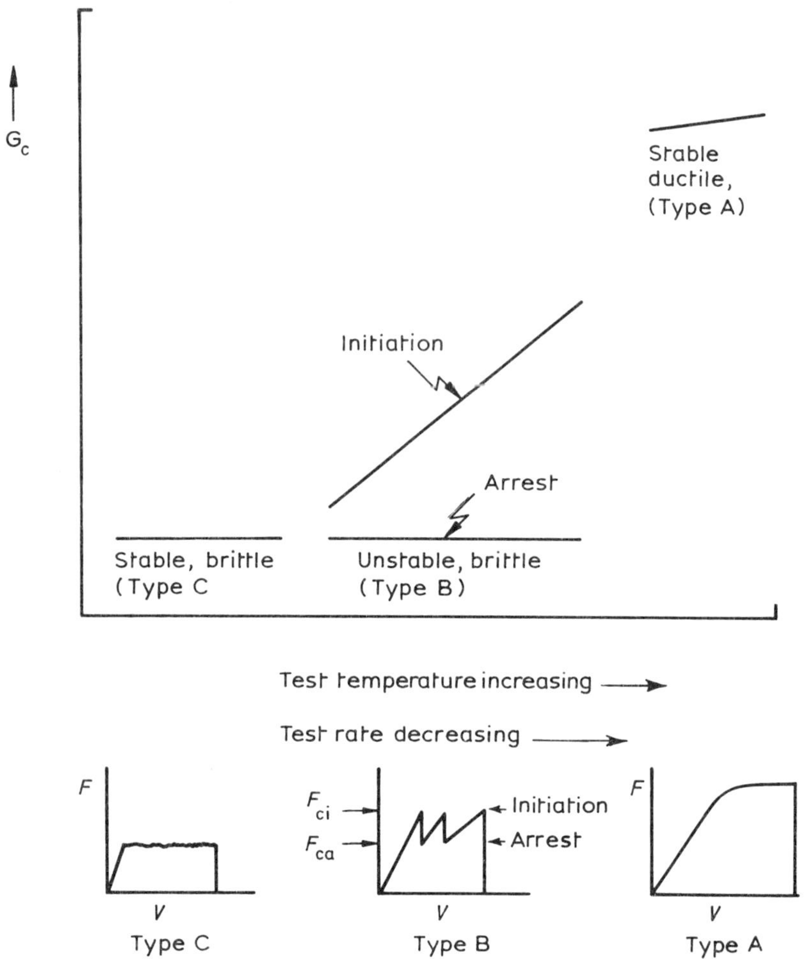

Figure 7.23 Schematic representation of the variation of the fracture energy, G_c, as a function of test temperature or test rate for structural adhesives. Typical load versus displacement traces are indicated for each type of crack growth for a test geometry where $\partial C/\partial a$ is independent of crack length, e.g. TDCB or double-torsion. (Note that for rubber-toughened adhesives the transition from type B to type A is less marked than for the more brittle, unmodified adhesives.)

propagation is constant. However, unlike stable brittle propagation discussed in (i), a relatively high value of G_c or K_c for propagation is now required and the fracture surfaces are far rougher and torn in appearance, indicating a far more ductile fracture process. Therefore, this type of crack growth is often observed at the highest test temperatures when the yield stress of the adhesive is relativly low: severe crack-tip plasticity and blunting result which give a ductile failure and stabilize the crack growth. Most recently, Kinloch *et al.* [153] have successfully quantitatively modelled this type of crack growth using a meniscus-instability model.

Whilst the above are the basic types of crack growth it should be noted that the exact transitions between each type is a strong function of adhesive formulation, test conditions and joint geometry. This obviously arises from the type of crack growth being controlled by the extent of crack-tip plasticity, which is dependent upon these factors. Further, in any test one may observe a mixture of the above types of crack propagation. For example, in rubber-toughened epoxies the crack may begin to grow in a stable ductile manner (type A) but at some point change to an unstable, brittle type of propagation (type B). The reasons for this are not currently fully understood.

7.6.3.4 Rate/temperature effects

Considering firstly the more simple, unmodified structural adhesives, the effect of test rate upon the measured adhesive fracture energy, G_{Ic}, and the type of crack growth is illustrated [108] in Fig. 7.24. As may be seen, one adhesive (DGEBA–TEPA) exhibits brittle stable crack growth over the complete range of test rates at room temperature whilst in the other the crack grows in a brittle, stable fashion at the higher rates but at slower rates propagates in an unstable brittle manner. In both cases the values of G_{Ic} are relatively low. These observations arise from the crack-tip micromechanisms. In both materials the extent of shear-yield deformations at the crack tip is relatively limited, so that the values of G_{Ic} are relatively low. However, the yield stress of the DGEBA–TEPA adhesive is relatively high under the complete range of test conditions so that the crack tip is comparatively sharp and crack propagation occurs in a stable brittle manner. The yield stress of the DGEBA–TETA adhesive is somewhat lower and decreases in value as the strain rate at the crack tip decreases, i.e. as the test rate is lowered. Thus, at room temperature with this adhesive, the crack propagates in a stable brittle fashion at the fastest test rates but at the slowest rates the crack tip is somewhat more blunt which leads to unstable brittle crack propagation; hence both initiation and arrest values are indicated on Fig. 7.24b. The effect of test temperature on these types of adhesives follows a similar pattern. At low temperatures stable brittle crack growth may be observed and as the temperature is raised unstable brittle cracking may be recorded. At the highest test temperatures there may be a relatively sudden increase in G_{Ic} as stable ductile crack growth occurs, although at the relatively high temperature needed to give this brittle/ductile transition the assumptions of linear-elastic fracture mechanics may no longer be valid. It should be noted that an artefact of the test technique that may arise is that at very low temperatures, i.e. below about −180 °C, there may be an apparent increase in G_{Ic} due to plasticization of the adhesive by the cryogen employed.

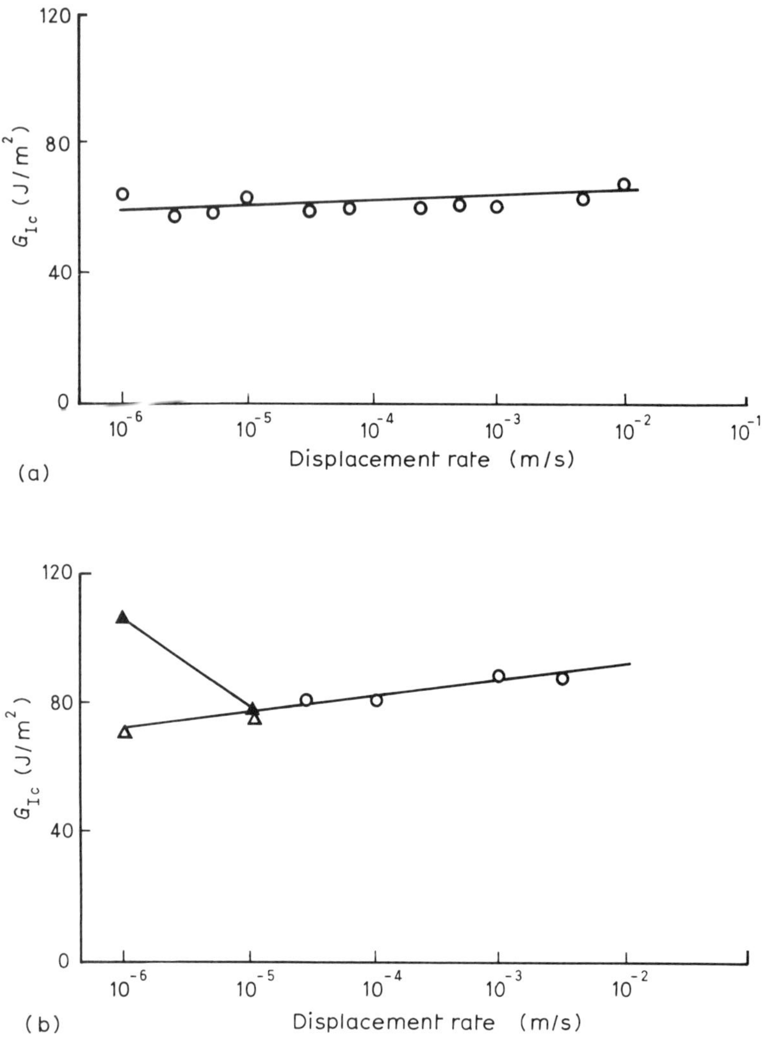

Figure 7.24 Effect of crosshead displacement rate on the measured fracture energy, G_{Ic}, and type of crack growth [108]. (a) DGEBA-TEPA adhesive; (b) DGEBA-TETA adhesive. ○ stable brittle crack growth (Type C); ▲ initiation and △ arrest of unstable crack growth (Type B). (Test temperature = 22 °C; TDCB joints; aluminium alloy substrates.)

Considering next the toughened structural adhesives, then a more complex picture emerges since the effect of the 'toughening additives' is to invoke crack-tip micromechanisms which increase the extent of energy dissipation. Thus, the energy-dissipation zone at the crack tip is larger, which increases the value of G_{Ic} or K_{Ic}, but this zone may now impinge against and be effected by, the substrates. As discussed in Section 7.5.3, if the extent of this zone is relatively large, then the measured value of the adhesive fracture energy, G_{Ic}, is dependent upon the test geometry and test conditions in a complex manner. This is illustrated in Fig. 7.25 where the adhesive fracture energy, G_{Ic}, is shown [154] as a function of temperature for two different rates of test. A TDCB joint geometry was employed consisting of mild steel substrates bonded with a rubber-toughened epoxy.

For the faster rate (Fig. 7.25a) the adhesive fracture energy, G_{IC}, increases steadily with increasing temperature and the bulk and the joint values are similar. The increasing degree of plastic deformation arising from multiple shear yielding and particle cavitation, and the associated blunting of the crack tip, may be seen from calculating the plastic-zone diameter, $2r''_{yc}$, from:

$$2r''_{yc} = \frac{1}{\pi} \frac{E_a G_{Ic}}{\sigma_{ay}^2} \tag{7.61}$$

Values of $2r''_{yc}$ are given as a function of temperature in Fig. 7.25a and, since the adhesive layer thickness, h_a, employed was about 0.7 mm, then over the complete temperature range, $2r''_{yc} < h_a$. Hence, as expected, the joint and bulk values of G_{Ic} are in good agreement. The type of crack growth observed was typically unstable brittle at the lower temperatures but with an increasing amount of stable ductile growth occurring as the test temperature was increased (i.e. type A → B).

However, when tested at the slower rate (Fig. 7.24b) the results are more complex. At temperatures below about 20 °C the bulk and the joint values are again in good agreement but above about 20 °C the joint values decrease. This may be understood by comparing the values of $2r''_{yc}$ and h_a. The value of $2r''_{yc}$ increases with increasing temperature until, at about 20 °C it is approximately equal to the adhesive bond thickness, h_a. The plastic zone size in the bulk specimen may, of course, grow still larger at higher temperatures and so the value of G_{Ic} (bulk) increases at higher temperatures. Following the discussions in Section 7.5.3.1, the decline in G_{Ic} (joint) values above 20 °C may be explained by the high yield stress substrates restricting any further growth of the plastic zone at the crack tip in the adhesive layer. Indeed, if above 20 °C the value of $2r''_{yc}$ in the adhesive layer is assumed to be constant and equal to the adhesive layer thickness, h_a, then G_{Ic} (joint) values may be deduced by recasting Equation 7.61 to:

$$G_{Ic}\,(\text{joint}) = \frac{2\pi r''_{yc} \sigma_{ay}^2}{E_a} \tag{7.62}$$

Values of G_{Ic} (joint) so calculated are shown in Fig. 7.25b and are in

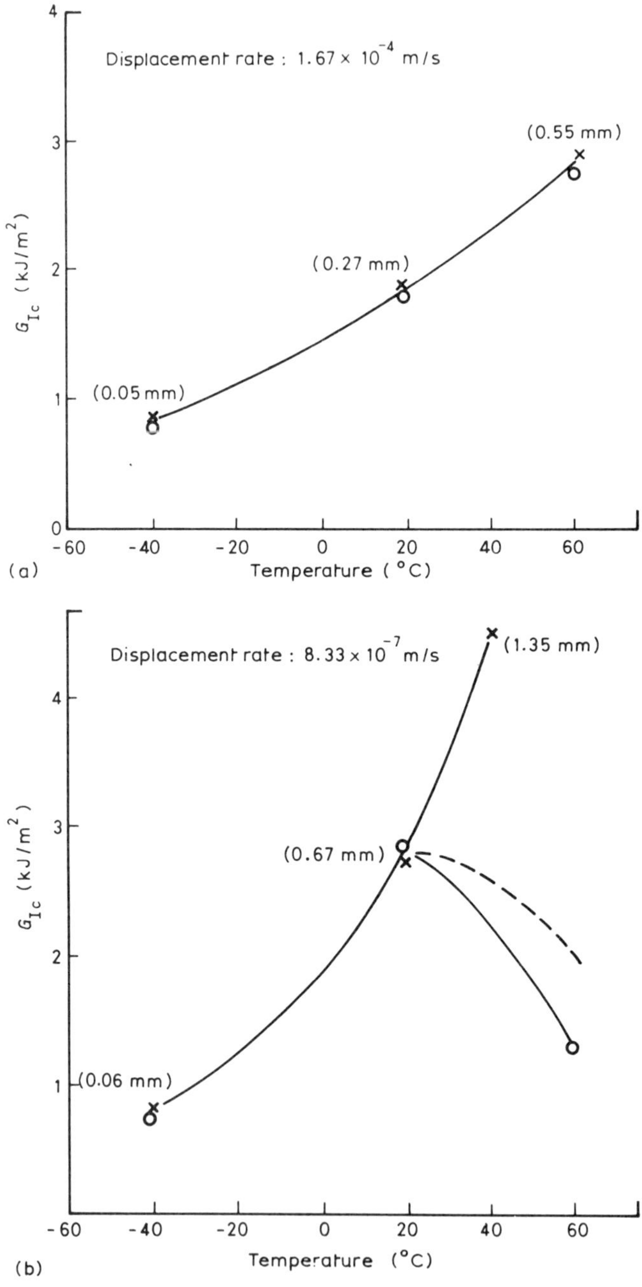

Figure 7.25 Effect of temperature on the fracture energy, G_{Ic}, for a toughened epoxy adhesive bonding steel substrates using a TDCB specimen [154]. ×, Bulk behaviour of adhesive; ○, adhesive joint behaviour; ⊃, predicted joint behaviour, from Equation 7.62. The plastic zone size, $2r''_{yc}$, calculated from Equation 7.61 is given in parentheses.

reasonable agreement with the measured values. This result demonstrates that G_{Ic} (joint) will also pass through a maximum for the test conducted at the faster rate (Fig. 7.25a) but will do so at a temperature exceeding 60 °C, i.e. when $2r''_{\mathrm{yc}} = h_{\mathrm{a}}$. Indeed, the interrelationship between temperature and rate has recently been emphasized by Hunston *et al.* [155]. These workers have developed a time–temperature equivalence argument for obtaining a master-curve of fracture energy, G_{Ic} (bulk) versus the effective rate of test for epoxy polymers tested below their glass transition temperature. A WLF-type equation was found to be inappropriate but the shift factors, a_{T}, followed an Arrhenius relationship over most of the range of test temperatures. Obviously, a similar time–temperature equivalence could be applied to the fracture of structural adhesive joints but would be more complex when the fracture behaviour is modified by the test geometry. Finally a comment about the type of crack growth which is observed. Essentially, the general pattern is the same as described in Fig. 7.23 but when the development of the plastic zone is restricted by the substrates, i.e. at slow test rates/high test temperatures/low h_{a} values, then there is a greater propensity for the crack to grow to a stable ductile manner compared to crack propagation in a bulk specimen.

Recently there has been considerable interest in the impact behaviour of structural adhesives and adhesive joints, largely generated by the increasing use of adhesives in the production of vehicles. Kinloch, Kodokian and Jamarani [163,164] have conducted instrumented impact tests on structural adhesives and adopted a fracture mechanics approach. Their work illustrates several interesting aspects. For example, it has been shown that such an approach may indeed result in 'material' characterizing parameters. However, impact tests are often dominated by dynamic effects and such effects must be taken into account if true 'material' characterizing parameters are to be determined. Nevertheless, when such effects are considered, it was found that the toughening micromechanisms that operate in the rubber-modified adhesives still do so even under impact conditions, and even at low test temperatures. However, as might be expected from the above discussions they are not usually as effective as under slow rates of test.

7.6.3.5 *Failure criteria*

It is evident from the above discussions that the parameters G_{Ic} and K_{Ic} do not provide a unique failure criterion but are usually functions of rate and temperature and, in some cases, also of joint geometry. However, a more general criterion may be formulated by considering the stress field at the tip of a crack. It has been shown [156] that, for a crack of tip radius, ρ_{t}, and length, a, the stress σ_{11}, normal to the axis of the crack at a small distance, r, ahead of the tip is given by:

$$\sigma_{11} = \frac{\sigma_{\mathrm{o}} a^{1/2}}{(2r)^{1/2}} \frac{(1+\rho_{\mathrm{t}}/r)}{(1+\rho_{\mathrm{t}}/2r)^{3/2}} \tag{7.63}$$

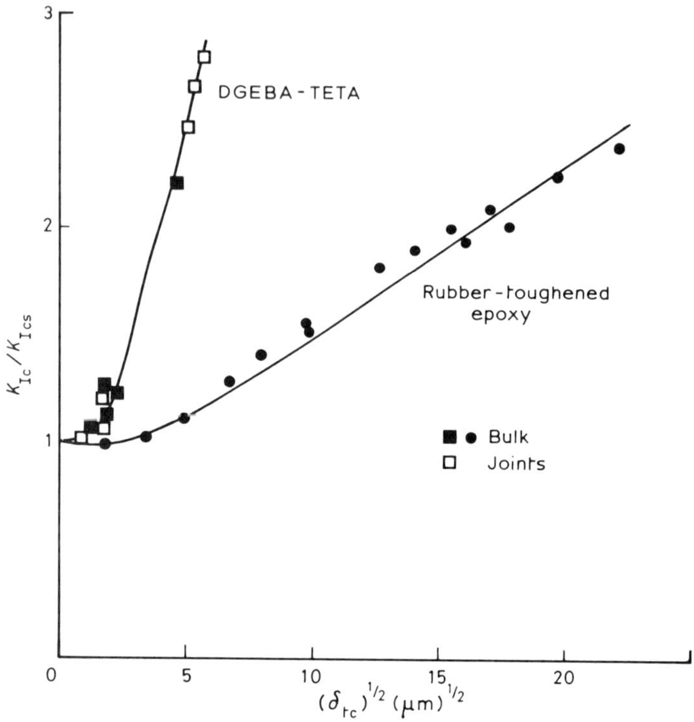

Figure 7.26 K_{Ic}/K_{Ics} ratio versus $\delta_{tc}^{1/2}$ for two epoxy compositions. The points are experimental results and the solid curves are theoretical results from Equation 7.67 [154].

where σ_0 is the applied stress. Now it has been postulated [157–159] that fracture occurs when a critical stress, σ_{tc}, is attained at a certain distance, r_{tc}, ahead of the crack tip; $\sigma_{11} = \sigma_{tc}$ and $r = r_{tc}$ and Equation (7.63) becomes:

$$\sigma_{tc} = \frac{\sigma_c a^{1/2}}{(2r_{tc})^{1/2}} \frac{(1+\rho_{tc}/r_{tc})}{(1+\rho_{tc}/2r_{tc})^{3/2}} \tag{7.64}$$

The measured stress intensity factor, K_{Ic}, at the onset of crack growth is related to the applied stress, σ_c, by (from Equation 7.18 since $Q = \pi^{1/2}$):

$$K_{Ic} = \sigma_c(\pi a)^{1/2} \tag{7.65}$$

and the propagation of a 'sharp' crack at a value of K_{Ics} may be interpreted as requiring the critical stress, σ_{tc}, to be attained at the distance, r_{tc}, such that:

$$K_{Ics} = \sigma_{tc}(2\pi r_{tc})^{1/2} \tag{7.66}$$

Thus, rearranging Equation 7.63 and substituting for Equations 7.65 and 7.66 yields:

$$\frac{K_{\mathrm{Ic}}}{K_{\mathrm{Ics}}} = \frac{(1+\rho_{\mathrm{tc}}/2r_{\mathrm{tc}})^{3/2}}{(1+\rho_{\mathrm{tc}}/r_{\mathrm{tc}})} \tag{7.67}$$

Finally Kinloch and colleagues [157,159] have shown experimentally that the value of ρ_{tc} is approximately the same as that for the crack opening displacement, δ_{tc}, and this is in agreement with the theoretical analysis of McMeeking [160]. Hence, the value of ρ_{tc} may be estimated from Equation 7.29.

The theory has been examined by measuring the ratio $K_{\mathrm{Ic}}/K_{\mathrm{Ics}}$ as a function of ρ_{tc}; where the values of K_{Ic} have been ascertained over a wide range of test rates and temperatures and the value of K_{Ics} has been taken to be the low temperature value, when the yield stress of the adhesive is high and the crack sharp. The results for two different adhesive compositions are shown in Fig. 7.26 and the theoretical lines have been fitted to the experimental points by selecting suitable values of the critical distance, r_{tc}, which is the only fitting parameter. The agreement between theory and experiment is good and values of the critical stress, σ_{tc}, and distance, r_{tc}, for various epoxy adhesives, obtained from bulk and adhesive joint specimens, are given in Table 7.6.

Thus, to summarize, the fracture data over a wide range of rates and temperatures may all be rationalized by the concept of a critical stress acting over a critical distance. The two parameters provide a unique failure criterion for the bulk fracture of adhesives and for the fracture of joints; although in the latter case it will be more complex when the development of the crack-tip deformation zone is affected by the presence of the substrates. Nevertheless, the concept of combining the critical stress and distance criterion with an analysis of the stress field in the adhesive layer as a function of joint geometry

Table 7.6 Values of critical stress, σ_{tc}, and critical distance, r_{tc}

Composition	Configuration	σ_{tc} (MPa)	r_{tc} (μm)	Reference
9.8 phr TETA	Bulk and joint	370	0.5	154,158
12.3 phr TETA	Bulk	300	1.1	158
14.7 phr TETA	Bulk	270	1.6	158
9.4 phr 3° amine	Bulk and joint	360	0.4	154,156
10 phr TEPA	Bulk and joint	495	0.1	154,156
5 phr piperdine	Bulk and joint	340	1.0	159,161
Rubber-modified epoxy	Bulk	200	10	159
Hybrid-particulate epoxy	Bulk	350	1.6	147

(a) Epoxy was a DGEBA resin.
(b) Hardeners: TETA, triethylenetetramine; TEPA, tetraethylenepentamine.

(see Fig. 7.14, for example) would appear to be a promising approach for extending the applicability of the proposed fracture criterion.

7.7 CONCLUDING REMARKS

This chapter has described the basic principles of the continuum fracture mechanics approach to the failure of adhesive joints. It is evident that from using such an approach it is possible to establish the interrelation between the measured fracture energy, the intrinsic fracture energy that is necessary solely for the rupture of intrinsic adhesion forces and the energy dissipated visco-elastically and plastically at the crack tip. From identifying such relationships, many of the observations concerning the effects of test rate and temperature and of joint geometry upon the measured fracture energy, G_c, or fracture toughness, K_c, may be rationalized and more fully understood. Also, in many circumstances, it is evident that values of G_c and K_c do provide unequivocal measures of the crack resistance of the material or joint and so enable quantitative joint design studies and service-life predictions to be undertaken. Hence, from the molecular level to the requirements of in-service quality assurance, the use of fracture mechanics has proved to be of assistance to the adhesives technologist.

However, it is also clear from the discussions in the present chapter that there are many areas in the application of fracture mechanics to the failure of adhesive joints which are far from fully understood. For example, the theoretical problems of dealing with cracks at, or near, bimaterial interfaces and the dependence of the measured values of G_c and K_c upon the joint geometry have yet to be completely understood and modelled. Thus, if the aims of predicting the short- and long-term behaviour of adhesive joints from the behaviour of the bulk adhesive material are to be realized, considerable work remains to be done.

The next, and final, chapter will consider the various types of loading conditions and environments which may adversely affect the service life of a bonded component. As will be seen, fracture mechanics has also made a most useful contribution to this important aspect of adhesives technology.

REFERENCES

1. Knott, J. F. (1974) *Fundamentals of Fracture Mechanics*, Butterworths, Guildford.
2. Broek, D. (1974) *Elementary Engineeering Fracture Mechanics*, Noordoff, Leyden.
3. Kinloch, A. J. and Young, R. J. (1983) *Fracture Behaviour of Polymers*, Applied Science Pub., London.
4. Williams, J. G. (1984) *Fracture Mechanics of Polymers*, Ellis Horwood, Chichester.
5. Griffith, A. A. (1920) *Phil. Trans. Roy. Soc.* **A221**, 163.

6. Orowan, E. (1948) *Rept. Prog. Phys.* **12**, 185.
7. Irwin, G. R. (1973) *Appl. Mater. Res.* **3**, 65.
8. Rivlin, R. S. and Thomas, A. G. (1953) *J. Polymer Sci.* **10**, 291.
9. Berry, J. P. (1961) *J. Polymer Sci.* **50**, 107.
10. Fenner, R. T., Kinloch, A. J., Thrusabanjong, E. and Williams, J. G. (1986) unpublished work.
11. Gent, A. N. and Kinloch, A. J. (1971) *J. Polymer Sci.* **A2**, **9**, 659.
12. Ahagon, A., Gent, A. N., Kim, H. J. and Kumagi, Y. (1975). *Rub. Chem. Technol.* **48**, 896.
13. Thomas, A. G. and Kadir, A. (1984) *J.Polymer Sci., Polymer Phys. Ed.* **22**, 1623.
14. Kinloch, A. J. and Tod, D. A. (1984) *Propellants Explosives Pyrotechnics* **9**, 48.
15. Andrews, E. H. (1974) *J. Mater. Sci.* **9**, 887.
16. Andrews, E. H. and Kinloch, A. J. (1973) *Proc. Roy. Soc.* **A332**, 385.
17. Andrews, E. H. and Kinloch, A. J. (1973) *Proc. Roy. Soc.* **A332**, 401.
18. Andrews, E. H. and Kinloch, A. J. (1974) *J. Polymer Sci., Symp.* **46**, 1.
19. Gent, A. N. and Schultz, J. (1972) *J. Adhesion* **3**, 281.
20. Westergaard, H. M. (1939) *J. Appl. Mech.* **A** June, 46.
21. Williams, M. L. (1959) *Bull. Seism. Soc. Am.* **49**, 199.
22. Erdogan, F. (1963) *J. Appl. Mech.* **30**, 232.
23. Rice, J. R. and Sih, G. C. (1964) *J. Appl. Mech.* **31**, 477.
24. Rice, J. R. and Sih, G. C. (1965) *J. Appl. Mech.* **32**, 418.
25. Malyshev, B. M. and Salganik, R. L. (1966) *Int. J. Fracture Mech.* **1**, 114.
26. England, A. H. (1965) *J. Appl. Mech.* **32**, 400.
27. Williams, M. L. (1984) *Adhesive Joints* (ed. K. L. Mittal), Plenum Press, New York, p. 703.
28. Comninou, M. (1977) *J. Appl. Mech.* **44**, 631.
29. Comninou, M. (1978) *J. Appl. Mech.* **45**, 287.
30. Comninou, M. and Dundurs, J. (1980) *Res. Mech.* **1**, 249.
31. Achenbach, J. D., Keer, L M., Khetan, R. P. and Chen, S. H. (1979) *J. Elasticity* **9**, 397.
32. Mak, A. F. and Keer, L. M. (1980) *J. Appl. Mech.* **47**, 816.
33. Williams, J. G. (1984) *Fracture Mechanics of Polymers*, Ellis Horwood, Chichester, p. 55.
34. Sawyer, S. G. and Anderson, R. B. (1972) *Eng. Fract. Mech.* **4**, 605.
35. Piva, A. and Viola, E. (1980) *Eng. Fract. Mech.* **13**, 143.
36. Wang, S. S., Mandell, J. F. and McGarry, F. J. (1978) *Int. J. Fract.* **14**, 39.
37. Sih, G. C. (1973) *Handbook of Stress-Intensity Factors for Researchers and Engineers*, Lehigh University, Bethlehem, USA.
38. Tada, H., Paris, P. and Irwin, G. R. (1973) *The Stress Analysis of Cracks Handbook*, Del Research Corporation, Hellertown, USA.
39. Rooke, D. P. and Cartwright, D.J. (1976) *Compendium of Stress-Intensity Factors*, HMSO, London.
40. Trantina, G. G. (1972) *J. Composite Mater.* **6**, 192.
41. Mai, Y. and Vipond, A. S. (1978) *J. Mater. Sci.* **13**, 2280.
42. Anandarajah, A. and Vardy, A. E. (1984) *J. Strain Anal.* **19**, 173.
43. Wu, E. W. and Thomas, R. L. (1965) *Proceedings of the 5th Congress on Rheology*, Vol. 1, University Park Press, Baltimore, p. 575.
44. Wang, T. T., Kwei, T. K. and Zupko, H. M. (1970) *Int. J. Fracture. Mech.* **6**, 127.
45. Arin, K. and Erdogan, F. (1971) *Int. J. Eng. Sci.* **9**, 213.
46. Arin, K. and Erdogan, F. (1972) *Int. J. Eng. Sci.* **10**, 115.
47. Cook, T. S. and Erdogan, F. (1972) *Int. J. Eng. Sci.* **10**, 677.

48. Erdogan, F. (1972) *Eng. Fract. Mech.* **4**, 811.
49. Erdogan, F. and Biricikoglu, V. (1973) *Int. J. Eng. Sci.* **11**, 745.
50. Lin, K. Y. and Mar, J. W. (1976) *Int. J. Fract.* **12**, 521.
51. Wang, S. S. and Yau, J. F. (1982) *Int. J. Fract.* **19**, 295.
52. Hong, C. C. and Stern, M. (1978) *J. Elasticity* **8**, 21.
53. Corten, H. T. (1972) *Fracture, An Advanced Treatise*, Vol. 7 (ed. H. Leiebowitz), Academic Press, New York, p. 675.
54. Erdogan, F. and Gupta, G. D. (1971) *Int. Sol. Struct.* **7**, 1089.
55. Lindley, P. B. (1971) *J. Inst. Rub. Ind.* **5**, 243.
56. Hata, T., Gamo, M. and Doi, Y. (1965) *Kobienshi Kagaku* **22**, 152.
57. Deryaguin, B. V. and Krotova, N. A. (1948) *Doklady Akad. Nauk. SSR* **61**, 849.
58. Kaelble, D. H. (1959) *Trans. Soc. Rheol.* **3**, 161.
59. Kaelble, D. H. (1960) *Trans. Soc. Rheol.* **4**, 45.
60. Kaelble, D. H. (1965) *Trans. Soc. Rheol.* **9**, 135.
61. Gent, A. N. and Hamed, G. R. (1975) *J. Adhesion* **7**, 91.
62. Gent, A. N. and Hamed, G. R. (1977) *J. Appl. Polymer Sci.* **21**, 2817.
63. Andrews, E. H. and King, N. E. (1976) *J. Mater. Sci.* **11**, 2004.
64. Danneberg, H. (1961) *J. Appl. Polymer Sci.* **5**, 125.
65. Williams, M. L. (1969) *J. Appl. Polymer Sci.* **13**, 29.
66. Williams, M. L. (1970) *J. Appl. Polymer Sci.* **14**, 1121.
67. Williams, M. L. (1973) *J. Adhesion* **5**, 81.
68. Bennett, S. J., DeVries, K. L. and Williams, M. L. (1974) *Int. J. Fract.* **10**, 33.
69. Anderson, G. P., DeVries, K. L. and Williams, M. L. (1974) *J. Colloid Interf. Sci.* **47**, 600.
70. Anderson, G. P., Bennett, S. J. and DeVries, K. L. (1977) *Analysis and Testing of Adhesive Bonds*, Academic Press, New York, p. 58.
71. Andrews, E. H. and Stevenson, A. (1978) *J. Mater. Sci.* **13**, 1680.
72. Bennett, S. J., Anderson, G. P. and Williams, M. L. (1970) *J. Appl. Polymer Sci.* **14**, 735.
73. Burton, J. D., Jones, W. B. and Williams, M. L. (1971) *Trans. Soc. Rheol.* **15**, 39.
74. Williams, M. L. (1972) *J. Adhesion* **3**, 1
75. Kunio, T. and Williams, M. L. (1970) *Proceedings of the International Symposium on the Science and Technology of Space*, Tokyo, Japan, p. 217.
76. Andrews, E. H., Majid, H. A. and Lockington, N. A. (1984) *J. Mater. Sci.* **19**, 73.
77. Johnson, K. L., Kendall, K. and Roberts, A. D. (1971) *Proc. Roy. Soc.* **A324**, 301.
78. Maugis, D. and Barquins, M. (1978) *J. Phys. D.* **11**, 1989.
79. Roberts, A. D. and Thomas, A. G. (1977) *Adhesion–1* (ed. K. W. Allen), Applied Science Pub., London, p. 141.
80. Fuller, K. N. G. and Tabor, D. (1975) *Proc. Roy.Soc.* **A345**, 327.
81. Mostovoy, S., Crosley, P. B. and Ripling, E. J. (1967) *J. Mater.* **2**, 661.
82. Mostovoy, S.and Ripling, E. J. (1966) *J. Appl. Polymer Sci.* **10**, 1351.
83. Crosley, P. B., Mostovoy, S. and Ripling, E. J. (1971) *Eng. Fract. Mech.* **3**, 421.
84. Mostovoy, S. and Ripling, E. J. (1975) *Fracture Characteristics of Adhesive Joints*, Final Report N00019-75-C-0271, Naval Air Systems Command, Washington, USA.
85. Brussat, T. R., Chin, S. T. and Ripling, E. J. (1977) *Fracture Mechanics for Structural Adhesive Bonds*, Report TR-77-163, Air Force Materials Lab., Dayton, Ohio, USA.
86. Mai, Y. (1975) *J. Adhesion* **7**, 141.
87. Francis, E. and Kinloch, A. J. (1986) Unpublished work.
88. Kinloch, A. J. and Young, R. J. (1983) *Fracture Behaviour of Polymers*, Applied Science Pub., London, p. 102.

89. Timoshenko, S. P. and Goodier, J. N. (1970) *Theory of Elasticity*, 3rd edn, McGraw-Hill, New York.
90. Andrews, E. H. and Stevenson, A. (1980) *J. Adhesion* **11**, 17.
91. Andrews, E. H. and Lockington, N. A. (1983) *J. Adhesion* **18**, 1455.
92. Parry, T. V. and Wronski, A. S. (1981) *Adhesion_5* (ed. K. W. Allen), Applied Science Pub., London, p. 1.
93. Chisholm, D. B. and Jones, D. L. (1977) *Exptl. Mech.* **17**, 7.
94. Vardy, A. E. and Anandaajah, A. (1984) *Proceedings of the 3rd International Conference on Numerical Methods in Fracture Mechanics*, University College, Swansea, p. 681.
95. Bascom, W. D., Cottington, R. L. and Timmons, C. O. (1977) *Appl. Polymer Symp.* **32**, 165.
96. Ripling, E. J., Mostovoy, S. and Patrick, R. L. (1963) *ASTM, STP* **360**, 5.
97. Trantina, G. G. (1972) *J. Comp. Mater.* **6**, 371.
98. Bascom, W. D., Timmons, C. O. and Jones, R. L. (1975) *J. Mater. Sci.* **10**, 1037.
99. Bascom, W. D. and Oroshnik, J. (1978) *J. Mater. Sci.* **13**, 1411.
100. Johnson, W. S. and Mall, S. (1985) *ASTM, STP* **876**, 189.
101. Dattaguru, B. Everett, R. A., Whitcomb, J. D. and Johnson, W. S. (1984) *J. Eng. Mater. Technol.* **106**, 59.
102. Ripling, E. J., Santner, J. S. and Crosley, P. B. (1983) *J. Mater. Sci.* **18**, 2274.
103. Kinloch, A. J. (1972) *Mechanics and Mechanisms of Adhesive Joint Fracture*, University of London, Queen Mary College.
104. Gledhill, R. A. and Kinloch, A. J. (1979) *Polymer Eng. Sci.* **19**, 82.
105. Mostovoy, S. and Ripling, E. J. (1969) *J.Appl. Polymer Sci.* **13**, 1083.
106. Kinloch, A. J. and Shaw, S. J. (1981) *J. Adhesion* **12**, 59.
107. Ebewele, R., River, B. and Koutsky, J. (1979) *Wood and Fibre* **11**, 197.
108. Gledhill, R. A., Kinloch, A. J., Yamini, S. and Young, R. J. (1978) *Polymer* **19**, 574.
109. Mostovoy, S. and Ripling, E. J. (1971) *J. Appl. Polymer Sci.* **15**, 611.
110. Bascom, W. D., Cottington, R. L., Jones, R. L. and Peyser, P. (1975) *J. Appl. Polymer Sci.* **19**, 2545.
111. Bascom, W. D. and Cottington, R. L. (1976) *J. Adhesion* **7**, 333.
112. Gent, A. N. and Hamed, G. R. (1977) *Polymer Eng. Sci.* **17**, 462.
113. Gent, A. N. and Hamed, G. R. (1978) *Plast. Rub. Proc.* **3**, 17.
114. Gent, A. N. (1981) *Int. J. Adhesion Adhesives* **1**, 175.
115. Gardon, J. L. (1963) *J. Appl. Polymer Sci.* **7**, 643.
116. Johnston, J. (1968) *Adhesives Age* **11**(4), 20.
117. Aubrey, D. W., Welding, G. N. and Wong, T. (1969) *J. Appl. Polymer Sci.* **13**, 2193.
118. Igarashi, T. (1975) *J. Polymer Sci., Polymer Phys. Ed.* **13**, 2129.
119. Yamamoto, Y., Yamakawa, S. and Tsuru, S. (1980) *J. Polymer Sci. Polymer Phys. Ed.* **18**, 1847.
120. Satas, D. and Egan, F. (1966) *Adhesives Age* **9**(8), 22.
121. Lake, G. J. and Stevenson, A. (1981) *J. Adhesion* **12**, 13.
122. Aubrey, D. W., Jackson, T. A. and Smith, J. D. (1969) *J. Inst. Rub. Indust.* **3**, 265.
123. Kendall, K. (1978) *J. Phys. D. Appl. Phys.* **11**, 1519.
124. Kaelble, D. H. and Ho, C. L. (1974) *Trans. Soc. Rheol.* **18**, 219.
125. Mulville, D. R., Hunston, D. L. and Mast, P. W. (1978) *J. Eng. Mater. Technol* **100**, 25.
126. Jonath, A. D. (1980) *Adhesion and Adsorption of Polymers* (ed. L. H. Lee), Plenum Press, New York, p. 175.
127. Bascom, W. D., Jones, R. L. and Timmons, C. O. (1975) *Adhesion Science and Technology*, Vol. 9B (ed. L. H. Lee), Plenum Press, New York, p. 501.

128. Williams, J. G. (1984) *Fracture Mechanics of Polymers*, Ellis Horwood, Chichester, p. 75.
129. Mall, S. and Kochhar, N. K. (1986) *Structural Adhesives in Engineering*, Institute of Mechanical Engineering, London, p. 71.
130. Brown, D. J., Windle, A. H., Gilbert, D. G. and Beaumont, P. W. R. (1986) *J. Mater. Sci.* **21**, 314.
131. Spencer, B. and Barnby, J. T. (1976) *J. Mater. Sci.* **11**, 83.
132. Jurf, R. A. and Pipes, R. B. (1982) *J. Comp. Mater.* **16**, 386.
133. Russell, A. J. and Street, K. N. (1985) *ASTM, STP* **876**, 349.
134. Ahagon, A. and Gent, A. N. (1975) *J. Polymer Sci., Polymer Phys. Ed.* **13**, 1285.
135. Kinloch, A. J. and Young, R. J. (1983) *Fracture Behaviour of Polymers*, Applied Science Pub., London, p. 376.
136. Kaelble, D. H. (1964) *J. Colloid Sci.* **19**, 413.
137. Kaelble, D. H. (1969) *J. Adhesion* **1**, 102.
138. Gent, A. N. and Petrich, R. P. (1969) *Proc. Roy. Soc.* **A310**, 433.
139. Aubrey, D. W. and Sherriff, M. (1980) *J. Polymer Sci., Polymer Chem. Ed.* **18**, 2597.
140. Aubrey, D. W. and Ginosatis, S. (1981) *J. Adhesion* **12**, 189.
141. Andrews, E. H., Khan, T. A. and Majid, H.A. (1985) *J.Mater. Sci.* **20**, 3621.
142. Kinloch, A. J. and Young, R. J. (1983) *Fracture Behaviour of Polymers*, Applied Science Pub., London, p. 43.
143. Beahan, P., Bevis, M. and Hull, D. (1975) *Proc. Roy. Soc.* **A343**, 525.
144. Kinloch, A. J. (1985) *Adv. Polymer Sci.* **72**, 45.
145. Kinloch, A. J. (1986) *Structural Adhesives: Developments in Resins and Primers* (ed. A. J. Kinloch), Applied Science Pub., London, p. 127.
146. Kinloch, A. J., Maxwell, D. A. and Young, R. J. (1985) *J. Mater. Sci. Letts.* **4**, 1276.
147. Kinloch, A. J., Maxwell, D. A. and Young, R. J. (1985) *J. Mater. Sci.* **20**, 4169.
148. Lange, F. F. (1970) *Phil. Mag.* **22**, 983.
149. Kinloch, A. J. and Young, R. J. (1983) *Fracture Behaviour of Polymers*, Applied Science Pub., London, p. 421.
150. Young, R. J. (1986) *Structural Adhesives: Developments in Resins and Primers* (ed. A.J. Kinloch), Applied Science Pub., London, p. 163.
151. Moloney, A. C., Kausch, H. H., Kaiser, T. and Beer, H. R. (1987) *J. Mater. Sci.* **22**, 381.
152. Kinloch, A. J., Shaw, S. J., Tod, D.A. and Hunston, D. L. (1983) *Polymer* **24**, 1341.
153. Kinloch, A. J., Gilbert, D. G. and Shaw, S. J. (1986) *J. Mater. Sci.* **21**, 1051.
154. Kinloch, A. J. and Shaw, S. J. (1981) *Developments in Adhesives–2* (ed. A. J. Kinloch), Applied Sci. Pub., London, p. 83.
155. Hunston, D. L., Kinloch, A. J., Shaw, S. J. and Wang, S. S. (1984) *Adhesive Joints* (ed. K. L. Mittal), Plenum Press, New York, p. 789.
156 Williams, J. G. (1980) *Stress Analysis of Polymers* 2nd edn, Ellis Horwood, Chichester, p. 350.
157. Kinloch, A. J. and Williams, J. G. (1980) *J. Mater. Sci.* **15**, 987.
158. Yamini, S. and Young, R.J. (1980) *J. Mater. Sci.* **15**, 1823.
159. Kinloch, A.J., Shaw, S. J. and Hunston, D. L. (1983) *Polymer* **24**, 1355.
160. McMeeking, R. M. (1977) *J. Mech. Phys. Sol.* **25**, 357.
161. Kinloch, A. J. (1985) Unpublished work.
162. Chai, H. (1986) *ASTM, STP* 893209.
163. Kinloch, A. J. and Kodokian, G. (1987) *J. Adhesion*, to be published.
164. Kinloch, A. J., Kodokian, G. and Jamarani, F. (1987) *J. Mater. Sci.* to be published.

CHAPTER EIGHT

The service life of adhesive joints

8.1 INTRODUCTION

Adhesive joints are frequently expected to perform satisfactorily under service conditions which include dynamically and statically applied loads and exposure to hostile environments such as water, organic solvents, etc; and, in many instances, combinations of these conditions may be experienced. It is therefore of prime importance for the adhesives technologist to be able to develop and recommend adhesive systems which will possess an adequate service life under the operating conditions which are to be experienced by the bonded structure. This, in turn, leads to the need to understand the mechanisms of failure and to develop test methods for (i) developing and selecting adhesive systems and (ii) predicting quantitatively the expected service life.

The present chapter is concerned with these aspects and the service life of adhesive joints subjected to dynamic and static fatigue will first be considered. As in previous chapters, if environmental attack is involved then such studies are discussed later when this aspect is addressed in detail. Indeed, the following section will then review the problem of environmental attack upon adhesive joints. This is one of the most important areas in adhesives technology and undoubtedly the one which gives the greatest concern to the design engineer. However, it will be shown how adhesive joints may withstand extremely hostile environments for very long periods of time, if due care and consideration is given to the choice of adhesive system and joint design.

8.2 FATIGUE

8.2.1 Dynamic fatigue

8.2.1.1 Introduction

Dynamic fatigue is the phenomenon of failure or fracture of a material, joint or structure under repeated or oscillatory loading. The importance of dynamic fatigue is that under fluctuating loads joints will fail at stress levels much lower than they can withstand under monotonic loading. Also, for a given alternating stress amplitude, they will fail in a much shorter time than the static fatigue time where a constant stress of the same magnitude has been applied. All types of material are liable to dynamic fatigue failure and it accounts for a great number of service failures. However, adhesives are generally regarded as possessing good dynamic fatigue properties compared to other methods of fastening such as rivets, spot welds and mechanical fasteners, and this is demonstrated by the data in Fig. 8.1, which is discussed below. This essentially arises from the adhesive giving a more even stress distribution in the joint compared to these other methods.

There is a large literature concerning the dynamic fatigue of polymeric

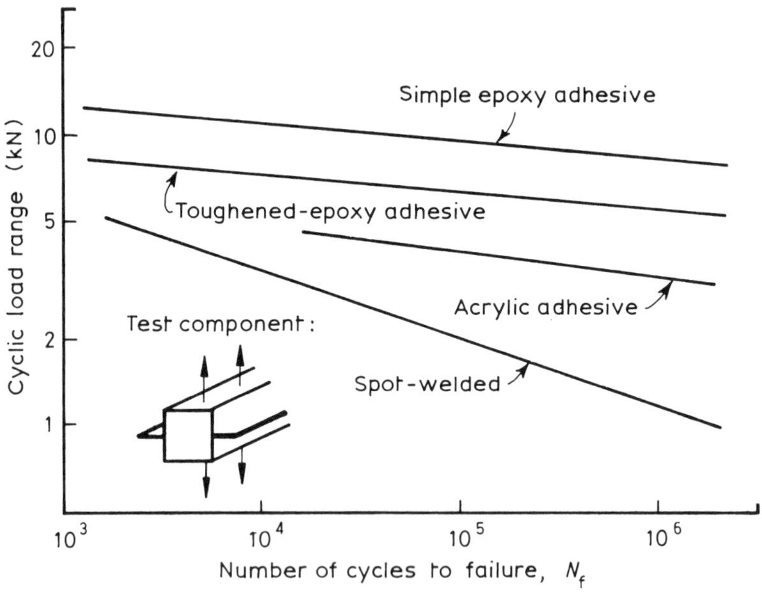

Figure 8.1 Dynamic fatigue properties of steel double-box hat structures either adhesively bonded or spot-welded [6].

materials and several reviews [1–3] have been published. Therefore, only the main features of the experimental and failure mechanisms will be covered in the present text before discussing the dynamic fatigue performance of adhesive joints. However, in this area there has only been a few detailed studies.

8.2.1.2 *Experimental considerations*

When designing dynamic fatigue tests the experimentalist has a wide range of variables to consider which contribute in different ways to the fatigue process. The following list is an extended version of that compiled by Andrews [4], namely:

1. A periodically varying load or stress system with a characteristic stress amplitude, σ_a [$\sigma_a = 1/2(\sigma_{max} - \sigma_{min})$];
2. A stress range, $\Delta\sigma$ ($\Delta\sigma = \sigma_{max} - \sigma_{min}$);
3. A corresponding fluctuating strain amplitude, ϵ_a;
4. A mean stress level, σ_m [$\sigma_m = 1/2(\sigma_{max} + \sigma_{min})$];
5. A mean strain level, ϵ_m;
6. A stress ratio, R [$R = \sigma_{min}/\sigma_{max}$];
7. A strain ratio, $\epsilon_{min}/\epsilon_{max}$;
8. A frequency, ω;
9. A characteristic wave-form (sinusoidal, square, etc.) for both the stress and strain;
10. The ambient and internal temperature of the specimen which in general will not be the same;
11. Environmental effects (see Section 8.3.5); and
12. The specimen geometry.

This list obviously presents the experimentalist with a formidable task if he wishes to study the fatigue behaviour in detail especially since, by their very nature, dynamic fatigue tests often occupy expensive equipment for long periods of time. Thus, complex service fatigue conditions are usually approximated by much simplified laboratory test programmes where many of the above parameters are held constant. This difficulty in reproducing expected service conditions emphasizes the importance of understanding the mechanisms underlying fatigue fracture. If they can be identified then the main parameters affecting the service life of bonded components are unlikely to be neglected when designing laboratory tests for the development and selection of adhesive systems and for life-prediction studies.

A number of different loading methods and specimen configurations have been devised and reported in the literature. For example, both the axially loaded butt joint [5] and the single overlap-shear joint loaded in tension [6] have been employed. These test geometries were described in Chapter 6 and the latter is used in the ASTM standard test for assessing the fatigue strength of adhesive joints: ASTM D3166-73, see Table 6.1. However, for the reasons discussed in Chapter 6, the results from these test geometries will be very dependent upon the exact geometry employed. Indeed, Hart-Smith [7] has commented that the problems associated with the stress concentrations in the single lap-shear joint may lead to some strange observations when this joint geometry is used to test adhesives under dynamic fatigue loading. For example, he has reported that such joints prepared with a commercial rubber-toughened epoxy adhesive may survive more than 10^7 cycles at a high frequency of 30 Hz but at a low frequency of about 10^{-4} Hz the joint might fail in only a few hundred cycles. The reason for this is to be found in the discussions in Section 6.4.4. The stresses and strains throughout the bonded overlap length of the typical ASTM single overlap-shear specimen, which uses a relatively short (typically 12.5 mm) overlap, are high. Hence, in the low-frequency tests the adhesive tends to creep quite markedly, not only near the ends but throughout the overlap. This progressive increase in the strain in the adhesive, due to the creep loads, results in joint fracture after a relatively few cycles. However, when tested at the high frequency, the load is always being removed before the adhesive has time to creep and the accumulated creep strain is low and the joint survives for a long time. As an aside, it is of interest to note that the actual bonded components were designed with a much longer length of bonded overlap, so that, even under high applied loads, the stresses in the adhesive in the central region of the overlap were low in magnitude and elastic in nature. Thus, even at low frequencies only the adhesive near the ends of the overlap will suffer creep, and when unloaded the elastic strains in the central region of the adhesive lead to the adhesive largely recovering any creep strains that may have been incurred. Therefore, the correct design of the bonded component may prevent the accumulation of creep deformation and the service life of the bonded component under dynamic loads may be good, whether a high or low test frequency is employed.

To overcome these problems of interpreting the results from such test geometries several different approaches have been adopted. Firstly, several workers [8–10] have used the thick-adherend shear-test specimen (Section 6.4.2) to obtain dynamic fatigue data which are more characteristic of the actual adhesive employed. Such data may, therefore, be used to rank various adhesive formulations and, possibly, to predict service life.

Secondly, several groups [6,11] have tested either sub-scale or full-scale bonded structures to determine their dynamic fatigue life. However, whilst testing a sub-scale or full-scale structure may give the designer confidence for a particular application it is often very expensive and does not generate basic design information for other different applications. An example of such data is shown in Fig. 8.1 and the test structures are double-box hat components which have been constructed as indicated with either one of several different types of structural adhesives or spot-welds. A typical format for describing the data is used; namely, the applied load (or stress) range is plotted against the number of cycles to failure, N_f. As may be seen, all the adhesively bonded structures perform better than the spot-welded structure. Also, with the adhesives studied, there is no significant advantage to be gained from the viewpoint of dynamic fatigue resistance by using the rubber-toughened epoxy compared to the simple epoxy adhesive. Indeed, generally with rubber-modified plastics there is no, or only at most a small, increase in the fatigue lifetime compared to unmodified materials. Finally, there is no indication of a value of the load range below which no signs of fatigue failure are detectable, i.e. a so-called endurance limit or threshold value; although such a limit might become apparent if longer term tests were to be conducted. Obviously, the presence of such a limit is of considerable use to the design engineer.

Thirdly, the concepts of continuum fracture mechanics (see Chapter 7) have been extended [8,12–14] to describe the rate of crack growth in a joint subjected to dynamic loading. The aims of adopting such an approach have been to gain an insight into both the kinetics and mechanisms of fatigue crack growth and to assess quantitatively the service life under fatigue loading. However, before describing the mechanical fatigue behaviour of adhesives the other main failure mechanism, thermal failure, will be briefly considered.

8.2.1.3 Thermal fatigue failure

A perfectly elastic material will remain at the temperature of the test environment throughout a dynamic fatigue test, since no energy is dissipated in the specimen. However, as discussed in Chapter 6, polymeric adhesives are viscoelastic and exhibit mechanical hysteresis even at moderate strains and at larger strains plastic flow is commonly observed. Thus, under cyclic loading some of this inelastic deformation energy will be dissipated as heat in each cycle. This will result in the adhesive's temperature rising until the heat generated per cycle is equal to the heat dissipated as a result of conduction, convection and radiation. The temperature may rise quickly to relatively high values, e.g. exceeding the glass transition or crystal melting temperatures, or stabilize at some particular value. Thus, the thermal fatigue mechanism may not necessarily involve fracture of the joint into separate pieces, but may occur due to excessive plastic deformation. The temperature rise and rate of increase will depend upon (i) the dimensions of the adhesive and the joint, (ii) the loss factor, thermal conductivity and specific heat of the adhesive and substrates, (iii) the test temperature, (iv) the applied frequency and (v) the stress and strain amplitudes. For example, the resulting temperature increase will rise with increases in the loss factor of the adhesive (hence, with respect to thermal fatigue effects, rubber-toughened adhesives may be inferior compared to more simple formulations) and with increases in the frequency and the maximum stress or strain experienced on each cycle. In connection with the last factor, neither deflection- nor strain-controlled tests tend to lead to runaway thermal failure, since, as the adhesive, or substrates, soften then the resulting stresses actually decrease. However, such runaway failures are far more likely in the more common stress-controlled dynamic tests since now as the adhesive, or substrates, soften the resulting strains steadily increase.

It should also be noted that even if any temperature rise upon dynamic fatigue is insufficient to cause thermal failure the properties of the adhesive, and possibly the substrates, will obviously be affected. Indeed, increases in the local temperature may cause blunting of the crack tip, due to the adhesive's yield stress being lowered, and so actually improve the mechanical fatigue resistance of the joint. Thus, thermal effects during dynamic fatigue are often important but frequently complex to interpret.

8.2.1.4 Mechanical fatigue failure

The other main mechanism of failure is mechanical fatigue which involves the initiation of a crack and its subsequent propagation. Most of the interest has focused upon the growth of a crack which has microscopic dimensions, since it is this phase of the mechanical fatigue mechanism which usually controls the fatigue life. In any event, by ignoring the crack initiation process, which is difficult to model, the calculated lifetime will be a conservative estimate, a situation favoured by the design engineer.

The most extensive study of using a fracture mechanics approach to describe fatigue crack growth in adhesive joints is that reported by Mostovoy and Ripling [12]. They employed TDCB specimens (see Table 7.1) which consisted of aluminium alloy substrates bonded with various structural adhesives, largely based upon epoxy resins. Some typical results obtained in a dry environment (r.h. 20%) are shown in Fig. 8.2. where the rate of crack growth, $\mathrm{d}a/\mathrm{d}N$, per cycle is plotted against the range of adhesive fracture energy, ΔG_{I}, that was imposed, i.e.:

$$\Delta G_{\mathrm{I}} = G_{\mathrm{Imax}} - G_{\mathrm{Imin}} \tag{8.1}$$

There are several points of interest. Firstly, as has been found for other polymers [3], over much of the range of the experimental data the crack growth rate may be expressed by:

$$\frac{\mathrm{d}a}{\mathrm{d}N} = A_{\mathrm{f}}\Delta G^{q} \tag{8.2}$$

where, in the present experiments, ΔG is equivalent to ΔG_{I}. The parameters A_{f} and q are constants but their values typically depend upon material variables, temperature, frequency, stress ratio and environment. The value of A_{f} varies

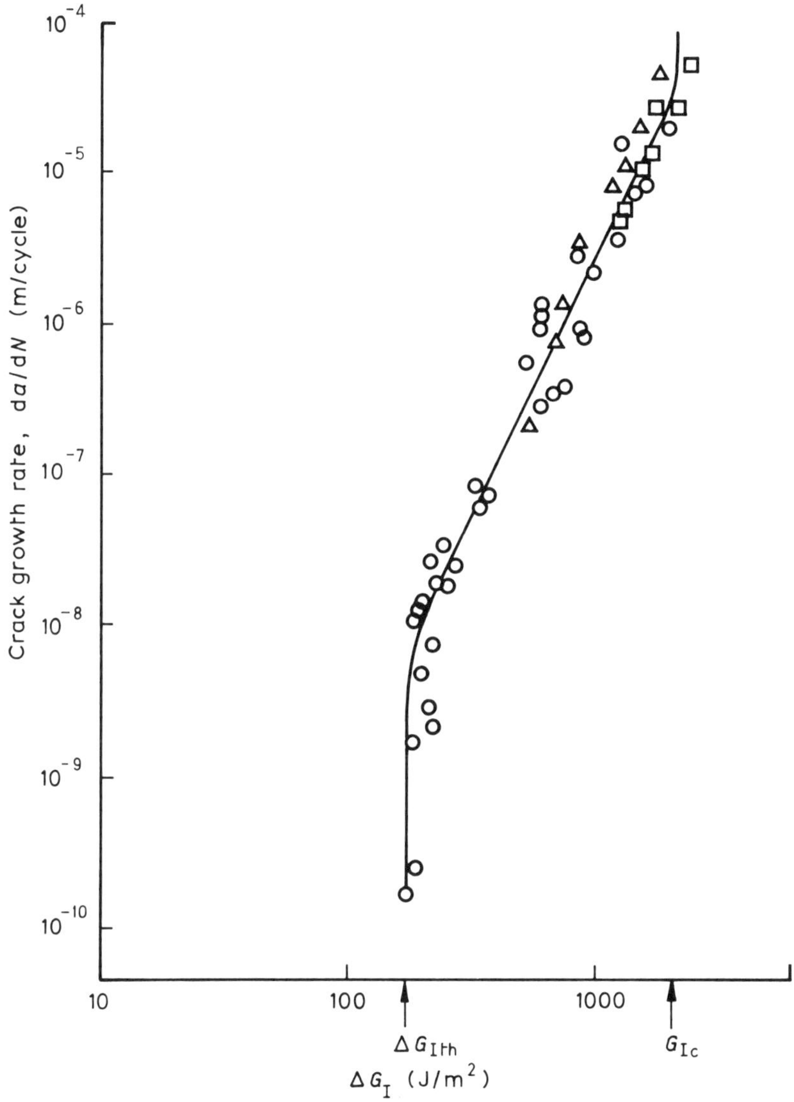

Figure 8.2 Crack growth rate, da/dN as a function of ΔG_I for an epoxy adhesive bonding aluminium alloy substrates for different test frequencies [12]. ○, 3 Hz; △, 0.25 Hz; □, 0.01 Hz.

between about 10^{-14} and 10^{-16} and q between 2 and 5 (da/dN is in m/cycle and ΔG_I in J/m^2). Secondly, however, the data shown in Fig. 8.2 and in other studies reveals that the relationship between da/dN and ΔG_I is actually sigmoidal in shape. Crack growth rates are found to decrease to very low values as ΔG_{I} approaches some limiting threshold value, ΔG_{Ith}, and to increase to very high values as G_{Imax} approaches the typical G_{Ic} value for crack growth under short-term monotonic loading conditions. Note that for the data shown in Fig. 8.2 the value of ΔG_{Ith} is only about 280 J/m^2 whilst the value of G_{Ic} is about 3500 J/m^2. For the seven different adhesives examined the ratio of $\Delta G_{\mathrm{Ith}}/G_{\mathrm{Ic}}$ varied from about 0.01 to 0.25, although in some cases it was difficult to define a true value of ΔG_{Ith}. This illustrates clearly the major effect that the application of an alternating, rather than a monotonic or static, load has upon the rate of crack growth in adhesive joints, and indeed in all materials. Thirdly, for the range of test frequencies used to obtain the results shown in Fig. 8.2, the frequency has no significant effect on the crack growth rate recorded at a particular ΔG_{I} value. However, in other instances, the crack growth rate has been found to decrease as the frequency is increased. This might arise from more localized heating occurring at the higher frequencies, since this will cause blunting of the tip of the advancing crack and so slow down the rate of crack advance.

Mall *et al.* [13] and Liechti and Lin [14] have examined the rate of crack growth under dynamic fatigue conditions when a mixed-mode adhesive fracture energy, $\Delta G_{\mathrm{I/II}}$, is applied. The former authors used the cracked lap-shear specimen (see Table 7.1) and prepared joints employing commercial structural adhesives to bond carbon fibre composite substrates. They reported that fatigue failure basically occurred by crack growth through the adhesive layer and that Equation 8.2 described the data successfully, with the value of ΔG being taken to be ΔG_{I}, ΔG_{II} or ΔG_{T} ($\Delta G_{\mathrm{T}} = \Delta G_{\mathrm{I}} + \Delta G_{\mathrm{II}}$). However, there were no indications of a threshold value, ΔG_{th}, being attained. Liechti and Lin [14] used four different geometries to study the mixed-mode cyclic fatigue in aluminium alloy/rubber-toughened epoxy joints. They concluded that the rate of crack growth in the adhesive layer is controlled by Equation 8.2 with ΔG being equivalent to ΔG_{T}, rather than some more complex, empirical combination of ΔG_{I} and ΔG_{II} as discussed in Section 7.5.4. Finally, there was no strong evidence of a threshold value, ΔG_{th}, below which dynamic fatigue crack growth would not occur although there was some indication that such a parameter might indeed exist and that the value of ΔG_{Tth} would be about 400 J/m^2, the value of G_{Ic} being approximately 3000 J/m^2.

The aspect of dynamic fatigue crack growth in adhesive joints is obviously of major importance but little work has been reported on gaining a fundamental understanding of the mechanisms involved, establishing design criteria and developing quantitative methods for estimating service life. For example, the nature of the damage which accumulates at the crack tip, and so causes the continual loading/unloading process to be so deleterious, has yet to be identified. Also, the effects of changing the various experimental parameters upon the crack growth rate equation are virtually unknown. Finally, it is possible to use data such as that shown in Fig. 8.2 to predict the long-term service life via the integration of Equation 8.2. This has been done successfully in the case of fatigue failure of rubbers [1,15] but has yet to be employed, and verified, in the case of adhesive joints.

8.2.1 Static fatigue

8.2.1.1 Introduction

Static fatigue, or creep rupture, is the phenomenon of fracture which occurs some time after the application of a constant load. The applied stress is usually lower than that required to cause fracture under monotonic loading conditions but, as shown later, there is no clear distinction between static fatigue and monotonic loading since in the latter the stress is continuously applied and a finite time must elapse between application of the stress and fracture. Again such experiments have been conducted in the presence of a hostile environment, but these will be considered later in Section 8.3.5.

8.2.1.2 Life-prediction studies

Only a few detailed studies have been reported on the static fatigue of adhesive joints and, even fewer, on modelling the data in order to predict the long-term behaviour of stressed joints from short-term data. Some workers [16–19] have concentrated on studying the single lap-shear joint whilst others [20,21] have used precracked specimens and adopted a fracture mechanics approach.

Lewis *et al.* [16,17] and Wake *et al.* [18,19] have examined in some detail the concept of an 'endurance limit' for static fatigue tests, i.e. a value of the applied stress below which joint failure will not occur. Such a concept is obviously of considerable benefit to the design engineer and is analogous to that discussed above for dynamic fatigue failure. Indeed, Wake *et al.* have considered this concept both for the case of simple static fatigue and for the case when the joint is first subjected to a number of load cycles and then the residual tensile fracture stress measured.

Considering the experimental and theoretical results that have been reported, Lewis and co-workers [16,17] have subjected single lap-shear joints, employing various structural adhesives bonding aluminium alloy, to a constant rate of loading and measured the joint strength and the time to failure, t_f. They progressively increased the loading rate and used an analysis first suggested by Prot to deduce an endurance limit. Essentially from their short-term tests they calculated that an endurance limit would be about 35 to 45% of the short-term joint strength values, but suggested that for design purposes a value of 25% would be more appropriate. These values appear, at first sight, to be very reasonable figures. (They would, of course, be lower if any hostile environment, such as high humidity, was also present.) However, Wake *et al.* [19] have recently pointed out that the assumptions made by Lewis *et al.* are fallacious and that any agreement between their calculated values and general design practice (i.e. to take 30 to 40% of the short-term strength and then have a safety factor of 2 to give a design value of 15 to 20%) is probably a fortunate coincidence.

Wake *et al.* [18] have found that for single lap-shear joints subjected to a static load a plot of the applied static stress versus log t_f is linear, and similar observations have been reported [22,23] for the bulk fracture of various polymeric materials. Following the theories of Zhurkov and Tomashevsky [22], such data may be modelled by an activated rate process, i.e. one involving the surmounting of a potential energy barrier where the activation energy is lowered by the presence of an applied stress. This theory can yield values of the activation energy for fracture and these values may be used to predict the long-term fracture behaviour under an applied load. This has been discussed [22–24] in depth elsewhere with reference to the fracture of polymers and,

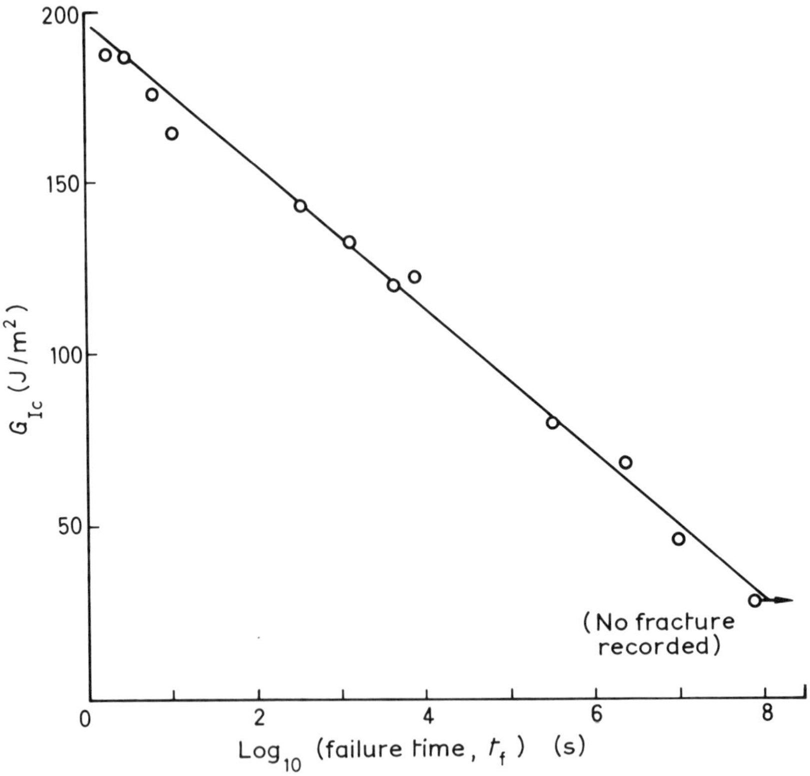

Figure 8.3 Relationship between the applied adhesive fracture energy, G_{Ic}, and the joint failure time, t_f [20].

since it has not been really developed as yet for adhesive joints, the interested reader is referred to these previous publications for further details. One interesting aspect of this theoretical analysis is that it does actually predict an endurance limit, but its value is extremely small and represents a negligible stress level from the engineering viewpoint. However, this does not prove that a significant value of the endurance limit stress does not exist, at least for some cases. On this subject the present author would agree with Wake *et al.* [19] who concluded that, from both theoretical and experimental considerations, whether an endurance limit really does exist or not has yet to be firmly established.

Gledhill and Kinloch [20] have used a fracture mechanics approach to study the static fatigue behaviour of joints which consisted of aluminium alloy substrates bonded with a simple epoxy adhesive. They employed a TDCB specimen which was precracked with a sharp crack and loaded joints to different levels and awaited events. Some of their results are shown in Fig. 8.3 and, as may be seen, they found a linear relationship existed between the applied adhesive fracture energy, G_{Ic}, and the log of the time to failure, t_f; the failure time decreased as the value of G_{Ic} increased. Note that there is no evidence that, within the time-scale investigated, there is a lower endurance limit to the value of G_{Ic} below which static fatigue failure is not observed. Now, the failure time represented an incubation period; the crack was not observed to propagate until the very end of the experiment, at the instant of fracture, when it propagated very rapidly. They suggested that these results could be interpreted and modelled by postulating a fracture criterion based upon the requirement of a critical crack-opening displacement, δ_{tc}. Hence, crack propagation does not occur whilst $\delta_t < \delta_{tc}$. By combining Equations 7.29 and 7.35 and using:

$$\sigma_{ay} = \epsilon_{ay} E_a \qquad (8.3)$$

they derived:

$$\delta_{tc} = \frac{1}{\epsilon_{ay}} \frac{G_{Ic}}{E_a} \qquad (8.4)$$

where G_{Ic} is the time-dependent applied fracture energy, E_a is the time-dependent modulus and ϵ_{ay} the yield strain which is approximately constant. Values of G_{Ic} and corresponding values of E_a are plotted in Fig. 8.4 and yield a linear relation over the entire time-scale of about seven decades over which the static fatigue experiments were conducted. Thus, as time passes, the value of σ_{ay} and E_a decrease because of the viscoelastic nature of the polymeric adhesive, and the crack opening displacement, δ_t, steadily increases and fracture occurs when $\delta_t = d_{tc}$. The value of δ_{tc} is about 4.5 μm and provides a unique failure criterion; until this critical value is attained no crack extension

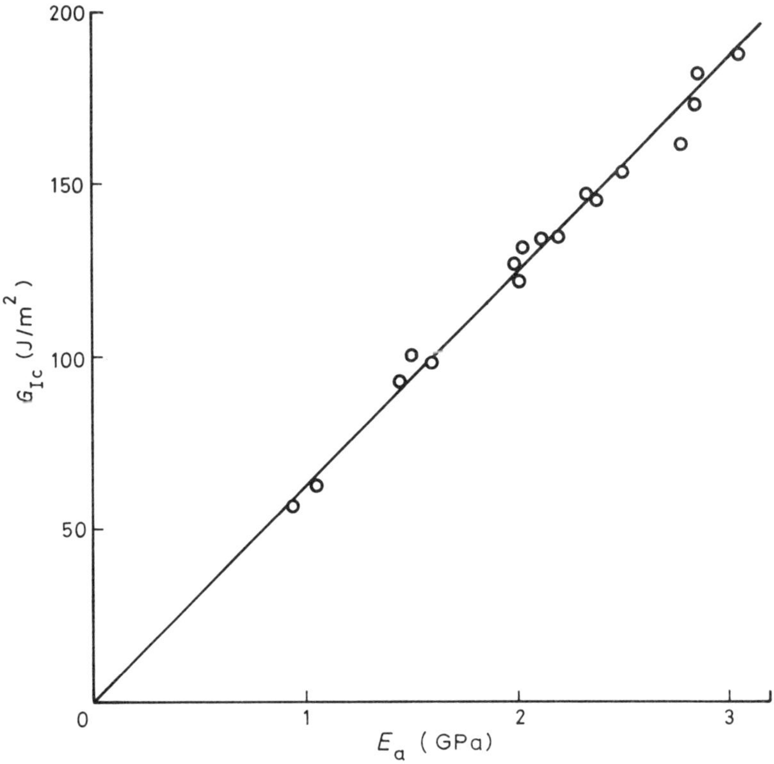

Figure 8.4 Applied adhesive fracture energy, G_{Ic}, as a function of the corresponding modulus, E_a, for the epoxy adhesive whose static fatigue behaviour is shown in Fig. 8.3 [20].

occurs. It is of interest to note that subsequent work by Kinloch *et al.* [21] found that in some adhesives the time-dependence of the yield stress was such that extensive crack-tip plasticity and blunting occurred prior to static fatigue fracture. This resulted in the stress concentration at the crack tip being reduced to such an extent that the tip stress level necessary for crack extension is not attained and thus this mechanism may assist greatly in delaying or even forestalling static fatigue failure.

8.3 ENVIRONMENTAL ATTACK

8.3.1 Introduction

One of the most important requirements of an adhesive joint is the ability to retain a significant proportion of its load-bearing capability for long periods under the wide variety of environmental conditions which are likely to be encountered during its service life. Unfortunately, for the reasons outlined later, one of the most hostile environments for joints involving high-surface energy substrates, e.g. metals, glasses and ceramics (see Section 2.4.2), is water and this is, of course, one of the most commonly encountered. Hence, in many applications which often involve some of the most critical uses of adhesives technology the joints are subjected to an environment which also happens to be one of the most potentially damaging. Indeed, it is probably true to state that the most important challenges facing the adhesives community are (i) to develop adhesives systems which possess excellent long-term durability, especially systems acceptable to the general engineering industries, and (ii) to develop test methods and models to predict the service life from short-term experiments, and thereby convince the potential user that an adequate durability will be realized.

The present section will discuss the aspects of environmental attack with reference to a wide range of environments although, because of its technological importance, the effect of moisture will be highlighted. Some general observation will be noted first before considering the mechanisms and kinetics by which such attack may occur. The effects of an applied stress on the mechanisms and kinetics will then be reviewed and, finally, the prediction of the service life from short-term data will be considered. Throughout this section, as in previous chapters, the importance of considering the joint as a 'system', i.e. the type of substrates, of surface pretreatment, of primer and of adhesive employed, will be emphasized. This is especially important when considering environmental attack, since all these factors may greatly effect the mechanical performance of the adhesive joint in a given environment.

Finally, it should be noted that the discussions will largely, although not exclusively, centre upon those environments/adhesive systems where the interfacial regions are attacked. This is not to say that the adhesive itself, or the substrates, may not be attacked by the ingressing liquid, and if this does occur then the joint may well be appreciably weakened. However, the general materials literature, including that specializing in adhesives chemistry (see Chapter 5), should enable the adhesives user to select an adhesive type which will itself be sufficiently durable and, also, to ensure that the substrate materials themselves will withstand the service environment. The more assidious problem, and the far more difficult to design against, is that of the environment attacking the interfacial regions, and it is this aspect on which the present discussions will therefore concentrate.

8.3.2 General observations

8.3.2.1 Mechanical performance and the service environment

In essence, it has been found that the mechanical properties of a bonded component may deteriorate upon exposure of the joint to its normal operating environment. Indeed, in some instances the loss of load-bearing capacity has

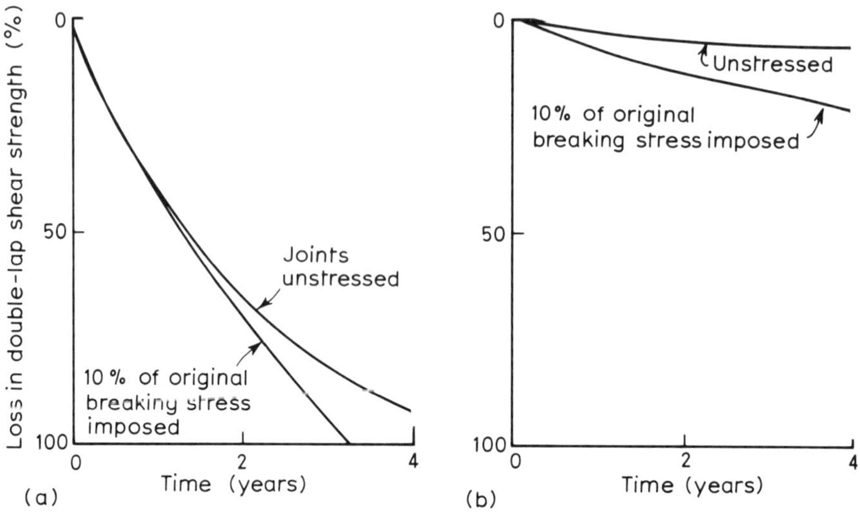

Figure 8.5 Effect of outdoor weathering on the strength of epoxy-polyamide/aluminium alloy (chromic acid etched pretreatment) structural joints [29]. (a) Hot/wet tropical site; (b) hot/dry desert site.

been so marked that premature joint failure has occurred at very low load levels and within a small fraction of the required service life.

For example, in the case of structural adhesive joints consisting of metallic substrates bonded with adhesives such as phenolic, epoxy or acrylic resins, then empirical laboratory investigations established [25–28] many years ago that water, either in the liquid or vapour form, is the most hostile environment that is commonly encountered and in natural outdoor climatic trials it is invariably the presence of moisture which is responsible for environmental attack rather than, say oxygen or ozone. Also, if aircraft applications are considered, then moisture usually presents more of a problem than de-icing fluids, hydraulic oil or kerosene. Further, the rate of environmental attack by water is greater if the temperature is relatively high or the joint is subjected to stress. These aspects are illustrated in Fig. 8.5 which imparts an appreciation of the extent of the problem, although it is one of the worst cases known to the author. It is evident that the hot/wet tropical environment was by far the more hostile and that the presence of an applied load increased the rate of loss of strength. Indeed, the joints exposed under this loading at the tropical site fell apart shortly after three years had elapsed.

8.3.2.2 Locus of joint failure

As discussed in previous chapters, the locus of joint failure of adhesive joints when initially prepared is usually by cohesive fracture in the adhesive layer, or possibly in the substrate materials. However, a classic symptom of environmental attack is that after such attack, the joints exhibit some degree of apparently interfacial failure between the adhesive (or primer) and the substrate. The extent of such apparently interfacial failure increases with time of exposure to the hostile environment. In some instances it is not accompanied by corrosion and the substrates appear clean, whilst in other instances the substrates may be heavily corroded. However, as will be shown later, first appearances may be deceptive. For example, whether the failure path is truly at the interface, or whether it is in the oxide layer, or in a boundary layer of the adhesive or primer (if present), is a matter of some debate and controversy. Also, the presence of corrosion on the failed surfaces does not necessarily imply that it was a key aspect in the mechanism of environmental attack. In many instances corrosion only occurs once the intrinsic adhesion forces at the adhesive/substrate interface or the oxide layer itself have been ruptured by the ingressing liquid: the substrate surface is now exposed and a liquid electrolyte is present so that post-failure corrosion of the substrate may occur.

Thus, a visual assessment of the fracture surfaces of a joint which has been subjected to environmental attack does not indicate exactly the locus of failure and techniques such as Auger and X-ray photoelectron spectroscopy are needed to identify precisely the locus of failure. However, it does indicate that we should concentrate our attention on the interfacial regions if we are to understand and prevent such environmental failure of adhesive joints.

8.3.2.3 Parameters affecting environmental resistance

The following parameters have been found to affect joint durability:

(a) Environment The presence of a hostile environment is the essence of the problem. In the case of thermoplastic adhesives then the adhesive itself may be appreciably weakened upon being exposed to a range of organic solvents, acids or bases. As commented above, the literature (see Chapter 5) on the chemistry of the particular polymeric adhesive should be consulted to ensure that this will not be a problem when selecting the adhesive. In the case of thermosetting adhesives, then most are hydrolytically stable in the presence of most organic solvents, water and moderate strengths of acids and bases. However, at relatively high temperatures, e.g. 70 to 100 °C, many of these types of adhesives will be susceptible to hydrolysis and show appreciable loss of mechanical properties as their molecular weight is decreased. In normal usage such temperatures are not very often attained and so this mechanism of attack is not commonly observed.

In the case of attack in the interfacial regions of the joint, then water is the most damaging environment, for the reasons discussed below, and the one on which the present review will concentrate. However, whether neutral water or with a pH below (acid) or above (basic) seven in value or a salt solution are the most damaging environments depends upon the exact adhesive system being investigated. For example, the rate of hydrolysis of interfacial primary bonds with some degree of ionic character will typically be a function of the pH of the environment. Also, if corrosion of the substrate is a primary mechanism then the use of a salt-spray environment may greatly increase the kinetics of the process. The most obvious advice that can be offered to the prospective investigator who wishes to undertake accelerated durability testing is to attempt to identify the mechanisms of failure and/or test in an environment which approaches that of the actual service condition.

(b) Concentration of water Essentially the higher the activity of the water present in the environment then the more rapid and greater is the degree of attack on the interfacial regions of the joint. Furthermore, there often appears to be a minimum concentration of water below which no environmental failure, or at least no significant attack over a comparatively long time-scale, occurs. Although this value is very dependent upon the adhesive system being employed. In practice in Northern Europe climates this critical concentraton of water translates to requiring a relative humidity above about 60 to 80% for attack to be observed. Thus, for most indoor applications, such as bonding office furniture, in Northern Europe the problem of environmental attack by water is minimal. However, if such bonded structures are to be exported to a hot/wet climate, e.g. the Far East, then attention must be given to selecting an adhesive system which has adequate durability.

These points are illustrated [30] in Fig. 8.6 where, as may be seen, the adhesive joints exposed to 55% r.h. at 20 °C do not lose any strength, unlike those immersed in water. Indeed, subsequent work revealed that even after about ten years the joints kept at 55% r.h. were as strong as the day that they were made and that appreciable weakening due to attack by ingressing moisture only occurred, with this particular adhesive system, above about 75% r.h.

(c) Temperature Increasing the temperature of the environment increases the rate of strength loss. Therefore, using a somewhat higher exposure temperature than may be experienced in the actual bonded structure is a typical method of accelerating the rate of attack in an accelerated laboratory test. However, as commented above, although many adhesives are themselves relatively resistant to hydrolysis at moderate temperatures, this is not always the case at elevated temperatures. For example, using some types of epoxy adhesives, when the joint is immersed in water at, say, 70 to 100 °C then hydrolysis and loss of bulk strength may well occur, particularly at long times of immersion. However, such exposure temperatures are not reached during normal outdoor aging and only rarely in certain applications. Therefore, unless the application demands that these temperatures be withstood, there is an obvious danger in using very high temperatures, e.g. a boiling water test, in laboratory aging trials in order to accelerate the rate of attack by moisture upon the joint, since, at such high

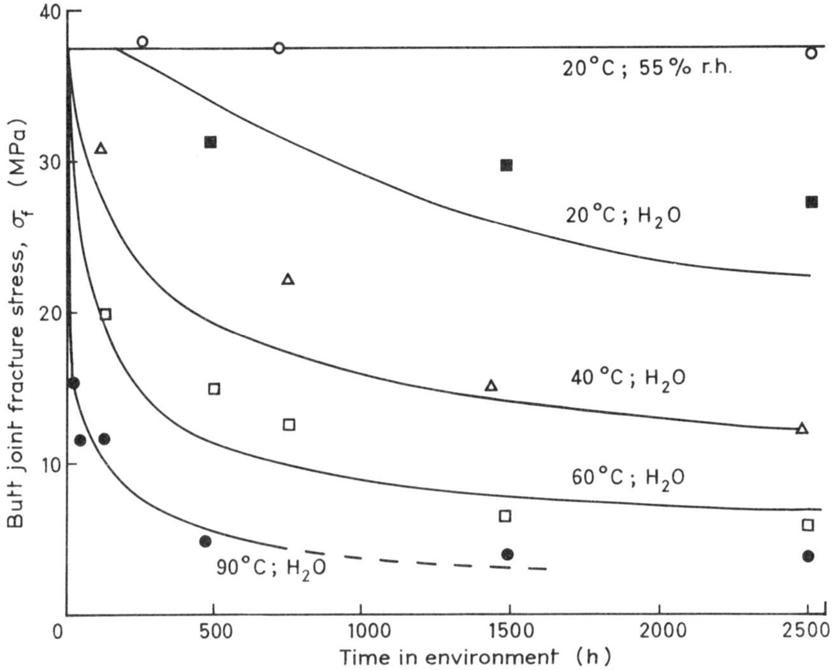

Figure 8.6 Fracture stress of epoxy/mild steel (abraded solvent-cleaned pretreatment) butt joints as a function of the time exposed to various environments; all fracture tests conducted at room temperature [30].

temperatures the mechanism of joint failure may be via a mechanism which is never observed during the service life of the bonded structure. The aim of any accelerated test must be to simulate the same mechanism(s) of failure as seen in normal usage but cause them to occur more rapidly; not to introduce different mechanisms which may lead to completely misleading conclusions being reached.

(d) Adhesive type The chemical type of adhesive employed can affect both the mechanism and rate of environmental failure and this arises for several reasons [31].

Firstly, as previously mentioned, if the service environment chemically attacks the adhesive to any significant extent, the joint may well be appreciably weakened. If high temperatures are avoided, then most modern adhesives generally have good resistance to chemical attack by moisture, although a few, like the cyanoacrylates (Section 5.2.2), for example, are quite susceptible to hydrolysis even at room temperature.

Secondly, the service environment may interact physically with the adhesive. The physical interaction of the adhesive with ingressing water results in plasticization of the adhesive [32–35]. All organic adhesives will absorb water to some extent and their mechanical behaviour will reflect this interaction. For adhesives which are not highly polar the effect will be very small but for the more polar adhesives, especially when they also contain polar additives such as polyamides or acrylonitrile rubber as hardeners and/or toughening agents, then the effect may be appreciable. For example, in the case of an epoxy-polyamide structural adhesive, the effect of exposure of the cured bulk adhesive to an environment of 43 °C and 97% r.h. for 1000 h has been found [32] to be quite marked, as might be anticipated from the chemical nature of this material. The effect of the absorbed moisture was to decrease its complex dynamic modulus from 2.19 to 0.37 GPa and decrease its glass transition temperature, T_g, from about 140 to about 100 °C. The original properties of the adhesive could be restored by removing the water by drying, demonstrating that the effects were due to physical interactions: the diffusing water disrupting the hydrogen bonding between the molecular chains of the adhesive. Whether plastication of the adhesive layer by the ingressing water actually affects the strength of the joint is difficult to predict. It may be recalled from Chapter 6 that, for many joint geometries, a decrease in the adhesive modulus may decrease the stress concentrations in the joint and, all other factors being equal, lead to an *increase* in joint strength. Similarly, using fracture mechanics tests, often the first observation made when the joint is initially exposed to moisture is that the toughness of the joint increases somewhat: plasticization of the crack-tip region by moisture lowers the adhesive's yield stress, blunts the crack tip and so increases the apparent toughness of the joint [36]. Nevertheless, since the T_g of the adhesive may have been appreciably lowered, the high temperature performance of the joint will obviously have been decreased. It is also sometimes found that some of the strength loss suffered by adhesive joints may be regained upon drying and this recoverable portion is usually ascribed to loss of joint strength due to plasticization of the adhesive. However, whilst with some types of adhesive plasticization by water may influence the mechanical properties of the joints, the preceding comments reveal that it is the interfacial regions which represent the critical area.

Thirdly, the adhesive type does appear to influence the environmental stability of the interfacial regions. For example, especially noteworthy is the often-superior joint durability shown, in the case of bonding aluminium alloys, by some of the older phenolic-based structural adhesives, compared to the more modern epoxy-based adhesives. This does not arise from the phenolics possessing lower diffusion or solubility coefficients with respect to water than the epoxies [153]. Indeed, many such phenolic adhesives are more permeable to moisture. However, it may arise from the ability of the phenolics to establish stronger, more stable intrinsic interfacial forces, as discussed later. Alternatively, it may arise from the pH of residues from phenolic-based adhesives being acid and therefore not attacking and weakening the aluminium oxide, unlike the more alkaline extracts from some of the epoxies. It is believed

[37,38] that epoxies which are hardened with curing agents such as dicyandiamide are particularly likely to affect the aluminium oxides in this manner, since any unreacted hardener may react with water to form alkaline products which are then leached out and react with the alumiunium oxide. However, the epoxy types are now generally preferred on the basis of their lower cure-temperature/pressure requirements and usually superior toughness and, hence peel strengths. Therefore, to obtain the very long service life in the hostile environments experienced by adhesive joints in modern aircraft, the aerospace community has developed improved surface pretreatments for epoxy adhesives, as discussed below.

(e) Substrate type Joints to high-energy substrates such as metals, glasses and ceramics present the main problem. Joints to rubbers, plastics, GRP and CFRP are far less susceptible to environmental attack by water; that is not to say that such joints never suffer from such attack, but when this does occur it is usually

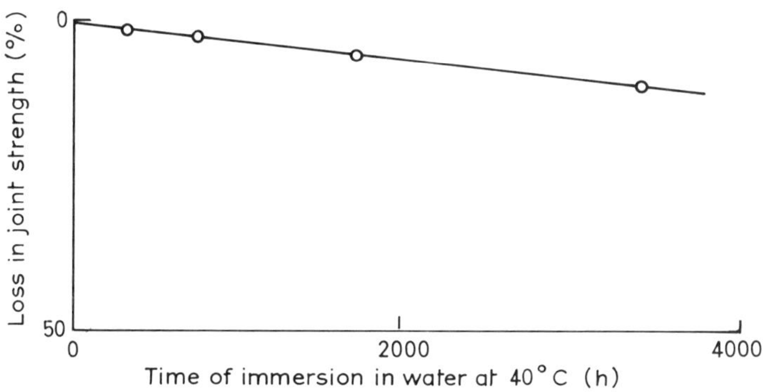

Figure 8.7 Effect of immersion in water at 40 °C on the tensile lap-shear strength of glass fibre reinforced plastic (GRP) substrates bonded with a two-part acrylic adhesive [40]. The surface pretreatment was simply an abrasion/solvent-cleaning process.

the substrate itself, e.g. the composite material, which is itself attacked more readiliy and rapidly than the bulk adhesive or the interfacial regions [39,40]. To illustrate this point, the good durability that may be attained with GRP substrates which have merely been abraded and solvent-cleaned prior to bonding with a two-part acrylic are indicated in Fig. 8.7.The fundamental reasons underlying these observations are discussed later.

(f) Substrate surface pretreatment This is an extremely important factor. As described in Section 4.3, to ensure initially reasonably strong joints in the case of bonding high-energy substrates it is generally sufficient to remove surface contamination, weak oxides, etc. which may act as weak boundary layers. However, to produce durable joints it is also necessary to form stable oxides which are 'receptive' to the adhesive and to establish strong, stable intrinsic forces at the adhesive (or primer) and oxide interface. There is broad agreement with respect to the pretreatments which impart the best durability but, as will be seen later, there is still much speculation concerning how these pretreatments actually lead to good environmental resistance.

Primers are often used to help ensure an adequate service life and, indeed, the presence, and the chemical type, of any primer may play a crucial role in ensuring a long service life for the adhesive joint. For example, it will be shown later that the use of organometallic primers may be extremely beneficial and many commercial 'corrosion-inhibiting' primers have been developed for use with epoxy aerospace adhesives; following the above comments, it is probably not just a coincidence that many of the latter types of primers contain phenolic-based constituents. Examples of the effect of substrate surface pretreatments on joint durability were shown in Figs 4.16 and 4.17 and Table 4.15; further examples are illustrated in Fig. 8.8.

These results are representative of the technology typically employed by the aerospace community to ensure good joint durability, where the pretreatment

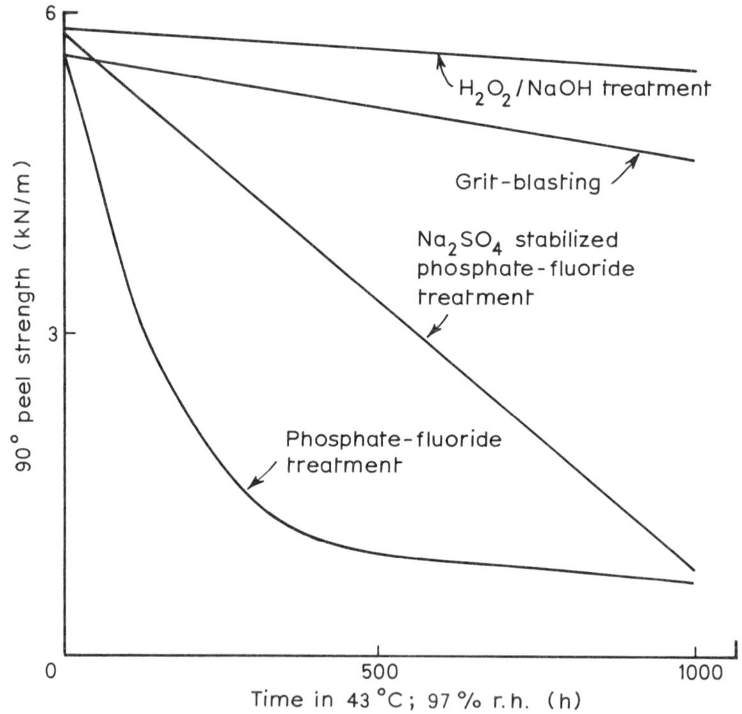

Figure 8.8 Peel strength of hot-curing epoxy/titanium alloy joints as a function of the surface pretreatment of titanium substrates (titanium 6Al 4V alloy). (Note that the various pretreatments have a relatively small effect upon the initial, unaged joint strength.)

Table 8.1 Strength and durability tests on aluminium alloy subjected to various ('non-aerospace') pretreatments [41]

Pretreatment	Single lap-shear strength (MPa)†		
	Initially	In salt spray at 43 °C	
		8 weeks	20 weeks
Alodine 407/47‡	21.5	16.3	13.7
Accomet C§	20.2	17.4	14.4

† A one-part hot-curing epoxy adhesive was employed.
‡ A phosphate/chromate conversion coating (from ICI plc).
§ A 'no-rinse' coating is produced by applying a chromate/silica concentrate (from Albright and Wilson Ltd).

(a) In all cases the pretreated substrate surfaces were coated with a protective press lubricant onto which the adhesive was applied.

(b) It was also found that shaping and forming the aluminium alloy prior to bonding did not significantly affect the joint strengths and the oxide layers were thin enough to spot weld through.

consists of many different stages and is generally undertaken in large chemical tanks and the adhesive (or more usually a primer) is applied to the freshly treated surface within 4 h. This technology is however, not very acceptable to the general engineering community. Therefore, there have been considerable efforts devoted to developing primers which may be applied to simply abraded, or even oily, metal surfaces and to the development of pretreatments which may be undertaken by the materials supplier. The former are discussed below and an example of the latter is shown [41] in Table 8.1. The basic philosophy with this latter type of process is that the materials supplier will undertake the pretreatment of the metal and then apply either a primer, a protective lubricant, etc. which will protect the surface during storage and transport. (A lubricant may also serve to enable the metal to be pressed into the required shape and the adhesive is selected to displace and absorb the lubricant when it is applied.) Thus, the adhesives user need undertake no pretreatment, but when he undertakes the bonding operation the resulting adhesive system is such that the required level of environmental resistance for the joint is assured.

Finally, attention should be drawn to the balance between the necessary service life required and the economics of the adhesive bonding operation. For example, if you are constructing a civil aircraft then a service life of 20 operating years and 60 000 flying hours in some of the very worst environments, which also involve dynamic fatigue loads, may be required. So durability is of prime importance and complex multistaged pretreatments are necessary and an accordingly high cost must be paid. However, for other applications, even outdoors, the requirements are often less demanding. For example, (i) the environment may be less severe, such as only Northern European, (ii) no dynamic loads may be present and, indeed, any applied stresses may be very low and (iii) the lifetime demanded may be only a few years at most. Thus, a less complex, and less costly pretreatment, may be perfectly adequate. This balance of cost against performance is a familiar one to the design engineer but the lack of engineering data in the area of adhesives makes it a difficult one to address with confidence.

(g) Applied stress As illustrated in Fig. 8.5, the rate of strength loss will generally be faster if a tensile or shear stress is present, albeit an externally applied stress or internal stresses induced by adhesive shrinkage (incurred during hardening) or by adhesive swelling (due to water uptake). Such stresses render interfacial molecular bonds more susceptible to environmental attack by lowering the free energy barrier that must be crossed if the bond is to change from an unbroken to a broken state, i.e. as mentioned before in the case of static fatigue (Section 8.2.2) they lower the activation energy for, and so increase the rate of, bond rupture. Stress also probably increases the rate of diffusion of the ingressing medium. These aspects are considered in more detail later.

(h) Test methods and joint design Since it is the interfacial regions in which environmental failure usually initiates, a test method or joint design which has a relatively high stress concentration at, or near, the interface will tend to reveal durability effects more readily. The standard peel test or the more recently developed 'Boeing' wedge-test (Fig. 8.9) have greater sensitivities towards environmental attack than, say, the conventional single lap joint. Indeed, the Boeing wedge-test has been found to be particularly useful as a quick and convenient quality control technique for assessing the adequacy of the substrate surface pretreatment with respect to joint durability, and some data obtained using this test method is shown [42] in Fig. 8.10. In a quality control laboratory a value of crack growth, Δa, would be specified which must not be exceeded after a given exposure to a hot/wet environment, so giving a pass/fail criterion. However, the Boeing wedge-test can give very misleading results if the adhesive creeps excessively, since the local stresses are then relieved. Indeed, many workers have found it to be only useful as a simple quality control tool, and even then only with certain adhesive types. Nevertheless, Cognard [43] has used the wedge-test to undertake more basic investigations and, when considerable attention is paid to several critical aspects of the test procedure such as employing adhesives which do not creep excessively and obtaining reproducible initial cracks, it may be a very useful test method.

As is evident from the data discussed above, there is no test method or test environment which is generally accepted as a standard throughout the adhesives community and with many adhesives users. Some rely solely upon assessing the loss of joint strength employing single lap-shear joints. Others prefer single lap joints with hole(s) drilled through the centre to increase the rate of water

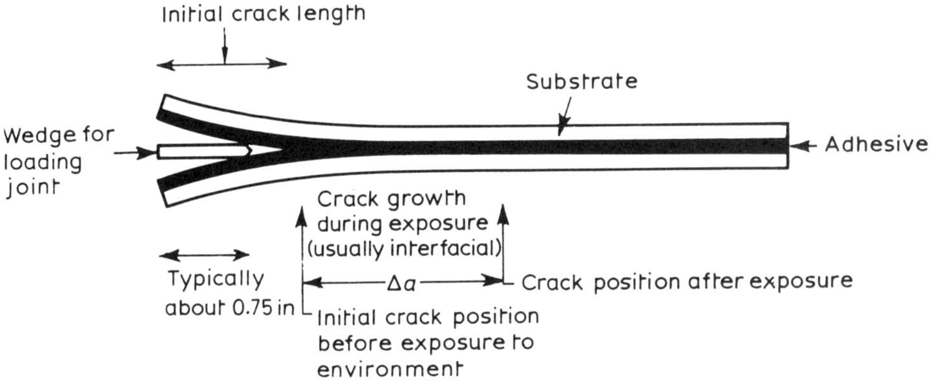

Figure 8.9 The Boeing wedge-test for assessing joint durability. Typical exposure to hot/wet environment is 1 h, or longer, at 50 or 60 °C, 100% r.h. Environmental crack growth, Δa, is then determined, which is a measure of joint durability.

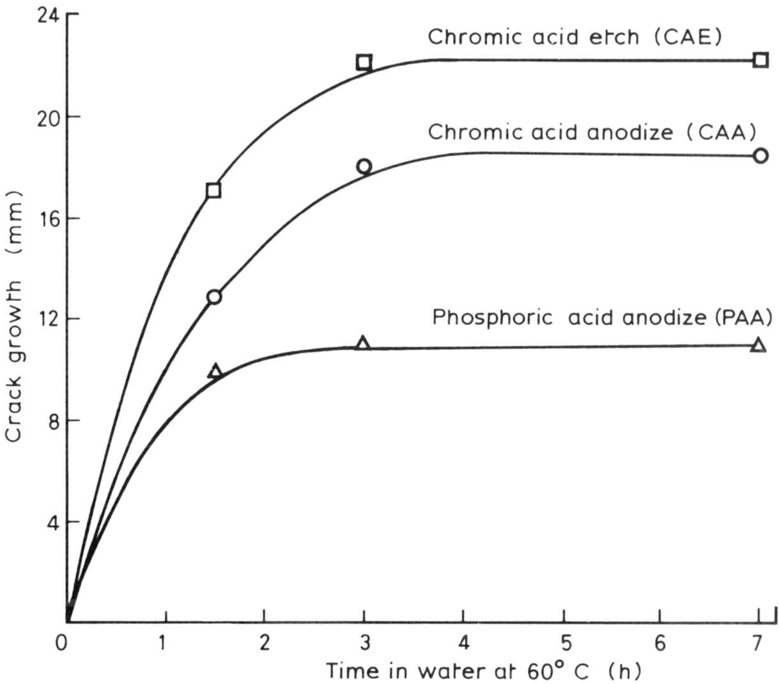

Figure 8.10 Crack growth versus time in environment of water at 60 °C from the Boeing wedge-test. Epoxy/aluminium alloy joints with three different pretreatments used for the aluminium alloy substrates [42].

penetration and, possibly, the rate of environmental degradation; yet others employ this method and then apply a constant load and assess either the time to failure or the residual strength after a given exposure time. In other studies the TDCB geometry (see Table 7.1) has been used, again with an applied load but now with the rate of crack growth being measured. A final example is the use of dynamic fatigue tests in the presence of the hostile environment. Unfortunately, with all these test techniques an adhesives technologist really has to use his experience from previous and currently in-service adhesive systems to decide whether any new development is worthy of using in an application or, more typically, worthy of further investigation; for example, whether it would be a good investment to conduct natural aging trials on a full-scale bonded structure. There is an obvious need for more meaningful test techniques and, further, tests which will enable the service life to be quantitatively predicted.

8.3.3 Mechanisms of attack

8.3.3.1 Introduction

The interfacial regions have been highlighted as the region where moisture may attack adhesive joints and the various mechanisms of such environmental attack which have been postulated to explain the observations recorded in the previous section will now be considered. It should be emphasized that they should not necessarily be viewed as competing mechanisms. Certainly there is ample evidence that no single mechanism can explain all the different examples which have been reported and which one is operative in any situation does greatly depend upon the exact details of the adhesive system and the service environment.

8.3.3.2 Interface stability

8.3.3.2a Locus of failure That environmental attack may indeed result in true interfacial failure between the adhesive and substrate has been established by examining the locus of joint failure of joints consisting of mild steel substrates bonded with an epoxy adhesive using Auger and X-ray photoelectron spectroscopy (XPS). Some of the data are shown [44] in Table 8.2 and are the results of XPS analysis for carbon on the mild steel substrates (i) prior to bonding using an epoxy adhesive, (ii) after bonding and 'dry' fracture and (iii) after bonding, exposure to a hot/wet environment and then fracture. To prevent the fresh fracture surfaces being contaminated by atmospheric contaminants the fracture experiments were conducted inside the ultra-high vacuum system of the X-ray photoelectron spectrophotometer. As may be seen, the substrate control, although freshly prepared, is covered by a carbonaceous layer and the 1*s* carbon photoelectron peak at 285.0 eV may be associated with elemental

Table 8.2 XPS analysis for carbon on mild steel substrates [44]

Specimen	Chemical state: Initially	After ion erosion of: 5 nm	After ion erosion of: 13 nm	Comments
Substrate control	285.0 eV; 19.7%	284.8 eV; 8.5%	284.5 eV; 4.2%	Hydrocarbon contamination from atmosphere; disappears on ion erosion
After 'dry' joint fracture	286.4 eV; 26.3%	284.0 eV; 20.0%	284.3 eV; 20.1%	Polymeric carbon which degrades on ion erosion. Indicates failure in adhesive
After environmental attack	284.7 eV; 2.0%	— —	— —	Negligible hydrocarbon contamination. Indicates failure at interface

(a) Binding energy in electron volts; atomic percentage present.

carbon or hydrocarbon. On ion bombardment the carbon concentration decreases indicating removal of the carbonaceous contamination. As discussed in Section 2.6.2, the typically polar adhesives used in industrial applications will displace and absorb this contamination and so make interfacial contact with the underlying oxide. After joint preparation and fracture, without joint immersion in water, the locus of joint failure was visually assessed as cohesive in adhesive and this was 'confirmed' by the XPS analysis. The carbon 1*s* peak at 286.4 eV was considered to be indicative of polymeric carbon in intimate contact with the substrate and on ion bombardment the polymer was degraded to lower molecular weight species. Finally, after joint preparation and fracture after joint immersion in water the mild steel substrate surface had virtually no carbonaceous material present. This demonstrated that no significant amount of epoxy adhesive remained on the oxide after environmental attack. Further, since analysis of the other (apparently adhesive) fracture surface showed no traces of oxide, the locus of failure for these joints after environmental attack was therefore considered to be truly interfacial.

8.3.3.2b Thermodynamic considerations The earlier discussions revealed that attack by moisture could adversely affect the strength of joints and the above studies clearly demonstrate that it may be the interface which is weakened. The obvious question is why is an interface, which can withstand comparatively high stresses when initially prepared, should be so unstable in the presence of liquids such as water?

Now the intrinsic stability of any adhesive/substrate interface in the presence of a liquid environment may be assessed from the thermodynamic arguments advanced by Gledhill and Kinloch [30]. From Section 3.5.2.2 it will be recalled that the thermodynamic work of adhesion is defined as the energy required to separate unit area of two phases forming an interface. If only secondary forces (e.g. van der Waals) are acting across the interface which, as discussed in Chapter 3, is believed by many to be the main mechanism of adhesion of most adhesives, then the work of adhesion, W_A, in an inert medium may be expressed by:

$$W_A = \gamma_a + \gamma_s - \gamma_{as} \tag{8.5}$$

where γ_a and γ_s are the surface free energies of the adhesive and substrate, respectively, and γ_{as} is the interfacial free energy. In the presence of a liquid (denoted by the suffix 'l') this expression must be modified and the work of adhesion W_{Al} is now given by:

$$W_{Al} = \gamma_{al} + \gamma_{sl} - \gamma_{as} \tag{8.6}$$

where γ_{al} and γ_{sl} are the interfacial free energies between the adhesive/liquid and substrate/liquid interfaces, respectively.

For an adhesive/substrate interface the work of adhesion, W_A, in an inert atmosphere, for example, dry air, usually has a positive value, indicating thermodynamic stability of the interface. However, in the presence of a liquid the thermodynamic work of adhesion, W_{Al}, may well have a negative value indicating the interface is now unstable and will dissociate. Thus, calculation of the terms W_A and W_{Al} may enable the environmental stability of the interface to be predicted.

From the concepts discussed in Section 2.4.1.3 and expressed in Equation (2.17) the values of W_A and W_{Al} may be calculated from:

$$W_A = 2(\gamma_a^D \gamma_s^D)^{1/2} + 2(\gamma_a^P \gamma_s^P)^{1/2} \tag{3.12}$$

$$W_{Al} = 2[\gamma_{lv} - (\gamma_a^D \gamma_{lv}^D)^{1/2} - (\gamma_a^P \gamma_{lv}^P)^{1/2} - (\gamma_s^D \gamma_{lv}^D)^{1/2}$$
$$- (\gamma_s^P \gamma_{lv}^P)^{1/2} + (\gamma_a^D \gamma_s^D)^{1/2} + (\gamma_a^P \gamma_s^P)^{1/2}] \tag{8.7}$$

where γ^D and γ^P are the dispersion and polar force components, respectively of the surface free energy and γ_{lv} is the surface free energy of the liquid.

Some examples of values of W_A and W_{Al} are shown in Table 8.3 and the generality of this concept is illustrated by reference to environments other than water.

For those interfaces where there is a change from a positive to a negative work of adhesion then this provides a driving force for the displacement of

Table 8.3 Values of W_A and W_{AI} for various interfaces and environments

Interface	Work of adhesion (mJ/m^2) Inert medium: W_A		In liquid: W_{AI}	Evidence of interfacial debonding after immersion of unstressed joints?	Reference
Epoxy adhesive/ferric oxide (mild steel)	291	Ethanol:	22	No	30
		Formamide:	−166	Yes	30
		Water:	−255	Yes	30
Epoxy adhesive/aluminium oxide	232	Water:	−137	Yes	45
Epoxy adhesive/silica	178	Water:	−57	Yes	45
Epoxy adhesive/CFRP	88 → 99	Water:	22 → 44	No	46,47
Vinylidene chloride-methacrylate copolymer/polypropylene	88	Water:	37	No	48
		Sodium *n*-octyl sulphate: (3.5% wt solution)	1.4	No	
		Sodium *n*-dodecyl sulphate: (0.5% wt solution)	−0.9	Yes	
		Sodium *n*-hexadecyl sulphate: (0.05% wt solution)	−0.8	Yes	

adhesive on the substrate surface by the liquid. It is therefore to be expected that, if the joint is subjected to such an environment, there will be a progressive encroachment of debonded interface. This will have the effect of progressively reducing the joint strength and also of progressively changing the locus of failure to interfacial between the adhesive and substrate. This is exactly what has been observed in practice. The thermodynamic approach also reveals that high-energy substrates, e.g. metal oxides and ceramics, will be the most difficult materials for the adhesives technologist to bond successfully with respect to ensuring a long service life in the presence of moisture and that fibre composites should generally present little problem in this respect, again in accord with industrial experience.

Finally, it should be noted that the thermodynamics as stated in Equations 8.5 and 8.7 take no account of intrinsic interfacial forces arising from primary bonds or mechanical interlocking. Further, they provide no information on the kinetics of the failure mechanism. These aspects are discussed later.

8.3.3.2c Interfacial secondary forces The relations derived above may be used to predict, for a given substrate and liquid environment, the required values of γ_a^D and γ_a^P of the adhesive if adhesive/substrate interface stability is to be maintained in the presence of the environment with only secondary forces acting across the interface. Hence, the surface chemical properties of the adhesive needed for interface stability may be calculated and then employed in the adhesive selection process.

This concept is illustrated in Fig. 8.11 for a poly(ethylene terephthalate) substrate and a mild steel (ferric oxide) substrate with, in both cases, water as the hostile environment. Values of γ_a^D and γ_a^P of the various adhesives may be measured, as described in Chapter 2, or extracted from the literature (see Table 2.3); for example, considering a styrene–butadiene rubbery adhesive the values are 27.8 and 1.3 mJ/m^2, respectively, and for a typical epoxy adhesive they are 41.2 and 5.0 mJ/m^2, respectively. Hence, it is evident that these (and most other) adhesives will form an environmentally water-stable interface with the poly(ethylene terephthalate) substrate but an unstable interface with mild steel. Indeed, the data confirm that if only secondary molecular forces are acting across the interface then water will virtually always desorb organic adhesives, which typically have low surface free energies of less than about 60 mJ/m^2, from a metal oxide surface. Hence, for such interfaces, stronger intrinsic adhesion forces must be forged which are more resistant to rupture by water.

8.3.3.2d Interfacial primary forces An approximate indication of the contribution to interface stability from the additional establishment of primary chemical bonds across the interface may be deduced by taking the chemical bond energy to be of the order of 250 kJ/mol and assuming a coverage of 0.25 nm^2 per adsorbed site. This yields an intrinsic work of adhesion of +1650

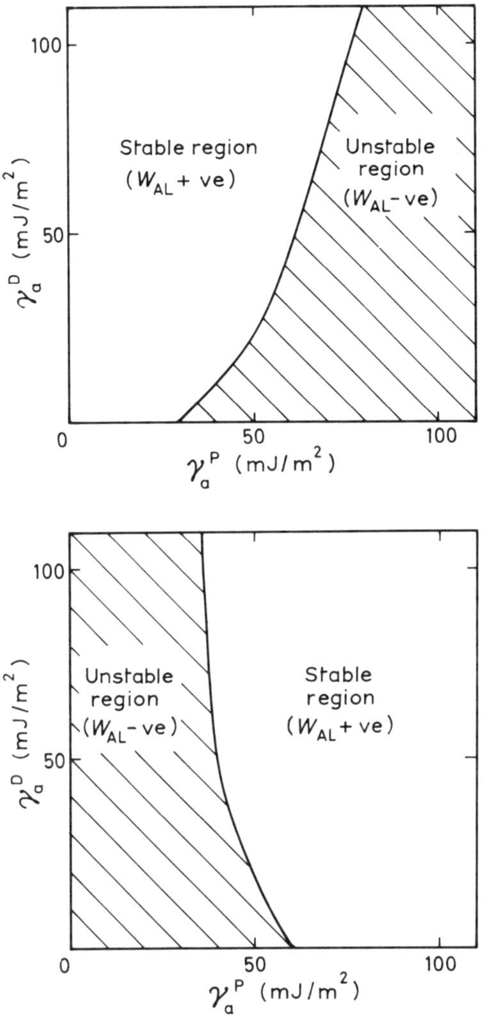

Figure 8.11 Predictions of interface stability, as a function of γ_a^D and γ_a^D values of potential adhesives, for: (a) poly(ethylene terephthalate) substrate and water environment; (b) Fe_2O_3 substrate and water environment.

mJ/m^2 and from energetic considerations it would be unlikely that water would readily displace such a chemisorbed layer, although it should be noted that more basic information on the nature of the interfacial forces and reaction mechanisms is required before more definitive calculations and predictions can be undertaken.

One technique for establishing interfacial primary bonds has been the use of organometallic silane-based primers, as discussed in Sections 3.5.4.2 and

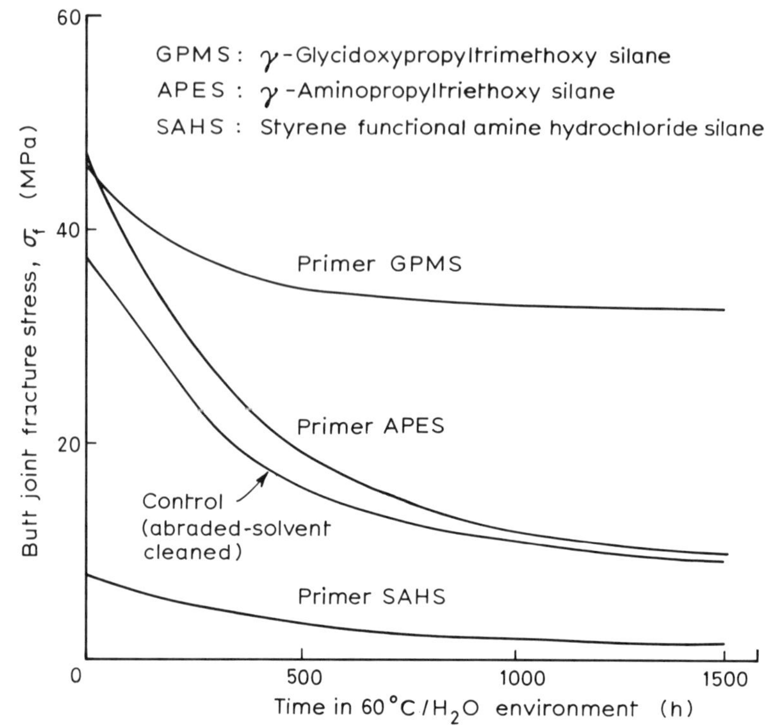

Figure 8.12 Durability of epoxy/mild steel joints employing various silane-based primers [49].

4.3.5.2. Gettings and Kinloch [49] have investigated a range of silane-based primers deposited onto mild steel substrates which resulted in joints possessing different durabilities, as shown in Fig. 8.12. The one silane primer which was found to be effective was the γ-glycidoxypropyltrimethoxysilane and it was only for this primer that there was evidence, from static secondary ion mass and X-ray photoelectron spectroscopy, that the primer had polymerized to give a polysiloxane and had established primary chemical bonds, probably –Fe–O–Si–, across the interface (see Figs 3.13 and 3.14). Undoubtedly, both of these aspects are important if the presence of the primer is to increase the durability of the joint. If the former does not occur then the silane layer, which is usually thicker than a monolayer, will act as a low molecular weight, weak boundary layer and if no primary chemical bonds are established then the thermodynamics would predict that the joint durability will not be significantly increased. The type of primary bonds could not be precisely identified from the data but they are probably mainly covalent in nature but with some degree of ionic character. Similar studies [50] have been conducted on stainless steel and the same picture has emerged, but with this substrate the chemical bonds appeared to be both –Fe–O–Si– and –Cr–O–Si–. More recently, inelastic electron tunnelling (IET) spectroscopy has been used [51,52] to explore further the structure and nature of bonding of the silane primers. It is probably true to state that the results have not been as positive as had been hoped at first, but nevertheless some interesting observations have been made. For example, it has been shown that the triethoxysilanes are relatively chemically unreactive on the metal surface whilst the trimethoxysilanes appear to react chemically. The evidence for this latter conclusion was that the –Si–O–C– peaks were strong in the infrared (IR) spectra of the bulk trimethoxysilanes but were weak or absent in the IET spectra. (Both IR and IET spectroscopy record vibrational spectra, but the latter does so of a very thin absorbed layer.) This may be seen in Fig. 8.13 which shows the IR spectrum of the bulk material and the IET spectrum of the primer adsorbed onto aluminium oxide for the γ-glycidoxypropyltrimethoxysilane. This particular silane was found to be the most reactive of those studied lending general support to the SIMS and XPS studies mentioned previously.

Silane-based primers have also been found to be very effective in increasing the environmental resistance of joints prepared from glass substrates [45,53] and moderately successful in the case of aluminium alloy [45,54,55] and titanium alloy [56,57]. However, there is a great deal about the details of their structure and the nature of their interactions with the substrate and adhesive that is not currently understood, as discussed in Section 4.3.5.2. For example, although the silane primer is shown to react with the adhesive in the schematic representation in Fig. 3.14 this is certainly not necessary from thermodynamic considerations: simply secondary bonding at the adhesive/silane interface would be adequate to ensure the environmental stability of this interface to water. However, it is likely that often an interpenetrating network is established at this interface. Thus, the role of the organic moiety on the silane molecule is not clear. Secondly, it has been found that the age [58] and pH [57,59] of the silane primer solution prior to being applied to the substrate affect the resulting durability of the joints. These factors appear to influence the structure of the polysiloxane adsorbed onto the substrate and the nature of the primer/substrate interfacial interactions.

It should be noted that in many instances the silane is not used as a primer but rather is added to the adhesive. Generally, this is not so effective at increasing the subsequent joint durability [58], probably because the silane molecules first have to diffuse to the interface before they can react and operate in the manner described above. Obviously, they must reach the interface before the adhesive hardens.

Locus of failure studies have shown [44] that, although a silane primer may increase interface stability and so enhance the service performance of the bonded component, the polysiloxane is now the weakest part of the joint and

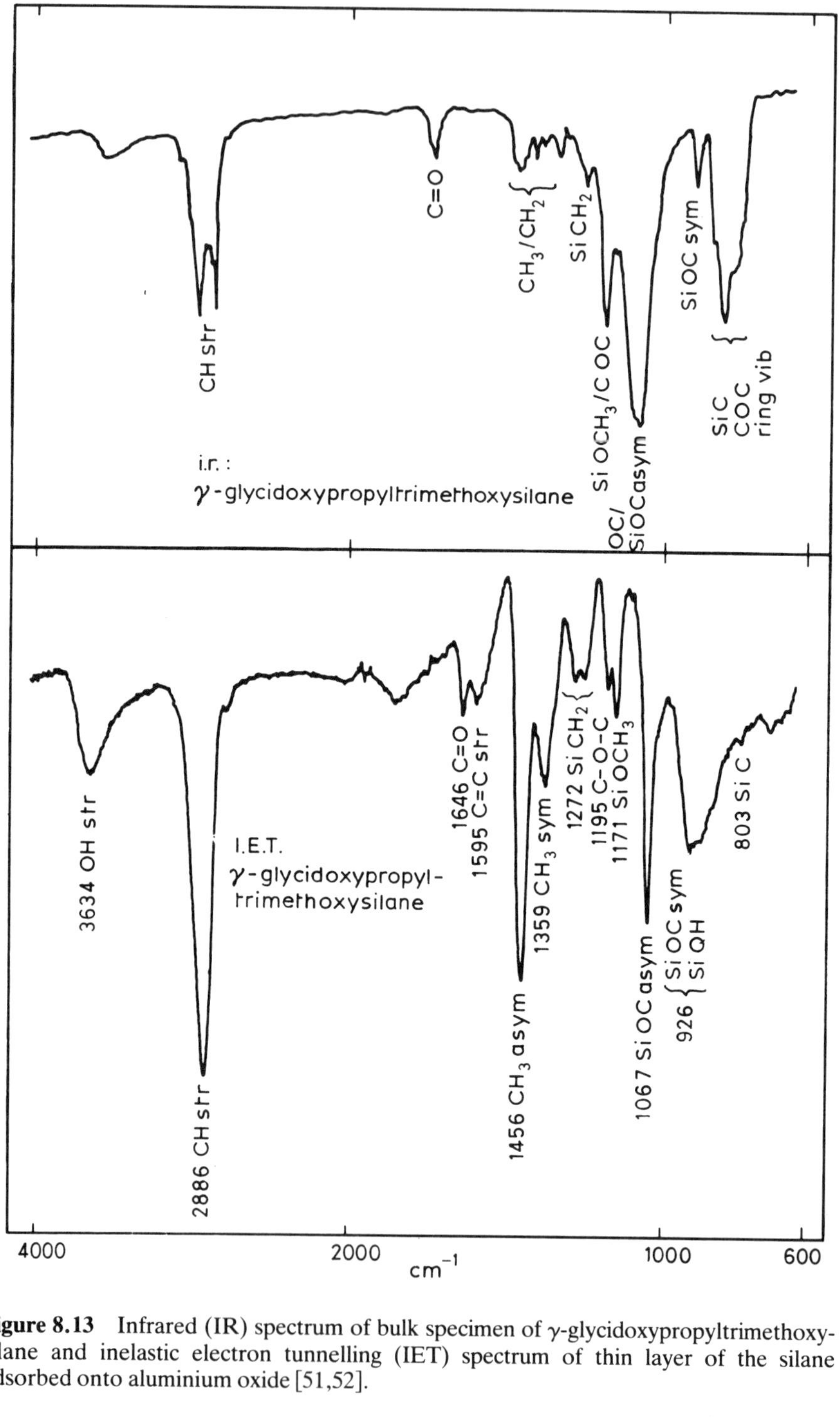

Figure 8.13 Infrared (IR) spectrum of bulk specimen of γ-glycidoxypropyltrimethoxysilane and inelastic electron tunnelling (IET) spectrum of thin layer of the silane adsorbed onto aluminium oxide [51,52].

environmental failure occurs by cohesive fracture of this layer. The mechanism of attack probably involves hydrolysis of the –Si–O–Si– bonds and so, to improve further the service life, attention needs to be focused on increasing the intrinsic resistance of the silane-based primers. A variety of techniques offer considerable promise [60] in achieving this objective and include enhancing their reactivity by *in situ* plasma polymerization (see Section 4.3.6) and the deposition of highly ordered layers via the Langmuir–Blodgett [61] or Sagiv [62,63] methods.

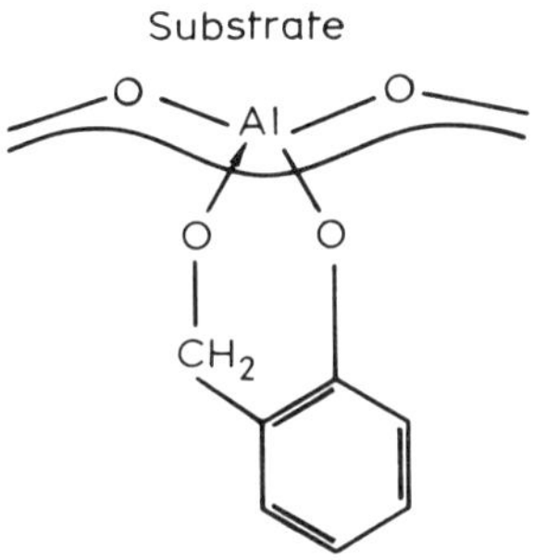

Figure 8.14 Schematic diagram of possible interfacial chelate structure for a phenolic-based primer or adhesive adsorbed onto an aluminium oxide substrate [69].

Other types of primers are also thought to increase the durability of adhesive joints by establishing strong, stable intrinsic adhesion forces to the metallic substrate. One important example is that of phenolic-based primers, and adhesives, and for many years it has been suggested [64] that the long-time, high-temperature conditions for hardening such types of resin are also precisely those that give the maximum probability of forming ether linkages between metal oxide surfaces and the organic resin. However, the existence of such intrinsic interfacial bonding has yet to be proven, although there is considerable indirect evidence for the chemisorption of phenolic derivatives onto aluminium oxide, typically obtained using model compounds. Mention has been made of the work of Chu *et al.* [65] in Section 3.4.5.1, and other work by Brockmann [66], Kollek [67], Turner *et al.* [68] and Zaporozhskaya *et al.* [69] has shown that phenolic derivatives may chemisorb onto aluminium alloys and that the type of bonding is usually via some form of molecular complex (Section 3.5.5). For example, when the molecular structure of the phenolic derivative permits, a chelated ring complex has been suggested, as drawn schematically in Fig. 8.14. Indeed, for such relatively ionic bonds to be stable in the presence of moisture the formation of a chelate is probably essential: a chelate will require both attachments to the substrate to be broken simultaneously to displace effectively the adsorbed molecule and the energy required for this compared, to that for a monodentate ligand, is likely to be far higher. Obviously, in a polymeric phenolic of this type a large number of such chelate structures would be established and, in addition, the high molecular weight polymeric material would possess an adequate cohesive strength. Further, the fact that only certain structures of phenolic derivatives may form such chelate bonds, as well as possessing adequate cohesive strength, may explain why only certain types of phenolic resins are effective as adhesives and primers from the viewpoint of giving good joint durability. So again we return to these two important aspects in the interfacial chemistry: strong, water-stable intrinsic adhesion and adequate cohesive strength in the boundary layer of primer or adhesive.

Similar arguments to the above may be advanced to explain the increase in durability that is observed with other types of chelating primers such as the benzotriazoles on copper substrates [70] and the mercaptoesters on steel substrates [71]. An example of the latter type is shown in Fig. 8.15. Finally, it is

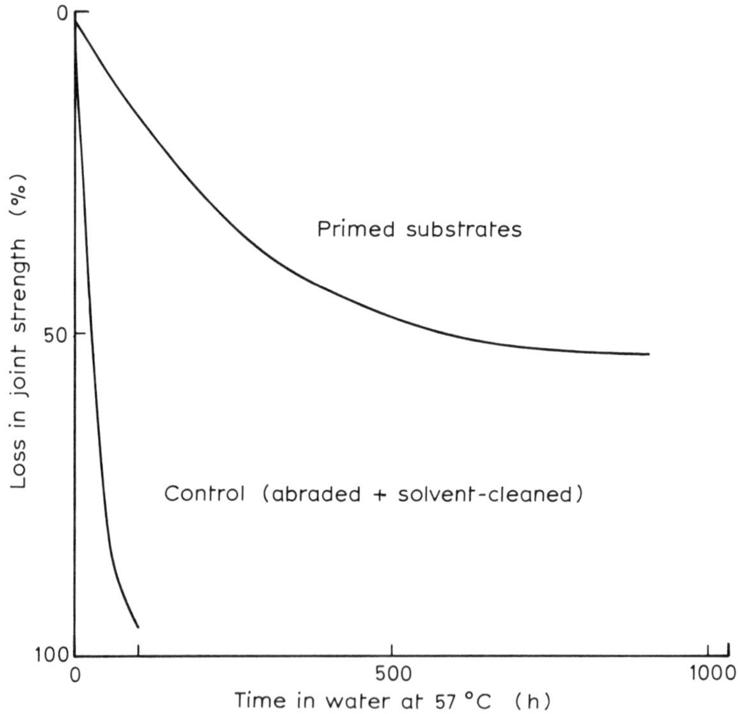

Figure 8.15 Effect on the durability of epoxy/carbon steel joints of a pretreatment based upon preactivation of the substrate with aqueous ammonium citrate followed by application of hexanetriol trithioglycolate as a primer [71].

noteworthy that Waite *et al.* [72] have found that the adhesive used by the marine mussel *Mytilus edulis* is based upon a polyphenolic protein which consists largely of hexa- and decapeptide repeats that contain amongst others the unusual amino acids Dopa, 3-hydroxyproline and 4-hydroxyproline. The presence of Dopa in the repeats was considered to be of prime importance for ensuring durability of the mussel's adhesive in seawater, since it contains a phenolic moiety which is capable of chelating to metal surfaces, on which the mussel frequently abides quite happily. Also, the Dopa is capable of cross-linking the adhesive and thereby increasing its cohesive stability and strength. Thus, this family of mussel appeared to have known well before the present generation of adhesives technologists that there are two prime requirements for establishing water-stable interfaces.

8.3.3.2e Conformational changes Another mechanism which may increase the stability of the interface is when the adsorption of a relatively large molecule involves a conformational change (see Section 2.4.3). It has been reported [73] that if changes in the conformation of the polymer chains are necessary then this will not in general lower the tendency to adsorb, but will greatly retard the process. Further, the necessity for such changes will even more delay the process of desorption in the presence of a solvent for the adsorbed polymeric molecule. The contribution of such a mechanism to the stability of adhesive/substrate interfaces has yet to be ascertained.

8.3.3.2f Mechanical interlocking The role that mechanical interlocking plays in determining the adhesion of materials was discussed in Section 3.2 and, on the level of simple keying of the adhesive into the substrate, was found to represent a generally inadequate mechanism for explaining the many experimental observations. A similar picture now emerges with respect to using this theory to explain why different substrate surface pretreatments lead to joints possessing different joint durabilities.

It is generally agreed [e.g. 74,75] that there is little correlation between the macro-roughness of the substrate and joint durability, and the poor performance of grit-blasted surfaces of aluminium alloy (Fig. 4.17) clearly shows this statement to be true. Nevertheless, mechanical interlocking, but on a micro-scale has recently been suggested [76] to be a major mechanism in increasing interface stability for both aluminium and titanium alloy substrates. For example, on phosphoric acid anodized surfaces of aluminium (see Figs 8.16 and 8.17) it may be shown theoretically [77] from Equation 2.26 and experimentally [78] that the adhesive, and most certainly a typical low-viscosity primer, will penetrate virtually to the bottom of the pores. Therefore, some workers believe that micro-mechanical interlocking will be established and represent a major reason for the excellent joint durability that is attained with this pretreatment. However, the experimental evidence does not support the

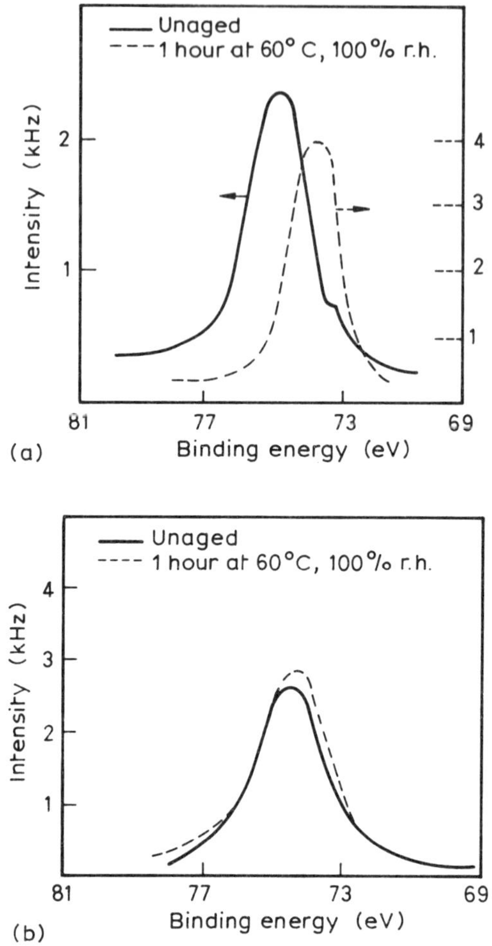

Figure 8.16 X-ray photoelectron spectroscopy analysis of aluminium alloy surfaces (Al 2*p* peak) before and after aging [86]. (a) Chromic acid etch (CAE) pretreatment; (b) phosphoric acid anodized (PAA) pretreatment.

mechanism of micro-mechanical interlocking as a major factor in influencing joint durability. For example, this mechanism does not explain the well-documented effects on the joint durability of (i) the adhesives chemistry, (ii) the nature of the environment and its pH, (iii) the presence of very thin layers of adsorbed primers, etc. (whether they are organometallic- or phenolic-based primers, or of other types of inorganic primers [79,80] or hydration inhibitors [81,82]). Also, it is at variance with (iv) the observation [83] that very large differences in joint durability may result from pretreated substrates where both the macro- and micro-roughnesses may be very similar, and (v) the intuitive

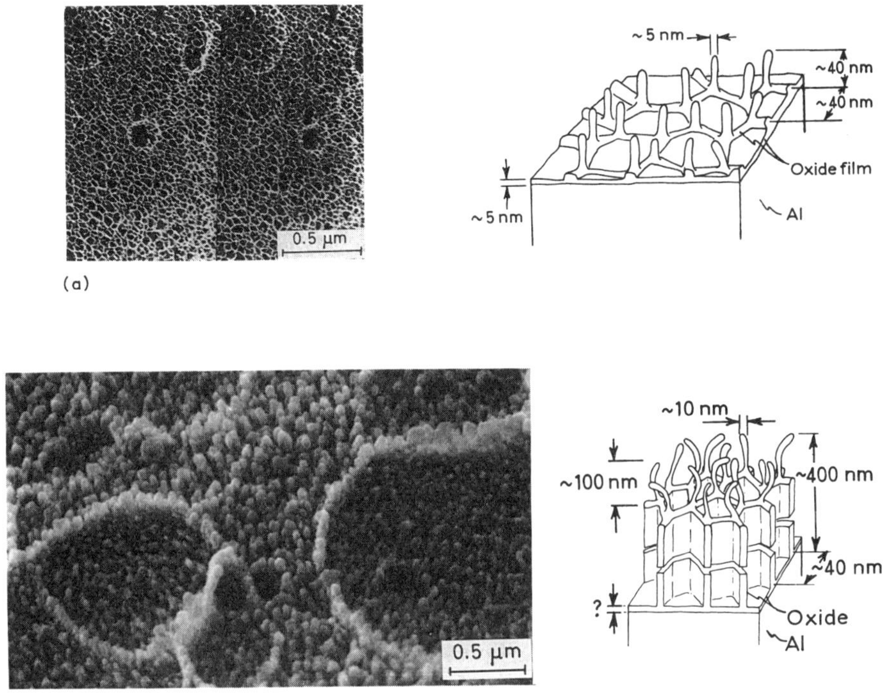

(b)

Figure 8.17 Scanning electron micrographs [76,97] and isometric drawings [76] of oxides on aluminium alloy. (a) Chromic acid etched (CAE); (b) phosphoric acid anodized (PAA).

feeling that if the interfacial molecular forces are completely destroyed by the ingressing moisture then, upon the joint being loaded, the resulting stress concentrations in the adhesive 'fingers' at the oxide surface would be extremely large and such an interface, now with only mechanical interlocking present, would fracture at low applied stress.

Thus, to summarize, it is generally agreed that it is often beneficial for good initial joint performance to have a substrate surface with a relatively rough topography; provided that the oxide is mechanically sound and the adhesive or primer can penetrate the pores of any porous oxide (the latter is difficult on some anodic-produced oxides, such as that formed by sulphuric acid anodizing, but various etches may be used to establish larger pores [84]). The reasons for relatively rough surfaces possibly being beneficial with respect to the initial joint strength were discussed in Chapters 2 to 4. However, establishing mechanical interlocking at the interface, even on a micro-scale, does not appear to represent a major contribution to increasing the durability and service life of adhesive joints.

8.3.3.3 Oxide stability

8.3.3.3a Introduction The stability of the oxide may be considered from two different aspects: subtle changes in the nature of the oxide layer which mechanically weaken the oxide and gross corrosion of the substrate which often involves the oxide layer. The former is considered below and the latter in the following section.

As discussed below, the classic example of the problem of oxide stability upon exposure of bonded joints is with aluminium and its alloys and this aspect has therefore been investigated in detail by the aerospace community. Particularly, the fundamental micro-mechanisms have been studied in order to explain observations such as those shown in Fig. 8.10, which reveals the effect of three common aerospace surface pretreatments upon the subsequent durability of the adhesive joints. The three treatments which have been studied in some detail are chromic acid etch (CAE), chromic acid anodize (CAA) and phosphoric acid anodize (PAA), and details of the processes were given in Section 4.3.4.5.

8.3.3.3b Locus of failure Work by Kinloch *et al.* [42] has clearly established that in the case of aluminium alloys the locus of joint failure after environmental attack may be through the aluminium oxide layer. They used XPS to analyse the fracture surfaces of various joints where the substrate had been subjected to different pretreatments. Prior to exposure to a hot/wet accelerated test environment, the joints all failed by cohesive fracture in the adhesive layer but after exposure to such an environment some pretreatments led to joints where the strength had decreased significantly and the locus of joint failure was largely through the oxide layer. This may be seen from the results shown in Table 8.4. The chromic acid etched (CAE) aluminium alloy joint, which gave the poorest joint durability (see Fig. 8.10), has a very high level of aluminium on both halves of the broken joint and a similar result is seen in the case of the chromic acid anodized (CAA) aluminium alloy. Also, the carbon level is very low on what visually appeared to be the 'adhesive' side of the joint, especially when some, if not most, of this carbon has come from atmospheric contamination deposited after the fracture event. Thus, joint failure has occurred largely through the oxide layer which has been weaked by the moisture which has ingressed into the joint. For the phosphoric acid anodized (PAA) aluminium alloy joint, the extent of failure within the oxide appears to have decreased, while the degree of cohesive fracture in the adhesive, albeit close to the interface, has increased. This treatment resulted in the best durability. Finally, it is noteworthy that visual assessment of all the above joints would have suggested that interfacial failure had occurred, except for the phosphoric acid anodized case when there was some visual evidence of cohesive fracture in the adhesive layer, in addition to the areas which appeared to be interfacial failure regions.

Table 8.4 XPS analysis of fracture surfaces from aluminium alloy joints which have been prepared using various pretreatments prior to joint preparation and environmental exposure [42]

Specimen	Depth of ion erosion of (nm)	Atomic % Al	C
Chromic acid etched			
Metal side	0	23	17
	2	27	6.9
Adhesive side	0	21	22
	2	26	12
Chromic acid anodized			
Metal side	0	23	12
	2	30	4.1
	12	31	2.0
Adhesive side	0	20	23
	2	30	7.2
	12	32	3.8
Phosphoric acid anodized			
Metal side	0	17	25
	2	24	19
	12	25	22
Adhesive side	0	10	50
	2	13	50

There is, therefore, another requirement for the formation of environmentally stable joints in addition to those mentioned in the previous discussions on the stability of the interface, namely, that any oxide layer does not significantly weaken upon the ingress of the hostile environment and so form an *in situ* weak boundary layer, thereby leading to a loss of joint strength. Much of the current understanding in this important aspect of joint durability comes from the elegant work of Venables and co-workers [76,85].

8.3.3.3c Chemistry of the oxide layer Early evidence for the importance of the chemistry of the oxide layer comes from the work of Noland [86]. He reported that the oxide produced on aluminium alloy by a CAE process is unstable in the presence of moisture and postulated that the oxide changes to a weaker gelatinous type which is hydrated and is termed 'pseudo-boehmite'. Support for the transformation came from XPS analysis of the oxide surface before and after aging and Fig. 8.16a shows that a change in binding energy is observed for the aluminium 2*p* peak, indicating a change in oxide structure. Noland also examined epoxy/aluminium alloy joints after exposure to hot, humid conditions and reported that, in agreement with the data shown in Table 8.4, the locus of failure appeared to be in the mechanically weak, hydrated oxide layer.

In direct contrast, the joints made using a PAA pretreatment for the aluminium alloy were far more durable and he found that the oxide produced by the PAA treatment was basically far more stable in the presence of moisture, as may be seen from the XPS analysis in Fig. 8.16b. More recent work has sought to identify those features of the chemistry, or other aspects, of the oxide which are responsible for these observations.

Several workers [87–90] have detected, from using XPS and AES, that the PAA treatment leaves an appreciable concentration of phosphorus in the oxide and that this is present as phosphate, PO_4^{3-}, ions and is distributed [88,91] on all surfaces of the oxide, including the cell walls, ridges and pore bases. (The fact that the phosphates are more highly concentrated near the oxide's surface means that even the sensitive surface analytical techniques such as AES and XPS do not detect a very high level of phosphorus.) The importance of this result comes from the conclusions of some previous work [92] on anodic films which revealed that the presence of such phosphates may greatly reduce the rate of hydration of the oxide layer when exposed to hot/wet conditions. Intuitively, one might expect that the presence of chromates in the CAE and CAA produced oxide layers would have the same effect, but there are no significant and reproducible concentrations of chromium detectable in these oxides [87,89,90,93]. Hence, neither the CAE nor CAA pretreatments on aluminium alloys appear to deposit significant concentrations of chromates in the oxide layers and this may arise from the relatively large size of the dichromate ion.

8.3.3.3d Structure of the oxide layer The changes in the structure of the oxides on aluminium alloys upon attack by moisture have been studied using ellipsometry [76,85], scanning transmission electron microscopy [94,95], scanning electron microscopy [84,96,97] and selected area electron diffraction techniques [98–100]. The initial structure of the CAE and PAA oxides produced on aluminium alloys are shown in Fig. 8.17. Small fingers are shown on the uppermost surface of the oxide but it appears that whether this feature is present or not is a function of the exact details of the pretreatment process and the type of alloy used. The PAA oxide is far thicker and more porous than that produced by the CAE pretreatment. This is, of course, expected since only the former is an anodic process. The formation, and subsequent growth, of the pores in anodically produced oxides have been explained [101–103] in terms of a field-assisted dissolution mechanism.

According to this theory, pore initiation occurs at points in the barrier layer which are somewhat thinner, and where the current density is therefore relatively high, the non-porous barrier layer being the oxide structure which is produced first in the anodization process. Pore initiation always occurs after the anodization current density drops to a value close to the steady-state current density or, correspondingly, after the barrier layer has grown to a nearly steady-state thickness. Once the steady-state structure has been established, oxide growth continues by two simultaneous processes: (i) migration of Al^{3+}, OH^- and OH^{2-} ions through the existing barrier layer to form new oxide at the interface with the underlying metal, and (ii) oxide dissolution at the bases of the pores due to the higher current density in these regions, i.e. field-assisted dissolution. The applied voltage and the solvent characteristics of the electrolyte are the two key factors which determine the structure of the anodic layer on a metal substrate. For example, if the anodic oxide layer is insoluble in the electrolyte then once the barrier layer has grown to the limiting thickness for the voltage applied then no further oxide growth takes place. On the other hand, when this stage is reached in an electrolyte in which the anodic layer is at least partially soluble the above mechanisms may operate and so produce a porous oxide. Thus, the voltage-assisted oxide-dissolution characteristics of the electrolyte determine the thickness of the barrier layer that can be maintained for a particular voltage and so determine the related pore diameter and cell-wall thickness. Indeed, in a given system it is usually found that the steady-state barrier-layer thickness, pore diameter and cell-wall thickness are all directly proportional to the applied voltage.

Returning to the stability of the oxides produced by the commonly used aerospace pretreatment methods, then after exposure to a hot/wet environment the oxide structure changes markedly, as schematically shown [76,99] in Fig. 8.18. The original oxide produced by the CAE treatment (Fig. 8.17a) has increased greatly in thickness and the uppermost oxide structure now consists of this platelet structure, termed a 'cornflake' structure. This increase in effective thickness of oxide upon exposure to water is about threefold and, according to Vedder and Vermilyea [104], in agreement with the increase to be expected when aluminium oxide is converted to the hydroxide. Thus, Venables

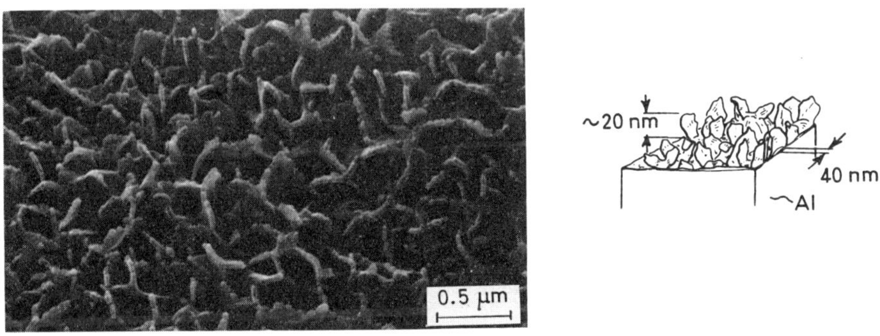

Figure 8.18 Scanning electron micrograph [97] and isometric drawing [76] of CAE-produced oxide after exposure to water and becoming hydrated.

and co-workers concluded that aluminium hydroxide was formed by hydration of the original CAE-produced oxide. The important practical aspect is that this hydrated oxide is mechanically weak and adheres only poorly to the aluminium metal substrate, so that, after exposure of a CAE pretreated joint to a hot/wet environment, joint failure may occur at a relatively low fracture stress through the oxide layer, as found in the locus of failure studies described previously. Furthermore, ellipsometry measurements on the rate at which the oxide increases its thickness has indicated that, compared to the CAE treatment, the PAA-produced surface takes far longer to hydrate and increase in thickness, presumably due to the inhibition of the hydration process arising from the presence of phosphate ions in the oxide layer.

The selected area electron diffraction studies also confirmed the above picture. They indicated that the oxides produced by all the pretreatments were largely amorphous in nature, although some evidence [100] for isolated areas of crystallinity on about the 5 mm scale has been reported. However, after exposure to water the oxides became more crystalline in nature and the ring pattern obtained was characteristic of hydrated aluminium oxide. The nature of the hydrated aluminium oxide is difficult to define exactly but Venables and co-workers concluded that it was probably a mixture of boehmite ($Al_2O_3 \cdot H_2O$) and pseudo-boehmite ($Al_2O_3 \cdot 2H_2O$); the latter is a hydrated form of aluminium oxide which contains somewhat more water than the perfectly crystallized boehmite ($Al_2O_3 \cdot H_2O$). However, if the substrate is left in the moist environment after this hydration stage a further change occurs as the oxide becomes further hydrated and forms the most advanced hydration state, i.e. bayerite, $Al(OH)_3$. This hydrated oxide typically grows on top of the boehmite structure.

Davis *et al.* [105] have studied the kinetics of the hydration of the PAA aluminium in detail and concluded that the rate-determining step in the hydration process is the slow dissolution of $AlPO_4$ in the outermost regions of the oxide, which exposes Al_2O_3 which quickly hydrates to the boehmite. However, more recently El-Mashri *et al.* [106] have suggested that the phosphorus is not largely concentrated at the surface of the oxide layer but is actually distributed more extensively throughout the oxide. This has led to them suggesting an alternative mechanism to that proposed by Davis *et al.* where the role of the phosphorus impurity is now to modify the internal packing structure and hence the susceptibility of the oxide to hydration and weakening.

Finally, it should be noted that the above work has been conducted on exposed oxide surface. It remains to be seen whether the mechanisms of oxide weakening are exactly the same when the oxide is covered by a primer or adhesive layer.

8.3.3.3e Hydration inhibitors The role of the phosphorus in apparently

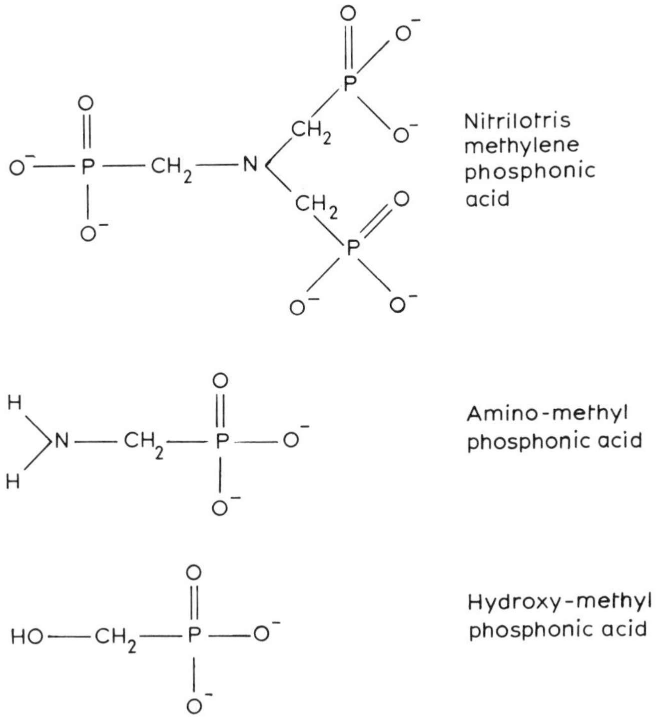

Figure 8.19 Schematic diagram of deprotonated hydration inhibitor molecules [85].

increasing the stability of the oxide on the aluminium alloy, by whatever detailed mechanism, has led to the development [81,82,107,108] of hydration inhibitors based upon phosphoric acid derivatives. Some of these types of inhibitors are shown in Fig. 8.19 and, especially when applied as monolayers, they are very effective in increasing the oxide stability, and resulting joint durability, on CAE pretreated aluminium alloys. It was concluded that, apart from decreasing the rate of hydration, the inhibitors should also be capable of establishing strong and stable intrinsic chemical bonds to both the aluminium oxide and the adhesive.

Two other points of interest are that it appears that such hydration inhibitors are only really effective on an aluminium alloy substrate which has already been pretreated by the CAE process. It would obviously be useful for the technologist if such simple primers were effective on a degreased or abraded surface. Secondly, the above work [82] also suggested that some silane-based primers could function as hydration inhibitors, so adding another possible mechanism whereby such organometallic primers might increase the service life of adhesive joints.

8.3.3.3f The role of other ions in the oxide layer Surface analytical studies have demonstrated [75,89,93,109–111] that ions other than phosphates may be present in the oxide layer on aluminium alloys and, unlike phosphorus, may often adversely affect the stability of the oxide layer. For example, Kinloch and Smart [89] examined a range of pretreatments using an aluminium–magnesium alloy and identified the elements present in the oxide layers using XPS. They found a correlation between high magnesium content, which was present as magnesium oxide (MgO), in the oxide layer and poor joint durability, as indicated in Table 8.5. However, whether this is a fortuitous correlation or of more general applicability has yet to be established. In any event this and subsequent work [75,112] clearly confirmed that a low level of magnesium is not the sole criteria for attaining good joint durability, but that good durability may be unattainable if a high concentration of magnesium is present in the oxide layer. Similar conclusions [93,111] have been reached with respect to the presence of copper in the oxide layer. These results emphasize that different alloy compositions will respond differently to any pretreatment process, and this has indeed been observed [90,113].

Table 8.5 Relationship between magnesium concentration in the oxide layer of aluminium alloy and subsequent joint durability [89]

Surface pretreatment for aluminium alloy	Al:Mg ratio in oxide layer†	Joint durability
Degreased	6:1	Very poor
Abraded	15:1	Moderate
CAE	57:1	Good
PAA	110:1	Excellent

† The ratio of Al:Mg in the bulk of the alloy was 53:1.

8.3.3.4 Substrate corrosion

Another mechanism of attack upon adhesive joints by moisture which may lead to loss of joint strength and greatly reduce the service life is that of relatively gross corrosion of the substrate. As remarked above, often, signs of gross corrosion are observed on the fracture surfaces of joints which have suffered environmental attack, but such corrosion may have frequently resulted after failure has occurred in the interfacial regions of the joint due to one of the above mechanisms. Thus, in such cases, corrosion of the substrates is a post-failure event, rather than a prime cause of environmental failure.

Nevertheless, there are instances where corrosion mechanisms may play the primary role in affecting the service performance of bonded joints. For example, clad aluminium alloys may present a particular problem and the reasons for this have been considered by Reil [114]. With clad aluminium alloys the electrode potential of the clad layer, which is a thin layer of pure aluminium, is generally higher than the base alloy. This choice is deliberate in that the clad material is selected to be anodic with respect to the base alloy so that in a corrosive environment the cladding will be consumed, thus protecting the base alloy. This mechanism is very effective in protecting the structure from surface corrosion such as pitting and Fig. 8.20 is a sketch showing how pitting penetration is restricted on clad aluminium, as compared to the same alloy without cladding. On the clad alloy pitting is less likely to occur due to the nature of the material and where pits do form and penetrate the clad surface, its anodic nature will cause the pit to grow laterally once the base alloy is reached, instead of penetrating into the base alloy.

However, whilst this mechanism of corrosion inhibition may be effective for exposed aluminium alloy structures, if one considers the mechanisms whereby clad aluminium alloy achieves its corrosion resistance then the clad layer is actually undesirable in the context of adhesive bonding; i.e. once the clad layer

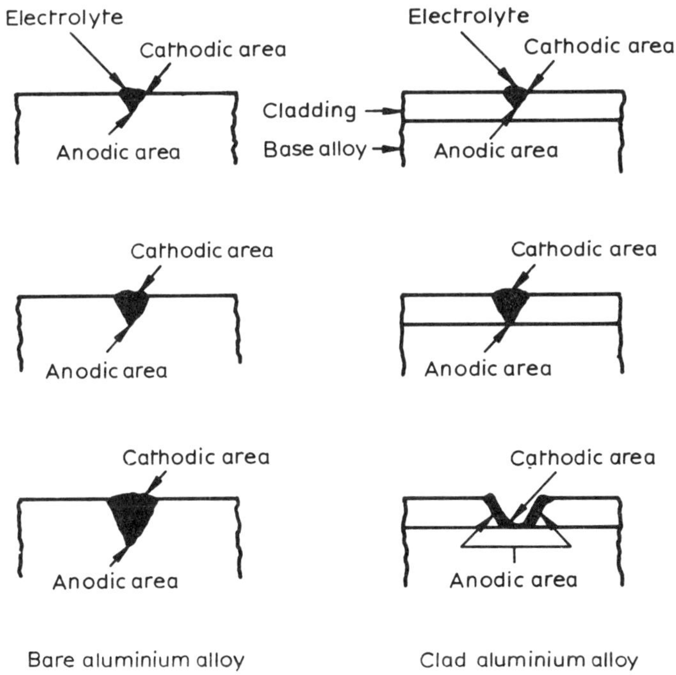

Figure 8.20 Progressive pitting of bare and clad aluminium alloys in a corrosive environment [114].

has disappeared the joint strength will not be very high! Figure 8.21 shows schematically how a galvanic cell may be established between cladding and underlying alloy with the progressive destruction of the interfacial regions. Once this galvanic action starts, the acidity and oxygen concentration factors come into play and progressive delamination occurs. Bascom and Patrick [115] have commented that in general the attachment of the adhesive to a layer of metal which is anodic with respect to the rest of the metallic system is always undesirable from the standpoint of corrosion resistance. Also, chemical components in the adhesive may diffuse into the electrolyte. For example, if the adhesive is an amine-cured epoxy any unreacted amine diffusing into the region of corrosion could affect or even control the local pH of the electrolyte.

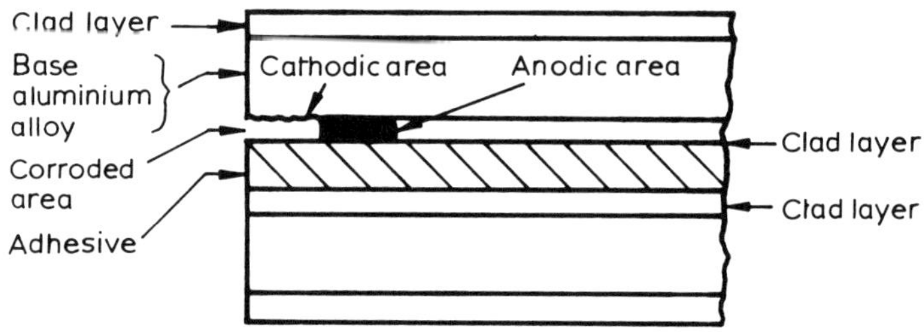

Figure 8.21 Schematic diagram of corrosion as a primary mechanism of environmental attack on clad aluminium alloy joints [114].

Indeed, it will be recalled from Section 8.3.2.3 that it has also been suggested that a reason for the often superior durabilities shown by phenolic-based adhesives and primers is that any extract is acidic rather than alkaline and does not, therefore, promote corrosion of the aluminium substrate.

In the USA the trend has been to move away from adhesive bonding to clad aluminium alloys, but where unclad alloys are bonded and used in areas exposed to corrosive environments any non-bonded, exterior surfaces must be protected by appropriate means, e.g. special primers and paint schemes, in order to limit surface corrosion. In Europe, clad aluminium alloys are still being widely used.

Also, whilst on the subject of corrosion it is of interest to note that in Europe the preferred pretreatment employed by the aerospace community is CAA rather than PAA. The CAA produces a much thicker oxide than the American developed PAA treatment, typically about 4 μm as opposed to 0.4μm. Hence, the CAA treatment offers far better intrinsic corrosion protection for the aluminium alloy. This reason, and the fact that historically the Europeans have developed and established confidence in the CAA pretreatment, has led to the continued use of the CAA process in Europe.

Other types of joints may also suffer loss of strength arising from corrosion of the substrate material and examples are certain types of rubber/steel joints exposed to seawater or salt-spray [116] *when* an electrochemical potential is present. (It is noteworthy that in electrochemically inert conditions the joints, which were bonded using proprietary adhesive/primer formulations, were very stable and showed no significant loss in strength even after three years immersion in seawater.) A general simplified scheme for the corrosion of steel under an organic adhesive or coating is shown [117] in Fig. 8.22. Water and oxygen, as well as some liquid electrolyte are required to establish an electrochemical cell at the adhesive/substrate interface. As a result of small differences between oxide structure or contaminant levels, the steel develops local anodic and cathodic areas, i.e. again small differences between electrochemical potentials are established. At the anodic sites iron dissolves:

$$Fe \rightarrow Fe^{2+} + 2e \tag{8.8}$$

and at the cathodic sites the major reaction is:

$$4e + 2H_2O + O_2 \rightarrow 4OH^- \tag{8.9}$$

i.e. oxygen is reduced, although other reactions are possible under other specific conditions. If the anodic and cathodic sites are in electrical contact with each other a corrosion current will flow. Thus, the liquid collected at the interface will become strongly alkaline and the Fe^{2+} ions will form $Fe(OH)_2$ by hydrolysis and by reaction with the cathodically generated hydroxyl ions; eventually corrosion products of trivalent iron are formed, e.g. FeOOH, Fe_2O_3 or Fe_3O_4 through a complex series of reactions. The observed loss of joint strength may occur from (i) the fact that the corrosion products are mechanically weak and cannot withstand significant applied loads, (ii) the iron oxides which are the corrosion products may actually dissolve under the very high pH created, (iii) the hydroxyl ions may cause the very rapid hydrolysis of the intrinsic molecular bonds at the adhesive/substrate interface, and (iv) the hydroxyl ions may cause the adhering organic polymer to saponify and so become mechanically weak.

Steels are often protected against corrosion by phosphating processes, which provide a mechanical barrier against aggressive species, or by galvanizing, which involves dipping the steel into molten zinc. The zinc coating has more resistance to natural weathering because its corrosion products form dense adhering layers which have low solubilities in water. The requirement to bond adhesively to these types of protected steel surfaces is growing and, from the author's own experience, two problems may arise. Firstly, the initial joint strength may be low due to the protective layer not firmly adhering to the underlying substrate or being insufficiently strong. Secondly, the protective layer may become mechanically weak as its nature changes as it is attacked by moisture, even though it is fulfilling its function of protecting the underlying

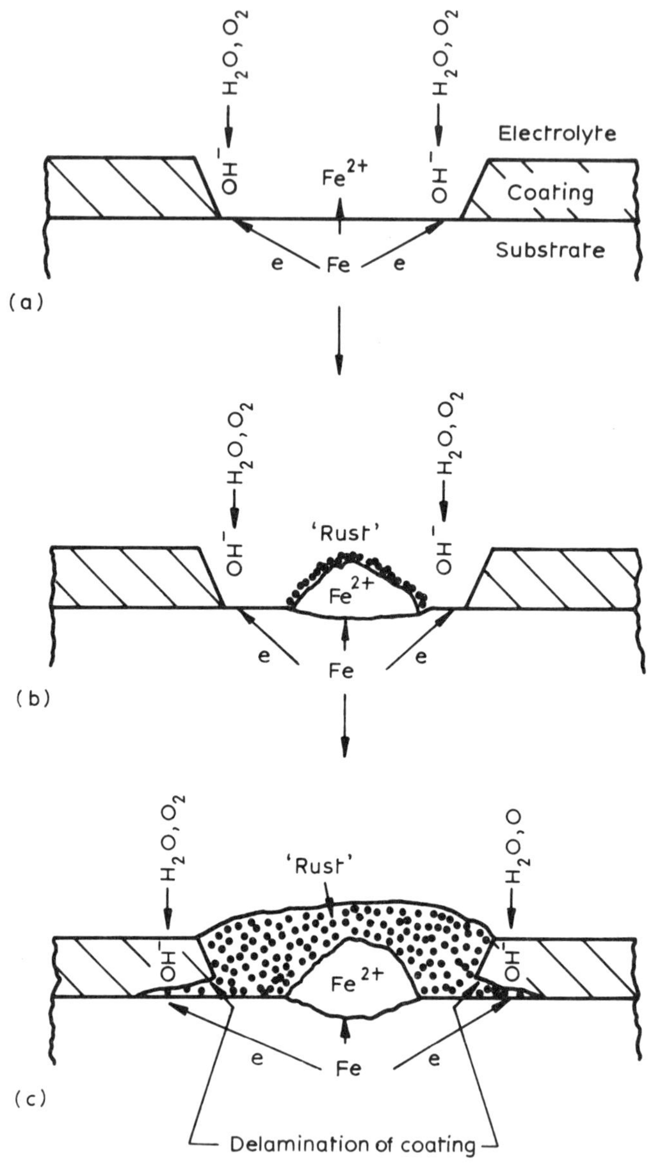

Figure 8.22 General schematic mechanism of corrosion as a primary mechanism of environmental attack upon an organic coating/steel interface (see text for details of delamination processes) [117].

steel substrate from corrosion. So in both cases the protective layer may act as a weak boundary layer and adversely affect joint performance.

Finally, to summarize, corrosion of the substrate has not been observed to be one of the most common mechanisms of environmental attack upon adhesive joints. Nevertheless, there are circumstances when it may be a primary mechanism and the examples of bonding to clad aluminium alloys and steel joints exposed to a seawater or salt-spray when electrochemical potentials are present are such instances. Several specific measures to prevent corrosion being a main mechanism have been mentioned and a general method is to ensure that all the edges of the joint are adequately protected by a suitable sealant, since the mechanism of corrosion attack will frequently initiate in adhesive joints at a raw, unprotected edge. However, it should be noted that all organic sealants are permeable to water and thus moisture will eventually reach the interfacial regions. A metal or glass 'sealant', free from pinholes, would offer complete protection but the methods of applying such a sealant are at present usually impracticable and costly.

8.3.3.5 Stability of the adhesive boundary layer

Early work [26] on the mechanisms of environmental failure concluded that hydrolysis in a boundary layer of the adhesive near the adhesive/substrate interface was a primary failure mechanism in the epoxy/aluminium joints investigated. However, subsequent studies have largely rejected this mechanism as a major cause of environmental failure, especially since in many cases the locus of joint failure after environmental attack has been clearly identified as occurring either at the interface or in the oxide layer. Of course, with the porous anodic oxides and assuming good wetting has occurred and led to penetration into the oxide pores, then fracture through the oxide region will inevitably cause some fracture in the adhesive (or primer); but this does not necessarily imply that the adhesive has been significantly weakened by the ingressing moisture.

However, when considering the stability of a boundary layer of adhesive, it should be borne in mind that there is firm evidence that the boundary layer of adhesive adjacent to the metal oxide surface may well possess a different chemical and physical structure compared to the adhesive away from the interface, and further, that these differences may depend upon the activity of the substrate surface resulting from the particular pretreatment employed. For example, it has been shown that a lower crosslink density [118] or a lower concentration of filler particles [119] or a separation [37] of the constituents of the adhesive or primer (due to the lack of penetration of relatively large molecules into the porous anodic oxide layer) may occur and so result in the boundary layer of adhesive being different to the material away from the interface. Thus, it is certainly theoretically possible that the boundary adhesive layer could be more or less susceptible to hydrolysis than the rest of the adhesive layer and that the degree of susceptibility would be a function of surface pretreatment.

More recently Brockmann and co-workers [37,78,120] have suggested that this mechanism might be important in the case of environmental attack upon aluminium alloy joints but at the present time the mechanism of oxide stability (Section 8.3.3.3) appears to be in better agreement with the experimental evidence or, in specific cases, that of gross corrosion of the substrate (Section 8.3.3.4).

However, one example where hydrolysis of the adhesive, near the interface, may indeed lead to loss of joint strength is where cathodic corrosion of the substrate takes place and leads to a highly alkaline local environment at the interface. This may then result in the hydrolysis and mechanical weakening of the adhesive adjacent to the substrate, as described in the preceding section. Hammond *et al.* [121] and Watts and Castle [122] have reported such mechanisms for the debonding of organic coatings on steel substrates. However, there is no suggestion that the boundary layer of the coating is initially weak, simply that it is at the interface that the local conditions are such that hydrolysis of the adhesive is possible.

Thus, the mechanism of environmental attack via hydrolysis and weakening of a boundary layer of adhesive, adjacent to the interface, arising from its different chemical or physical structure compared to the bulk adhesive has yet to be firmly established as an operative mechanism.

8.3.4 Kinetics of attack

The kinetics of environmental attack will be governed by the rate-determining step in the overall mechanism of failure and very little information exists on such details.

Kinloch and co-workers [30,123] have studied the kinetics of environmental failure for epoxy/steel joints where the mechanism of attack had been found to be via the displacement of adhesive on the oxide surface by the ingressing moisture, as descibed in Section 8.3.3.2. The rate of dissociation of the interface is shown as a function of the water immersion temperature in the form of an Arrhenius plot in Fig. 8.23 and several interesting features are evident. Firstly, for the three lowest temperatures a linear correlation exists which yields an activation energy for the displacement of adhesive by water of 32 kJ/mol. Secondly, the rate of interfacial debonding at an immersion temperature of 90 °C is far greater than would be expected from this activation energy. It was suggested that the main cause for this is that the glass transition temperature, T_g, of the adhesive is 85 °C and, at temperatures above this, segmental motions of the polymer chains greatly increase. This permits diffusing species to penetrate far more readily than would be predicted from

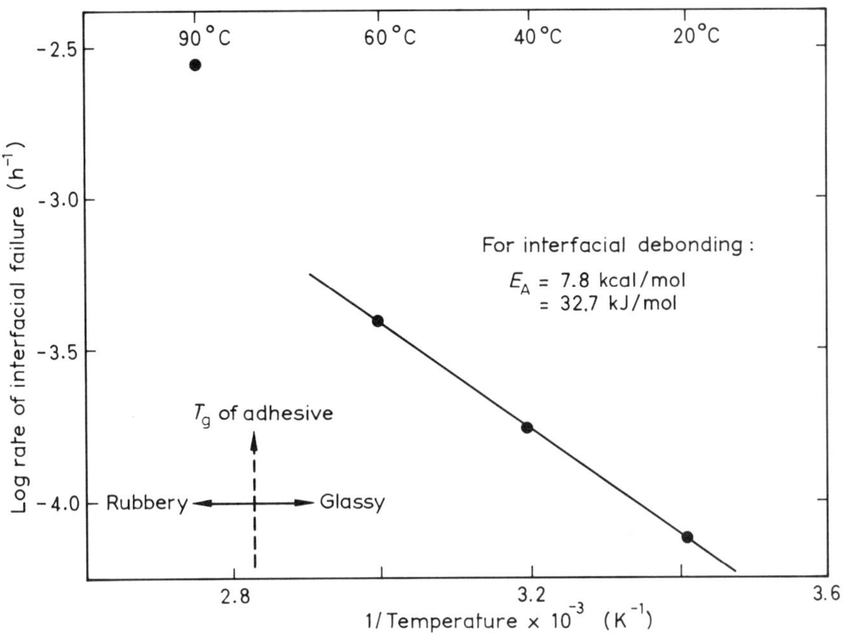

Figure 8.23 Rate of interfacial debonding of epoxy/mild steel joints versus reciprocal of the water immersion temperature [30].

experiments conducted below T_g. The independently measured coefficients for water diffusion through the bulk adhesive are also shown as a function of temperature in the form of an Arrhenius plot in Fig. 8.24. This data yields an activation energy of 36 kJ/mol, with the rate of diffusion at 90 °C being greater than expected from the lower temperature results. These data and values of activation energies clearly suggest that the rate of interface debonding in these joints is controlled by the availability of water at the interface, which in turn is governed by the diffusion of water through the adhesive.

However, in other adhesive systems the situation is more complex. For example, if an effective silane-based primer was used in the above joints then the rate of loss of joint strength would be far lower and the kinetics of attack would no longer be governed by diffusion of water. Indeed, water would be expected to diffuse faster through the polysiloxane primer by several orders of magnitude, although it is of interest to note that the solubility coefficient of water in the polysiloxane primer would be far lower than in the typical structural adhesive. Also, in the case of aluminium alloy joints a correlation [124,125] may be found between loss of joint strength and uptake of water into the adhesive layer; at least until the adhesive layer is saturated. This arises from both initially increasing with increasing time, but the exact relation will

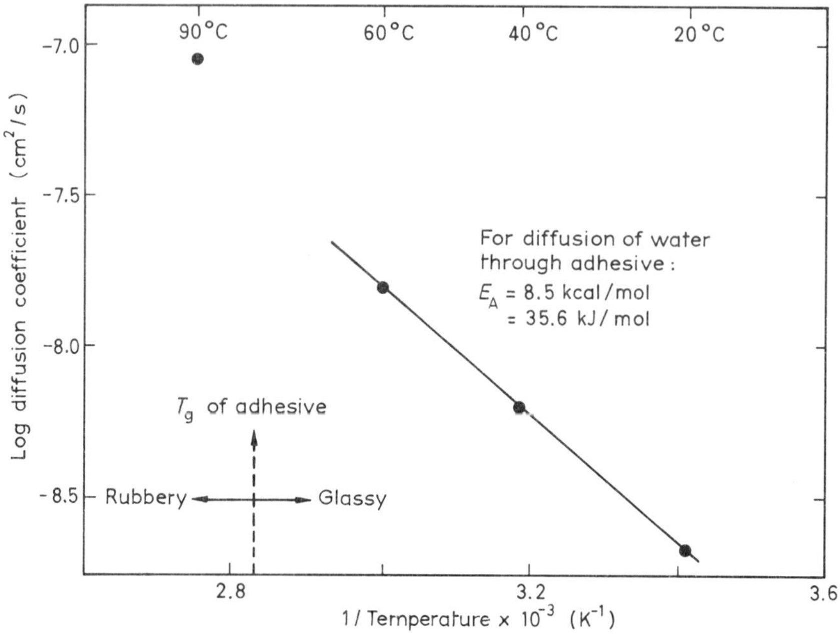

Figure 8.24 Diffusion coefficient, D_f, for water through the adhesive layer versus reciprocal of the water immersion temperature [30].

depend, of course, upon other factors such as choice of pretreatment. Indeed, there is no general relationship between the diffusion and the solubility coefficients of water in the adhesive layer and loss of joint strength, although obviously any feature which will slow down the rate of water ingress, all other factors being equal, will tend to increase the service life. Hence, the use of sealants around joint edges, designing long access paths for moisture, etc. will all be of help in ensuring durable joints. However, it must be generally accepted that water will ingress into the joint eventually and therefore the adhesive system must be selected to possess sufficient stability in the presence of moisture.

Returning to the detailed kinetics of attack, for most adhesive joints the actual rate-determining steps have yet to be precisely identified and their nature will undoubtedly depend upon the particular adhesive system being studied, although, it will be recalled that, for the mechanism of joint attack via hydration and weakening of the aluminium oxide layer (Section 8.3.3.3), Davis *et al.* [105] have proposed that the rate-determining step is the slow dissolution of the $AlPO_4$ layer on the outermost surface of the PAA oxide. This then exposes Al_2O_3 which rapidly hydrates to the mechanically weak boehmite form of aluminium oxide.

8.3.5 The role of stress

As commented earlier, the rate of loss of strength of a joint under environmental attack will usually be faster if a tensile or shear stress is present [45,126–130], albeit an externally applied stress or internal stresses induced by adhesive shrinkage (incurred during cure) or by adhesive swelling (due to water uptake) [131,132]. Such stresses render primary and secondary molecular bonds more susceptible to environmental attack and also probably increase the rate of diffusion and the solubility in the adhesive of the diffusing medium [133–135].

Of particular interest is whether there is a lower limit to the applied stress below which no environmental attack is observed and this has been investigated in some detail using a fracture mechanics approach [36,45,136,137]. The TDCB geometry (see Table 7.1), consisting of aluminium alloy beams bonded with an epoxy adhesive, has been widely employed and the commonly used experimental technique has been to place this specimen under an applied load in the environment of interest and monitor the rate and locus of crack growth. The typical observation is that the cohesive in adhesive starter crack does not

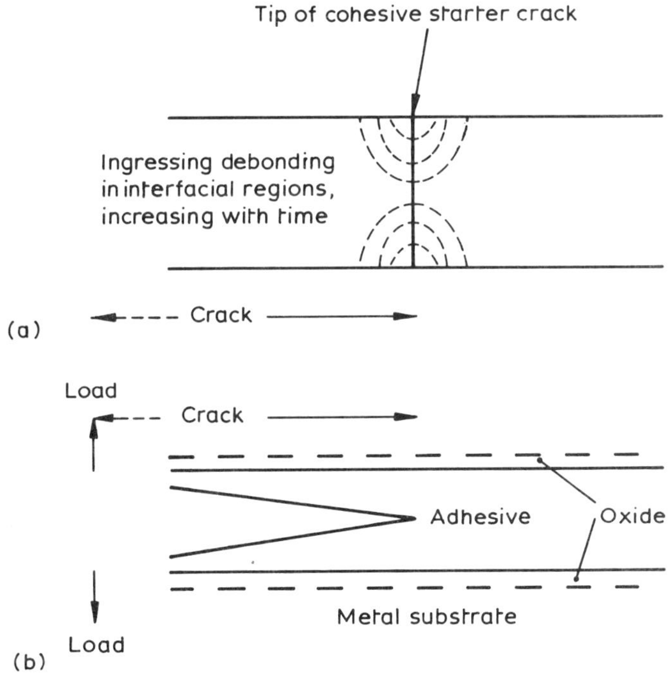

Figure 8.25 The mechanism of attack upon the interfacial regions in an adhesive joint resulting from the combined influences of stress and water. (a) Plan view; (b) side view.

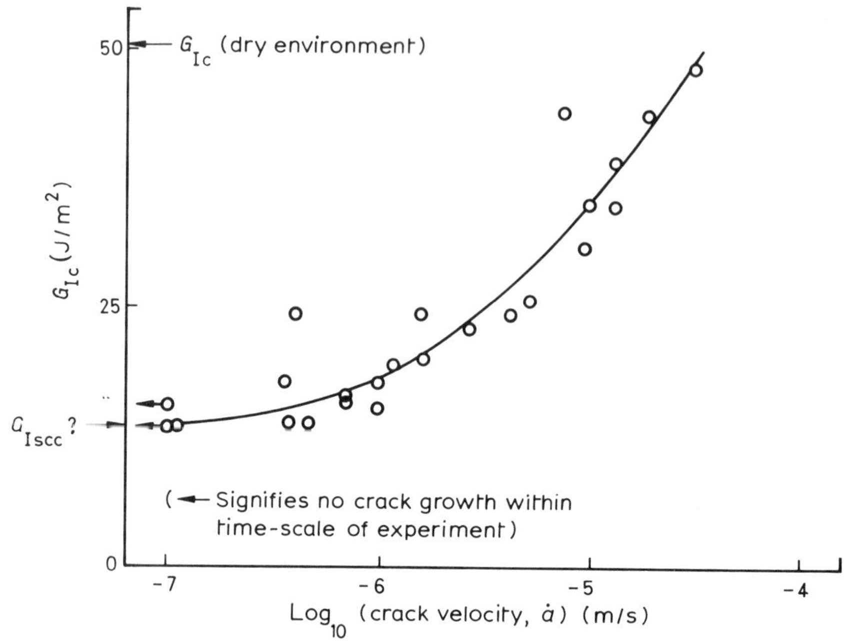

Figure 8.26 Adhesive fracture energy, G_{Ic}, versus crack velocity, $\dot{a}$, in water for epoxy–TEPA/aluminium alloy joints [36,136].

propagate but that new cracks initiate in the interfacial regions, as shown in Fig. 8.25. The new environmental cracks progress inwards from the edges of the joint but are limited to the vicinity of the interfacial regions directly above and below the position occupied by the tip of the initial cohesive starter crack, i.e. within the vicinity of the stress field generated by the loaded starter crack. The freshly created crack then propagates through the interfacial regions down the length of the joint. A typical relationship between the adhesive fracture energy, G_{Ic}, and the resulting crack velocity, $\dot{a}$, is shown in Fig. 8.26. As may be seen, the value required to cause crack growth in an aqueous environment is much lower than that needed in a relatively dry environment. However, the measured adhesive fracture energies required for these crack growth rates are, of course, much higher than the values of the thermodynamic works of adhesion given in Table 8.3. This is because, at the typical crack velocities that result, inelastic energy dissipative processes, e.g. plastic and viscoelastic deformations, occur in the regions of the adhesive surrounding the crack tip. The values of W_A and W_{AI} do not allow for any such processes.

From such data as that shown in Fig. 8.26 it was concluded that there was a minimum value of G_{Ic}, denoted G_{Iscc}, below which slow crack growth would not occur in aqueous environments. Such a minimum value implies that there is a

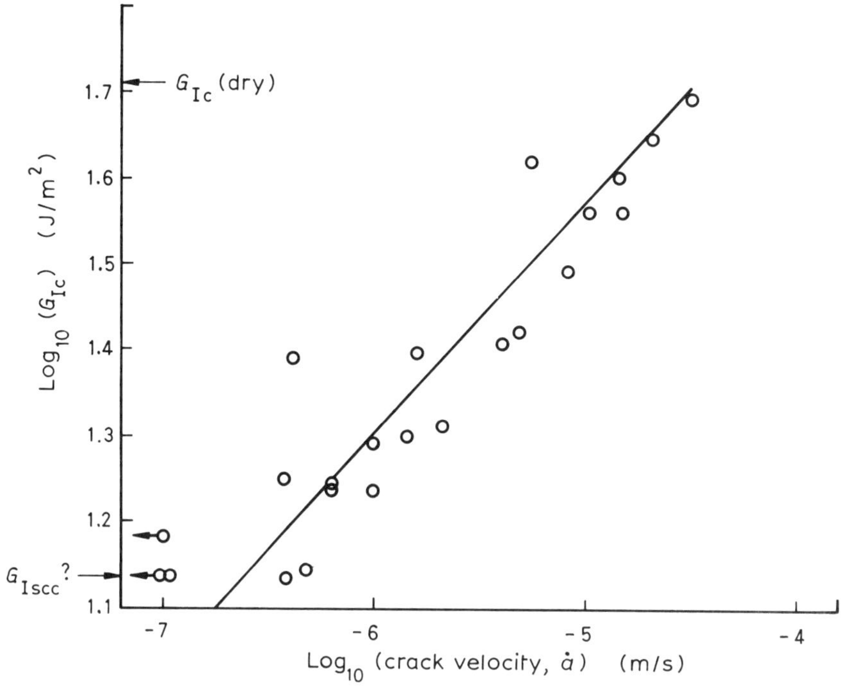

Figure 8.27 Existence of a minimum G_{Iscc}? Data from Fig. 8.26 replotted as $\log_{10}G_{Ic}$ versus $\log_{10}$ crack velocity.

stress level below which no environmental attack will occur. However, the situation is more complex than the data shown in Fig. 8.26 initially implies.

Firstly, the time-scales over which the experiments have been conducted to obtain such plots as shown in Fig. 8.26 are insufficiently long to be able to state with confidence that a true minimum value of G_{Ic}, i.e. a G_{Iscc} value, has been attained and replotting the data in the form of Fig. 8.27, with a logarithmic G_{Ic} scale, demonstrates this point. Secondly it is well established that joints under no externally applied stress may still suffer environmental attack, see for example Figs 8.5, 8.6 and 8.8. Although, it is of interest to note that Cherry and Thomson [138–140] have recently also argued that the presence of stress is essential for environmental attack but consider that in the absence of an externally applied stress, internal shrinkage stresses provide the necessary strain energy requirements. However, apart from the above comments, their data cannot readily explain that (i) hot-cure adhesives have the highest shrinkage stresses but often result in the most durable joints and (ii) any shrinkage stresses will probably be rapidly diminished by stress relaxation processes in the adhesive, accelerated by plasticization by water, and swelling stresses due to water adsorption.

However, from theoretical considerations of the mechanisms of attack a clearer picture emerges. Recalling Equation 7.8, the adhesive fracture energy, G_{Ic}, may be expressed by:

$$G_{Ic} = G_o \Phi_v(\dot{a}, T, \epsilon) \qquad (7.8)$$

If the case of environmental attack upon the adhesive/substrate interface is considered (hence from Equation 7.9, $i = 1$) this equation may be rewritten as:

$$G_{Ic} = G_o\,(\text{interfacial})\,\Phi_v(\dot{a}, T, \epsilon) \qquad (8.10)$$

where G_o (interfacial) is the intrinsic interfacial failure energy. Now in some environments the work of adhesion, W_{Al}, may be considerably reduced. For example, for an epoxy/aluminium oxide interface in carbon tetrachloride the values of W_A and W_{Al} (carbon tetrachloride) are about 230 and 90 mJ/m^2, respectively, and these values may be equated to the respective G_o (interfacial) values. Thus, one would predict that the value of G_{Ic} needed to drive the crack along the interface in carbon tetrachloride would be lower than in dry air, as indeed observed [141]. This lowering of G_{Ic} for crack growth in environments for $W_A > W_{Al} > 0$ has also been seen by other workers [142,143]. Further, a lower limit to the value of G_{Ic}, i.e. a G_{Iscc} value, would exist. The value of G_{Icss} will depend upon the value that $\Phi_v(\dot{a}, T, \epsilon)$ attains when deformation energy losses at the crack tip are at their lowest. For model rubbery adhesives the value of $\Phi_v(\dot{a}, T, \epsilon)$ may tend to unity, so that $G_{Iscc} \cong W_{Al}$, but for other adhesives where plastic flow is involved, e.g. epoxy adhesives, the value of $\Phi_v(\dot{a}, T, \epsilon)$ may always be appreciable so that $G_{Iscc} >> W_{Al}$. However, in other environments W_{Al} may be negative, for example, for the epoxy/aluminium interface in water the value of W_{Al} is -137 mJ/m^2. As discussed in Section 8.3.3.2, this indicates the spontaneous displacement of the adhesive on the metal oxide by the ingressing liquid which would result in a layer of water at the interface, i.e. G_o (interfacial) $\rightarrow 2\gamma_{lv}(H_2O)$ and $\Phi_v(\dot{a}, T, \epsilon) \rightarrow 0$. Hence, the implication that $G_{Iscc} \rightarrow 0$. The other mechanisms of environmental attack are more difficult to analyse but it is unlikely that, for example, the loss of joint strength via the oxide hydrating and weakening (Section 8.3.3.3) would lead to a lower stress limit below which this mechanism was inoperative. Thus, really each case needs to be considered individually but it would appear that there is no general argument for an endurance limit, i.e. a stress level below which no attack is observed, and for structural joints exposed to water none appears to exist.

Finally, the effect of cyclic stresses applied in the presence of a hostile environment has been studied in a limited number of investigations [8,9,12]. As might be expected, such a dynamic loading may accelerate any environmental attack but, by careful attention to the choice of substrate surface pretreatment, long service lives for the joints may be attained.

8.3.6 Service life predictions

8.3.6.1 Basic model

To predict the service life of adhesive joints exposed to a hostile environment requires a knowledge of the mechanisms and kinetics of attack. In the relatively simple epoxy/abraded steel joints exposed to water the work described in previous sections has identified the mechanism as being the rupture of interfacial secondary forces, as predicted from a consideration of the thermodynamics of the system, and the kinetics as being governed by the diffusion of water through the adhesive to the interface. Thus, in this system one might hope to describe a model to predict quantitatively the service life, and Kinloch and co-workers [47,123] have proposed the following model.

The first stage is the accumulation of a critical concentration of water in the interfacial regions which must be exceeded for environmental attack to occur and the rate of attaining this critical concentration appears to be governed by the rate of diffusion of water through the adhesive. As remarked earlier, this requirement of a critical concentration appears to be a general one from the experimental observations of attack by moisture upon many different types of adhesive joints [30,124,125,144,145], although its value is somewhat dependent upon the particular adhesive system. It is not clear why such a critical concentration should exist. It may arise from the absorbed water forming clusters within the adhesive, and the careful studies conducted by Comyn and colleagues [146,147] on the diffusion of water in adhesives shows that clustering of water molecules may occur. Only the molecularly dispersed water molecules would contribute to the process of environmental attack. Alternatively, when the mechanism of environmental attack is via hydration of the oxide then the critical water concentration may correspond to the critical level needed to form the particular hydrate, a well-documented phenomenon [148].

The second stage involves a loss of integrity of the interfacial regions. In the particular case of the epoxy/mild steel joints this arises from the rupture of interfacial secondary bonds. In other adhesive systems, however, any of the mechanisms discussed above could be envisaged.

The third stage concerns the ultimate failure of the adhesive joint. However, for the joint to fracture or lose an appreciable amount of its original strength upon subsequent testing it is usually not necessary for the weakening of the interfacial regions to have proceeded completely through the joint. From basic fracture mechanics considerations only a relatively small environmental crack is required to have developed before a substantially decreased failure time, under a constant load test, or a diminished joint fracture stress is observed. Indeed, with many joint geometries subjected to an imposed load and moisture, catastrophic failure will occur when the environmental crack, which is growing by the mechanisms outlined above, attains a critical length.

8.3.6.2 Life predictions

To use the above model to predict the durability of the epoxy/mild steel joints (see Fig. 8.6) first requires a knowledge of the rate of diffusion of water into the adhesive layer in the joint. The diffusion and solubility coefficient of water into the bulk adhesive was measured and the diffusion process was also found to be of the concentration independent Fickian type [147]. Further, work [147,149,150] has also shown that the diffusion of water into the adhesive layer may often be modelled by assuming Fickian diffusion. Support for this comes, for example, from studies which have employed tritiated water and so enabled the water concentration in the joint to be directly measured; good agreement with the calculated water concentration profile was obtained.

Thus, by assuming concentration-independent Fickian diffusion the concentration, C_t, of water at time, t, in an adhesive layer of square cross-section, b^2, involving two-dimensional diffusion (no water may penetrate through the metallic substrates) in an infinite bath is given by [151,152]:

$$C_t/C_\infty = 1-(1-C_t^1/C_\infty)^2 \tag{8.11}$$

and where C_t^1, the concentration of water at time, t, in an adhesive layer resulting from one-dimensional diffusion may be calculated from:

$$\frac{C_t^1}{C_\infty} = 1-\frac{4}{\pi}\sum_{n=0}^{\infty}\frac{(-1)^n}{(2n+1)}\exp[-D_f(2n+1)^2\pi^2t/b^2]\cos[(2n+1)\pi x/b] \tag{8.12}$$

where C_∞ is the equilibrium concentration of water, b is the width of the adhesive layer, x is the distance from the layer's outer edge given $0 \geq x \geq b/2$, and assuming that the initial concentration of water in the adhesive layer is zero. A typical relation between concentration of water, C_t, at any point in the adhesive layer at a given time for immersion in water at 60 °C is shown in Fig. 8.28. Upon comparing the data shown in Fig. 8.28 with the loss of joint strength results (see Figs 8.6 and 8.29) then several interesting features are apparent. Firstly, at 60 °C, for example, the joints have lost almost all their original strength after 2500 h immersion. Thus, the concentration profiles of water shown in Fig. 8.28 demonstrate that the attainment of the equilibrium concentration, C_∞, in the adhesive is obviously not a necessary requirement. Secondly, the time taken to achieve complete saturation of the adhesive layer is relatively long, i.e. about 2.5 years and, whilst this is obviously a function of the diffusion coefficient and geometry, it is not an untypical time-scale. Thirdly, as discussed previously, a critical minimum water concentration appears to be necessary; at 20 °C and 55% r.h. the joints suffered no environmental attack, even though of course the adhesive still absorbed water up to an equilibrium concentration, i.e. 0.75 g of H_2O per 100 g of adhesive.

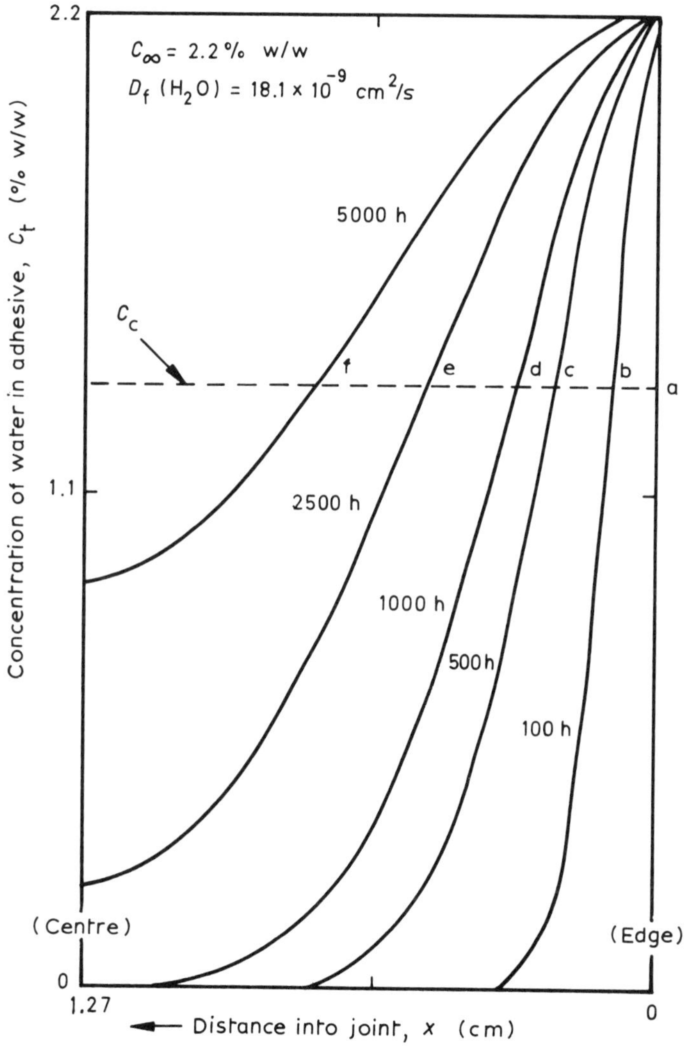

Figure 8.28 Water concentration, C_t, in the adhesive layer in a butt joint as a function of distance, x, into the joint for joints immersed in water at 60 °C. After an immersion time of 100 h the environmental crack length, a, would be the distance 'ba', after 500 h it would be 'ca', etc. [123].

The next aspect of the service life prediction model is to use a continuum fracture mechanics analysis to relate the growth of debonded interface, due in the present joints to rupture of interfacial secondary bonds, to the fracture stress of the joint. For a butt joint this relation is:

$$\sigma_f \cong \left(\frac{E_a G_c}{a}\right)^{1/2} \tag{8.13}$$

Thus, the values of a, E_a and G_c now need to be ascertained. If the hypothesis of a critical water concentration, C_c, for the debonding mechanism to operate is valid, then at a point where the concentration, C_t, in the adhesive exceeds C_c there will exist a relatively sharp boundary between the outer weakened regions of the interface and the inner regions, as yet unaffected by environmental attack. The extent of debonding, or incipient interfacial debonding, that has occurred during environmental exposure is equivalent to the environmental crack length, a, and may be calculated from the water concentration profiles, as shown in Fig. 8.28. The constant value of C_c may be estimated from identifying the activity level of the aqueous environment at which attack is first observed or by interpolation; for the present example it was found to have a value of 1.35 g/100 g. Considering the appropriate values of E_a and G_c, upon testing of the environmentally aged joints the environmentally induced crack was subsequently found to propagate through the adhesive layer. This implies that, to a first approximation, the appropriate values of E_a and G_c are those for the bulk adhesive; i.e. values of 2.8 GPa and 0.4 kJ/m^2, respectively, although an obvious refinement would be to deduce the values for the adhesive plasticized by the appropriate water concentration.

Values of the fracture stress, σ_f, of the joints were deduced from Equation 8.13 as a function of the temperature and time of environmental exposure prior to testing. The theoretically predicted values are shown in Fig. 8.29 as the solid curves and may be compared to the experimentally measured strength values, shown as individual points. The agreement is generally very good. This analysis has also given good agreement when employed to predict the residual fracture strength of epoxy/steel joints used in full-scale bonded structures and subjected to hot/wet tropical exposure.

Thus, it is to be hoped that predictive models may be developed for other adhesives systems and used to estimate quantitatively the service life of bonded components exposed to hostile environments.

8.4 CONCLUDING REMARKS

The present chapter has emphasized the types of loading conditions and environments which may be experienced by a bonded structure and which may greatly reduce the service life of adhesive joints unless consideration is given to

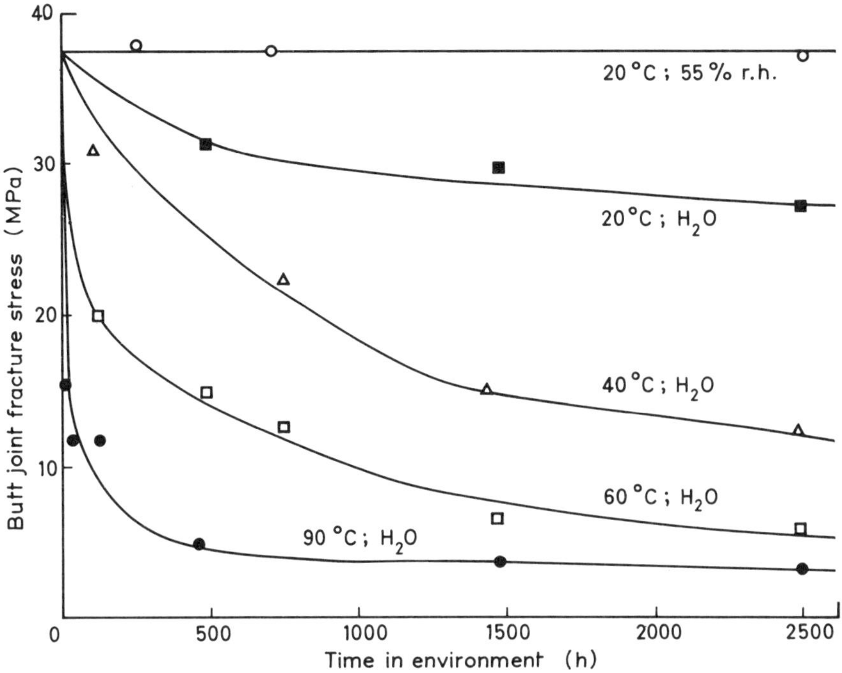

Figure 8.29 The fracture stress, σ_f, of butt joints at 20 °C of epoxy/mild steel joints versus time in environment. Points are experimental measurements and the solid curves are theoretical predictions from the model described in the text [47,123].

all aspects of the material selection and design of the bonded component. It has concentrated on some of the most damaging ways any structure may be loaded and some of the most hostile environments, but it has clearly shown that when careful attention is given to such aspects, and the correct decisions are made, then adhesive joints are capable of surviving under some of the most arduous conditions for long periods of time.

In discussing the above it has been only natural to concentrate on some of the most demanding applications for adhesives, like the aerospace industry. However, it should be emphasized that very few other applications are quite as demanding, either in terms of the applied loadings, the nature of the environments which may be encountered or in the very long service lives which are required. For less demanding applications the critical balance which the adhesives technologist and design engineer must make is that between attaining the necessary service life but at an economically viable cost.

Thus, there is an urgent need to develop adhesive systems which result in bonded components which have the required level of performance, especially an adequate service life. However, the technology so developed must be acceptable both in terms of general engineering practice and economics, when compared to other fastening techniques. Indeed, this is undoubtedly one of the main challenges which currently face those working in the area of the science and technology of adhesion and adhesives.

REFERENCES

1. Lake, G. J. and Lindley, P. B. (1966) *Physical Basis of Yield and Fracture*, Conference Proceedings Institute of Physics, London, p. 176.
2. Hertzberg, R. W. and Manson, J. A. (1980) *Fatigue of Engineering Plastics*, Academic Press, New York.
3. Kinloch, A. J. and Young, R. J. (1983) *Fracture Behaviour of Polymers*, Applied Science Pub., London, p. 197.
4. Andrews, E. H. (1969) *Testing of Polymers IV* (ed. W. A. Brown), Interscience, New York, p. 237.
5. Imanaka, M., Kishimoto, W., Okita, K., Nakayama, H. and Shirato, M. (1984) *J. Comp. Mater.* **18**, 412.
6. Jones, T. B. and Williams, N. T. (1986) *SAE International Congress and Exposition*, Detroit, Michigan, USA, p. 1.
7. Hart-Smith, L. J. (1981) *Developments in Adhesives–2* (ed. A. J. Kinloch), Applied Science Pub., London, p. 1.
8. Marceau, J. A., McMillan, J. C. and Scardino, W. M. (1977) *Sci. Adv. Mater. Process Eng. Series* **22**, 64.
9. Romanko, J. and Knauss, W. G. (1981) *Developments in Adhesives–2* (ed. A. J. Kinloch), Applied Science Pub., London, p. 173.
10. Althof, W. (1984) *Adhesive Joints* (ed. K. L. Mittal), Plenum Press, New York, p. 659.
11. Thrall, E. W. (1979) *Bonded Joints and Preparation for Bonding*, NATO-AGARD Lecture Series No. 102.
12. Mostovoy, S. and Ripling, E. J. (1975) *Adhesion Science and Technology 9B* (ed. L. H. Lee), Plenum Press, New York, p. 513.
13. Mall, S., Johnson, W. S. and Everett, R. A. (1984) *Adhesive Joints* (ed. K. L. Mittal), Plenum Press, New York, p. 639.
14. Liechti, K. M. and Lin, C. (1986) *Structural Adhesives in Engineering*, Institute of Mechanical Engineering, London, p. 83.
15. Kinloch, A.J. and Young, R. J. (1983) *Fracture Behaviour of Polymers*, Applied Science Pub., London, p. 396.
16. Lewis, A. F. (1972) *Adhesives Age* **15**, 38.
17. Lewis, A. F., Kinmouth, R. A. and Kreahling, R. P. (1972) *J. Adhesion* **3**, 249.
18. Wake, W. C., Allen, K. W. and Dean, S. M. (1979) *Elastomers: Criteria for Engineering Design* (ed. C. Hepburn and R. J. W. Reynolds), Applied Science Pub., London, p. 311.
19. Allen, K. W., Smith, S. M., Wake, W. C. and van Raalte, A. O. (1985) *Int. J. Adhesion Adhesives* **5**, 23.
20. Gledhill, R. A. and Kinloch, A. J. (1976) *Polymer* **17**, 727.
21. Gledhill, R. A., Kinloch, A. J. and Shaw, S. J. (1979) *J. Mater. Sci.* **14**, 1769.
22. Zhurkov, S. N. and Tomashevsky, E. E. (1966) *Physical Basis of Yield and Fracture*, Conference Proceedings, Institute of Physics, London, p. 200.
23. Kinloch, A. J. and Young, R. J. (1983) *Fracture Behaviour of Polymers*, Applied Science Pub., London, p. 43.

24. Andrews, E. H. and Reed, P. E. (1978) *Adv. Polymer Sci.* **27**, 1.
25. Kerr, C., MacDonald, N. C. and Orman, S. (1967) *J. Appl. Chem.* **17**, 62.
26. Orman, S. and Kerr, C. (1971) *Aspects of Adhesion–6* (ed. D. J. Alner), University of London Press, p. 64.
27. Sharpe, L. H. (1969) *Appl. Polymer Symp.* **3**, 353.
28. Albericci, P. (1983) *Durability of Structural Adhesives* (ed. A.J. Kinloch), Applied Science Pub., London, p. 317.
29. Cotter, J. L. (1977) *Developments in Adhesives–1* (ed. W. C. Wake), Applied Science Pub., London, p. 1.
30. Gledhill, R. A. and Kinloch, A.J. (1974) *J. Adhesion* **6**, 315.
31. Minford, J. D. (1983) *Durability of Structural Adhesives* (ed. A. J. Kinloch), Applied Science Pub., London, p. 135.
32. Butt, R. I. and Cotter, J. L. (1976) *J. Adhesion* **8**, 11.
33. Morgan, R. J. and Mones, E. T. (1979) *11th National SAMPE Tech. Conf.* 218.
34. Brewis, D. M., Comyn, J. Shalash, R. J. A. and Tegg, J. L. (1980) *Polymer* **21**, 357.
35. Jurf, R. A. and Vinson, J. R. (1985) *J. Mater. Sci.* **20**, 2979.
36. Ripling, E. J., Mostovoy, S. and Bersch, C. F. (1971) *J. Adhesion* **3**, 145.
37. Brockmann, W. (1986) *Structural Adhesives in Engineering*, Institute of Mechanical Engineering, London, p. 61.
38. Kollek, H., Brockmann, W. and Henneman, O. D. (1985) *Progress in Advanced Materials and Processes* (eds G. Bartelds and R. J. Schliekelmann), Elsevier Science Pub., Amsterdam, p. 83.
39. Parker, B. M. (1978) *Proceedings of the International Conference on Jointing in Fibre Reinforced Plastics*, Butterworths, Guildford, p. 95.
40. Bowditch, M. R. and Stannard, K. J. (1982) *Composites* **13**, 298.
41. Selwood, P. G., Maddison, A. and Sheasby, P. G. (1984) UK Patent Application, 2,139,540A.
42. Kinloch, A. J., Welch, L. S. and Bishop, H. E. (1984) *J. Adhesion* **16**, 165.
43. Cognard, J. (1986) *J. Adhesion* **20**, 1.
44. Gettings, M., Baker, F.S. and Kinloch, A. J. (1977) *J.Appl. Polymer Sci.* **21**, 2375.
45. Kinloch, A. J., Dukes, W. A. and Gledhill, R. A. (1975) *Adhesion Science and Technology 9B* (ed. L. H. Lee), Plenum Press, New York, p. 597.
46. Kaelble, D. H. (1974) *J.Appl. Polymer Sci.* **18**, 1869.
47. Kinloch, A.J. (1979) *J. Adhesion* **10**, 193.
48. Owens, D. K. (1970) *J. Appl. Polymer Sci.* **14**, 1725.
49. Gettings, M. and Kinloch, A. J. (1977) *J. Mater. Sci.* **12**, 2511.
50. Gettings, M. and Kinloch, A. J. (1979) *Surf. Interf. Anal.* **1**, 189.
51. Brewis, D. M., Comyn, J. Kinloch, A. J., Oxley, D. P., Pritchard, R. G., Reynolds, S. and Werrett, C. R. (1984) *Surf. Interf. Anal.* **6**, 40.
52. Comyn, J., Kinloch, A. J., Horley, C. C., Mallik, R. R., Pritchard, R.G., Reynolds, S. and Werrett, C. R. (1985) *Int. J. Adhesion Adhesives* **5**, 59.
53. Andrews, E. H., Pingsheng, He and Vlachos, C. (1982) *Proc. Roy. Soc.* **A381**, 345.
54. Patrick, R. L., Brown, J. A., Cameron, N. M. and Gehman, W. G. (1981) *Appl. Polymer Symp.* **16**, 87.
55. Boerio, F. J. and Gosselin, C. A. (1981) *Proceedings of the 36th Annual Technical Conference*, SPI Reinforced Plastics/Composites Institute, Section 2G.
56. Schrader, M. and Cardamone, J. A. (1978) *J. Adhesion* **9**, 305.
57. Boerio, F. J. and Dillingham, R. G. (1984) *Adhesive Joints* (ed. K. L. Mittal), Plenum Press, New York, 541.
58. Kinloch, A. J. and Gledhill, R. A. Unpublished work.

59. Boerio, F. J. and Williams, J. W. (1981) *Appl. Surf. Sci.* **7**, 19.
60. Kinloch, A. J. and Cave, N. (1986) unpublished work.
61. Srivastava, V. K. (1973) *Physics of Thin Films* (ed. G. Hass), Academic Press, London, p. 311.
62. Netzer, L., Iscovici, R. and Sagiv, J. (1983) *Thin Solid Films* **99**, 235.
63. Netzer, L., Iscovici, R. and Sagiv, J. (1983) *Thin Solid Films* **100**, 67.
64. Bolger, J. C. and Michaels, A. S. (1968) *Interface Conversion for Polymer Coatings* (eds. P. Weiss and G. D. Cheever), Elsevier, New York, p. 4.
65. Chu, H. T., Eib, N. K., Gent, A. N. and Henriksen, P. N. (1979) *Advances in Chemistry Series 174* (ed. J. L. Koenig), American Chemical Society, Washington DC, p. 87.
66. Brockmann, W. (1977) *Adhesives Age* **20**(6), 30.
67. Kollek, H. (1985) *Int. J. Adhesion Adhesives* **5**, 75.
68. Turner, R. J., Brown, N. M. D. and Walmsley, D. G. (1981) *Vacuum* **31**, 603.
69. Zaporozhskaya, E. A., Ginsberg, L. V. and Donstov, A. A. (1980) *Vyskomol. Soedın Ser. A* **22**, 1222.
70. Park, J. M. and Bell, J. P. (1982) *Adhesion Aspects of Polymeric Coatings* (ed. K. L. Mittal), Plenum Press, New York, p. 205.
71. DeNicola, A. J. and Bell, J. P. (1983) *Epoxy Resin Chemistry*, American Chemical Society, Washington DC, p. 119.
72. Waite, J. H., Housley, T. J. and Tanzer, M. L. (1985) *Biochemistry* **24**, 5010.
73. Silberberg, A. (1984) *Polymer Adsorption and Dispersion Stability* (ed. E. D. Goddard and B. Vincent), American Chemical Society, Washington DC, p. 161.
74. Minford, J. D. (1977) *Adhesives Age* **20**(9), 41.
75. Kinloch, A. J., Bishop, H. E. and Smart, N. R. (1982) *J. Adhesion* **14**, 105.
76. Venables, J. D. (1984) *J. Mater. Sci.* **19**, 2431.
77. Packham, D. E. (1982) *Adhesion Aspects of Polymeric Coatings* (ed. K. L. Mittal), Plenum Press, New York, p. 19.
78. Brockmann, W., Henneman, O. D. and Kollek, H. (1984) *Adhesive Joints* (ed. K. L. Mittal), Plenum Press, New York, p. 469.
79. Pike, R. A. (1985) *Int. J. Adhesion Adhesives* **5**, 3.
80. Pike, R. A. (1986) *Int. J. Adhesion Adhesives* **6**, 21.
81. Davis, G. D., Ahearn, J. S., Matienzo, L. J. and Venables, J. D. (1985) *J. Mater. Sci.* **20**, 975.
82. Matienzo, L. J., Shaffer, D. K., Moshier, W. C. and Davis, G. D. (1986) *J. Mater. Sci.* **21**, 1601.
83. Assefpour-Dezfuly, M., Vlachos, C. and Andrews, E. H. (1984) *J. Mater. Sci.* **19**, 3626.
84. Arrowsmith, D. J. and Clifford, A. W. (1985) *Int. J. Adhesion Adhesives* **5**, 40.
85. Davis, G. D. and Venables, J. D. (1983) *Durability of Structural Adhesives* (ed. A. J. Kinloch), Applied Science Pub., p. 43.
86. Noland, J. S. (1975) *Adhesion Science and Technology 9B* (ed. L. H. Lee), Plenum Press, New York, p. 413.
87. Solomon, J. S. and Hanlin, D. E. (1980) *Appl. Surf. Sci.* **4**, 307.
88. Sun, T. S., McNamara, D. K., Ahearn, J. S., Chen, J. M., Ditchek, B. and Venables, J. D. (1980) *Appl. Surf. Sci.* **5**, 406.
89. Kinloch, A. J. and Smart, N. R. (1981) *J. Adhesion* **12**, 23.
90. Poole, P. and Watts, J. F. (1985) *Int. J. Adhesion Adhesives* **5**, 33.
91. Thompson, G. E., Furneau, R. C., Wood, G. C., Richardson, J. A. and Goode, J. S. (1978) *Nature* **272**, 433.

92. Hunter, M. S., Towner, P. F. and Robinson, D. L. (1959) *Proc. Am. Electroplaters Soc.* **46**, 220.
93. Sun, T. S., Chen, J. M., Venables, J. D. and Hopping, R. (1978) *Appl. Surf. Sci.* **1**, 202.
94. Venables, J. D., McNamara, D. K., Chen, J. M., Sun, T. S. and Hopping, R. L. (1979) *Appl. Surf. Sci.* **3**, 88.
95. Young, R. J., Assefpour-Dezfuly, M., Andrews, E. H. and Kinloch, A. J. (1983) *Inst. Phys. Conf. Series* **68**, 347.
96. Arrowsmith, D. J. (1986) *Trans. Inst. Met. Finish.* **64**, 6.
97. Davies, R. J. and Kinloch, A. J. (1987) *Adhesion '87 Conf.* York University, 7–9 Sept. 1987. Plastics and Rubber Inst., London.
98. Kinloch, A. J. and Young, R. J. (1984) Unpublished work.
99. Venables, J. D., McNamara, D. K., Sun, T.S., Ditchek, B., Chen, J. M. and Hopping, R. L. (1979) *Structural Adhesives and Bonding*, Technology Conference Association, El Segundo, CA., p. 12.
100. El-Mashi, S. M. and Forty, A.J. (1981) *Inst. Phys. Conf. Series* **61**, 395.
101. Hoar, T. P. and Mott, N. F. (1959) *J. Phys. Chem. Sol.* **9**, 97.
102. O'Sullivan, J. P. and Wood, G. C. (1970) *Proc. Roy. Soc.* **A317**, 511.
103. Xu, Y., Thompson, G. E. and Wood, G. C. (1985) *Trans. Inst. Met. Finish.* **63**, 98.
104. Vedder, W. and Vermilyea, D. A. (1969) *Trans. Faraday Soc.* **65**, 561.
105. Davis, G. D., Sun, T. S., Ahearn, J. S.and Venables, J. D. (1982) *J. Mater. Sci.* **17**, 1807.
106. El-Mashri, S. M., Jones, R.G. and Forty, A. J. (1983) *Phil. Mag.* **48**, 665.
107. Hardwick, D. A., Ahearn, J. S. and Venables, J. D. (1984) *J. Mater. Sci.* **19**, 223.
108. Hardwick, D. A., Ahearn, J. S., Desai, A.and Venables, J. D. (1986) *J. Mater. Sci.* **21**, 179.
109. Smith, J. A. and Martinsen, W. E. (1973) *Ceram. Bull.* **52**, 855.
110. McDevitt, N. T., Baun, W. L. and Solomon, J. S. (1976) *J. Electrochem. Soc.* **123**, 1058.
111. Datta, A., Pethrick, R. A. and Affrossman, S. (1982) *J. Adhesion* **15**, 13.
112. Kinloch, A. J. (1982) *Adhesion–6* (ed. K. W. Allen), Applied Science Pub., London, p. 95.
113. Moulds, R. J. (1984) *Int. J. Adhesion Adhesives* **4**, 23.
114. Reil, F. J. (1971) *SAMPE J.* **7**(4), 16.
115. Bascom, W. D. and Patrick, R. L. (1974) *Adhesives Age* **17**(10), 25.
116. Stevenson, A. (1985) *Int. J. Adhesion Adhesives* **5**, 81.
117. van Ooij, W. J. (1985) *Industrial Adhesion Problems* (eds D. M. Brewis and D. Briggs), Orbital Press, Oxford, p. 87.
118. Comyn, J., Horley, C. C., Oxley, D.P., Pritchard, R. G. and Tegg, J. L. (1981) *J. Adhesion* **12**, 171.
119. Allen, K. W., Alsalim, H. S. and Wake, W. C. (1972) *Faraday Spec. Disc.* **2**, 38.
120. Brockmann, W. (1983) *Adhesion Aspects of Polymeric Coatings* (ed. K. L. Mittal), Plenum Press, New York, p. 265.
121. Hammond, J. S., Holubka, J. W., DeVries, J. E. and Dickie, R. A. (1981) *Corrosion Sci.* **21**, 239.
122. Watts, J. F. and Castle, J. E. (1983) *J. Mater. Sci.* **18**, 2987.
123. Gledhill, R. A., Kinloch, A. J. and Shaw, S. J. (1980) *J.Adhesion* **11**, 3.
124. Brewis, D. M., Comyn, J. and Tegg, J. L. (1980) *Int. J. Adhesion Adhesives* **1**, 35.
125. Brewis, D. M., Comyn, J., Cope, B. C. and Moloney, A. C. (1980) *Polymer* **21**, 1477.
126. Kreiger, R. (1973) *SAMPE Q* **5**, 643.

127. Minford, J. D. (1978) *Adhesives Age* **21**(3), 17.
128. Bascom, W. D. (1979) *Adhesives Age* **22**(4), 28.
129. Bascom, W. D., Gadomski, S. T., Henderson, C. M. and Jones, R. L. (1977) *J. Adhesion* **8**, 213.
130. Pocius, A. V., Wangsness, D. A., Almer, C. J. and McKown, A. G. (1984) *Adhesives Chemistry* (ed. L. H. Lee) Plenum Press, New York, p. 617.
131. Weitsman, Y. (1977) *J. Comp. Mater.* **11**, 378.
132. Sargent, J. P. and Ashbee, K. H. G. (1980) *J. Adhesion* **11**, 175.
133. Gillat, O. and Broutman, L. J. (1978) *ASTM, STP* **685**, 61.
134. Fahmy, A. A. and Hurt, J. C. (1980) *Polymer Comp.* **1**, 77.
135. Morgan, R. J., O'Neal, J. E. and Fanter, D. L. (1980) *J. Mater. Sci.* **15**, 751.
136. Mostovoy, S. and Ripling, E. J. (1969) *Appl. Polymer Sci.* **13**, 1083.
137. Kinloch, A. J. and Shaw, S. J. (1981) *Developments in Adhesives–2* (ed. A. J. Kinloch), Applied Science Pub., London, p. 83.
138. Cherry, B. W. and Thomson, K. W. (1977) *Adhesion–1* (ed. K. W. Allen), Applied Science Pub., London, p. 251.
139. Cherry, B. W. and Thomson, K. W. (1977) *Fracture Mechanics and Technology* (ed. G. Sih), Sijthoff and Nordhoff, The Netherlands, p. 723.
140. Cherry, B. W. and Thomson, K. W. (1980) *Int. J. Polymeric Mater.* **8**, 13.
141. Mai, Y. W. (1975) *J. Adhesion* **7**, 141.
142. Gent, A. N. and Schultz, J. (1972) *J. Adhesion* **3**, 281.
143. Carre, A. and Schultz, J. (1985) *J. Adhesion* **18**, 171.
144. DeLollis, N. J. (1977) *Natl. SAMPE Symp. Exhib.* **22**, 673.
145. Brewis, D. M., Comyn, J., Cope, B. C. and Moloney, A. C. (1981) *Polymer Eng. Sci.* **21**, 797.
146. Brewis, D. M., Comyn, J. and Tegg, J. L. (1980) *Polymer* **21**, 134.
147. Comyn, J. (1983) *Durability of Structural Adhesives* (ed. A.J. Kinloch), Applied Science Pub., London, p. 85.
148. Glasstone, S. and Lewis, D. (1964) *Elements of Physical Chemistry*, Macmillan and Co., London, p. 382.
149. Althof, W. (1979) *Aluminium* **55**, 600.
150. Brewis, D. M., Comyn, J., Moloney, A. C. and Tegg, J. L. (1981) *Eur. Polymer J.* **17**, 127.
151. Crank, J. (1956) *Mathematics of Diffusion*, Oxford University Press, Oxford.
152. Carslaw, H. S. and Jaeger, J. C. (1959) *Conduction of Heat in Solids*, Oxford University Press, Oxford.
153. Brewis, D. M., Comyn, J. and Tredwell, S. T. (1987) *Int. J. Adhesion Adhesives* **7**, 30.

Glossary

MAIN NOTATIONS

English alphabet

Lower case symbols

a	crack length
a_T	shift factor
$\dot{a}$	rate of crack growth
b	width
b'	area fraction of cohesive in adhesive failure
d	thickness of the substrate
d_c	value of d above which no plastic deformation occurs in the substrate
d_i	thickness of the inner substrate
d_o	thickness of the outer substrate
d_m	moment arm
f	function
h_a	thickness of the adhesive layer
h_{am}	thickness of the adhesive layer when the fracture energy is at a maximum
h_c	discharge gap
i	area fraction of interfacial failure
$k_1 \rightarrow k_8$	constants
l	length
l_a	overlap length
l_e	moment of inertia of flexible substrate cross-section
l_p	penetration depth
m	constant
m_b	constant
m_p	distance
n	constant
n^{AB}	number of moles of acid–base pairs per unit area
p	vapour pressure
p_a	atmospheric pressure
p_r	pressure

p_{rc}	pressure at onset of crack growth
q	constant
r	distance
r, θ, z	(subscripts) radial, circumferential and axial cylindrical co-ordinates respectively
r_c	radius of curvature
r_f	roughness factor
r_o	radius
r_{oi}	inner radius
r_{oo}	outer radius
r_{tc}	critical distance
r_y	radius of plastic zone at the crack tip
r_{yc}	radius of plastic zone at the crack tip at crack growth
r'_y	plane-strain plastic zone radius at the crack tip
r''_y	plane-stress plastic zone radius at the crack tip
s	area fraction of cohesive-in-substrate failure
t	time
t_c	time of contact for the interdiffusion of rubbery polymers
t_f	time to failure
v	displacement
v_s	velocity
w	weight
x, y	(subscripts) cartesian coordinates

Upper case symbols

A	area
A_c	energy required to separate electrical double layer
A_{aa}, A_{bb}, A_{ab}	attractive constants between molecules
C	elastic compliance ($= v/F$)
C_c	critical concentration of water
C_t	concentration of water in the adhesive layer at time t
C_∞	equilibrium solubility coefficient of water in the adhesive
C_1, C_2	constants in WLF equation
C^A, C^B	acid–base constants
D	depth
D_b	bending stiffness (per unit width)
D_d	constant
D_f	diffusion coefficient
E	Young's or tensile modulus
E_a	Young's or tensile modulus of the adhesive
E'_a	apparent Young's or tensile modulus of the adhesive, i.e. in a butt joint

E_s	Young's or tensile modulus of the substrate
E^A, E^B	acid–base constants
F	applied force or load
F_c	applied force at the onset of crack propagation
F_f	applied force at fracture ($2F_f$ in the case of a double lap joint)
G_a	shear modulus of the adhesive
G_c	fracture energy
G_{cm}	maximum value of the adhesive fracture energy, i.e. when $h_a = h_{am}$
G_{scc}	a possible minimum mode I fracture energy below which no crack growth will occur in a hostile environment
G_o	intrinsic fracture energy
G_{max}	maximum value of the fracture energy
G_{min}	minimum value of the fracture energy
ΔG	fracture energy range during dynamic fatigue cycling ($= G_{max} - G_{min}$)
ΔG_{th}	threshold value of ΔG, below which no dynamic fatigue crack growth will occur (subscripts I, II or III to the above indicate mode of failure)
G_s	shear modulus of the substrate
ΔH^{AB}	enthalpy of acid–base interactions
ΔH_v	molar heat of vaporization
I	tunnelling current
K	stress intensity factor
K_c	stress intensity factor at the onset of crack growth, or fracture toughness
K_{ic}	interfacial fracture toughness (subscripts I, II or III to the above indicate mode of failure)
L	constant
M	molecular weight
M_e	bending moment (per unit width)
M_p	moment of peel force (per unit width)
$\mathbf{N}$	Avogadro's number
N_f	number of cycles to failure
N_c	number of chains crossing the interface
P	peel energy, or peel force per unit width (also referred to as the peel strength)
Q	geometry constant
Q_i	geometry constant for cracks at, or near, an interface (subscripts I, II or III to the above indicate mode of failure)
R	stress ratio
$\mathbf{R}$	gas constant
S	equilibrium spreading coefficient

T	temperature
T_g	glass transition temperature
T_q	applied torque
ΔT	temperature difference
U	stored elastic energy
U_d	applied or input strain energy density (i.e. strain energy per unit volume)
U_{dc}	applied strain energy density at onset of crack growth
U_{df}	applied strain energy density at fracture
U_{dh}	hysteresis, or loss, strain energy density
U_{dhc}	value of U_{dh} for crack growth
V	molar volume
V_c	discharge potential
W_d	work done
W_A	thermodynamic work of adhesion
W_A^{AB}	thermodynamic work of adhesion arising from acid–base interactions
W_A^P	thermodynamic work of adhesion arising from polar interactions
W_{Al}	thermodynamic work of adhesion in the presence of a liquid
W_F	free energy change for formation of interface

Greek alphabet

Lower case symbols

α	peel angle
α_m	constant
$\Delta\alpha_t$	difference in linear thermal expansion coefficient
β	constant
β_p	constant
γ^D	dispersion force component
γ^P	polar force component
γ_a	surface free energy of the adhesive
γ_{ae}	maximum elastic shear strain of the adhesive
γ_{af}	shear fracture strain of the adhesive
γ_{al}	interfacial free energy of adhesive/liquid interface
γ_{as}	interfacial free energy of adhesive/substrate interface
γ_c	critical surface tension
γ_{ij}	shear strain
γ_{lv}	surface free energy, or tension, of the liquid surface free energy of a material
γ_s	surface free energy of the solid
γ_{sl}	interfacial free energy of the solid/liquid interface

γ_{sv}	surface free energy of solid substrate in the presence of vapour
γ^{AD}	dispersion contribution to the surface free energy of 'acid'
γ^{BD}	dispersion contribution to the surface free energy of 'base'
γ^{P}	polar contribution to surface free energy
δ_s	solubility parameter
δ_t	crack opening displacement at the crack tip
δ_{tc}	crack opening displacement at the crack tip at the onset of crack growth
ϵ	normal strain
$\dot{\epsilon}$	strain rate
ϵ_a	strain amplitude
ϵ_{af}	tensile fracture strain of the adhesive
ϵ_{ay}	tensile yield strain of the adhesive
ϵ_d	dielectric constant
ϵ_m	mean strain level
ϵ_z	axial normal strain
ζ	bimaterial constant
η	viscosity
η_c	stress concentration
θ	contact angle (but polar coordinate in Chapter 7)
θ_a	advancing contact angle
θ_d	dynamic contact angle
θ_f	contact angle on a rough surface
θ_r	receding contact angle
θ_s	contact angle on a smooth surface
θ_t	contact angle at time t
θ_∞	contact angle at infinite time
κ	bending moment factor
κ^*	rotation factor
κ_p	factor in peel force analysis
λ_c	extension ratio
ξ	constant
ρ	density
ρ_t	crack tip radius
ρ_{tc}	crack tip radius at fracture
σ	normal, or direct, stress
σ_a	stress amplitude
σ_{af}	tensile fracture stress of the adhesive
σ_{ay}	tensile yield stress of the adhesive
σ_f	tensile fracture stress of the joint
σ_{ij}	components of the normal, or direct, stress
σ_{im}	maximum stress the interface can carry
σ_m	mean stress in dynamic fatigue cycling

σ_{o}	applied tensile stress
σ_{p}	boundary cleavage stress
σ_{r}	radial stress
σ_{sy}	tensile yield stress of the substrate
σ_{tc}	critical stress at crack tip
σ_{z}	axial stress
σ_{θ}	circumferential stress
$\Delta\sigma$	stress range in dynamic fatigue cycling
τ	shear stress
τ_{ae}	Shear stress of the adhesive at its elastic limit
τ_{af}	shear fracture stress of the adhesive
τ_{ay}	shear yield stress of the adhesive
τ_{ij}	components of the shear stress
τ_{o}	applied shear stress
ν	Poisson's ratio
ν_{a}	Poisson's ratio of the adhesive
ν_{s}	Poisson's ratio of the substrate
ϕ	scarf angle
χ	constant
ψ	energy dissipated in viscoeleastic and plastic deformations at the crack tip
ω	frequency

Upper case symbols

Γ	surface concentration of adsorbed vapour
Δ	dimensionless coefficient
Π_{s}	spreading pressure
Φ	Good–Girifalco interaction parameter
Φ_{v}	mechanical loss function

MAIN ABBREVIATIONS

AES	Auger electron spectroscopy
ASTM	American Society for Testing and Materials
ATR	attenuated total reflectance
BS	British Standard
CAA	chromic acid anodized
CAE	chromic acid etched
CASING	crosslinking by activated species of inert gases
CFRP	carbon fibre reinforced plastic
DCB	double cantilever beam specimen
DGEBA	diglycidyl ether or bisphenol A-based epoxy resin

ESCA	electron spectroscopy for chemical analysis
EVA	ethylene/vinyl acetate
FEP	fluorinated ethylene–propylene copolymer
GRP	glass fibre reinforced plastic
IET	inelastic electron tunnelling spectroscopy
IR	infrared
LEFM	linear elastic fracture mechanics
NDT	non-destructive testing
PAA	phosphoric acid anodized
PET	poly(ethylene terephthalate)
PCTFE	poly(chlorotrifluoroethylene)
PTFE	polytetrafluoroethylene
RTV	room-temperature vulcanizing
SBR	styrene–butadiene rubber
SBS	styrene–butadiene–styrene block copolymer
SIMS	secondary ion mass spectroscopy
TDCB	tapered double cantilever beam specimen
TETA	triethylenetetramine
TEPA	tetraethylenepentamine
WLF	Williams–Landel–Ferry
XPS	X-ray photoelectron spectroscopy, also termed ESCA
r.h.	relative humidity

Author Index

Page numbers in *italics* indicate references cited in figures; numbers in **bold** indicate where a full reference is to be found.

Subject Index

Page numbers in *italics* refer to figures